Intelligent Systems Reference Library

Volume 241

Series Editors

Janusz Kacprzyk, Polish Academy of Sciences, Warsaw, Poland

Lakhmi C. Jain, KES International, Shoreham-by-Sea, UK

The aim of this series is to publish a Reference Library, including novel advances and developments in all aspects of Intelligent Systems in an easily accessible and well structured form. The series includes reference works, handbooks, compendia, textbooks, well-structured monographs, dictionaries, and encyclopedias. It contains well integrated knowledge and current information in the field of Intelligent Systems. The series covers the theory, applications, and design methods of Intelligent Systems. Virtually all disciplines such as engineering, computer science, avionics, business, e-commerce, environment, healthcare, physics and life science are included. The list of topics spans all the areas of modern intelligent systems such as: Ambient intelligence, Computational intelligence, Social intelligence, Computational neuroscience, Artificial life, Virtual society, Cognitive systems, DNA and immunity-based systems, e-Learning and teaching, Human-centred computing and Machine ethics, Intelligent control, Intelligent data analysis, Knowledge-based paradigms, Knowledge management, Intelligent agents, Intelligent decision making, Intelligent network security, Interactive entertainment, Learning paradigms, Recommender systems, Robotics and Mechatronics including human-machine teaming, Self-organizing and adaptive systems, Soft computing including Neural systems, Fuzzy systems, Evolutionary computing and the Fusion of these paradigms, Perception and Vision, Web intelligence and Multimedia.

Indexed by SCOPUS, DBLP, zbMATH, SCImago.

All books published in the series are submitted for consideration in Web of Science.

Anthony L. Brooks

Editor

Creating Digitally

Shifting Boundaries: Arts
and Technologies—Contemporary
Applications and Concepts

 Springer

Editor
Anthony L. Brooks 🆔
CREATE—Section for Media Technology
Campus Aalborg (Human Machine
Interaction/The Center for Applied
Game Research)
Department of Architecture, Design
and Media Technology; The Technical
Faculty of IT and Design
Aalborg University
Aalborg, Denmark

ISSN 1868-4394   ISSN 1868-4408 (electronic)
Intelligent Systems Reference Library
ISBN 978-3-031-31359-2   ISBN 978-3-031-31360-8 (eBook)
https://doi.org/10.1007/978-3-031-31360-8

This Springer imprint is published by the registered company Springer Nature Switzerland AG
The registered company address is: Gewerbestrasse 11, 6330 Cham, Switzerland

Preface

This book's preface shares how the book was conceived, researched, and put together, as well as linking it to the editor's credentials. Over 100 abstract texts were submitted proposing chapters for the volume that were weeded down to the selections herein. Up front, I thank all the authors for their time, efforts, and patience in achieving their publication, hopefully something that they are proud to have alongside others in this book and maybe even on their bookshelves in a physical form. The journey to finalize the book was arduous with issues arising and causing delays that need not be shared in a preface. In this opening, I introduce by acknowledgement what has been a big part of my *Creating Digitally* life and endeavor for the last 20 years or so that led to this volume.

The title *Creating Digitally* acknowledges Aalborg University (AAU), and especially those people involved in the Medialogy (Medialogi in Danish) education from over the years across campuses in three regions of Denmark. Both this book and the Medialogy education have a common central focus upon creativity and technology. In acknowledging as such I inform that this book had been planned to include my own penned chapters on Medialogy to share what it is alongside some of its untold stories and history. In doing so, the original plan was thus an attempt toward supporting the positioning of the education in its rightful place alongside others where comprehension of denotation is unquestioned. My authoring agreement from Aalborg University in sharing the Medialogy story and history reflects that at the time of my transition to emeritus, in early fall 2023, credentials included that I was the last remaining employee from the originating team behind the education's forming, thus many stories are known by me solely. However, or course, these are not all encompassing, but offer insight that many others involved are unaware of and can reflect their own sides. I was heavily involved in the birth of Medialogy, in fact, being involved much earlier than others on the initial AAU employment and student rosters for Medialogy's initial delivery in the Danish higher education system. This is posited after initially meeting and discussing on the concept with the "founding father" of Medialogy Professor Jens Arnspang when I was first artist/researcher in residence at CAVI—the Centre for Advanced Visualization and Interaction—located at Aarhus University around the turn of the millennium. Jens visited with around eight of his

PhD students and I presented my research to them all. Some would later visit my applied research projects[1], notably at the Centre for Rehabilitation of Brain Injury, Copenhagen University[2].

At the time of the initial meeting, my research had already been responsible for receiving sizable national (Danish) and international (European) funding grants for my sole designed projects. My research led to being invited onto the coordination board leading the "European Network in Intelligent Information Interfaces (i3net)"— a project funded under the European Commission's 5th Framework Program (5FP) for Long Term Research associated to market requirements and citizens needs aligned to the emerging information society. Via my i3net role I comprehended 5FP's identifying, in the mid-90s, how European (this was pre-Brexit) industry lagged the rest of the world being both weak in terms of hardware and software products and stagnant in terms of computer services. Further, it questioned employment, exclusion, and culture (and more) relative to an identified low level of investment (per employee) that had direct consequences on the adoption of the new technologies. The emerging vision from the program was investment for developing a European information society, able to match traditional humanistic and social values to satisfy the expectations of the citizens for an improved quality of life, economic growth, and employment[3]. A goal was to compete with those other global powers who were leading in these aspects, i.e., America and Asia-Pacific.

It can thus be argued how Medialogy was formed to offer an education and, eventually, apropos under AAU's flagship Problem—Based Learning (PBL—or POPBL: Project–Oriented Problem—Based Learning) toward promoting scholarly research, knowledge, and experiences that would provide, albeit minuscule in comparison, an effective contribution to the creation of a European Information Society—one of the key objectives for the 5FP[4]. Just as any form of education targets student vocations post their higher education studies, Medialogy was conceived with diligent consciousness to support graduating students in their future endeavors.

From subsequent discussions between Jens and myself, my understanding was that, following its initial conceptualizing, the education form proposed to various universities around the last years of the 1990s had no adopters. It was hinted at the time that this lack of uptake was due to Medialogy's distancing from traditional topics and subjects in title and structure. This alongside its hybrid untraditional form sitting on the cusp between the arts and sciences and its focus upon relations to the human and society as central by targeting, within the education, courses on understanding humans, for example, in their sensing, perception, and cognition aligned to their (designed for) use of products. The designing of human interactions; creation of hardware prototypes; and related programming of software and the questioning of related aesthetics, ethics, and justifications became integral to the holistic

---

[1] For example, HUMANICS 1 & 2 https://www.researchgate.net/publication/237769859 https://vbn.aau.dk/ws/portalfiles/portal/18596205/pdf.

[2] https://link.springer.com/chapter/10.1007/978-1-4757-5569-5_17.

[3] https://www.ercim.eu/publication/Ercim_News/enw29/chasseriaux2.html.

[4] https://cordis.europa.eu/programme/id/FP5.

entity that Medialogy was to become following our meetings. By incorporating these elements as a synthesized offering, it can be reflected how management traditionalists within Danish higher education establishments may have been provoked, indifferent, and alienated to not adopt. Those involved at the time, and even ensuing, may reflect how Medialogy was avantgarde and ahead of its time through its potential targeting of future challenges as identified by "The Information Society Forum" set up by the "European Commission" in February 1995 to consider social, societal, and cultural aspects. In its form sculptured and fine-tuned following our exchanges, Jens eventually found sympathetic ears and Medialogy eventually began in the small city campus of Esbjerg on the Danish southwest coast under Aalborg University. Timing was right, as the twentieth century had just transitioned into the twenty-first century, and contemporary entities within education, research, and industry—especially within the fields of information and communications technologies—grew and became more intertwined in a transdisciplinary fashion. This development aligned with the European Commission's creation of a Future and Emerging Technologies (FET) call under its Information and Communication Technologies (ICT) programs where it reiterated how an increased synergy between the different players in the information society: researchers, industry, service providers, and users were necessary. Accordingly, since conception, the Medialogy education design targeted external collaborations for student projects across sectors of interest and influence.

Within my own research in the preceding period, projects titled Human Interactive Communication Systems (HUMANICS)[5]; The World Is As You See It (TWIAYSI); and Creating Aesthetically Resonant Environments for Handicapped, Elderly, and Rehabilitation (CAREHERE) all received sizeable funding support that led to the creation of a family of published patents with myself as inventor and a spin-out company called Personal Interactive Communication Systems (PERSONICS). All entities targeted social, societal, and cultural human-centered computing with goals to impact, benefit, and educate—utilizing creativity and technology—within contexts unfamiliar with such interventions. My (Creating Digitally) research focus was on helping therapists to advance (re)habilitation interventions and their patients' experiences and improvement whereby patients were across the spectrum of age, abilities, and situation. Thus, I targeted the most challenging cases (aligned to abilities) as well as those less challenging (more able) with the understanding from experiences that the created digital systems could be adapted to each individual or group as determined by profile(s)—including needs, desires, preferences, and targeted outcomes from interventions (as advised by medical professionals). In the 1990s, sensors and multimedia were prevalent in their advancement alongside software that could easily map (or route and adjust) data sourced from a human by sensors to manipulate multimedia. The manipulation of the multimedia opened a channel for direct feedback such that within an optimally tailored interactive system (according to a patient's profile,

---

[5] HUMANICS was my design for an "At-Home Rehabilitative Exercise" sensor-based system for stroke patients, investigated at Denmark's leading clinic for acquired brain injury rehabilitation while PERSONICS targeted adaptive systems that built upon HUMANICS incorporating video games in (re)habilitation alongside empowered creativity.

etc.,) a typical process chain changes from human feedforward (afferent to efferent to motoric to multimedia) to a situation where the feedback takes over leading within the afferent efferent neural feedback loop closure cycle. A control of feedback to feedback controlling motion if you like (think Hendrix). This has been known to drive a participant to expand their motoric gesticulations as opposed to a more traditional situation where consciousness may restrict.

The sensors originally conceived and created for my own research were an infrared emitter and receiver replacing a light bulb on a cheap swan neck lighting product. A single neck/sensor eventually transitioned to become a three-headed version, thus enabling a person to manipulate multimedia in the form of musical compositions (e.g., three sounds) or color visuals (RGB/HSV) or navigate in gameplay. Such manipulated audiovisuals included early interactive games (made in Macromedia Flash) that were played via recognized gestures. Accordingly, I am acknowledged as one of the first to use such channels of feedback (directly responding to feedforward actions) within healthcare (re)habilitation in this way. The data generated within the interactive environment can also be used for analysis. The same systems used in the healthcare (re)habilitation interventions were used in my art works, e.g., interactive installations and stage performances. Both entities I found to be cross-informing whereby the (re)habilitation interventions informed the art interventions that concurrently informed the (re)habilitation interventions. This cross-informing I relate to creative thinking. Thus, when I read articles, such as the 2016 titled "Person-Centered Multimedia Computing: A New Paradigm Inspired by Assistive and Rehabilitative Applications"[6], I look back at the period introduced herein, the projects, the Medialogy education, and smile at having been avantgarde in pioneering and contributing in the way I did.

Thus, the originally planned authoring of Medialogy in this volume was toward a history garnered from experiences I encountered, thus not all encompassing. However, a need for the education's history was identified by me and many others following discussions after identifying how texts describing the education had fallen short over the years, typically in delimited conference papers or journal article forms. Typically, such authoring was penned by those not having the credentials and such a history innate to Medialogy as myself. This was especially so when I identified similarity to a contemporary education, I considered aligned to Medialogy titled as "Human-Centered Computing (HCC)"[7] which "studies the design, development, and deployment of mixed-initiative human-computer systems. It is emerged from the convergence of multiple disciplines that are concerned both with understanding human beings and with the design of computational artifacts" additionally, "Human-centered computing (HCC) has emerged as a major subfield of computational science and emphasizes the understanding of human behavior, needs, and expectations in the

---

[6] S. Panchanathan, S. Chakraborty, T. McDaniel and R. Tadayon, "Person-Centered Multimedia Computing: A New Paradigm Inspired by Assistive and Rehabilitative Applications", in *IEEE MultiMedia*, vol. 23, no. 3, pp. 12–19, July-September 2016, doi: https://doi.org/10.1109/MMUL.2016.51.

[7] https://en.wikipedia.org/wiki/Human-centered_computing.

design and development of technologies". Subsequently, literature states how "HCC principles and methods have served researchers well, the increasing need for individualized solutions warrants a person-centered approach" thus leading to "Person-Centered Multimedia Computing" as argued by Panchanathan et al. (see footnote). In this cited work, referencing Mozaffarian et al (2015)[8], it states how "Motor rehabilitation has gained substantial traction as an application area of multimedia computing. More than 795,000 Americans are diagnosed annually with having experienced a stroke, resulting in costs of over US$34 billion per year. The rise of ubiquitous technology capable of sensing and responding to human behavior has granted an increased sense of autonomy to individuals in a wide variety of motor rehabilitation programs. Researchers are exploring new ways in which multimedia systems can interact with these users in both clinical and nonclinical environments". Thus, aligned to this train of thought and statistics, and from my own research of approximately four decades on this very subject, it seemed obvious that there were business opportunities to create solutions to such challenges and this is how Medialogy students were educated to think, and many started their own spin-out companies from their education projects—especially in Esbjerg where a dedicated business support framework networked with industries was created. A number of these spin-out companies, consisting of students I had supervised, focused upon motor rehabilitation training and well-being implementing technologies such as Virtual Reality, thus following in their supervisor's footsteps. The education was a great success as was many of the students' companies. I consider myself fortunate to having been a part of the Medialogy team for over two decades since being a member of the founding team of the education, wherein it's been acknowledged my input has had significant influence pre- and post- it's realization. A TED talk[9] is online by Professor Jens Arnspang informing on Medialogy wherein he acknowledges my input. My Medialogy chapters planned for this book were eventually placed aside due to the unexpected mass of responses received by authors from around the world wishing to contribute to this topic of "Creating Digitally" and related "Shifting Boundaries: Arts and Technologies—Contemporary Applications and Concepts". There were only so many pages planned. Thus, my Medialogy authoring will follow as will a book on my selected published works, rather than leaving for another to (mis)interpret posthumously. Suffice to say for now that readers have ahead in this book over 20 chapters from digital creatives living around the world who share their stories and histories as I introduce with Medialogy. There is diversity in topic and meaning within the chapters and as a whole the result is a volume that I am proud to be editor of.

In closing, fondly, I recall how early in the Medialogy education the first-semester students in Esbjerg read an article of my nickname being Mister Beam[10] due to the

---

[8] D. Mozaffarian et al., "Heart Disease and Stroke Statistics—2015 Update: A Report from the American Heart Association", *Circulation*, vol. 131, no. 4, pp. 29–322, 2015.

[9] 'Medialogy—bridging science of nature with creativity and art | Jens Arnspang | TEDxEAL'— https://www.youtube.com/watch?v=KnGKGvDnhwg.

[10] Brooks, A. (2000). Mr. Beam. Journal of the European Network for Intelligent Information Interfaces http://www.i3net.org/ser_pub/services/magazine/march2001/i3mag10.pdf pp. 2–6.

sensor beams invented and implemented in my research, which I showed in class. They picked up on this nickname and transitioned it to become "Mister Medialogy" that they informed me was due to their experiencing my obvious commitment following hearing of my preconceptual and initial inputs to the education from Professor Jens Arnspang in his lectures: That was a tag I carried with a smile over many years even though I believed it more fitting for Jens who in meeting changed my life…. I hope that you dear reader have such good fortune as I did. Enjoy your read…

Aalborg, Denmark   Anthony L. Brooks

Contents

Chapter 1
New Media Arts—The Thinking Space for Digitality

Monika Fleischmann and Wolfgang Strauss

Abstract With their works, the authors describe their approach to the digital and their artistic thinking as a mixture of analog and digital. They show how media art shapes *digitality* and illustrate strategies of interactivity with and without the use of *artificial intelligence*. Their *groundbreaking* works (Peter Weibel, 2012) each reveal new aspects of the digital. Since the late 1980s they have been realizing these works as artist-scientists in research institutions, where they created their own structures. They co-founded *ART + COM* in Berlin and headed art and technology departments in major German research institutions such as the *GMD Institute for Media Communication* and the *Fraunhofer Institute for Artificial Intelligence*. The motif of the *Denkraum* [thinking space] - a space filled with data - runs like a thread through all projects. How can one encourage one's own thinking and feeling despite all-encompassing automation? Based on an understanding of cultural hacking as a form of participation, the concept of the *thinking space* is presented as a counter-aesthetic to an inhuman digitality.

1.1 Emphasis

According to *Plato*, thoughts are the only real thing in the world. His allegory of the cave contains thoughts about truth and reality. Here we learn that truth is at home in the world of ideas. It is a matter of ascending from the sensually perceptible shadow world of ephemeral things, which is juxtaposed with an underground cave, into the purely spiritual world of ideas and thoughts. With the *Virtual Reality* (*VR*) installation *Home of the Brain* (1989), the authors depict the world of ideas as a networked computer brain. *VR* is the representation and simultaneous perception of an apparent reality and its physical properties in an interactive virtual environment

M. Fleischmann (✉) · W. Strauss
Center for Art and Media—ZKM, Karlsruhe, DE, Germany
e-mail: monika@fleischmann-strauss.de

W. Strauss
e-mail: wolfgang@fleischmann-strauss.de

that is computer-generated in real time. The authors, however, reverse the paradigm of *Virtual Reality*. Not what we see is real, but what we think is the decisive factor: *Real Virtuality*.

In order to gain access to this *Real Virtuality*, the authors develop *performative interfaces* that evoke the viewer's own thoughts. *Performative* means not only speaking and acting, but also an inner speaking that is connected to a physical presence or even a data body. The *Performative Interface* draws attention to the unknown. Not (only) the virtual environment is important, but the thoughts of the viewer. This hypothesis is further developed by the authors over the decades with each new work created. Interactivity does not simply aim to create a generative aesthetic, for example by altering the digital image through manipulations. Rather, interactivity is meant to evoke the viewer's own thoughts as an expression of a neural aesthetic. The resulting data is not recorded. What is important is the process of interactivity that is recorded and visualized once exemplarily in the work *Energy-Passages*. The *Industrial Interface* and the *Performative Interface* created by artists are fundamentally different. The essence of the industrial interface is *what you see is what you get*. The performative interface, on the other hand, turns things around and reveals what is hidden: *what you get is what you didn't see.*

1.2   ART + COM—An Interdisciplinary Media Lab in West-Berlin (1987–1992)

How did it all start? In the mid-1980s, nobody thought that everyone would have a computer at home in the future. Exploring the computer as a tool and a medium for communication was a rather futuristic goal. But, in *West Berlin* there was already a nucleus of digital media. *Urban Studies* Professor *Edouard Bannwart* led a research project on data communication between art academies in *Berlin*, *Braunschweig* and *Kassel*. In *Berlin*, he gradually brought in other digital and analog experts. Some, like *Fleischmann and Strauss*, came from the *University of the Arts*, others from *Mental Images*, a Berlin based software firm, from the *Technical University* and some associated with the *Chaos Computer Club*. Discussions went on all night long until a common language interlinking the disciplines were found. One year later, in 1988, they established *ART + COM* as an institute for the design and development of digital media in the form of a *registered association*. *Berkom, Deutsche Telekom's* research department, generously supported them. The IT industry provided them with hardware free of charge. They wrote their first research proposals.

At that time, remarkable scenarios were conceived as part of the long-term research projects *New Media in Urban Planning* and *The Digital Model House*. The interactive city simulation *Berlin-Cyber City* (1989)[1] took visitors into the past and the future of Berlin after the fall of the Wall. This was followed by one of the

---

[1] *Berlin-Cyber City* (1989-90) by *Monika Fleischmann* and *Wolfgang Strauss*, *ART + COM*, www.youtube.com/watch?v=BYmnVTvGssg&t=2s

world's first artistic VR installations *Home of the Brain* (1989),[2] which took the visitor inside the virtual New National Gallery. Both parts of the interrelated works illustrated the idea of a virtual information landscape that follows the seamless transition from outside to inside as in *Alice in Wonderland* by *Lewis Carroll* or in *Power of Ten* by *Charles & Ray Eames*. From the bird flight perspective over the city into *Mies van der Rohe's New National Gallery* and into the virtual exhibition about four thinkers and further into their houses of thought.

Here, the basic idea of a consistent world-related information structure is already emerging, as it was realized a short time later in the *ART + COM* project *TerraVision* (1993)[3] using satellite data and finally introduced as the mass product *Google Earth* (2007)[4] and *Google Earth Timelapse* (2021).[5] The legal dispute between *ART + COM* and *Google* on this topic—partly fictional—became public through the *Netflix* movie *The Billion Dollar Code*.[6]

Nevertheless, the two idea-generating projects—Berlin Cyber City and Home of the Brain—were much further ahead of the *Google Earth* adaptation at their core. Both were created as virtual walk-in thought spaces. No longer producing an image drawn by a computer as in the early computer art, but spaces that established communication and participation.

The audience of *Berlin-Cyber City* took part in virtual city walks, exploring and reconstructing the empty center of *Berlin* through their memories whilst discussing wishes for the future of the reunited city. The *Cyber City* interface, an interactive table with an aerial photo, brings visitors from *East* and *West Berlin* into conversation. The goal of the project was to stimulate a public discussion. *Berlin-Cyber City* opened up a space for communication and action. In contrast, *Home of the Brain,* on the other hand, presented the positions of four thinkers on the impact of digitalization. The audience gained access to the work by moving through the virtual buildings of thought. *Home of the Brain* created a learning space through which one could move virtually with VR devices to look and stimulate one's own thoughts.

1.2.1   *Home of the Brain (1990–91)—Philosophers Houses*

The *VR* installation *Home of the Brain* not only reflects the new medium, but also the media discourse itself is exhibited here as a philosophical debate. The virtual

---

[2] *Home of the Brain - Virtual Reality* installation (1989-91/Video 1992) by *Monika Fleischmann* and *Wolfgang Strauss earned them a Golden Nica at Ars Electronica in Interactive Art (1992).*
    https://www.youtube.com/watch?v=uMAA6GYLJPU

[3] *TerraVision* (1993-94) by *Axel Schmidt, Gerd Grüneis, Pavel Mayer, Joachim Sauter* and many others at *ART + COM.* https://www.youtube.com/watch?v=zBMJVgi8vm8

[4] *Europe Tour Google Earth* 2007, https://www.youtube.com/watch?v=wAtu3aGpNs8,

[5] *Our Cities | Timelapse in Google Earth* (2021) https://www.youtube.com/watch?v=v74_mf2usc0

[6] *CRE222 Terravision 2021.*
    https://cre.fm/cre222-terravision?podlove_template_page=episode-transcript-html

environment features the houses of philosophers, represented as metaphorical buildings of thought, dedicated to the ideas of *Vilém Flusser, Marvin Minsky, Paul Virilio* and *Joseph Weizenbaum*. Their opposing positions on digital culture become visible and audible as a battle of words—a heated debate. Art historian *Oliver Grau* calls the *Home of the Brain* (1991) "one of the earliest memory spaces that represented a completely new form of public space—that of global computer networks. ... A modern version of a *Stoa*, it offers a simulated, highly symbolic information space in which to engage in a metaphorical discourse about the ethical and social implications of new media technology (Fig. 1)".[7]

Using data gloves and data glasses, visitors enter the virtual space, inhabited by the thoughts of four media theorists. They had been chosen as important and controversial thought leaders in *Digital Culture* and *AI*. Their voices can be heard, they ask questions, and debates are triggered virtually as one moves through the virtual space. In the hypermedia architecture, the thinkers' voices become audible through the participants' movement. Within texture maps of light, shadow and sound examples, statements, sentences, and words build an interactive real-time landscape. The visitor moves through this virtual narrative and literally dives into the philosophers' minds. The audience observes this audio-visual scenario on a large projection screen. The navigator becomes the storyteller, while the audience takes on the role of the choir by commenting as in the theatre of antiquity.

The theoretical discourse on new media, which normally takes place in books, shifts from an interview into this virtual environment, e.g. *Flusser's* quote can be heard in text fragments: "People are getting worse, but technology is getting better". *Minsky's* view that there is no difference between the real you and your digital clone forms the narrative basis for the cognitive mood of virtual space. Even with its few text fragments (4 × 4), *Home of the Brain* allows for complex arrangements and combinations of sentences. The natural voices fill the virtual room and, together with the portraits of the protagonists, have an identity forming effect. The quality of the visual atmosphere is enhanced by the first-time use of special image synthesis in a 3D environment that normally appears graphically flat due to standard rendering methods. In contrast, in *Home of the Brain* the virtual space is bathed in physically calculated light that casts shadows and makes individual objects glow using a specially developed radiosity algorithm. The spatial impression is reminiscent of *Giorgio de Chirico's Pittura Metafisica* (Metaphysical Painting). The ability to simulate magical light scenes is an effective way of conveying attention and meaning.

On the occasion of the *Geneva Telecommunications '91*, an early version of *Home of the Brain* was sent virtually from *ART + COM* in *Berlin* to colleagues in *Geneva* via *ISDN* data lines. At both locations, the participants equipped with *data glove* and *data glasses* moved through the virtual environment and tried—successfully— to shake the virtual hand of the remote counterpart. Surprisingly, the colleagues

---

[7] Oliver Grau, "Immersion and Interaction. From circular frescoes to interactive image spaces". In: medienkunstnetz.de/themes/overview_of_media_art/immersion/scroll/

Fig. 1 *Home of the brain* (1990–91)—Media theory as a walk-in thinking space

in *Berlin* were able to recognize the participants in *Geneva* by their gestures and movements.[8]

Home of the Brain transfers the concept of the *Stoa* as a public place of encounter to a virtual environment. In doing so, the authors address the concept of telepresence

[8] Fleischmann, Monika; Strauss, Wolfgang: Staging of the Thinking Space. From Immersion to Performative Presence. In: Paradoxes of Interactivity, by Uwe Seifert, Jin Hyun Kim, Anthony Moore (eds.), transcript Verlag Bielefeld 2008. S. 268.

and refer to *Hole in Space* (1980) by *Kit Galloway & Sherrie Rabinowitz.*[9] While in *Hole in Space* people in *Los Angeles* greet each other by waving at a real department store window in *New York*; in *Home of the Brain* participants from *Geneva* and *Berlin* meet in a virtual place. *Home of the Brain* was awarded the *Golden Nica for interactive Art* of the *Prix Ars Electronica 1992*[10] for the organization of multimedia information as an interactive narrative, for interaction, navigation, and telepresence in virtual space.

However, the classic virtual reality interface with data glasses and data glove excludes the viewer from the physical world. In contrast, *touch,* and *tangible user interfaces (TUI)* offer the potential to support sensory perception and cooperative situations. The interactive sensor of the *data glove* had already been removed for *Berlin-Cyber City* and used as a tangible object to navigate across the table. Exploring intuitive interfaces that do not require long explanations and do not keep the participant stuck in front of the computer had been and continues to be an interest of interactive interface design.

New interface concepts soon became counter designs to virtual reality, the latter of which cuts off from reality and replaces it with a computer-generated world. Instead of immersing ourselves in *Virtual Realities*, we at *ART + COM* wanted to remain in the everyday environment that can be experienced with all our senses in the newly reunified city of *Berlin.*[11] More than 30 years later, however, *Virtual Reality* as well as *Mixed Reality* is now omnipresent in contemporary art and has established itself as an independent—now affordable—technology and visual genre.

1.3   VisWiz—Visual Wizzards at the GMD: The German National Research Center for Information Technology (1992–95)

In the same year that the *Prix Ars Electronica* was awarded (1992), the authors followed the call from *Wolfgang Krueger,* a former colleague and physicist at the *GMD, the German National Research Center for Information Technology* in *Sankt Augustin,* and *Bernd Girod,* a computer scientist at the *Academy of Media Arts (KHM)* in *Cologne to join the institutes.* They share their time between the newly emerging *Visual Wizzards* research group at *GMD* as visiting scholars, and as first artist fellows at the newly founded *KHM*, in rapid succession becoming permanent members of the

---

[9] Telecollaborative Art Projects Of Electronic Café International Founders Kit Galloway & Sherrie Rabinowitz. Overview Of A Quarter Century Of Pioneering Artistic Achievements 1975–2000 http://www.ecafe.com/museum/history/ksoverview2.html.

[10] *Prix Ars Electronica* 1992 / Jury Statement / Interactive Art https://webarchive.ars.electronica. art/en/archives/prix_archive/prixJuryStatement.asp%3FiProjectID=2581.html

[11] The art of the thinking space — a space filled with data. Wolfgang Strauss and Monika Fleischmann. Image Science, ZKM | Center for Art and Media, Karlsruhe, Germany. In: Digital Creativity 2020, VOL. 31, NO. 3, 156-170. https://doi.org/10.1080/14626268.2020.1782945

VisWiz research team. In the aftermath of *Krueger's* sudden death, *Dennis Tsichritzis*, president of the *GMD along with the scientists within the group* voted for *Monika Fleischmann* to take over the leadership of the research group. *Wolfgang Strauss* joined as artistic research director. In this way, the scientists were able to continue the work they had begun as doctoral students or postdocs, working with the artists on joint artistic-scientific projects. During this time many lasting works, tools, and interfaces were created by this interdisciplinary team, composed of disciplines such as computer science, physics, mathematics, art, and architecture. One of the authors' iconic masterpieces from this period is the two-part interactive work: *Liquid Views* (1992) as a media mirror of contemporary *Narcissus,* and *Rigid Waves (1993),* a metaphor for the story of *Narcissus and Echo.*

1.3.1 *Liquid Views (1992)—Narcissus' Digital Mirror. Ovid's Metamorphosis as a Real-Time Wave Algorithm*

Liquid Views (1992) is an interactive installation showing a virtual water surface that reflects the viewer's image. A touch screen with a miniature camera simulates water and mirrors the image of the approaching viewer. As soon as the viewer touches the screen, algorithms are activated to amplify the artificial waves, and the viewer's reflection dissolves. In this way, the integrity of the looking glass is disrupted. As soon as the viewer stops touching the surface, the water becomes tranquil, and the reflection reappears. With this work, the audience participates in a *Narcissus* experience that can be captured in virtual memory as a kind of automatic selfie, so to speak (Fig. 2).

The unpredictability of the artificial wave algorithm in *Liquid Views* symbolizes the Internet. Whoever touches the water leaves traces that can be traced back. People are seen by others without realizing it. The mirrored face of the participant is presented on a large projection, turning the introverted gaze into a public spectacle. The interacting people look up and perceive themselves from a different perspective. It is like a glance from the outside, behind the mirror, into another world. In *Lewis Carroll's Through the Looking-Glass* (1871), *Alice* steps through a mirror into a world of dreams. *Orpheus* also passes through a liquid mirror in *Jean Cocteau's* film *Orphée* (1949).

Liquid Views connects the real world with the virtual world, as an interplay of image within image. In each of the more than 30 countries where *Liquid Views* was exhibited since 1992, cultural differences were noticed in the way visitors interacted with the installation. In *Mexico*, reverently; in *Los Angeles,* playfully; in *Madrid*, many couples kissed in front of the water mirror; in *Paris*, some men didn't think they were beautiful enough to see themselves. In *Japan*, people compared the *Narcissus* mythology to the story of the little sun *Amaterasu*. Today, the situation is no longer observed and reflected upon as it was back then. Since 2007, everyone immediately pulled out their smartphone and takes photos of themselves. Now the moment is captured for one's own social media archive. Reflection is postponed until later.

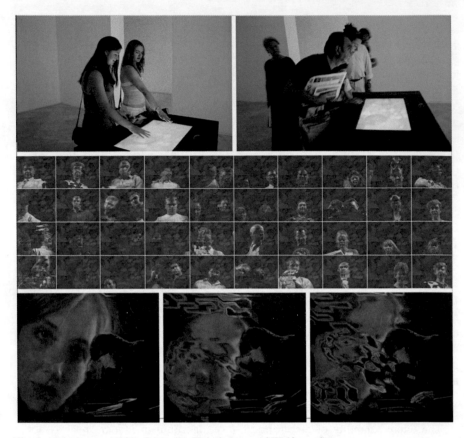

Fig. 2 *Liquid views* (1992) © Monika Fleischmann and Wolfgang Strauss

Liquid Views anticipates the *selfie* moment, in which the image and the self are increasingly difficult to parse. Here, the digital medium is camera, mirror and archive in one. It records us, mirrors us and stores us. Screen and camera are on one and the same level. The structure of *Liquid Views* from 1992 already anticipates partly the interface of the *Apple iPhone* from 2007. The greatest missed opportunity was that no patent office, neither in Germany nor in EU Brussels understood the concept, or was prepared to patent the authors' touch screen camera. But in the end, it's not just about invention when you want to bring something new into the world, but also about a mass-compatible application. Of course, at this early stage, the hardware was still extremely expensive. It was only 15 years later with the smartphone that this combination—supplemented by the voice function—became an internationally marketed mass medium. Back in 1992, *Liquid Views* was a homemade, high-resolution multi-touch screen with an integrated mini camera as the interface that simulates a water surface and mirrors the viewer's image. In 2022, an online version was developed that runs on a *tablet,* making it a lot easier to exhibit.

At the beginning of the digital age in the mid 1990s, establishing a lab for *Art and Science*[12] meant leading a research team and collaborating to develop breakthrough tools and concepts. Our research focused on interfaces for the senses. These are body-related interfaces that work with and address or respond on sensual perception. Each new application prototype and artistic-scientific work was usually presented to an invited audience at the much acclaimed annual *GMD Schlosstag* (Castle Day). *Liquid Views* was first exhibited at *GMD's Birlinghoven Castle* (1992), and internationally at *Siggraph's Machine Culture* show in *Anaheim* (1993). Sadly, in the excitement of this new digital era, few had the foresight to protect intellectual rights. It was all the more difficult to protect artistic authorship and scientific patents. Naturally, the ultimate goal was to create interests for further investments, however in those days, the *German* state, its politicians, universities and companies did not believe in the future technology of the Internet and that it could go beyond scientific operations. *MP3*, the world standard for audio coding, was sold to a *US* company by *Fraunhofer* Research after the patent expired and they no longer expected any added value. It was later bought back at a high price and the not inconsiderable proceeds are now used to finance further research.

In those early days, there were many obstacles into succeeding with digital trans-formation, innovative prototypes, and interactive art. There is a difference in attitude to the way *Europe* and the *US* view their investments in pioneering new technolo-gies. Whilst both are forward looking, the *US* is willing to take a longterm view on investing in innovative research projects, but with a more assured anticipation of commercial success, providing continual legal expertise and a watertight patent. A classroom example of this scenario, documented in the *Netflix* production 'The Billion Dollar Code',[13] tells of the legal dispute between *Europe's ART + COM* (*Terravision*) and the *US' Google* (*Google Earth*). A story—told with some artistic liberty—but showing clearly the inspiring *European* vision of the participating artists and scientists of *Terravision* fifteen years earlier and the *American* entrepreneurial talent in bringing *Google Earth* to a global market.

Other noteworthy new interfaces and installations include the *Responsive Work-bench*[14] *(1995)*, a successor to the interactive *Berlin-Cyber City* table installation. It was selected for its outstanding contribution at *Siggraph* in *Orlando* in 1994, and further showcased for several weeks at the *Disney EPCOT* center. *EPCOT* stands for *Experimental Prototype Community of Tomorrow* and within the *Walt Disney World Resort* in *Orlando* proactively presents *New Prototypes for the Future*, as in the case of *ART + COMs' Terravision* or *GMDs' Responsive Workbench* (Fig. 3).

---

[12] From 1992-1995 *Monika Fleischmann* is head of *VisWiz* and - together with *Wolfgang Strauss* - the artistic-scientific director of the research group. This involves 8 scientists and is, so to speak, the predecessor research group of *MARS*.

[13] *'The Billion Dollar Code': The battle over Google Earth* https://www.dw.com/en/the-billion-dollar-code-the-battle-over-google-earth/a-59433376.

[14] *The Responsive Workbench: A Virtual Work Environment. Wolfgang Krueger* et al. In Journal "IEEE-Computer", 1995.

Fig. 3 *Responsive workbench, airial drawing, virtual balance, virtual studio TV* © Monika Fleischmann and Wolfgang Strauss

1.3.2 Responsive Workbench (1993)—Thinking with Your Hand

The *Responsive Workbench,*[15] an interactive 3D VR workstation for urban planning and medicine, created at *GMD* under the direction of *Wolfgang Krueger,*[16] was a further development based on the interactive table by *Wolfgang Strauss* at *ART + COM*. The *Responsive Workbench* combines analog and digital working methods for the first time. Computer-generated stereoscopic images are projected onto the horizontal display of a table surface via a projection and mirror system and viewed through shutter glasses. Control is via stylus or voice recognition and designed as a collaborative workbench for real-time 3D display. A remote-controlled workbench for two hands and two viewers is being developed at *Stanford University* in collaboration with *Pat Hanrahan*. Drawing with the hand in the air—comparable to *Picasso's light drawings*—also emerged from this collaboration and was later implemented. *Steven Schkolne* from *CalTech* presents this intuitive design interface in his talk *Tracing the line of thought on the Responsive Workbench*[17] at the memorable *CAST01* at *Schloss Birlinghoven*. It means "*thinking with your hand*", as *Strauss* puts it. Fifteen years later, this invention is distributed as 3D VR software by *Google* under the name *Tilt Brush* (2016). The success of such projects relied heavily not only on the complicated project management, but also on a collaborative team and its passion for the cause, in this case the *Responsive Workbench* (1993) had already received the award for their inspiring work at *Ars Electronica* back in 1994. Oddly, *Strauss*, who designed the table, and *Fleischmann's* instrumental project management have

---

[15] Responsive Workbench http://netzspannung.org/database/responsive-workbench/

[16] Even before Wolfgang Krueger's death, the project management was continued by Monika Fleischmann together with Wolfgang Strauss, Christian A. Bohn and Bernd Froehlich and in collaboration with Pat Hanrahan at Stanford. Froehlich was assigned Krueger's Fellow invitation to Stanford.

[17] Steven Schkolne, Peter Schroeder, Tracing the Line of Thought on the Responsive Workbench, at CAST01, Schloß Birlinghoven, Sankt Augustin, Germany, 2001. www.academia.edu/68014217/Cast_01_Living_in_mixed_realities_netzspannung_org_event_conference_proceedings_Monika_Fleischmann_Wolfgang_Strauss_21_22_September_2001

not been mentioned in most publications, this reflecting the apparent chasm in the collaboration between art and science and onesided recognition.

1.3.3 The Virtual Balance (1995)—Looking with Your Feet

It requires body balance but no fine movements to get through virtual spaces and extreme perspectives with the *Virtual Balance*. The performer stands on a balance platform and navigates through a three-dimensional landscape only by shifting body weight. It was therefore not surprising that physically disabled people were particularly motivated to increase their range of motion with the help of the virtual world. Beneath their feet, the ground became an interactive surface, and even their partially limited body movements became a control tool through the virtual reconstruction of the archaeological excavations of an *ancient Roman* settlement near *Xanten* in *western Germany*. Navigating with this body-oriented interface is like *looking with your feet*. The body itself becomes the interface.

The weighing sensors detect every shift in weight and convert it into position, movement and navigation through the 3D landscape. The result was such an intense form of sensory perception that visitors could fully immerse themselves in the virtual environment even without VR goggles. This was seen at the first public exhibition in *Luxembourg* (1995), when three members of *The Grand Ducal Family of Luxembourg* flew through the virtual temples, and the 12-year-old son in particular moving very skillfully over the landscape and into virtual houses or on the ground. They felt immersed in a virtual reality without excluding the real world.

The sense of the senses in the virtual environment belong to the object of investigation in the work on new human-machine interfaces, which was deepened by the repeated cooperation with the neurologist *Hinderk M. Emrich*. Tactile perception includes not only touch perception, but also exploratory perception. Sensing one's own body in the virtual environment—here with the help of the *Virtual Balance* floor plate—observably expanded body awareness. During the presentation at *CeBit '96* in *Hanover*, *Emrich* got carried away with dance-like movements and discovered an inspiring perspective on the virtual world.

Sensing one's own body in the difference between one's own and another's body establishes bodily self-awareness. The sense of touch is our basic sense: ultimately, it is only through the experience of touch that we learn to perceive and classify visual impressions. In a small series of experiments, people with physical disabilities were enabled to experience the transition between real and virtual 'worlds'. We could observe that they felt more motivated to navigate in the virtual environment through the tangible exploration of this world. According to neuroscientist *Antonio Damasio*: I feel, therefore I am. It had to do with the fact that they could move there almost more skillfully than performers without disabilities, and because no difference could be seen in their virtual avatar compared to that of "normal" people.

1.3.4 Virtual Striptease—First Networked Virtual Studio TV Production (1995)

For the *International Video Art Award* ceremony organized by *ZKM Karlsruhe* and *Suedwestfunk Baden-Baden (SWF)* in 1995, the authors conceive a broadcast with audience participation. With their teams they realize the performance and the corresponding digital set design for the virtual stage. For the live production of the show, the signals from the studio camera in *Baden-Baden* are connected via the high-speed network 300 km away with the virtual backdrops in the *GMD* computer pool in *Sankt Augustin* via a *Telekom broadband transmission (ATM)*. The entire huge TV studio in *Baden Baden* is set up as a *Bluebox* stage for 150 spectators. On the large screen opposite the stage, the audience watches the actors move through the virtual scenery. The moderator welcomes *Heinrich Klotz*, the founding director of the *ZKM*, and both sit on an invisible bench in a kind of wireframe backdrop. *Klotz* congratulates the individual *ZKM* award winners in different digital scenarios that refer to the respective award-winning work. Media artist *Bill Seaman*, who receives the main prize, finds his way around the stage and the virtual scenario after orienting himself on the large screen on the opposite wall, which shows the final image as for the TV viewer at home.

The audience sits on *swivel chairs* and has the choice of alternately watching the real action on the blue stage or the rendered TV image on the large projection opposite. One of the three performances is a *Virtual Striptease*.[18] Two performers, wearing blue catsuits under their dresses, slowly disappear onto the blue stage and become one person as they shed their clothes. The entire performance is a great success. The audience is fascinated.

This and other work catapults the authors and their research group to the forefront of artistic-scientific IT and media art scene. They are invited to show their work at international fairs such as *Siggraph*, *Imagina*, *ISEA* and in museum exhibitions such as the *ZKM* in *Karlsruhe*, the *Centre Pompidou* in *Paris* or the *MOMA* in *New York*. For the authors, the foundation of a new laboratory for media art is in the air, one that is even more self-determined in its orientation to their own research interests.

1.4 MARS—the Media Arts Research Studies at GMD (1995–2001) and at Fraunhofer (2001–2012)

The *MARS - Exploratory Media Lab*,[19] initiated and directed by *Monika Fleischmann* and *Wolfgang Strauss* since 1997, designed and developed research prototypes, artistic projects, tools, interfaces and events intersecting art, science and technology.

---

[18] Virtual Striptease - Video Only (1995) by Monika Fleischmann & Wolfgang Strauss (YouTube Chanel) https://www.youtube.com/watch?v=2Qg1tQYv-PQ,

[19] MARS stands as acronym for Media Arts Research Studies.

The experimental research lab invented *"tools for the art of tomorrow"*[20] in teams of architects, artists, designers, computer scientists, art and media scientists. They asked how people interact with technology and about life in a networked society such as how is *"Living in Mixed Realities"*[21] and they designed future models of networked knowledge and communication spaces. The aim was to present artistic approaches to media technology, to promote media-theoretical and practical skills in art, culture and education, and to critically reflect on technological developments in terms of their impact on people and the environment. With its focus on integrating *IT* research and media art, the *MARS* research group was unique in *Germany*.

The *MARS Lab* started in 1997 with three long-term projects on interfaces and community:

(1) The *EU* funded *eRena—electronic arenas for art and entertainment* (1997–1999) explored various tracking techniques such as computer vision and electro field sensing. A multi-user camera tracking system for the stage space was developed for the networked performance installation *Murmuring Fields* (1997–99), as well as a novel interface, the *PointScreen* for touchless navigation.
(2) The *BMBF* funded *CAT—Communication of Art & Technology* (1998–2006) developed *netzspannung.org* for and with the media art community, an online archive as a community platform for digital culture. This was preceded by the *CAT*[22] feasibility study (1997–98), written with input from the international media art community.
(3) The *eCulture Factory*[23] financed by the *Senate of Economy* in *Bremen* (2005–2008). The objective was to install a think tank by bringing people and labs together to create a digital culture of innovation in Bremen: Transfer of research results into the corporate sector.

At peak times, the *MARS Lab* consisted of more than 25 employees of different disciplines, nationalities and genders. Since the end of the 1990s, some employees have also been working at home in *Berlin, Frankfurt, Zagreb* or *Trieste* from time to time for conceptual or family reasons. For more than a decade, *MARS* has acted as a kind of experimental learning and teaching institution for digital media in Germany, when there were still hardly any degree programs in art colleges or universities.

---

[20] MARS is working in the eRENA consortium with ZKM and other research institutions during an intensive three-year research project, funded by the European Commission's i program, to develop a new generation of tools for planning and production for arts and entertainment in the twenty-first century. https://www.erena.kth.se/intro.html.

[21] Living in Mixed Realities: cast01. press release 08/06/2001. https://idw-online.de/en/news37787.

[22] CAT - Communication of Art and Technology. Machbarkeitsstudie für ein Kompetenzzentrum Kunst, Kultur und Neue Medien, Band 1 + 2. Monika Fleischmann, Wolfgang Strauss, 1997-98, GMD German Research Center for Information Technology, Sankt Augustin, Germany.
www.researchgate.net/publication/357781624_CAT_-_Communication_of_Art_and_Techno logy_Machbarkeitsstudie_fur_ein_Kompetenzzentrum_Kunst_Kultur_und_Neue_Medien_Band_ 12_Monika_Fleischmann_Wolfgang_Strauss_1997-98

[23] eCulture Factory, a temporary lab for digital culture and interactive media in Bremen and part of the MARS Lab. http://eculturefactory.de/CMS/index.php?id=337.

MARS worked on the digital culture and asked about how the Internet influences human life and the entire planet.

The *MARS Lab* was scientifically evaluated four times between 1999 and 2007. Not only was the excellence of the various projects with advanced technology always highlighted, but also the diversity of the staff and the gender sensitivity of the management. From the first scientific evaluation report on the *Arts & Technology Research Department MARS* (1999): "We are impressed with the diversity and impact of many projects that blend technology and art into unique forms of expression. These works are embodied variously as sculpture, multimedia content, sensor-based mechanical systems. What may at first seem like frivolous uses of advanced sensing technology, backed up by computation, real-time graphics, and computer-generated sound, are in fact at the forefront of what is variously known as *ubiquitous computing*, *pervasive computing*, *perceptual user interfaces*, and *mixed reality environments*."[24]

MARS was one of four departments of the *GMD Institute for Media Communication* with 240 employees. In 2001, the federal government decided to close *the GMD, the German Research Center for Computer Science*, with its basic research and to merge it with the *Fraunhofer Organization for Applied Research*. For reasons of efficiency, the former *GMD Institute for Media Communication* and the *Institute for Artificial Intelligence* were also merged under new management. At *Fraunhofer*, different standards apply to the acquisition of project funds. Orders from industry count above all. And although *MARS*, with public research funds, is one of the most successful departments financially and in terms of content and has received excellent international evaluations, it no longer fits in with the plans of the new institute management and was being wound up.

This marked the end of the national and international impact of a special research group on digital creativity. So that not everything ends up in the wastepaper basket, the artistic works and tools were handed over to the *ZKM* on permanent loan, where parts of them are exhibited repeatedly. One important work from this period is the platform *netzspannung.org*, which has been documenting the history of creative digital work at universities in German-speaking countries for almost a decade.

1.4.1 Netzspannung.org—One of the First Media Art Online Archives (1998–2010)

The Internet platform *netzspannung.org* is a comprehensive archive of the media theoretical discourse, artistic work and new mediation strategies for digital culture. A key feature is the transdisciplinary consideration of art, design, theory and information technology as well as the provision of online teaching and learning modules. The site has thus succeeded in establishing an interdisciplinary information pool that

---

[24] 1999 Scientific Evaluation of the *GMD - Institute for Media Communication*, Arts & Technology research division MARS - Media Arts Research Studies headed by *Monika Fleischmann*, Dec. 23, 1999, https://www.fleischmann-strauss.de/Evaluation

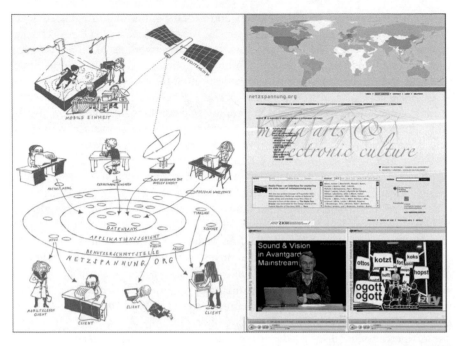

Fig. 4 *Netzspannung.org* sketch 1999, homepage of the platform 2001. Maps: origin of users 2002, With the *Hyper media Tele-Lecture* (2003) *Dieter Daniels* accesses the two networked databases in his lecture © Monika Fleischmann and Wolfgang Strauss

appeals equally to humanities scholars and computer scientists, artists and designers, agencies and IT companies.[25]

In the first decade of the 2000s, *Netzspannung.org* has become a steadily growing online archive documenting digital culture in Germany. One of the main motives for this project was to document the numerous interactive media artworks and thus save them from disappearing because the hardware and software were no longer available. At least what was created during the period of active operation of the platform should be documented. Most museums were not interested in digital art because it was too technically complex. Even today, they rarely have competent staff nor the necessary resources (Fig. 4).

---

[25] Cf. *Monika Fleischmann, Wolfgang Strauss,* 2005: On the Development of netzspannung.org – An Online Archive and Transfer Instrument for Communicating Digital Art and Culture. In: Present Continous Past(s). Media Art. Strategies of Presentation, Mediation and Dissemination, edited by Ursula Frohne, Mona Schieren, and J.F. Guiton, p 162-173. Wien: Springer. www.aca demia.edu/23731707/Fleischmann_M_Strauss_W_On_the_Development_of_netzspannung_org_ An_Online_Archive_and_Transfer_Instrument_for_Communicating_Digital_Art_and_Culture

The *CAT* study (1997–98)[26] for the conception of the online platform, later named netzspannung.org, started with an e-mail survey. Artists, scientists and theorists from the field were asked about the need for international networking of digital culture. At the same time, the survey served community building within the international media art scene. The numerous responses supported the funding of the project by the *German Federal Ministry of Research and Education*. The concept envisaged creating several distributed stages (server areas) and bundling information on media art design and education. The idea was to create digital culture as a platform for various actors and providers and to network with partners like *Medienkunstnetz.de* or *ADA*—the *Archive of Digital Art*, to create a global network for media art. Visionary was the *MARS Lab's* idea of distributed servers that collectively create cross-border knowledge repositories. Not a central archive but a networked platform archive was to be created that would bundle distributed knowledge on media art. This idea of networked archives was later implemented on *netzspannung.org* in collaboration with *MedienKunstNetz.de*. The MARS Lab invited art historian *Dieter Daniels* to give a hypermedia lecture based on the archives' material. He spoke about *Sound & Vision in Avantgarde & Mainstream*[27] and was able to access materials from both archives during his talk. Unfortunately, this one experiment has remained, because one feared—without reason—that all content would only be found at *netzspannung.org*. The fear of losing personal recognition through collaboration with others is sometimes stronger in the academic community than the prospect of shared greater success. Therefore, it is always beneficial to encourage individual achievement while not neglecting collaboration between partners.

Finally the concept of the platform was presented at the *Memoria Futura* symposium (1999) organized for this purpose, and the first results were presented in a live workshop with the *MARS* online video conferencing system *i2tv* on site in *Sankt Augustin* in a shared network with *New York* and *Toronto,* creating another mixed reality situation that became the norm more than 20 years later with *ZOOM* and other conferencing systems.[28] Participants at the time included *Regina Cornwell, Christina von Braun, Derrick de Kerckhove, Mihai Nadin, Gerfried Stocker, Nadja Thalmann, Joseph Weizenbaum*. The venue for the debates on virtuality was the *GMD's Birlinghoven Castle* in *Sankt Augustin*, where the baroque portraits on the walls, were confronted with the discussion about equally idealized cloned images of human beings, prompting heated arguments.

---

[26] CAT—Communication of Art and Technology Machbarkeitsstudie für ein Kompetenzzentrum Kunst, Kultur und Neue Medien, Band 1 + 2 Monika Fleischmann, Wolfgang Strauss, 1997-1998. www.academia.edu/67875616/CAT_Communication_of_Art_and_Technology_Machbarkeitss tudie_f%C3%BCr_ein_Kompetenzzentrum_Kunst_Kultur_und_Neue_Medien_Band_1_2_M onika_Fleischmann_Wolfgang_Strauss_1997_1998

[27] "Sound & Vision in Avantgarde & Mainstream" by Dieter Daniels. http://netzspannung.org/tele-lectures/hypermedia-lectures/

[28] The i2tv System: A Mixed Reality Communication Interface Jasminko Novak, Monika Fleischmann, Wolfgang Strauss, Predrag Peranovic, Christoph Seibert. www.researchgate.net/public ation/228853605_The_i2tv_System_A_mixed_reality_communication_interface

Two years later *netzspannung.org* was presented at the legendary conference *CAST 01—Living in Mixed Realities*,[29] which took place 10 days after the terrorist attack on the World Trade Center on September 11, 2001. The meeting was characterized by an intense atmosphere and the team spirit of the participants, who hardly wanted to part even after 3 days. The conversations were increasingly influenced by current world politics and sought answers to the terrible event.

The conference was aimed at anyone working at the intersection of art, culture, new technologies, and media studies. The theme of the conference was exciting for a large audience and several hundred attended. "What does it mean to live, play and work in a world shaped and perceived through digital media, networks, and architectures of real and virtual space? How can the development of complex communication spaces, living environments, and economic models be designed as an interplay of technological, social, and artistic forces, as Mixed Realities of Art, Science, and Technology?".[30]

Despite the terror attack, there were hardly any cancellations and prominent speakers appeared such as *Rebecca Allen, Roy Ascott, Maurice Benayoun, Jay D Bolter, Sabine Breitsameter, Hiroshi Ishi, Steve Mann, Laurent Mignonneau, Christa Sommerer, Tamiko Thiel, Jill Scott, Victoria Vesna* and many others. New Yorkers like *Bill Buxton* and *Natalie Jeremijenko* cancelled for all too understandable reasons. Who wanted to board a plane at that time anyway? Nevertheless, more than 250 international participants from the arts and sciences attended the conference at the *GMD's Science Park in Birlinghoven,* along with several hundred online participants.

All lectures were recorded with the *MARS Lab's* new *Mobile Unit System* (2001) and broadcasted live on the Internet. Highlights included the launch of the *online media lab netzspannung.org* and the *Digital Sparks* initiative. The annual student competition held from 2001 to 2008, was realized directly in its first year 2001 using the platform's infrastructure. The entire conference, the competition and all future activities were documented online. In contrast to monopolistic platforms such as *Facebook* (2004), *netzspannung.org* differed in its decentralized structure, which corresponds to a cooperating community with individual input/output data.

The platform for interactive media art and teaching with digital media became an instrument for research, reflection and mediation of electronic culture. More than 3,000 work descriptions (texts, images, videos) for interdisciplinary education have been archived and are still accessible to the public for free. There are curated channels such as *Media Art Research, Tele-Lectures, Media Art Learning, Digital Sparks* and there is *netzkollektor*, the open channel for the community to integrate their projects. The platform technology has been designed to connect to other archives. *Netzspannung.org* was presented at media art festivals and exhibitions. With the

---

[29] cast01 // Living in Mixed Realities. Conference on artistic, cultural and scientific aspects of experimental media spaces. Sept 21–22, 2001, Sankt Augustin (Bonn, Germany). First website http://netzspannung.org/version1/cast01/

[30] CAST01//Living in Mixed Realities. (2001) Artistic, Cultural and Scientific Aspects of Experimental Media Spaces. Monika Fleischmann + Wolfgang Strauss (eds). www.academia.edu/129 0856/CAST01_Living_in_Mixed_Realities_2001_Artistic_Cultural_and_Scientific_Aspects_of_ Experimental_Media_Spaces_Monika_Fleischmann_Wolfgang_Strauss_eds_

Mobile Unit (2001) lectures from selected partner events were broadcast live on the Internet and to auditoriums of German universities as *Tele-Lectures* (2002) to allow a chat feedback. The streaming of the lectures was connected with the simultaneous archiving of different formats. It was a first model of a video portal, as *Youtube* (2005) later became.

One of netzspannung.org's partners was the *Burda Akademie*, which is backed by the major German publisher *Burda Media*. For the successful and always crowded *Burda* lecture series *Iconic Turn*[31] at the University of Munich, the netzspannung.org system was switched to higher bandwidths to cope with the large number of online accesses and to ensure high transmission quality as soon as prominent speakers such as video artist *Bill Viola*, film director *Wim Wenders* or architect *Norman Foster* came on stage to talk about the digital image from their point of view. Then the lectures also had to be broadcast to the entrance hall and even to the square in front of the university.

Around 150,000 monthly visitors used the free access to the multimedia database for studies, research and teaching already in the first years and still after almost 10 years. In addition to the collection and processing of current media art projects, one focus was on learning and teaching with digital media. This, as well as the consideration of contributions from the community in the *netzkollektor*, the open channel, distinguishes the platform from others.

Fairly early on, the *MARS Lab* explored, and tested archiving interfaces specifically suited to solving the problems of a growing platform. For this purpose, novel search and discovery tools were developed, the *Tools for Knowledge Discovery*, that were recognized as outstanding in the years to come. In its active period (2001—2010), the platform is a chronicler of media art and media art education in the German-speaking world. Due to a lack of funding, netzspannung.org was handed over to the ZKM | Center for Art and Media in 2010 and archived there as a *Virtual Machine*. Today, some things on the platform would need to be "fixed" in order for videos and documents to become fully accessible again.

The late *ZKM* Director *Peter Weibel* comments: "Netzspannung.org and its groundbreaking interfaces create cognitive and data-related structures. What has been created here is a model for an international educational infrastructure that deserves imitation" (Peter Weibel 2012)[32]. *Knowledge discovery tools* such as the *Semantic Map*, the *Matrix* or the *Media Flow* help people intuitively discover content and connections in the archive.

---

[31] Iconic Turn - The new image of the world. Burda Academy in Munich with netzspannung.org, as a cooperation partner, 2002-2003 http://netzspannung.org/tele-lectures/series/iconic-turn/index.xsp?lang=en

[32] Inter-Facing the Archive. The Media Art Portal netzspannung.org, 2012 https://zkm.de/en/event/2012/06/inter-facing-the-archive.

1.4.2 Faces of the Archive

It should be noted that the best way to achieve something new is to really know the subject and its problems. In this case, *MARS* had to deal with growing amounts of data and develop and deploy AI algorithms when there were hardly any applications to learn from. It was about reformulating the search. Basically, there are two ways to access electronically stored data: through a precise, targeted search or through imprecise, untargeted rummaging around. Searching assumes that people know what they are looking for, that they can formulate their interest, and can narrow or broaden the definition as needed. Rummaging or browsing, on the other hand, is about viewers being guided and inspired by what is presented to them. Finding embodies the principle of serendipity. And here we come to the semantic relations through text analysis, which lets us find out more using *AI* to train an artificial neural network (*Kohonen*, 1995[33]).

A digital archive that is nothing more than an index box with individual data sheets makes little sense. It is just a classic archive. Instead, the database of the media art archive is represented as an interconnected network that can take on interface formats that differ from the traditional listing. Various *Knowledge Discovery Interfaces* were being developed next for a test data set of 500 documents, which demonstrates different processes for unlocking information. S*emantic Map*, *Media Flow* or *Digital Sparks Matrix* are browsers that address a new paradigm of self-learning.

Two versions were always designed for different needs, an Internet screen version and a large visual-auditory interactive spatial installation for various situations at home or in the museum. *Mapping the archive of netzspannung.org*, the *Knowledge Discovery Tools* were exhibited at *ZKM* in the form of three installations with great response (Fig. 5).[34]

1.4.3 Semantic Map—A Navigation System for the Data Space (1999–2002)

The *Semantic Map*[35] does not represent a digital archive as a list of search terms. Rather, it relationally locates archive entries and makes semantic correlations visible. This *Self-Organizing Map (AI)* uses semantic text analysis to (1.) uncover content

---

[33] Kohonen, Teuvo. Self-Organizing Maps. In: Springer Series in Information Sciences (Vol. 30). Berlin, Heidelberg, New York: Springer. pp 51–75, 1995. https://link.springer.com/book/10.1007/978-3-642-97610-0

[34] YOU_ser 2.0: Celebration of the Consumer. Further development of the successful show YOU_ser: The Century of the Consumer, which has already been shown at the ZKM | Medienmuseum since October 2007. Video-Trailer: zkm.de/de/media/video/youser-20-celebration-of-the-consumer, http://www.eculturefactory.de/CMS/?s=Aktuell&c=Ausstellungen

[35] Strauss, Wolfgang, and M. Fleischmann. 2006. Semantic Map for Knowledge Discovery. http://www.eculturefactory.de/CMS/index.php?id=371.

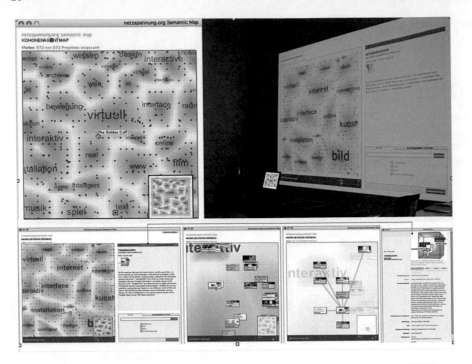

Fig. 5 *Semantic map.* Web version and spatial installation © Monika Fleischmann and Wolfgang Strauss

relationships between texts, (2.) identify similarities, and (3.) group them into thematic clusters. The *Semantic Map* shows in which conceptual environment a document is located and which other documents are most similar to it. The documents of the individual works appear as nodes in a network of thoughts, themes and practices that expands with each new database entry. The viewer discovers information through differentiation and thus becomes a co-producer of meaning. Therein lies the performative act of viewer and interface. Searching and finding, overview and detail are combined in the interface on one and the same level. It is an operational image that evokes movements of thought.

So, what happens when one studies the semantic map? New thought patterns emerge. Astronomer and *Leonardo* editor, *Roger Malina*, aptly compared this interface to a *telescope for viewing and evaluating the data cosmos.*[36] The *Semantic Map* transforms the database archive from a static information store into an intelligent and navigable knowledge network. It seems as if the data know about each other, which

---

[36] Cf. Wolfgang Strauss, Monika Fleischmann, Cultivation of Knowledge Spaces. In: ARTMEDIA VIII, Paris, October 2002. http://archive.olats.org/projetpart/artmedia/2002eng/te_mFleischmann. html

is reminiscent of a vision by *AI* researcher *Marvin Minsky*: *Can you imagine that they used to have libraries where the books didn't talk to each other?*[37]

1.4.4 MediaFlow—Thoughts in the Flow (2005)

Thoughts in the flow, that's what the *MediaFlow* Interface stands for. The entire content of an archive—here from *netzspannung.org*—is visible at a glance. Two parallel streams of images and words run along the projection wall. The large-format browser makes the contents of the online archive visually accessible. Images of projects and people, including titles, authors, and keywords from the entire archive stream pass by the viewer. In a text-to-speech process, the terms are whispered by computer voices. The *MediaFlow* creates the impression of a walk-in data room. The motif of the flow sets static and passive amounts of information in motion. Images and words are awakened to a visual and acoustic sphere and take on a narrative position. A large touchscreen interface translates images and words into scrollable text bands ready for selection. Semantically related documents, the database entries, are visually linked and highlighted on the big screen.

The formal structure of the *MediaFlow* installation is reminiscent of *Aby Warburg's* art historical concept of a thinking space—his *Mnemosyne Picture Atlas*. In the 1920s, he developed a collection of photographs, documents and texts arranged on many panels into a thinking space. These wooden panels were not meant to appear in a fixed order, but rather to continually evoke new insights by arranging and rearranging the documents. Instead of *Warburg's* variable image panels, the *MediaFlow* consists of streams of information, but can also be freely selected and rearranged.

Art historian *Daniel Becker* elaborates: "*MediaFlow* like *Mnemosyne* are correspondingly (…) fragmentary and therefore only the source or origin of a body of knowledge; the actual formation of knowledge arises through decisions in the interaction, reception or (re)configuration of the existing content by the user over the course or, or better, in the flow of time. Similar to *Warburg*, but algorithmically controlled, the fluid interface allows for arrangement and rearrangement, thus initiating a dialogue between viewer and content. This method explicitly allows for mind games, and so the digital archive becomes a space for thought movement. (…) The work of *Fleischmann and Strauss* thus enriches the discourse around collections of knowledge, because here it is not about—like a classic archive search, which is connected with previous knowledge - about targeted searching, but about browsing and finding. This affirmation of media-savvy *surfing* in the data pool of an online

---

[37] Minsky M. (1990) in an interview next to: The Future of Fusion of Science, Art and Psychology. Ars Electronica Symposium 1990: Natural Intelligence - Artificial thinking. https://monoskop.org/images/4/47/Ars_Electronica_Facing_the_Future_A_Survey_of_Two_Decades_1999.pdf

Fig. 6 *MediaFlow* Browser and Installation (2007) © Monika Fleischmann and Wolfgang Strauss and *Aby Warburg* Denkraum (1925)

database is inscribed in a long history of concepts for storing, archiving and collecting data (Fig. 6)."[38]

The *MediaFlow* is a visual finding tool and a research browser used to get a quick overview of the contents in the archive and direct access to individual documents. Through the openly visible creation of relationships between works, texts and lectures, a form of learning is practiced as thinking in contexts. Part of the concept

[38] Becker, Daniel. 2017. Atlas oder Orakel? zkm.de/de/magazin/2017/01/atlas-oder-orakel English: Atlas or oracle? zkm.de/en/magazine/[node:blog-post-date:value:custom:Y]/[node:blog-post-date:value:custom:m]/atlas-or-oracle.

of the Knowledge Archives[39] exhibition at *Edith-Ruß-Haus for Media Art* (2008). In Oldenburg, Germany was to use the *MediaFlow* interface for teaching and learning. Between the large installation screens were intimate research stations in the exhibition space, set up like workstations for archivists. Students visited the installation with their teachers to deepen their knowledge of contemporary media art and theories. They browsed the *MediaFlow*, learned about comparative works, and gained insight into the media art they were discursively discovering.

Due to the flexible selection possibilities, small ad hoc lectures were created, not least by asking questions of comparison like this one by *Joseph Weizenbaum*: "What is the difference between the virtual hand and the hand of a beautiful woman?" Following *Weizenbaum's* method, individual findings were compared and analysed. Again, it's the viewer's own thinking that counts, and that's the added value of knowledge discovery tools.

1.4.5 Digital Sparks Matrix (2001–08) on the Web and on PointScreen (1996–2002)

Digital Sparks, the student media art & design competition at German-speaking universities, is also a tool for learning and teaching on netzspannung.org. Using a web *DVD* and a matrix of overall 450 project images as the navigation interface, the viewer can select a video clip of the respective competition project, similar to a visual jukebox, and then proceed to the informative project sheet on the Internet. The *DVD* was printed in an edition of 1000 copies and distributed as a teaching aid to research partners at participating universities. The visual matrix, whose individual image sizes are adjustable, proved to be a suitable form for the presentation of hundreds of projects. A virtual magnifying glass serves as a cursor with which the viewer glides over the individual matrix fields to select one of the project videos. The matrix allows for clear browsing and the dynamic magnifying glass for detailed viewing (Fig. 7).

PointScreen (1997–2002), a new interface technology developed by the authors, is suitable for the presentation of the collection in the context of a museum. PointScreen allows the control of screen objects by touchless body gestures. Inspired by the *Theremin* (1920), the first electric musical instrument that can be played without touch, the *MARS Lab* invented and produced an *Electro Field Sensing* interface technology. The position, direction of movement and speed of e.g. hand gestures can be used for navigating the digital display.

The *Electro Field Sensing* (*EFS*) system does not *see* the hand, but it *detects* it. The body is sensed by the body's own electrostatic field. In contrast to optical tracking, which has to be programmed for specific gestures, the *PointScreen* is a *fuzzy* interface. It is not necessary to learn sign language, but only to perform a movement. This ultimately means that everyone can communicate performatively

---

[39] Wissensarchive/Knowledge Archives - Online Platforms, Monika Fleischmann and Wolfgang Strauss, Edith-Ruß-Haus for Media Art, Oldenburg, Germany, 2008. https://bit.ly/3z10goQ

Fig. 7 *Matrix* Interface. Selection of *Digital Sparks* Award Winners: Nora Krug *How to Bow*, Julius Popp: *Bit Fall*, Jana Linke *Click and Glue*, Markus Kison: *Roermond Ecke Schönhauser*, Martin Frey: *Cabboots*, Ralf Baecker: *Rechnender Raum* © Monika Fleischmann and Wolfgang Strauss

with their own individual gestures. The system reacts so finely that *Leon Theremin*, the Russian physicist and inventor of the analog *Theremin*, built *EFS* mines during *World War II* that were effective even when birds flew (Fig. 8).

The development phase of *PointScreen* technology gave rise to experimental applications[40] such as the *Energy Meter* (1997), an analog re-imagining of the *Theremin* that displays the body's energetic state through three flickering, colored light bulbs. The MARS Bags (1998), electrified sounding handbags in various shapes, were

---

[40] Experimentelle Arbeiten und Prototypen, die in diesem Zusammenhang seit 1997 am MARS Lab der Autoren, am Fraunhofer Institut für Medienkommunikation entwickelt wurden, sind u.a. der Energy Meter, die MARS Bag, das MARS Field und Dynamic Capacities. http://netzspannung.org/about/mars/projects/

Fig. 8 *PointScreen* navigations (2006); *Lev Thermen/Leon Theremin* playing the *Theremin* 1917; *Energy Meter* and *Mars Bags* (1998)—using analog electro-field sensing © *Fleischmann & Strauss*

created for a performance and turned out to be a theft-proof object with an alarm trigger. With the *PointScreen*, the authors anticipate a future gesture-based form of interaction. It is only ten years after these first prototypes that interactivity through gestures such as wiping and swiping across the *Smartphone* (2007) becomes dynamic and marketable. *PointScreen's* digital *Electro Field Sensing* technology was developed starting in 1997 and finally patented in the *US* in 2007 as a *Gesture-based input device for a user interface of a computer*.[41] The patent offices in *Munich* and

---

[41] FleischmannM07a] Fleischmann, Monika; Strauss, Wolfgang; Li, Yinlin; Groenegress, Christoph "Gesture-based input device for a user interface of a computer", US Patent 7,312,788, December 25, 2007. Hardware patent on electrostatic/capacitive non-contact digitizer, similar in principle to a theremin, to sense in-air gestures. Requires use to stand on on a grounding pad or reference pad. Hardware for iPoint device from Fraunhofer Research Association. Includes one claim relating to detecting termination of a gesture by not moving. In: History of Pen and History of Pen and Gesture Computing: Annotated Bibliography in On-line Character Recognition, Pen Computing, Gesture User Interfaces and Tablet and Touch Computers (https://doi.org/10.13140/2.1.3018.8322) http://users.erols.com/rwservices/pens/biblio07.html

Brussels—with whom negotiations lasted 10 years—did not find the touchless inter-face principle patentable. While in the US technology is equated with progress and innovation, in Europe an innovation is usually not even recognized.

Another work by the authors mentions energy in the title, is not based on the *EFS* principle, but is to be understood metaphorically. It is about measuring and mapping people's mental energy.

1.4.6   Energy Passages—A Public Thinking Space (2004)

Energy Passages (EP) is an interactive visualization of the daily news in public space. The title derives from an understanding of language as mental energy. In front of the *House of Literature*, a cultural institute in *Munich*, murmurs can be heard, and catchwords—reduced to nouns through semantic mapping—can be seen flowing on the floor along the building. Artificial voices read out individual terms, creating a poetic soundscape. Passers-by walking through the city unexpectedly find them-selves in a Wittgenstein-like (1952)[42] *Language Game Performance*. A projection is reflected on the rain-soaked stone ground and the words seem to glitter and float above the floor. The flow of words combined with the auditory sphere invites visitors to participate.[43] (cf. Fleischmann & Strauss 2013) (Fig. 9).

By superimposing virtual entities with physical space, the installation creates an expanded urban reality. Sound scientist *Holger Schulze* notes, that "we could have the impression all the current news streams were running through ourselves".[44] (Schulze 2005) This could be called the crucial moment of entry into the staging. Visitors are wondering, pause or walk through the field of light and sound. The flowing movement, the murmuring words stimulate the audience to check what is going on here. Obviously, the only way to find out is by choosing one of five hundred words appearing on a touchscreen or by speaking into a microphone. The selected term is highlighted in the flow of words along with five semantically related words—the so-called *nearest neighbors*. Relationships between individual terms may look arbitrary but are calculated by the underlying semantic classification system.

The selection process encourages visitors to engage in a kind of inner dialogue as they consider the meaning of the terms. They realize a certain word and discover their own associations to it. Thus, a visitor finds the word *Faust* and is delighted to

---

[42] Wittgenstein, Ludwig: *Philosophische Untersuchungen.* Frankfurt am Main: Suhrkamp, 2003, p. 249.

[43] Monika Fleischmann and Wolfgang Strauss. *"Energy Passages – Knowledge as interactive spec-tacle."* In: *The city and the others.* Virus 09 Brasil. 2013 http://www.nomads.usp.br/virus/virus09/ ?sec=3&item=1&lang=en

[44] Schulze, Holger. On Taking Back An Artifical Separation. Intermodalperception in Energie-Passagen by Monika Fleischmann and Wolfgang Strauss. Introduction to the Panel: Sound Art Visual, Transmediale 05, 2005. http://netzspannung.org/about/mars/projects/pdf/energie-passagen-2005-1-en.pdf

Fig. 9 *Energie passagen/energie passages* streetview © Monika Fleischmann and Wolfgang Strauss

discover *Goethe's* drama in the flow of words, when in fact the newspaper has used this term to report on a politician banging his fist on the table.

The original text of the newspaper is projected onto the visitor pavilion as soon as a term is selected. The information pavilion is part of the whole staging. In this way, the participants and the audience recognize the contextual reference to the chosen term, but also tell each other their own memories or associations. Via microphone and touchscreen, visitors inspire each other with words that become visible in the flow. Some literally throw words around and play with others as if in scenic dialogues. In a sense, they use the voice interface as an intercom, and so, in addition to the encounter with digital words in the flow, analog conversations arise outside the technical system. Visitors become participants in a performance of varying durations and co-authors of emerging narratives that go far beyond the current newspaper article. It is a communication between people through a digital-analog interweaving of image, text, sound, voice and gestures. Participants become data performers. (Fleischmann & Strauss 2008)[45] Because of the billions of combinations based on the *urn model* calculation (cf. Johnson 1977),[46] the interaction process is unlimited, and its duration is determined only by the participants' desire to engage in this *Language Game Performance*.

[45] Fleischmann, Monika and Strauss, Wolfgang: Staging of the Thinking Space. From Immersion to Performative Presence. In: Paraoxes of Interactivity. Eds. Uwe Seifert, Jin Hyun Kim, Anthony Moore. Bielefeld: transcript Verlag, 2008, pp. 266-281.

[46] Johnson, Norman L. and Samuel Kotz: Urn models and their applications: an approach to modern discrete probability theory, by. Pp xiii, 402. New York: John Wiley & Sons 1977s.

Actually, the point is to awaken one's own associations and thus contribute to a dialogue in public space. Through their various attempts to decipher, the audience engages in conversation. Unlike computer generated interactivity here true social interactivity unfolds. Something rare in public. Philosopher Peter Matussek comments as follows: *"It is not about throwing around technoid text fragments like what the hypertext cult celebrated excessively, but filling the remaining gaps using in an odd way, smoothly operating automatisms. This especially due to the strikingly harmonious, however, fragmentary recomposing of what we have latently retained as consumers of the news. In this way of staging life is emphatically breathed into scripture. It becomes vivid not just because of the bare motion of pictures itself, but because of its media practice to stage performative readings in which text and reader equally participate in a constructive way. Additionally, the compensatory flare-up makes the motion of pictures continuously flirt with its own demise. In the contemporary silicon age, the future of scripture rather lies within visual, sculptural and architectural forms of expression than in secondary orality. This is exactly what the installation makes the visitors sensitively experience"* (Matussek, 2004[47]).

An important part of the four-week installation is the statistical evaluation of the visitors' interactions with the newspaper's keywords. How are topics like economy, politics, culture and education weighted and described in the newspaper? And what is the interest of the audience? What terms did the visitors find and select? The five most frequently mentioned terms of the daily newspaper in these weeks were: Germany, years, percent, millions, people. The *Sueddeutsche Zeitung/South German Newspaper* focuses on economics and politics in its reporting and marginalizes culture and education. In this way, the newspaper shapes public consciousness away from arts, culture and education. In contrast, the choice of words by some four thousand visitors to the *Energy Passages* over four weeks reveals a certain mood in the city, with emotionally charged terms such as *cost, parents,* victim, *love, food, girls, happiness.* By analyzing the language of the newspaper and the activities of the participants, two different realities of the city are recorded and reproduced in the statistics as an image of the city—an economic and a poetic reality. A color-coded bar chart, illustrated as an abstract cityscape, distinguishes different sections. It shows where and how frequently the terms occur. Most frequently the economy (purple), often in politics (turquoise), some in culture (yellow), and the least in education (red). Pragmatic media language versus personal choice and imagination. The result shows the state of the discursive energy of the audience during this period. The installation is an energy meter and comparable to Google's Nowcast, albeit on a different scale. Although unintentional, with 4000 visitors over four weeks, the installation is at least as meaningful as a standard survey with 1000 participants (Fig. 10).

The sociologist *Sherry Turkle* describes *Energy Passages* as a true evocative object. *"The notion of a spatial experience of the discourse of the news within a*

---

[47] Matussek P. (2004), Kein technoides Herumgewürfle mit Textfragmenten … / It is not about throwing around technoid text fragments" (translated by the authors). Available from: http://www. energie-passagen.de/presse2.html.

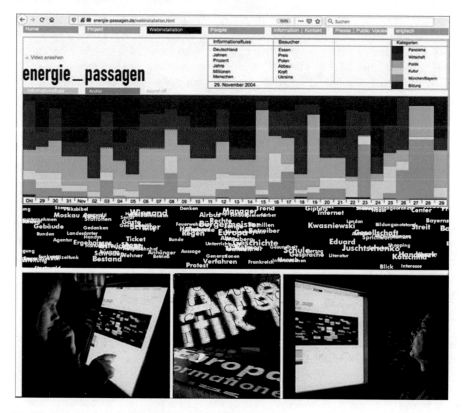

Fig. 10 *Energie passagen/energy passages.* Statistical image of the city, total flow of words, participants in conversation © Fleischmann & Strauss

city space and the possibility of deconstructing the newspaper captures the fragmentation of how media is experienced by citizens in a culture of simulation. It thus mirrors and concretizes an important cultural and political moment, turning it into an object for reflection." (Turkle 2004).[48]

The interactivity of the system is extended through the communication between visitors and the idea of a thinking space as a collective work becomes evident. With *Energy Passages* a participatory staging is created in public space with techniques such as semantic data analysis, interactive visualization, and atmospheric visual soundscape. The authors thus contribute their previous experiences as a synthesis of knowledge about digitality. Staged as a flow of information, the installation became a walk-in information browser that invites visitors to stroll through the news. All activities are stored, evaluated and visualized as energy images of the city during the four-week runtime. The *Mixed Reality* stage on the public square in front of the *House of Literature* in *Munich* is a thinking space for the flow of thoughts and a staging

---

[48] Turkle, Sherry. *"A true evocative object."* http://energie-passagen.de/presse2_engl.html (2004).

of people's interest. (cf. FS 2010)[49] In *Energy Passages* for the first time a neural aesthetic appears as next-generation interactivity, so to speak. What the authors call a thinking space becomes evident as a contemporary form of social interaction in public. The authors refer to *Vilém Flusser's* concept of the passage as a journey in which individual elements are traversed as fragments of a larger context in order to merge individual impressions as parts of a whole into a tangible image (Findeisen 2004).[50]

1.5 Digitality for Good or Bad

Digitality brings numerous benefits, including greater efficiency and better communication. Technology can be a great equalizer by enabling access to commerce and public services. Yet, transparency or control of it can fall by the wayside, given that personal data evolves into raw materials and merchandise. Date of birth, postal code or mobility and health data—our data is collected in bulk on the internet and sold at a profit. It has become an asset for new businesses.

Only gradually the underlying dark meaning of buzzwords such as *Internet of Things* or *Smart things for Smart People* is becoming clear. Smart means data trading, a smart car means that data is being sold. Data trade is trade with human data. Data trade is human trafficking.

Many *industrial actions* and *digital innovations* must be understood as an attack on our civil culture, and defense strategies must be developed. We are in an information war! This is not just about cyber-attacks or fake news. Like the music industry, the games industry is commercially successful and recognized. But games are by no means just harmless or educational. They also have a malicious component, especially when they entice constant consumption. People are ripped off fraudulently, as in gambling. This is just as criminal as the consequences of fake news and disinformation. When there is no way out, something has to change radically, and we have to become visionaries.

The *United Nations* has declared the current decade the *Decade of Restoration* (2021–2030) for our ecosystems. *Reimagine, Recreate and Restore*[51] was the credo of the Virtual Youth Forum organized by the *UN* and *Generation Restore*[52] called for action. The *Ecosystem Restoration Playbook* provides an introduction to the range of actions that can slow and halt ecosystem degradation and promote their recovery.[53]

---

[49] Fleischmann, M.; Strauss, W. Energie-Passagen: Die Stadt lesen und (be) schreiben. 2010. https://publica.fraunhofer.de/entities/publication/12f77b06-6db8-475e-9d5e-5097d97544ad/details

[50] Findeisen, Hans-Volkmar: Auf und davon. 2004. In: DIE ZEIT 2004 Nr.12. https://www.zeit.de/2004/12/Passagen-Alpen

[51] https://www.unep.org/es/node/29554#, https://www.youtube.com/watch?v=C_sBtDOdg7U.

[52] https://www.youtube.com/hashtag/generationrestoration

[53] https://www.decadeonrestoration.org/publications/ecosystem-restoration-playbook-practical-guide-healing-planet.

Likewise, the *Ars Electronica* Festival and the *Johannes Kepler University* invited 2022 to *Linz* for a thought experiment: "What would our lives look like if we had already mastered the crises of our time?" Answers were sought by outstanding experts from art, science and civil society, an ongoing *process* open to the public and worthy of imitation as "Planet B is not the second chance for another place where we can continue as before, it is the cipher for the indispensable, new and in many forms completely different life and action on this only planet that exists for us."[54]

Culture and care, a mindful way of thinking and acting, all of which must become the unifying element for innovation, education, the environment, sustainability, and new technologies. The Internet is offering a paradigm shift in thinking about communication and knowledge. But until recently, there was no public awareness of the problem of surveillance through social media or the breakdown of privacy or data protection. To counteract the increasing digitization of the world with its *smart* objects scanning through us, it needs thinking and acting people. As a counter-aesthetic to a mindless and inhuman digitality, the authors attempt to formulate a searching movement of thought, based on the concept of the thinking space.

The focus of the media art works listed above is both the interactivity between participant and system, and the media-mediated communication between the participants.[55] Such kind of communicative interaction opens up new ways of thinking. If it succeeds, the transition from the imaginary to action also becomes visible and the development from a representative to a relational and performative space can be experienced. "*Fleischmann and Strauss place the viewers of their works in such an environment. Consequently, viewers become integral elements of these works, as if intelligently working components. As a result, the perceptual experience of Fleischmann's and Strauss' installations becomes transgressive; It is an activity performed in one environment but having an impact on another, and the results are fed back to the interactors acting, establishing an evolving context of interaction, motivating their further behavior and thus the structure of an interactive work event is co-designed.*" (Kluszczyński 2011, 13).[56] The authors understand the interactivity of the performative interface as a contemporary strategy to "intervene aesthetically in the internationally operating media industries and create a third space between the poles of fusion" (cf. Spielmann 2010),[57] namely a thinking space.

Various spaces overlap in the hybrid thought space of the performative interface: the physical, the mental and the virtual space. Thinking space here means creating a place for collaborative thinking as a medium for a reconfiguration of thinking. A

---

[54] The Highlights of the 2022 Ars Electronica. Welcome to Planet B - A different life is possible! But how? Press release 29.8.2022 https://ars.electronica.art/press/en/2022/08/28/das-sind-die-highlights-der-ars-electronica-2022/

[55] Video documentations of the projects of Monika Fleischmann & Wolfgang Strauss on youtube https://www.youtube.com/MonikaFleischmannWolfgangStrauss

[56] Kluszczynski, Ryszard W., "Living between reality and virtuality. Remarks over the works of Monika Fleischmann and Wolfgang Strauss." In *Performing Data* [DVD, catalog]. Ed. Krzysztof Miekus. Warsaw: National Centre for Culture, 2011, p. 13 www.academia.edu/23731708/Fleischmann_Monika_Strauss_Wolfgang_Performing_Data

[57] Spielmann, Yvonne. Hybridkultur. Frankfurt am Main: Suhrkamp, 2010.

striking difference becomes apparent between the use of the target-oriented industrial interface and the performative interface. The essence of the industrial interface is *what you see is what you get*. The performative interface, on the other hand, turns things around and uncovers what is hidden: *what you get is what you did not see*. Being amazed at something that becomes visible through one's own actions creates a new space for thinking.

The *Performative Interface* represents the enactment of a particular property. Namely, it encourages participants to respond to the evocative language of the interface. The evocative interface invites to act, and only then does the interactive work and thus a personal experience emerge. Interactivity is not about carrying out instructions to operate an interactive system. Rather, it is about exploring the interactive work and thereby making it appear. While one strategy represents a stimulus-reaction scheme, the other way of dealing with the work opens up a playing field for improvisation and interpretation.

As the current world situation shows, everything—but really everything—needs to be rethought. It is time to finally establish transdisciplinary content and courses of study at schools and universities to find sustainable solutions to the problems of *digitalization, climate change, global food,* and *land use,* to name just a few. Just like adults, children and young people need to be involved in tackling the world's problems. Their interests and their future are at stake and it is necessary to invest in the education of thinking and feeling and the appreciation of the mind. Not only languages, but also programming languages are contemporary cultural techniques. In the rearview mirror appears a figure like Se*ymour Papert*, the father of Logo, the first programming language for children. What would it be like if everyone—including senior citizens—learned a little programming and thus understood digitality at its core?

Chapter 2
Pick My Brain: Thoughts on Anatomical Representations of the Human Inner Body from the Renaissance, Baroque Paintings, and Book Illustrations of the Modern Age to Current Medical Imaging Visualizations

Danne Ojeda

Abstract In this essay, I take the reader through a set of theatrical scenarios or mise-en-scènes, while focusing on the analysis of 3 essential artistic masterpieces and their entwined compositional theatrical strategies. In referring to the structure of a theatrical script in a text in 3 Acts, I am also suggesting a backdrop to my own work Vanitas book series (Figs. 2.15, 2.16, 2.17, 2.18, 2.19 and 2.20) further discussed in the coda of this paper.

2.1 Act 1. Pictorial Bodies: Anatomy Lesson's Mise-en-Scènes

The external representation of the human anatomy has been a *leitmotif* in the Histories of Art. Examples range from drawings or paintings that represent a life model, a person who poses in a studio setting,[1] and that epitomize the exquisite surface and features of the human epidermis from the (self)portrait type to the variety of nudes in all artistic genres, styles, and orientations. This obsessive representation of the body—in art and design—develops in parallel to scientific discoveries that aid re-imagining how the morphology of the human inner anatomic structures contour

---

[1] Other examples might be the paintings that resulted from the interaction of 'the author and his model'.

D. Ojeda (✉)
Nanyang Technological University, Singapore, Singapore
e-mail: danneojeda@ntu.edu.sg

© The Author(s), under exclusive license to Springer Nature Switzerland AG 2023
A. L. Brooks (ed.), *Creating Digitally*, Intelligent Systems Reference Library 241,
https://doi.org/10.1007/978-3-031-31360-8_2

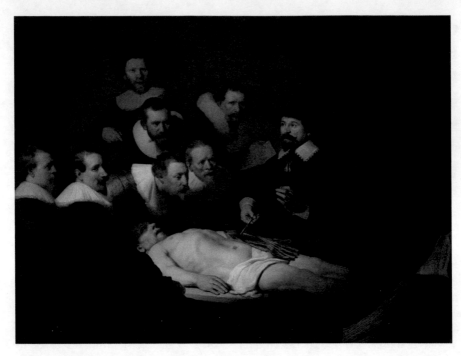

Fig. 2.1 Rembrandt van Rijn, *The Anatomy Lesson of Dr. Nicolaes Tulp*, 1632. Mauritshuis, The Hague. Collection Mauritshuis Museum, The Netherlands. © Mauritshuis Museum, The Hague

the outer anatomical figure. In other words, whenever scientific practice reported breakthrough findings, there would be a consequence in the artistic realm.[2]

Art historian Bob Haak [14] pointed to *The Anatomy Lesson of Dr. Nicolaes Tulp* (1632) by Rembrandt van Rijn (Fig. 2.1), as one of the first occurrences in which the dissection of a human body was part of an anatomy lesson. *The Anatomy Lesson* is a reference to the frequent public scientific demonstrations that took place in the seventeenth century,[3] where a deceased human body—usually that of a criminal, as permitted by the law—was 'examined' post-mortem in front of an audience with medical-pedagogical aims and objectives. All in all, they sought to prove therapeutic professionals' theories in favour of a more accurate clinical understanding of how flesh and bones scaffold us.

---

[2] The fact that the internal anatomical images of the body usually appear aesthetically unpleasant to the gaze because they deal more with the tenets of the abject, might have been an important factor as to why inner human anatomical structures are rarely protagonists in artistic masterpieces.

[3] Anatomical dissections became possible in the fourteenth century [7, p. 27]. However, the study of anatomy in medicine remained based on readings of Latin translations from Arabic, rather than a science based on direct observation [7, p. 34].

In this context, it is worth a digression to refer to Professor Francis Barker's analysis of *The Anatomy Lesson* which has a visible poststructuralist backdrop: Michel Foucault, Jacques Derrida, Jacques Lacan and, the Marxism of Louis Althusser. And perhaps, because of the former, there is also a strong bond between 'text' and 'power' in the form of a pictorial drama. Thus, aided by the 'play theme' of an anatomy lesson—pedagogically and highly illustrative in nature—, this painting evidences the paradigmatic change in mentality of the Cartesian Era, where knowledge production was intrinsically tied to rational thinking.

In the mise en scène of *The Anatomy Lesson*, Rembrandt depicts a public demonstration which is theatrical in nature, and moreover allusive to the form of a 'tragedy' to judge by the post-mortem destiny of its main deceased character. Barker writes: [2].

> Its neck broken (…) the chest is thrust forward in *rigor mortis*: Adriaan Adriaanszoon, also known as Aris Kindt, is less than thirty-six hours death (…). A life has just been extinguished; but a body is being made to signify, and to signify according to a mode of representation to which the body is only accessible on pain of a kind of death, actually for this petty thief from Leiden, more comprehensively—and more literally— exercised by the regime of subjection which is beginning to be practiced in this painting (…)

> It is, after all, a ghastly tableau, close to tragedy, which was in fact produced in a theater in Amsterdam before an audience who had paid for their tickets, who stand where we stand as we observe the painting, and whom the anatomist Tulp, his left hand raised in indication of speech addresses across this public body.

Let's note how the notion of theatricality and re-presentation are intertwined as if they are seen as part of an infinity mirror. First, there is a reference to a real-life event, a quotidian medical practice of the epoch, that is represented in the fictional setup of the painting. In this latter context, and with the textual citation in mind, there is a visible double scenario: *The Anatomy Lesson* does take place, first, in a theatre in Amsterdam; and second, in the 'ghastly tableau' of the painting. In between both scenarios there is a delay, or a gap from the 'real' event to the representational space of the painting whereby, all 'actors-characters' have the chance to appear rearranged, as if in a script or a sentence, or spaced by commas and semicolons, in a desirable pose-indicative-of-a-narration that underlines the characters' personal relevance and position in the medical guild.[4]

---

[4] We witness a visual memoir of a scene that acts as a singular empowerment pose, where all males doctors are represented at work, and where their legitimation in society for the sake of human knowledge when conducting such dissection acts, will supersede the relevance and accuracy of the post-mortem act itself. A proof of the latter is, for instance, that this painting showcases anatomical inaccuracies, which stresses the fact that the joints, vessels, bones, and muscles' true depiction was not the central aspect of the canvas. The latter is also subtly handled by Rembrandt's majestic dominion of light and compositional elements, which leaves the dissection practice in dark colours and in the shadow zone of the painting.

While Barker's passage can be related to the notion of theatricality and re-presentation as part of a double mise-en-scène (as explained above); it does equally involve its audience. For theatricality to exist, it is fundamental the existence of a premeditated setup of an event is displayed for the sake of the spectator [5, 11, 12]. Thus, in *The Anatomy Lesson* there is a double exposure *for* the spectator. We witness a reenactment as Barker's recalls, 'before an audience who had paid for their tickets', and later, in the realm of the painting, and for the gaze of the beholder. In this sense, theatricality, together with its literal meaning of double presentation, does parallelly imply the distortion of a postponed account of a piece of 'reality' that is presented to a viewer and filled up with layers of interpretations and para-texts.[5]

Now, let's focus on the diverse acts on the stage and their possible meanings. Speaking of the 'real' event depicted in *The Anatomy Lesson* and the portrayal of the characters staged by Rembrandt, Barker [2] asserts that the 'drama' is re-presented in the name of scientific scholarship, whereby a person metamorphoses into a persona: a sort of stereotype of 'flesh made word'. In his view, 'the carnality of the body' dissipates to be reconstructed or better revised 'in writing at a distance from itself'.

> As the flesh is de-realized, representation, which becomes at last representational, is separated from it and puts in train a mode of signification for which, to borrow a word from Derrida, the body has become supplementary. Neither wholly present, nor wholly absent, the body is confined, ignored, exscribed from discourse, and yet remains at the edge of visibility, troubling the space from which it has been banished.
>
> (…) no eye within the painting sees the body. The scientific gaze, the perspective of natural philosophy, may be organized around the corpse, but in order not to see it. The lines of sight can be easily traced. They return from the surgeons to we who look (how real we are) or they run to the anatomy atlas open at the foot of the corpse.

Flesh dematerialized in the act of becoming text is one of Barker's major annotations which is evidenced in the painting with the paradoxical absence of the corpse to those present at the scene. '*No eye within the painting sees the body*'. Like a ripple in a pond, we might substitute the corpse by a water drop fading in its resonating waves. The first ripple that emphasizes the de-materialization of the body that Rembrandt re-enacted in *The Anatomy Lesson of Dr. Nicolaes Tulp*, is accentuated by the gaze of three doctors surrounding the corpse. The doctor's eyesight frames a book that lays wide open at the bottom right of the painting, which is believed to be Andreas Vesalius's *De humani corporis fabrica libri septem* (1543), a testimony and paradigm

---

[5] One of these para-textual interpretations is for example, what I see as a remarkable symbolism that emanates from the pictorial treatment of the diseased body. *The Anatomy Lesson…* also 'speaks' about the mortality of the human being—emphasized by the brightness and almost white corpse's epidermis tints. This treatment might remind the viewer about the unfortunate condition of pariah of its protagonist, which bears certain resemblances to the pale aesthetics of the descents of Christ iconographies.

of anatomic representations.[6] Thus, the eyesight of the characters disregards the exposition of the corpse to focus on its textual re-composition and illustrative re-birth (This analysis will be continued in Act 2 of this text).

The second ripple is formed by the other three doctors that surround those first three enthralled by the book. They stare at the viewer. These second lines of eyesight acknowledge that the viewer is being somehow interrogated as an active participant in completing Rembrandt's magistral mise-en-scène.

The Anatomy Lesson continues in its intermittent ripples. The tableau is a theatrical re-presentation, a *déjà vu*, whereby the spectator might assist to fill the gaps of the magistral move from the pictorial to the illustrational. The apex of 'flesh made word' [2] seems to be the transition—yet another theatrical act—from the invisibility of the pictorial displayed corpse; to the spreads of soft paper bodies articulated by Vesalius's *De humani corporis fabrica*. All in all, in the name of scientific illustration and knowledge construction and dissemination. This call for attention that the characters of the painting anticipate, takes the audience outside the realm of the painting: 'flesh made word' focalizes the space of writing, and the re-presentational domain that a book inhabits favoring the renaissance of paper bodies.

2.2 Act 2. Paper Bodies: The Body-Corpse Unbound

It is 1543—one of the most paradigmatic years in the history of science—and two innovative and complementary books about the anatomy of 'bodies' are published: *De humani corporis fabrica* (On the Fabric of the Human Body) by the Flemish-born Andreas Vesalius (1514–1564) [Figs. 02–04], Basel; and *De revolutionibus orbium coelestium* (On the Revolutions of the Celestial Spheres), by Nicolaus Copernicus (1473–1543), and printer Johann Petreius, Nuremburg. While the *Fabrica* set the path to comprehensive knowledge about the morphology of *human bodies*; *De revolutionibus*, explained the anatomy of *celestial bodies*, revealing Copernicus's ground

---

[6] Referring to the book that appears in the painting some authors (A.B. De Vries, Magdi Tóth-Ubbens, W. Froentjes, 1978) stated that: 'In view of the subject being taught, it is nevertheless quite possible that the painter intended to suggest Vesalius's *Fabrica*. This would, moreover, be in line with the portrayal of Dr. Tulp as *Vesalius Redivivus*… it is obvious that the book has a part to play in the composition of the work. (A.B. De Vries, Magdi Tóth-Ubbens, W. Froentjes. (1978). *Rembrandt in the Mauritshuis*. Springer Netherlands. p. 100).

Another reference to *De humani corporis fabrica* was established by the Mauritshuis Museum in its catalogue description of this painting: 'The frontispiece of a standard work on human anatomy, *De humani corporis* fabrica (1543) by the German anatomist Andreas Vesalius (1514–1564), contains a portrait of the author holding a dissected arm. It is tempting to conjecture that Tulp may have wanted to emulate his illustrious predecessor and therefore asked Rembrandt to portray him while working on an arm.' (*The Anatomy Lesson of Dr Nicolaes Tulp*. (n.d.). Mauritshuis Museum. Retrieved January 15, 2022, from https://www.mauritshuis.nl/en/our-collection/artworks/146-the-anatomy-lesson-of-dr-nicolaes-tulp/.

The text is an adaptation of P. van der Ploeg, Q. Buvelot, *Royal Picture Gallery Mauritshuis: A princely collection*, The Hague 2005, pp. 146–148.).

breaking heliocentric theory which has been amply recognized as one of the pillars of modern science. The contextual backdrop of this editorial yin and yang combination, is the authors' and publishers' rational (and illustrative) thinking examining their own nature, what humans are made of, and how they position themselves as part of a complex system of universe of bodies.

All my thoughts are now concentrated on the *Fabrica*, the book painterly cited in *The anatomy lesson* by Rembrandt, to continue with some reflections on the re-presentations and dis-locations (perhaps disassociations) of human anatomic features, and its theatrical and symbolic connotations that lead to create my own series of works discussed at the coda of this paper.

Scientific illustration in books thrived in the sixteenth century with the invention of printing, the moveable type, and the use of engravings. If there is a book whose illustration caused a rétro-alimentation from the scientific imaging to the artistic field (and vice versa) via its prints and woodcut reproductions, it is *De humani corporis fabrica* [Figs. 02–04]. The *Fabrica* is considered a seminal book of its kind: Vesalius was both the author (as an anatomist[7]), and the supervisor of the woodcut illustrations, being also a draughtsman himself. Medical historian Gernot Rath spoke of the *Fabrica* as the 'foundation of modern medical science which is characterized by systematic observation of case' [22, p. 354]. Vesalius, based on the analysis that derived from his own dissection practice, dared to correct anatomical errors, challenging the Galenic view prevalent at the time, and debunking past beliefs about human anatomical architecture that had been previously accepted for more than a millennium[8] [17, 22]. With this, the *Fabrica* emerged as a novel series of anatomy lessons and as a vocabulary of anatomical forms that would set the bar for the grammar used by future generation of scientists and artistic studies.

In terms of the visualization of the *Fabrica*'s anatomical syntax, besides Vesalius himself, other draughtsmen's identities have not been fully established. However, researchers Douglas J. Lanska and John Robert Lanska refer to well-known artists from the Venetian school of Italian Renaissance painters such as Titian (Tiziano Vecelli, ca. 1488/1490–1576), and his student Jan Steven van Calcar (c. 1499–1546), as well as Domenico Campagnola (c. 1500–1564). These artistic collaborations were influential in defining an illustration style in which the practice of dissection appeared aestheticized and emulating fine art pictorial conventions of the times.

While browsing through the illustrations of Vesalius's book, it is moreover inevitable to think of Leonardo da Vinci's drawings on the human anatomy as they bear an extreme resemblance. Leonardo da Vinci (1452–1519) has been named one of the most significant artist-anatomists of his time. Da Vinci admitted he had dissected

---

[7] Vesalius was trained as a Galenist. He also edited some of Galen's texts as referred to by authors Douglas J. Lanska and John R. Lanska. He later started practising dissections when he was appointed to a lectureship in surgery and anatomy in Padua in 1537 [17, p. 45].

[8] Lanska & Lanska refer to the limited available anatomic illustrations found in the Middle Ages. Those available were "stylized representations of writings in the Galenic–Arabic canon. Because classical Greek medical lore had been preserved in Arabic writings, illustrations were necessarily absent in the Arabic sources because dissection and pictorial representation of the human figure were prohibited by Islamic law" (Lanska & Lanska [17, p. 34] citing Choulant [7]).

more than thirty bodies to understand the real 'nature' of human representation. Following these practices, da Vinci sketched a series of human anatomical studies in layers and in three-dimensional forms. Some of his drawings of skulls, for example, are also shown sectioned or in cutaway views.[9] In his anatomical drawings, we are presented with features unseen by the human eye such as muscles unwrapped and hanging from the body as flaps—in a similar morphology as the illustrations depicted by Vesalius and his collaborators. Carmen C. Bambach, curator in the Department of Drawings and Prints at The Metropolitan Museum of Art, New York, has said that these representational techniques transcended da Vinci's lifetime to finally be seen in the *Fabrica* although none of da Vinci's anatomical drawings and annotations were published in his lifetime [1]. Additionally, da Vinci died when Vesalius was five years old, which made it improbable for him to have benefited from the illustrations published in the *Fabrica* twenty-four years after da Vinci's death [9, pp. 12–13]. Hence, the similarities of these two authors' styles, suggests that there was undoubtedly aesthetic feedback in the Italian Renaissance regarding the treatment of human anatomical representations, when visualizing the innermost *fabrica* of the human body.

Having established the contextual backdrop of the *Fabrica*, we will now link it to the theatrical mise-en-scène compositional strategies we have been weaving throughout this text.

Elizabeth Burns [5] stated that theatricality takes place when a behavior is purposedly showcased as unnatural to underline 'rhetorical and authenticating conventions' of the epoch [11]. Thus, I argue that the *Fabrica* has been carefully *designed* to fulfil this prophecy by Burns, and later by **Josette** Féral in the interface of the following two aspects linked to theatricality: (1) A sort of 'troubling corporeality', a notion described by Roland Barthes [3, p. 27] which is associated with the intentional duplicity of the actor's body (E.g. 'Don Juan's son be played by a girl'), and to what Féral calls 'ostension' [11], the process of showing-off specific poses and behaviors that are seen by the spectator as over-acting; and (2) the technology of the book as a paper bound and sectioned material, with its own compositional edited framework, whereby paper bodies mise-en-scène make its spectacular appearance in diverse layers and para-textual surfaces.

(1) The *Fabrica* appears indifferent to mortality. Its 'troubling corporeality' [3, p. 27] is evidenced by a seemingly deceased body, a skinned corpse that plays out its opposite: a living person. Thus, human anatomical re-presentations are sometimes exhibited in a *skeleton mode* (Fig. 2.2), or in an afterlife style that follows canonical artistic representations of human depictions of the epoch.

---

[9] See da Vinci, L. (1519). *The muscles of the shoulder and arm, and the bones of the foot c.1510–11.* [Illustration]. Royal Collection Trust.
https://www.rct.uk/collection/919013/the-muscles-of-the-shoulder-and-arm-recto-the-muscles-of-the-shoulder-and-arm-and; and da Vinci, L. (1489). *Verso: The skull sectioned 1489* [Illustration]. Royal Collection Trust. https://www.rct.uk/collection/919058/recto-the-cranium-sectioned-verso-the-skull-sectioned.

Fig. 2.2 *Humani corporis ossium caeteris quas sustinent partibus liberorum suaque sede positorum ex latere delineation.* Illustration attributed to Jan Steven van Calcar, a student of Titian.—Cf. Choulant, L. Original image in: Vesalius, Andreas, 1514–1564. De humani corporis fabrica (Basileae: Ex officina Joannis Oporini, anno salutis reparatae 1543. Mense Iunio), page 164. © Image courtesy of the Historical Medical Library of The College of Physicians of Philadelphia

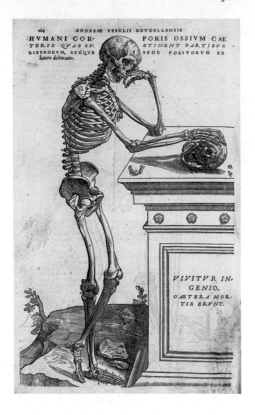

The reader-spectator witness a new vision of body-muscles (Fig. 2.3), or body-bones illustrated as living creatures. These bodies-inside out have been literally 'taken for a walk' premiering new poses: standing, wandering, pensive about its own mortality (Fig. 2.2), or posing over-acting in idyllic landscapes[10]— as implausible vivisections. As Barthes noted in a fragment that seems written for this text, in the *Fabrica*, we witness 'the *theatre as a site of ultraincarnation*, in which the body is double, at once a living body deriving from a trivial nature, and an emphatic formal body, frozen by its function as an artificial object' (emphasis added [3, pp. 27–28]).

(2) But the *Fabrica* is also an object to thumb through, a 'paper body'. Thus, let's now shed light on to the second point earlier enunciated in this section, the realm of the book, its technology, anatomical, and compositional edited structure where Vesalius's paper bodies make its spectacular mise-en-scène. Hence, the technology of the book, its very foundation and production culture come

---

[10] See also *Secunda musculorum tabula*, and *Quinta musculorum tabula*, found respectively at: *The College of Physicians of Philadelphia Digital Library*, accessed January 9, 2022, https://www.cpp digitallibrary.org/items/show/2459; Calcar, Jan Steven van, 1499–1546?, *The College of Physicians of Philadelphia Digital Library*, accessed January 9, 2022, https://www.cppdigitallibrary.org/items/show/2254.

Fig. 2.3 *Prima musculorum tabula*. Illustration attributed to Jan Steven van Calcar, a student of Titian.—Cf. Choulant, L. Original image in: Vesalius, Andreas, 1514–1564. De humani corporis fabrica (Basileae: Ex officina Joannis Oporini, unno salutis reparatae 1543. Mense Iunio), page 181. © Image courtesy of the Historical Medical Library of The College of Physicians of Philadelphia

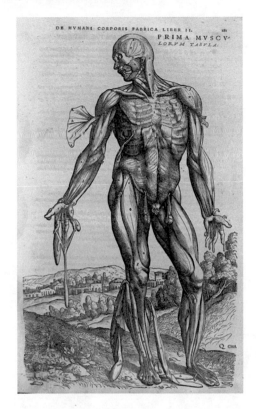

into play to ex-pose its content. The *Fabrica* has the intention of reaching out to an audience beyond the medical dominion while dispossessing the illustrative studies based on post-mortem dissections from its macabre allure. With the reader's gaze in mind, once again, there is the anticipation of an audience, and the consequent premeditation of an (editorial) scenario that is intentionally *fabricated*—the very theatrical undertone—which comes to justify the *Fabrica*'s paper bodies characteristics.

When turning its pages, anatomical organs are found scattered as if they were study annotations, ex-posed and sectioned like minute bodies; a reminder of the *flesh made word* Barker earlier anticipated. The book spreads are the scenario for a full catalogue of body fragments, overly dis-membered to be further scrutinized, read, memorized… The latter occurs as part of an intelligent hardcover book production, that it is also noteworthy for its skilful use of typographical elements, woodcut printing, and book page compositions (Fig. 2.4). The book offers para-textual elements or scenes within scenes, where the reader could find,—also attuned with the conventions of printing— detailed historiated initials across the book paragraphs, that imbricate their own narratives with the main plot described in the book sections. In a historiated initial H, for example (Fig. 2.5), we witness how there is an alternative plot to the main storyline illustrated on the page, that addresses former medical practices. Letter H is a pre-text

Fig. 2.4 *Prima figura V. capitis* (page 18). Illustration attributed to Jan Steven van Calcar, a student of Titian.—Cf. Choulant, L. Original image in: Vesalius, Andreas, 1514–1564. De humani corporis fabrica (Basileae: Ex officina Joannis Oporini, anno salutis reparatae 1543. Mense Iunio), page 18. © Image courtesy of the Historical Medical Library of The College of Physicians of Philadelphia

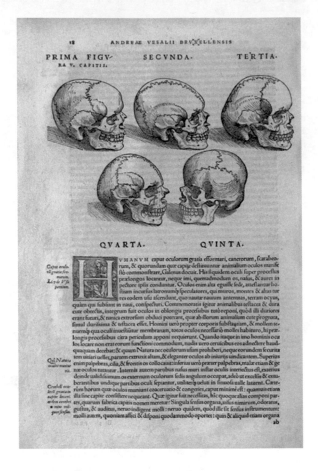

for demonstrating a trephining of the skull scene with hot iron plates, an instrument, and a procedure which are an outmoded tradition in modern surgery.[11] In so doing the book initials, masterly embody their fundamentals of historically contextualizing the subtopics of the *Fabrica*'s human anatomy display.

From the 'troubling corporeality' showed-off by the main skinned characters of the *Fabrica*, to the body fragments, or micro-scenes found in the historiated

---

[11] See Samuel W. Lambert: "The Initial Letters of the Anatomical Treatise, *De humani corporis fabrica*, of Vesalius," p. 19. in *Three Vesalian Essays to Accompany the Icones Anatomicae of 1934* (New York, Macmillan, 1952). The text reads: 'The small initial *H* shows one of the students of the *De fabrica* burning the scalp or the cranial calvaria with an actual cautery. The methods of protection against a too extensive burning by means of a cold metal plate with orifices to limit the application of the hot iron present an instrument called the plate, which is practically unknown in modern surgery'. Lambert also notes how some authors such as: 'Metzger, Chaigneau, and Rosenkranz all unite in calling the operation in letter H; a trephining of the skull instead of an operation by cautery'.

Fig. 2.5 Letter H, historiated initial. Fragment from Fig. 2.4. © Image courtesy of the Historical Medical Library of The College of Physicians of Philadelphia

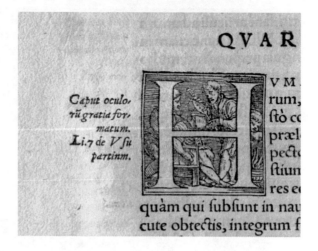

initials, the intertwined editorial mise-en-scènes underline the ambivalent meanings of re-presentation [as ex-hibition], and presentation [as ex-position],[12] as in 1 and 2 combined we have been analysing throughout this text. While the first loudly calls for the attention of the viewer inciting multifarious socio-cultural poses and readings, the latter one fragments the subject-matter, dislocating it.

2.3 Act 3. A Portrait in Absence: Pieter Claesz's *Vanitas Still Life* (Figure 2.6)

At that table, which is so familiar, yet so alien, I arrange some personal belongings… I possess them in a way that I might not have sufficiently used them. I stare at them, feel them, once and again in a hurry, or as of today with time to spare. And while trying to improvise an orderly composition, I leaf through the manuscript, and the room echoes the clacking of my fingers that thumb through those infinite handwritten pages, mumbling words, smelling its dark-deep-water-like-rancid-humidity; while admiring the sinuous glyphs of the pages, the effacing ink here and there… inhale… exhale…. And in a blink of an eye: a dry 'clack!'. The chalice hits the table, and my breath is abruptly suspended… Thank God! I gasp. Exhale… I have just finished savoring this velvety smooth wine, while playfully stacking the objects as if piling up a sandcastle. Inhale… and I glare at this dark glass that feels suddenly stranded between the table corner and a candle that has just finished breathing, which softly returns my face reflected in the chalice, like a boomerang… Exhale.

---

[12] I use the term 'ex-position' as explained by Hanneke Grootenboer, while discussing modern portraiture in How to Become a Picture [13, p. 323]. Inspired by Jean-Luc Nancy's thoughts on portraiture, Grootenboer writes: 'any sense of unveiling of (part of) the self can only take place on condition of its exposition, that is, the exposure or display of the self, in a sense an explanation or expounding of it, that takes place via an account of its exterior appearance (Jean-Luc Nancy, 'The look of the portrait', in Simon Sparks, ed., Multiple Arts: The Muses II, Stanford, CA, 2006, 220–41.).

Fig. 2.6 Pieter Claesz, Vanitas Still Life, 1630. Collection Mauritshuis Museum, The Hague, The Netherlands. © Mauritshuis Museum, The Hague

I imagine the above passage being a hypothetical description of Pieter Claesz, *Vanitas Still Life*, 1630 (Fig. 2.6); written by *a woman*: by me, or by her, the female that gazes, or who might have witnessed Claesz's still-life real table arrangements. I base this assumption first, on the intelligent reflections raised by Norman Bryson in his "Still Life and the 'Feminine' Space". Bryson writes:

> But to understand still life of the table one needs to take into account the asymmetry of the sexes with regard to the domestic space of which the table is the centre. In Dutch culture of the seventeenth century responsibility for the orderliness and good management of this space is regarded as a female rather than a male prerogative, and when still life of the table sounds the theme of disorder, asymmetry of the sexes is an important factor in the emotional nuancing of the scene. [4, p. 160]

With an astonishing monochromatic virtuosity Claesz, but also the woman who could have made possible these table arrangements in place of a bouquet of flowers, carefully re-positions what (s)he needs the viewer to see. She also foregrounds the objects, 'curating' an imperfectly meticulous *tableau* that is rather than represented, presented to us as a monumental sculpture with a book-pedestal that locates and fore-grounds the cranium it supports in the space of the table.

Her delicate dis-arrangement made me suspicious, and I wondered, what if the cranium that crowns Claesz's *Vanitas Still Life's* (1630) depicted scene, was a cranium of another of her kind: a woman's cranium? What if she was just taking care of 'her kind', yet in another moment, another place, as if an infinite mirror of cyclical occurrences? *To be* or *not to be*?

In suggesting that a *theatrical* space in Claesz's painting had been re-arranged or lived and experienced by a woman as hinted at by reading Bryson; I would like to additionally argue that in Claesz's *Vanitas Still Life*, the viewer witnesses the presentation of an anonymous portrait, *a portrait (in absence)* of an unidentified woman's depicted cranium that creates a double scenario, one which is inside the other.

To the question of what characteristics indicate that the cranium depicted in Claesz's *Vanitas Still Life* (1630) might have been from a woman, I turned for an answer to the scientific findings that engage comparative morphologic studies on craniums that lead to sex determination. There are two recent biological anthropology studies based on the analysis of the current European population, one that included the revision of CT scans of 103 persons from the southern French population [19][13]; and another that based its findings on the study of contemporary Czech population [6]. Both studies overwhelmingly concluded that morphological characteristics of the frontal bone and sinuses of craniums are essentially useful to estimate the sex of a person since there are verified differentiations between male and female frontal bones surfaces.[14]

The comparative study lead by Musilová et al. [19] determined that 'females... had more globular skulls, more convex foreheads with frontal eminence, and less prominent glabellar and nasal regions with lower and more vertically oriented premaxilla', as opposed to male craniums who have 'more elongated skulls with more sloped foreheads, more prominent glabellar areas, and more posteriorly sloped premaxillar regions'. Likewise, Čechová et al., concluded that 'the whole cranial surface was significantly different between males and females in size (form)' asserting that the skull skeletal elements are sexually dimorphic, and supporting the findings of Musilová et al. One might therefore resolve that the asymmetry of the craniums mentioned in both studies might—and I stress, only might—be an indication to suggest that the cranium that pinnacles Claesz's *Vanitas Still Life* belonged to a woman.

Hence, the suggestion that the cranium in Claesz's *Vanitas Still Life*, 'acts' as a portrait in absence of a woman is a plausible fact. Moreover, the cranium of a still life might be atemporal and asexual, specifically when a depicted cranium underlines the metaphor of a human being, which ultimately embraces both sexes.

Thus, I want to imagine that in Pieter Claesz's *Vanitas Still Life* (1630), we are in the presence of a female cranium that somehow proposes a different reading of this painting. The characteristics of this space are twofold. On the one hand, the beholder confronts a feminine space of an intimate mise-en-scène, that is the representation of a daily womanly activity set at the intimacy of a table. On the other hand, we find in this painting the representation of the cranium of another possible

---

[13] The CT scans were carried out for 52 males and 51 females whose ages ranged from 18 to 92, using 3D software systems: Rapidform, Avizo, Morphome3cs, and approaches in geometric morphometrics.

[14] Moreover, after the pelvic bone, there are several authors that consider the skull as the second-best bone structure for sex identification. See Krogman WM (1962) The human skeleton in forensic medicine. Charles C Thomas Publisher, Illinois; Pickering RB, Bachman D (2009) The use of forensic anthropology, Second edn. CRC Press, Boca Raton; and Byers SN (2015) Introduction to forensic anthropology, Fourth edn. Routledge, New York.

female that appears subtle, ex-centric, and ex-positioned, as a portrait in absence. The infinite mirror metaphor comes repeatedly into play: there is a woman that re-arranges the space of the table where, yet other women reminiscences appear to perhaps be immortalized on these sorts of carefully crafted theatrical spaces where none of these females are visibly identified by the objects we observe, but the cranium; which can be mistakenly appreciated as yet another of its kind, or even disguised as its opposite male sex.

And I will copy and paste my own words from a paragraph taken from Act 1 where I *italicize* the *orthographic changes*: there is a visible double scenario: *Vanitas Still Life* does take place, first, in a *homely set-up inhabited by the persons who have touched and carefully positioned the objects for the painting to be executed by Claesz*, and second, in the *frozen instant that immortalizes the* 'tableau' of the painting. In between both scenarios there is a delay, or a gap from the 'real' event to the representational space of the painting whereby, *the female* have the chance to *dis*-appear in a desirable *set of metonymic objects* that underlines the characters personal relevance and *ex*-position in *a domestic-ated space*.

It is in these interconnected (a)temporal and symbolic dimensions that another interpretation of the Still Life foregrounds—aside of the impermanence of life which is still noticeably present—that is the reaffirmation of the disguised feminine ex-centric, dis-located, re-presentational space, I reenact in my Vanitas-book series.

2.4 Coda. 'Bodies Without Organs', or Another (Book) Anatomy Lesson

What is the body without organs of a book? (…) As an assemblage, a book has only itself, in connection with other assemblages and in relation to other bodies without organs. We will never ask what a book means, as signified or signifier; we will not look for anything to understand in it. We will ask what it functions with, in connection with what other things it does or does not transmit intensities, in which other multiplicities its own are inserted and metamorphosed, and with what bodies without organs it makes its own converge. [10, p. 4]

2.4.1 Recollections

Andreas Vesalius's *De humani corporis fabrica libri septem* (1543) was extensively plagiarized and reproduced by Vesalius's successors despite efforts for protecting his work, aspect that helped to the profuse dissemination and generalization of its contributions to both the scientific and the artistic knowledge. This is why, I firmly believe that the *Fabrica*'s artistic illustrations, especially those presented in book 7, where the brain and the skull are depicted wide opened, must not have escaped the radar of the painters of the *Vanitas* genre.

De humani corporis fabrica libri septem, paved the way for the prolific use of craniums that later enthroned the *Vanitas* paintings. While assisting in the visual assimilation of the human skull as a recurrent desk paperweight, it also aided in neutralizing their morbid strangeness. Likewise, the *Fabrica*, assisted in the identification of the skull representation with a metaphor of 'life-rejection' [4] and mortality-warning for the sake of appreciating—and re-enunciating—human existence.[15] One of the *Fabrica*'s figures reads: *Vivitur ingenio, caetera mortis erunt* [Genius lives on, all else is mortal], that is a variant of a still-life, and the *Vanitas* genre (Fig. 2.2).

The peak of the *Vanitas* genre also coincided with the popularization of post-mortem dissection acts that only wealthy persons could afford to witness in person. It is well known that the University of Leiden included a *Theatrum Anatomicum*, which was among the first of its kind in Europe with the sole purpose of offering public dissections in wintertime. Built in 1594, it promptly became a tourist attraction until its demolition in 1821. In these anatomical theatres, it was common to find arranged human skeletons with texts about 'vanitas' and 'memento mori'.[16]

2.4.2   *To* Image *and the Self-imaginary*

The surge of Magnetic Resonance Imaging (MRI) in the 1980s,[17] incited a turn of perception and visualization of biological bodies, which also triggered an immediate effect in the art field. Long has passed since the outmoded public post-mortem human dissection scrutiny. Nowadays, via the MRI visualization technique, one witnesses an intelligent process-based machine that non-invasively *'renders the body nakedly public'* while this same human body is thinking, breathing, and resting alive. On this perspective, researcher Barbara M. Stafford wrote in her book *Body Criticism: Imaging the Unseen in Enlightenment Art and Medicine* [23]:

> The computer-mediated milieu *renders the body nakedly public.* (...) one result of the non-invasive imaging technologies in the area of medicine is the capability of turning the person inside out. If the late nineteenth century developed the photographic sounding of the living interior through endoscopy, gastroscopy, cystoscopy, and, most dramatically, X-rays, the late twentieth century revealed its dark core three-dimensionally through MRI projections.

---

[15] See Lanska and Lanska [17, p. 5]. "Book 7 of the Fabrica includes a systematic dissection of the brain, illustrated with woodcut prints through 12 sequential stages. As demonstrated in these prints, Vesalii was the first to accurately depict numerous brain structures, including the corpus callosum, thalamus, basal ganglia, and cerebral peduncles [17].

[16] See the account offered (in Dutch) by researcher Tim Huisman at the Leiden University online platform via this link: https://www.leidenuniv.nl/nieuwsarchief2/2286.html (Retrieved on 19 June 2020).

[17] MRI scanners were first commercialized in the 1980s. (See Ramani, R. (2018–12). Introduction to Functional MRI. In Ramani, R. (Ed.), *Functional MRI: Basic Principles and Emerging Clinical Applications for Anesthesiology and the Neurological Sciences.* Oxford, UK: Oxford University Press. Retrieved 23 Jun. 2020, from https://oxfordmedicine.com/view/10.1093/med/978019 0297763.001.0001/med-9780190297763-chapter-1.).

Using radio waves and magnetic fields, this technique for painlessly exploring morphology, nonetheless raises the spectre of universal diaphaneity [emphasis added].

In the context described by Stafford, one of my initial questions was: how does this representational turn that renders the 'person inside out' influence a contemporary work that engages both my art and design practice?

The *Vanitas* book series development started in 2013 as part of an academic-pedagogic research project entitled *D-SIGN-LAB: Research Experiments on Art, Design and Science, with a focus on Magnetic Resonance Imaging analysis*. The technique of MRI is described as a non-invasive scientific practise that scans and represents the human—and animal—body. The technology of MRI has been considered highly innovative in the medical realm for its accuracy in unveiling morphological details of the inside biological configurations for investigating the physiopathology of diseases.

Notably, MRI is to date the medical technique that offers more accuracy *to image* detail and complex bodily morphologies, whose benefits are used for diagnosis and further treatments in the medical domain. In medical argot, this representation technique is called '*to image*', which means to expose the invisible *fabric* of our bodies to the gaze of the beholder by representing its anatomic morphologies 'when electromagnetic radiation passes through it or is reflected from it' [18]. In this sense, scientific images are visualizations of technologically gathered and processed data, and the sole act of collecting, analysing, and exposing these images to the public, constitutes a novelty or a discovery insofar as these *imagins* are the plausible evidence of 'the making of scientific knowledge' [16].

How can the meaning of 'to image' or—the making of [scientific] representation—be understood in the context of MRI aesthetics when entering the realm of artistic production? A look into the etymology of the word 'image' (mimesis, idea) underlines the making of a likeness through similitude. To 'image' also invokes developing one's 'imagination' or the 'imaginary', which is said by Julia Cresswell to equal to 'picture to oneself' [8]. Following this entwined logic, *to image* might as well be a trope for self-reflection: the act of technologically and digitally 'dissecting' while slicing a body invisible to the eye, to later re-compose it as a sort of a (self)portrait.

To 'picture to oneself', the *Vanitas* book series engages at first with the idea that a book portrays its author's inner thought—which might be seen as representations of the author's mental, and physical matter. Thus, with the help of MRI clinical technology, my brain anatomic *imagins* became the book content in the *Vanitas* book series. For this, two MRIs of my brain were carried out. The brain was 'imagined' by software after a complex process that facilitated the 'readings' of brain's molecular frequencies. Hence, the brain was first computer generated, and finally materialized into a modular and sequential architectural paper 'body'. To *image* the brain and its matter, all layers resulting from the MRI, were printed sequentially and literally

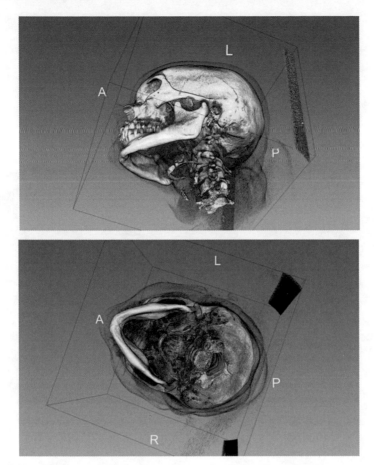

Figs. 2.7–2.10   Scan visualizations of Danne Ojeda's skull and brain, 2016

bound together with a common *fabric* following a manual bookbinding technique[18] as if wrapping or sculpting one's brain.

The result is a representation of the 'architecture' of a human brain, and the 'ineffable space' of the mind. The book size is similar to square to the human size of my own skull measurements (Figs. 2.7, 2.8, 2.9 and 2.10). This is perhaps the ultimate transposition of Barker's 'flesh made word' into 'flesh made paper', or what Deleuze and Guattari might have called 'plateaus that communicate with one another across microfissures, as in a brain.

---

[18] Book design technology also plays an important role in the series. A book-as-object is, among other things, a singular expression of the technological development of its time. The *Vanitas* book has been made possible, thanks to the medical devices that translated human anatomy into digital data; that was later processed by a combination of visualization software. Lastly, the resulting representations have been printed and bound into a book with the utmost technological as well as manual craft development present in the Netherlands in the printing and binding industries.

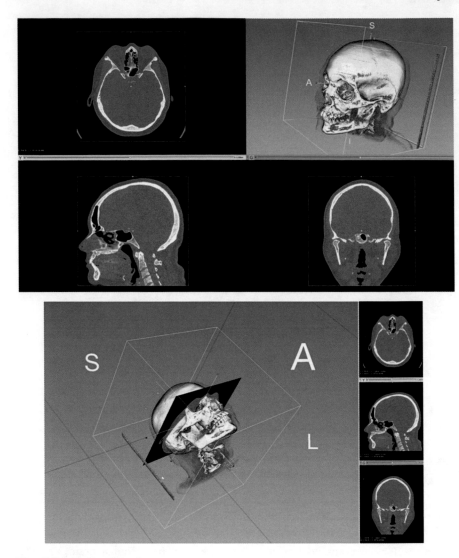

Figs. 2.7–2.10 (continued)

2.4.3 Paper Sculptures and the Vanitas Series

The work-in-progress of *Vanitas* was first presented to the public in the exhibition *Paper Brains.*[19] This show presented the initial *Vanitas* book *massing models.*

---

[19] The PAPER BRAINS exhibition "approaches the question of how the human brain has been represented as a subject of study throughout different epochs in prints and books. The exhibition hints at the multiplicity of 'readings' of the brain from scientific and artistic perspectives." More information can be found here: https://www.paperbrains.net.

Massing models were considered in search of an appropriate form to reflect the book initial ideas. The term *book massing model* or *paper massing model*[20] can aid us in understanding the book design ideation and creation process. A *massing model* is a term borrowed from architecture: 'A massing model is a dimensionally accurate summary of the fundamental exterior forms of a building. It is generally not hollow, but made of solid blocks. Conceptual mass models use basic solid shapes to verify the use of available space, whilst taking into consideration the client's requirements. Window openings in walls are generally not shown. Building detail is either left out entirely or summarized succinctly with a few relatively simple blocks.' When the concept of massing model is brought to the book design realm a *book massing model* would leave aside its tri-dimensional surface, features such as typographic applications or compositional layout to pay attention to *a mass of paper* as a whole (Figs. 2.11, 2.12, 2.13 and 2.14).

After these initial 'paper sculptures', I saw with clarity that *Vanitas* could not exist only as one, but as several book iterations to emphasize the different nuances I had first envisioned in this practice-based research project. In so doing, The *Vanitas* book series challenges the fundamentals of a book concept. It defies the idea of a unique book as a finite object by creating diverse final works that conform a series of explorations of the original book idea. In future exhibitions to come where *Vanitas* would feature, new varieties of the work might add to the current series. The *Vanitas* book series is ongoing. All its physical iterations contribute to the general book *imaging* or idea (Figs. 2.15, 2.16, 2.17, 2.18, 2.19 and 2.20). I could not express it better than Bryson [4, p. 11] when he asserted: '...the word *series* might be more useful here: still life paintings were made to enter the still life series. That series has no essence, only a variety of family resemblances. And it is not a *linear* series, like successive generations of computers or atomic reactors; rather the series (plural) regroup themselves around the individual work, the boundaries of the series fluctuate around each new case. It is a category, in other words, not only within reception and criticism, but within the historical production of pictures.'

2.4.4 A Portrait in Absence

If what the spectator witnesses in *Vanitas* by Pieter Claesz is a portrait in absence—the *feminine space* that while being re-arranged by another woman, enthrones a women's cranium—then I shall be several and continue these cyclical inhabitation. How would I imagine an artwork that will entangle all the possible painterly references and studies my mind consumes, and stacks as if in layers or remembrances? Shall I embody these other womanly spaces to continue with an *infinite mirror of cyclical occurrences* while attempting to *portrait* myself (if possible)?

---

[20] Kevin Matthews: Design Integration Laboratory (http://www.designlaboratory.com/courses/a222.f95/a222.f95.hwk6.html). Retrieved 25.11.2014.

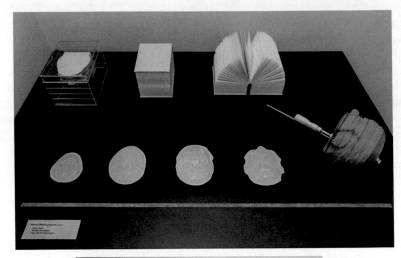

Figs. 2.11–2.14 Danne Ojeda, *Vanitas* (work in progress), © 2017. Pages and books prototypes, variable dimensions. PAPER BRAINS exhibition, Amsterdam, The Netherlands. *Photography* Stefanie Archer

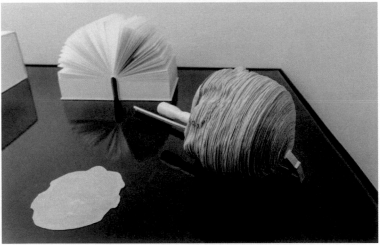

Figs. 2.11–2.14 (continued)

Figs. 2.15 and 2.16 Danne Ojeda, *Vanitas* (Skull), © 2018. Fragments. The brain scans derived from the MRI of the author are debossed onto the book inner pages to form the outer skull book form. 140 × 200 × 140 mm. *Photography* d-file studio

The *Vanitas* book series [21]²¹ reads in its all-encompassing explorations based on the polysemy of meanings set by the relationship between a book and a human skull, originally inspired by the *Vanitas Still Life* painting by Claesz, and by the multilayered historical connections analysed in the different acts of this text.²²

---

²¹ I am an art historian and a graphic designer. Although the combination of these fields is a constant leitmotiv seen throughout my entire work production, *Vanitas* is one of the pieces were most clearly both realms overlap.

²² See www.paperbrains.net. PAPER BRAINS. The exhibition, was part of a research for my creative work *Vanitas* that came to light a year later. PAPER BRAINS, was supported by the Book and Manuscript Studies and the Special Collections Library, University of Amsterdam and Nanyang Technological University, Singapore.

Fig. 2.17 *Vanitas* book spine (for the black, white, and gold **versions**). The spine of the book is adjusted to the proportions of the dimensions of the real skull of the author to make an equivalent between the book anatomy and the real human anatomy of the author. Consequently, the spine has the size of a cover. That is also why; the author uses to say that the spine is also the cover of the book

Beyond literally portraying part of my own inner anatomy, *Vanitas* is also an *autobiographical portraiture* that summarizes the areas of interest, I have been busy with since my upbringing [15]. First, it is a contemporary interpretation of the genre of still life paintings that comes from my background as an art historian, and my inclination towards a critical view of the historical artistic legacy. Also, in the process of taking my MRI tests, I too had to become a Still-life. I had to stay still while the tests lasted. So, the metaphor of (theatrically) staging a 'still life' became part of my own experience.

Second, there is my interest on the book-as-object that can actively express content beyond its texts by means of its formal characteristics. In so doing, I see my making process as an extension of my own thinking and writing. And here is where the notion of authorship is re-presented.

Third, The *Vanitas* book series is a project that examines the relationship between art, design, and the sciences, which has been a recurrent topic throughout my life. As my parents are medical doctors, I grew up surrounded by anatomical study books that made me aware of the visualization of body parts as something independent from human anatomy; almost like prosthetic clinical objects subjected to scrutiny. Later, this observation was reinforced with my own artistic studies on human anatomy when I was a visual art student and had to draw body parts endlessly.

In sum, *Vanitas* is built on the paradox of being one et al., not only because of the serial work it entails, but also by the content it re-presents. When I first imagined *Vanitas*, I had some reservations about part of my anatomy being exposed to the public. However, the final images printed in the book have no sign of my identity

Figs. 2.18 and 2.19 Danne Ojeda, *Vanitas* (Black), © 2018. 660 pages. 140 × 200 × 140 mm. Printed and bound in The Netherlands. *Photography* d-file studio

per se unless spelled out in this prose via the *imagins*'s connections with other texts, paintings, and illustrations. The book and its anatomy (which is also mine) are just what one sees: mere clinical pictures, or yet another portrait in the absence of my mind.

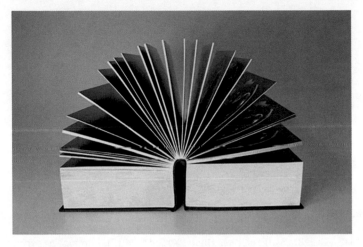

Fig. 2.20 Danne Ojeda, *Vanitas* (Gold), © 2018. 660 pages. 140 × 200 × 140 mm. Printed and bound in The Netherlands. *Photography* d-file studio

Acknowledgements This research is supported by the Ministry of Education, Singapore, under its Academic Research Fund Tier 1.

References

1. Bambach, C.: Anatomy in the renaissance. In: Heilbrunn Timeline of Art History. The Metropolitan Museum of Art, New York (2000). http://www.metmuseum.org/toah/hd/anat/hd_anat.htm. Accessed 15 Dec 2021
2. Barker, F.: The Tremulous Private Body. Essays on Subjection. University of Michigan Press, Michigan (1995)
3. Barthes, R.: Critical Essays (trans.: Howard, R.). University Press, Northwestern (1972)
4. Bryson, N.: Looking at the Overlooked: Four Essays on Still Life. Reaktion, London (1990)
5. Burns, E.: Acting. In: Theatricality. A Study of Convention in the Theatre and in Social Life, pp. 144–183. Harper & Row, New York (1972)
6. Čechová, M., Dupej, J., Brůžek, J., Bejdová, Š., Horák, M., Velemínská, J.: Sex estimation using external morphology of the frontal bone and frontal sinuses in a contemporary Czech population. Int. J. Legal Med. **133**(4), 1285–1294 (2019). https://doi.org/10.1007/s00414-019-02063-8
7. Choulant, L.: History and Bibliography of Anatomic Illustration: In Its Relation to Anatomic Science and the Graphic Arts (ed., trans.: Mortimer, F.). University of Chicago Press, Chicago (1920)
8. Cresswell, J. (ed.): Oxford Dictionary of Word Origins, p. 219. Oxford University Press, New York (2010)
9. Da Vinci, L., Keele, K.D., Roberts, J.: Leonardo da Vinci: Anatomical Drawings from the Royal Library, Windsor Castle. The Metropolitan Museum of Art, New York (1984)
10. Deleuze, G., Guattari, F.: A Thousand Plateaus: Capitalism and Schizophrenia. University of Minnesota Press, Minneapolis (1997)
11. Féral, J.: Foreword. SubStance **31**(2&3), 3–13 (2002). https://doi.org/10.1353/sub.2002.0025

12. Fried, M.: Absorption and Theatricality: Painting and Beholder in the Age of Diderot. University of California Press, Berkeley (1980)
13. Grootenboer, H.: How to become a picture: theatricality as strategy in seventeenth-century Dutch portraits. Art Hist. **33**(2), 320–333 (2010). https://doi.org/10.1111/j.1467-8365.2010.00746.x
14. Haak, B.: The Golden Age: Dutch Painters of the Seventeenth Century (ed., trans.: Willems-Treeman, E.). Harry N. Abrams, Inc., New York (1984)
15. Heller, S.: The Brain as an Open Book (2018). PRINT Magazine, 11 April 2018
16. Kelly, M.: Encyclopaedia of Aesthetics, 2nd edn. Oxford University Press, Oxford, New York (2014)
17. Lanska, D.J., Lanska, J.R.: Medieval and renaissance anatomists: the printing and unauthorized copying of illustrations, and the dissemination of ideas. Prog. Brain Res. **203**, 33–74 (2013). https://doi.org/10.1016/B978-0-444-62730-8.00002-5
18. Last, J.M. (ed.): A Dictionary of Public Health, 1st edn. Oxford University Press, Oxford; New York (2007)
19. Musilová, B., et al.: Exocranial surfaces for sex assessment of the human cranium. Forensic Sci. Int. **269**, 70–77 (2016). https://doi.org/10.1016/j.forsciint.2016.11.006
20. Ojeda, D.: Paper Brains, Exhibition Digital Archive (2017). https://www.paperbrains.net/. Accessed 08 Dec 2019
21. Ojeda, D.: Vanitas, Digital Archive (2018). https://www.vanitas-book.com/. Accessed 10 Dec 2019
22. Rath, G.: Charles Estienne: contemporary of Vesalius. Med. Hist. **8**(4), 354–359 (1964). https://doi.org/10.1017/S0025727300029823
23. Stafford, B.M.: Body Criticism: Imaging the Unseen in Enlightenment Art and Medicine. The MIT Press, Massachusetts (1993)

Chapter 3
A Framework of Networked Art as a Diagram that is an Image as a Map that is a Plan and that is a Space as a Territory

Garrett Lynch IRL

Abstract This chapter discusses the development of a framework for an artistic practice that employs networks, what I have termed as networked art. Instead of focusing
on how the framework is employed by demonstrating and discussing the practice
that is created through its use, which is discussed at length in my doctoral thesis,
I focus on the diagrammatic development of the framework. How the framework
evolved over a series of diagrams from a static image to an interactive model, from
a representative to a planning form and, most importantly, from a two-dimensional
to three-dimensional space. The development of the diagram, itself a manner of
employing a methodology of Practice as Research (PaR), specifically practice-led
research, has been instrumental to how I understand networked art is conceived,
functions and exists in relation to contemporary art practice.

Keywords Networked art · Practice as research · Diagram · Framework · Map ·
Plan · Space · Territory

3.1 How Did I Get Where?

For over twenty years I have worked with networks in my art practice. Initially, this
was exclusively technological and occurred within the space of the, still relatively
new, World Wide Web. The web was many things at that time, a new frontier [1],
a space, a site, a "new home of Mind" [2], a form, a medium "for production,
publication, distribution, promotion, dialogue, consumption and critique" [3] and of
course a message [4]. As the dot-com bubble burst and society awaited the expected
anarchy resulting from the Y2K bug, which never occurred, I felt aligned with a
group of artists who were *working in/on/with/of the web* and have been variously
called net.artists, internet artists and web artists.

G. Lynch IRL (✉)
Cork, Ireland
e-mail: garrett@asquare.org

© The Author(s), under exclusive license to Springer Nature Switzerland AG 2023   59
A. L. Brooks (ed.), *Creating Digitally*, Intelligent Systems Reference Library 241,
https://doi.org/10.1007/978-3-031-31360-8_3

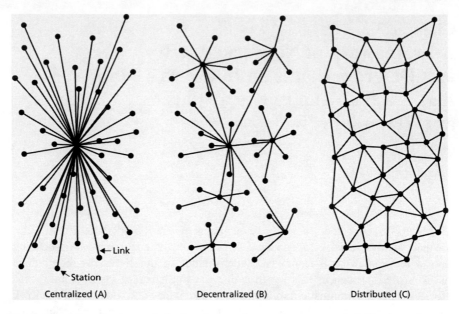

Centralized (A)   Decentralized (B)   Distributed (C)

Fig. 3.1 Paul Baran's proposed distributed network structure as a more reliable alternative to a centralised or decentralised network. Courtesy of the RAND Corporation

There was one significant problem. For all the hopes of the web as a decentralising technology[1] it did not decentralise artists from the historic centres of artistic milieu. As Manuel Castells had already noted concerning power structures [6], networks, such as the web, consolidated controlling power in existing major cities and relegated the mundane tasks it required for its preservation to geographically distant and/or isolated locations. This occurrence soon became manifest with the rise of services, such as Amazon Mechanical Turk and freelance service marketplaces. I am an Irish artist. Irish artists, at that time, were not at the forefront of being actively engaged with contemporary discourses in art, let alone considered technology as intertwined with those discourses. I was born on the geographical outskirts of Europe, not descended from a people with a long history of avant-garde practice; not from a city that overflows with art galleries; not living in a superpower where commerce and industry afford ample patronage for the arts or one where a government suppresses art to the point of it becoming an act of resistance. I was, therefore, artistically isolated in this new era of networks.

As a consequence, several years previously, I had already left my home to pursue my education and to be closer to that artistic milieu but it seemed as if I would

---

[1] These hopes were supported by scientific theory diagrammatically visualised by Paul Baran in his 1964 memorandum for the Rand Corporation [5] (Fig. 3.1). Figure one of the memorandum and its accompanying text explained that a centralised network would be vulnerable if its central node failed or was attacked. A decentralised, but in the long-term distributed, network would be more reliable.

always remain marginalised, regardless of actual geographic location. The ideology of an art practice of the web had for me, therefore, failed. Numerous twentieth-century antecedents, including cybernetics, systems theory, conceptual art, systems art, video art and social art had long developed and implemented ideas of a diversity of knowledge [7], systems, collectivity, institutional critique, such as that of the white cube gallery [8–10], the dematerialisation of art as object [11], the democratisation of media, tools and art for everyone [12]. An art practice of the web did little, seemed incapable, to effectively engage these ideas or move them forward.

3.2 And from There?

Technologies then did as they do, they evolved. Software companies were taken over while tools and mediums were abandoned, either caught in a proprietary no-mans land or repeatedly weaponised in hacking and excluded by web browsers. The vulnerabilities of art in/on/with/of the web as a technologically-enabled art form were exposed. As a result, the landscape of practice changed and the emphasis rapidly shifted from creating to preserving art and its required technologies; ironically something that had never been desired by many of its artists. If art of the web was rapidly becoming what seemed like a failed project, perhaps the dematerialised/digitised progeny of the pairing of conceptual art and digital technology, what was the way forward from here? How could there be a networked art practice or practises if they were technologically situated? Could such a practice be uncoupled from technologies that rapidly changed? Be wielded, worked, without the web and digitally disentangled?

At this point in time, I started to employ the research method of observation to form a survey of artists' practices, comparing them with my practice, and document my findings via a weblog.[2] The objective was to understand how artists were practising with the web, before and now following the dramatic changes described above. How they might now be exploring more broadly the internet and not simply the web region of the internet, its various platforms, spaces and protocols, or, in a much more extensive sense, networks as a whole and what that might incorporate as a practice. This is the line I traced from beginning to realisation.

In 2008 realisation came in the form of an invitation to give a guest lecture on my practice to fine art masters students at a university in England. I had, for some time, been attempting to devise an explanation for and identify a commonality between the variety of practices that my survey of artists' practices was steadily uncovering. At first appearance, these practices, what I was now calling networked art, seemed disparate when considered from a perspective of style, aesthetics, form and even

[2] The weblog Network Research (originally located at: http://asquare.org/networkresearch/) operated from June 2006 to November 2014. It is now archived at the Internet Archive's Wayback Machine (https://web.archive.org/web/20200713205948/http://asquare.org/networkresearch/).

technology. Yet they seemed to have a *vital element of connectedness* and, through this, became more than the sum of their parts.

I hoped to be able to express my explanation diagrammatically. The initial reasons for this were simple. As an artist, I primarily create visual work and the explanation I sought in my research, because this is what my search for a way forward for a networked art was becoming—practice-led research, ultimately needed to be possible to communicate to other artists. Diagrams in social, science, art and pedagogical contexts have long been used as visual explanations. These include but are not limited to examples such as: the Marshall Island Stick Maps, navigational aids to allow natives of the Marshall Islands to navigate the sea [13]; Paul Baran's diagrams of centralised, decentralised and distributed networks [5] (Fig. 3.1); the increasing number of diagrams that map the early days of the Arpanet network to mapping projects of the internet as it is now; Paul Klee's Pedagogical Sketchbook as a teaching aid for new artists [14]; François Molnár and François Morellet's diagram of a cycle of actions between artworks (oeuvre), creator, society and spectator exhibited within the context of the exhibition New Tendencies 2 [15] (Fig. 3.2); Joseph Beuys drawings on blackboards used in performance lectures [16]; Sol Lewitt's diagrammatic wall drawings that move from written instruction to visual form [17]; the Conceptual Framework of Art [18] (Fig. 3.3); Mark Lombardi's diagrams of political and economic power relations [19]; Suzanne Treister's diagrams of societal relationships [20] and Jeremy Deller's The History of the World as a map of social, political and musical relationships that influenced his practice [21] (Fig. 3.4). I reasoned, therefore, that a visual form of communication would, in a sense, speak more clearly to both artists and those with a technological background.

A large variety of diagrams uncovered in my research led me to understand how I could start to formulate my diagram and how it might function. The diagrams uncovered invariable consisted of lines as links, points, shapes, letters, numbers or text as nodes at the intersection of the links and frequently, although not always, employed triangular forms. A selection of these diagrams is illustrated in this chapter, interspersed with versions of my diagram as it developed. These should not solely be considered as individually illustrative of points or discussions recounted in the text

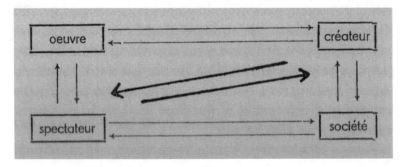

Fig. 3.2 François Molnár and François Morellet's cycle of actions with artworks (oeuvre), creator, society and spectator (1963). Courtesy of the artists

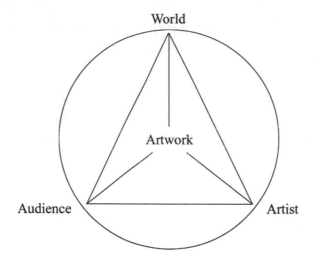

Fig. 3.3 The conceptual framework of art. Courtesy of NSW Education Standards Authority (NESA)

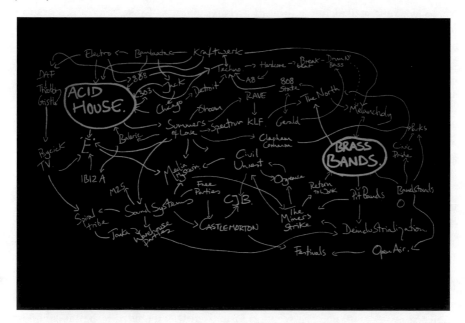

Fig. 3.4 Jeremy Deller, History of the World, 1996. Wall painting based on an earlier hand-drawn diagram by the artist that maps the social, political and musical relationships that influenced the artist's practice. Dimensions Variable. Installation view, Energy Flash, Museum of Contemporary Art Antwerp, 2016. Courtesy of the artist and The Modern Institute/Toby Webster Ltd, Glasgow. Photo by M HKA

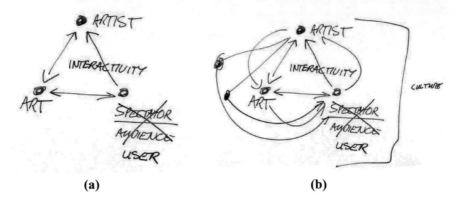

(a)   (b)

Fig. 3.5 (a) The framework of networked art whiteboard drawing, version one. The nodes of art, spectator or user, artist and interactivity as an interconnected triangle of lines. Courtesy of Garrett Lynch IRL. (b) The framework of networked art whiteboard drawing, version one modified. A more complex configuration with additional artworks contributing to culture. Courtesy of Garrett Lynch IRL

but collectively as depicting a form of a visual journey of research. In preparation for the lecture, I devised the first version of my diagram. It illustrated a relationship between an artist, spectator or user in the case of an interactive media artwork, and art, meaning an individual artwork, as a triangle of nodes interconnected by lines (Fig. 3.5). I presented it as a work-in-progress and invited discussion from students. The students' practice and background were largely not technological and what experience they had with technology was in no way related to networks or the internet. The diagram was successfully understood. In discussions following the lecture, the focus was concentrated on the social aspect inferred in the diagram. How an artwork would be created as a result of the juxtaposition of the nodes of artist, spectator or user and art.

Several points arose out of the presentation of my diagram. The first point was that the diagram was reductive. All diagrams are reductive but the point being made here was that the diagram only visualised one instance of a grouping of an artist, spectator/user and art. I had realised this already. There would most certainly be many networked artworks, there could be many spectators/users and even artists. Each artist could, for example, create many artworks or many artists could collaborate on an artwork. Depending on the specific artwork there could be one or many spectators/users implicated in it, which was already hinted at through the use of audience in the diagram. The scenario visualised was intended to be the simplest possible configuration of a networked artwork to communicate it yet be modular to allow scaling up towards more complex configurations. During the presentation, the diagram was adapted through the addition of more nodes, more artworks, on the left side of the diagram (Fig. 3.5) demonstrating how this issue would be resolved and how in turn artworks could be understood to contribute to culture.

The second point that quickly became clear was the unsuitable use of terminology to name spectatorship; a spectator, user or, collectively, an audience. Spectator and user are both individuals, singular nouns. The former is considered passive in an interactive sense while the latter is active. However, audience, generally also considered passive, is a group or collective noun. More precise terminology was needed to express this node in the diagram and if the diagram were to remain a visualisation of the *simplest possible configuration* of a networked artwork it needed to be a singular noun. Thirdly, and finally, perhaps the most problematic point that arose out of presenting and discussing the diagram was the use of the term art, an artwork, as a node. It became apparent that the diagram was not simply visualising a system or network within which a networked artwork participated as a node. A key distinction between my diagram and, for example, Molnár and Morellet's cycle of actions diagram (Fig. 3.2). My diagram as a whole was visualising a networked artwork. The nodes of artist, spectator/user and, what was poorly labelled, art collectively formed a networked artwork. This distinction had many implications that would take years to resolve.

3.3 Wayfinding

In attempting to find a way forward for networked art practice, to move beyond the confines of the web, discussions started to emerge out of some of the practices I had been observing. These discussions, articulated under the title Post-Internet Art, framed their objective as moving practice beyond the internet. However, in reality, this more often than not meant simply moving from the web to gallery contexts. Regardless of this, the intent was there. The practices seemed to be thinking in similar ways as I was, and in this, I understood it as a clear sign that there was an urgency to find a route forward. However, their understanding of that route was that it was with the provision that everything was now informed by the internet, what I considered a part, and not by networks, the whole.

Over the next ten years, I presented my ongoing research, my diagram, in various contexts such as conferences, publications and lectures. Most importantly, it became the central component of my doctoral research, which was completed in 2018 [22]. By this time, I was no longer comparing my practice to the practice of other artists through observation. I was now defining it through the use of the diagram as a framework for networked art practice. In researching Gilles Deleuze and Felix Guattari's conception of diagrams, itselfs based on Charles Sanders Peirce's writings that detail his conception of diagrams and continually incorporate diagrams, in particular of his triadic model of signs (Fig. 3.6), I uncovered a way to achieve this. The purpose of a diagram is, as Guattari has stated, "not to denote or to image the morphemes of an already-constituted referent, but to produce them" [23]. My diagram was not, and should not be, simply a representation or documentation of existing networked artworks. It should not function as an *image or map* of existing

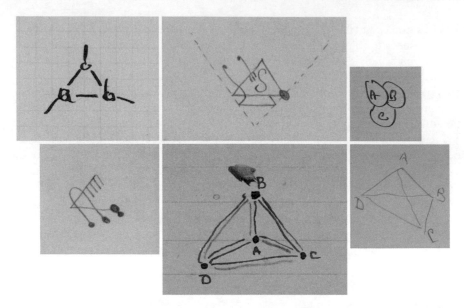

Fig. 3.6 A selection of Charles Sanders Peirce's hand-drawn diagrams from his notes in later life employing triangles, a Venn diagram of three sets of information and tetrahedrons. Courtesy of Houghton Library, Harvard

practice. As a framework, it must set out how networked artworks could be considered to be created.

I have now resolved all of the diagram's prior issues. The diagram consists of the four nodes of *artist*, *observer–user*, *artwork element* and *behaviour* (Fig. 3.7). Observer–user replaces the terminology of the node spectator or user employed in the first diagram and which was deemed unsuitable. The naming of observer–user takes its cues from the fields of science, art and media. Observer has long been used in science. However, here it is employed specifically in relation to systems theory and cybernetics discourses, particularly to second-order cybernetics, which specifies that while there may not be physical interaction between an observer and an observed system there has always been an active relationship [24]. This reframing of observer and the act of observation in science during the twentieth century echoes changes occurring in art discourses about audiences at the same time. These included a move away from a psychical distance [25] between audience and art, self and object, or what Roy Ascott terms the Renaissance paradigm of "the artist standing apart from the world and depicting it and the observer standing outside of the artwork and receiving this depiction" [26].

User in observer–user is employed concerning discourses in computing, media and how these have informed technological arts. It defines the change that occurs in an observer when they physically interact through technology with a media form influencing or changing it. Combined with the use of an en dash, observer and user are not intended to follow a rule of punctuation in language but are instead intended

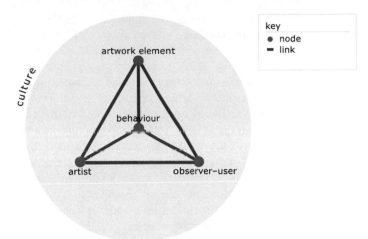

Fig. 3.7 The framework of networked art digital drawing, version two, consisting of four nodes of artist, observer–user, artwork element and behaviour. Courtesy of Garrett Lynch IRL

to form a neologism where the en dash as a range between figures visually recalls a *timeline* of how audience modes have changed over time and a *continuum* of movement from observer to user, observation to interaction occurs (Fig. 3.8). As such, observer–user is to be understood as a singular term and not a phrase.

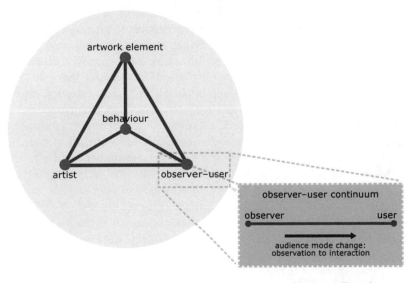

Fig. 3.8 The framework of networked art digital drawing, version two. The observer–user continuum indicates a change in audience modes from observation to interaction. Courtesy of Garrett Lynch IRL

The node artwork element replaces the node art employed in the first diagram. An artwork element is any discernible component of an artwork of any type, size etc. that is distinct from the artist, observer–user and behaviour. It is similar, in effect, to the term non-human in philosophy, which broadly consists of animal, vegetable, mineral and technologically manufactured objects. Human and non-human are often employed together to create a reductive binary opposition of subject and object, Self and Other. In recent years this binary opposition has been challenged in many ways [27], including by actor network theory employing non-human extensively in association with the terms actor or actant [28, 29], and has influenced discourses in contemporary art practices [30]. By replacing art with artwork element, it is clear that this new node no longer visualises an entire artwork but is instead an element of an artwork. The diagram, therefore, no longer visualises a system or network within which a networked artwork participates as a node. It has, instead, become a visualisation of a networked artwork—the simplest possible configuration of a networked artwork. This suggests two things. If only a single artwork element is contained within the diagram then a networked artwork must either be collectively formed by the four nodes of artist, observer–user, artwork element and behaviour and/or that the diagram is extended beyond its four nodes and triangular form to a more complex form in order to complete a networked artwork. Both suggestions are intended and left as options towards achieving simple or complex networked artworks.

Artist as a node positioned in the left corner of the diagram's triangle configuration, is, as it was always intended to be, representative of a human actor who initially creates a networked artwork. The term creation used here is intended to encompass typical creative processes of creating new nodes, such as an artwork element, as well as the techniques of positioning and juxtaposition of existing nodes, such as an observer–user or an already existing artwork element, of a networked artwork. The techniques of positioning and juxtaposition have been a common occurrence in art practices since the twentieth century and extend into various forms and modes including the readymade, systems art, appropriation, re-contextualisation, remixing, sampling, the mash-up, relational art and intersections between art practice and curation [31–33]. Since a networked artwork is, as noted above, collectively formed by the nodes of artist, observer–user, artwork element and behaviour each of these nodes is considered contained *within* a networked artwork. Therefore, for the artist and observer–user, nodes that are human actors which are active and can be physically interactive, being *within* allows a networked artwork to be performative.

In the new diagram performativity and all other forms of nodal relationships are presented as a fourth node called behaviour, which is centrally positioned between the nodes of artist, observer–user and artwork element. Behaviour replaces interactivity as the labelling of the space between nodes in the original diagram. By replacing interactivity with behaviour the new diagram indicates a shift from considering networked art within a new media art context and its discussions of interactivity to a wider systems theory and cybernetics context and their use of the term behaviour. In addition, the term behaviour is less specific about relationships between nodes, thereby allowing behaviour such as observation by observers as well as interaction by users and enabling parity between both parts of observer–user. Behaviour is distinct from other nodes. It is a quality of the relationship nodes have with each other and, therefore, only exists as a consequence of those nodes. As such, it is non-existent prior to a networked artwork existing, not just as a named node but in any sense of existing, and once a networked artwork is formed is emergent from a networked artwork's nodes.

3.4 "I Looked, and, Behold, a New World!" [34]

To visualise the distinction of behaviour from other nodes it is necessary to move and position it separately from other nodes. The nodes of artist, observer–user and artwork element are visually separated in the diagram by using two dimensions. These are all at once the X and Y of a graph, the length and breadth of paper on a horizontal plane that a diagram might typically be drawn on and the latitude and longitude of a map as seen from a top-down perspective. Behaviour cannot be sufficiently separated using these dimensions so an alternative solution is necessary. Version three of the diagram transposes it from two dimensions to three dimensions, from a triangle to a tetrahedron configuration, allowing behaviour to be separated onto another plane (Fig. 3.9). Crucial in understanding how the diagram as a triangle could become a tetrahedron, how earlier versions of the diagram could be interpreted as a top-down view of a tetrahedron and how behaviour could be understood as emergent from other nodes were two diagrams by Fritjof Capra [35] (Fig. 3.10). The first of these diagrams visualised the relationship between matter, form and process as a triangle and the second diagram visualised meaning as a result of that relationship emerging up or out of the triangle and transforming it into a tetrahedron. The plane containing the node behaviour in my diagram can be considered to be layered separately to the plane containing the other three nodes, however, it should not be considered to be on top or in front of those nodes. Instead, it can be considered to be emerging from the nodes and its position is removed or *away* from them, similar to how a vector in mathematics or programming indicates a direction away from a point. Behaviour is, in a sense, the meaning of the relationship between the nodes of artist, observer–user and artwork element.

Transposing the diagram from two dimensions to three dimensions is not simply a solution for visualising the node behaviour in relation to other nodes. It also constitutes several fundamental changes to the diagram. If the diagram now exists in three dimensions then it exists in the same type of space that we, artists and observer–users, occupy and essential to that type of space is time; a non-compulsory requirement in two dimensions. As discussed above, Deleuze and Guattari's interpretation of diagrams as producing and not denoting morphemes [23] allowed me to develop the diagram from simply representing or documenting existing networked artworks to setting out how they could be considered to be created. The diagram has in effect expanded from a role as *image* and *map* of existing practice, to incorporate

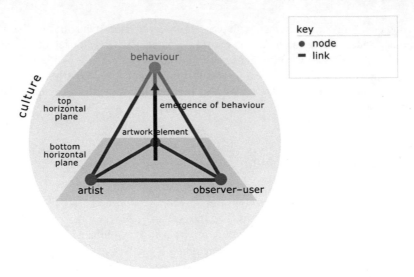

Fig. 3.9 The framework of networked art digital drawing, version three, as a three-dimensional tetrahedron allowing behaviour to be separated onto a different plane to other nodes. Courtesy of Garrett Lynch IRL

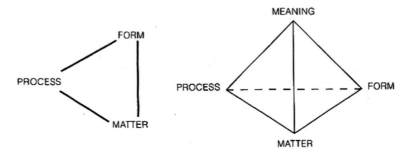

Fig. 3.10 Left, Fritjof Capra's diagram of matter, form and process visualised as a triangle and right, meaning emerging from the triangle to form a tetrahedron. Courtesy of Fritjof Capra

being a *plan* or *blueprint* for practice yet to be created. The integration of three-dimensional space and time in the diagram, however, also shifts it to become more similar to networked artworks than ever before. Space and time across both diagram and networked artworks are to be understood in similar ways. The diagram is no longer simply a representation of space, it is to be understood as a space.

Deleuze and Guattari's concept that diagrams are a means to plan implies that what will result from a diagram being implemented has both spatial and temporal implications. A plan or blueprint for a building, for example, will be used to construct a building at some point in the future. However, as emphasised by Simon O'Sullivan, diagrams are not simply about producing what will be, diagramming "can involve a re-engineering of the past (understood as resource and living archive) that will then allow a different kind of future to emerge" [36]. So diagrams are always to some degree representational or documentative of what has been. They can simultaneously engage the past, present and future to have a spatial impact and are an externalisation of understanding and intentionality. Deleuze and Guattari's concept of the diagram, therefore, needs to be understood as not excluding the ability to represent or document but that its priority *should* be as a means to plan. Furthermore, the intentionality behind the creation of diagrams that range from description to conception and the diverse fields they are employed within supports the dual purpose of diagrams representing and planning.

What complicates Deleuze and Guattari's illusive description of their concept of diagrams is Deleuze's ambiguous use of the term map as the French term *carte*, his manipulation of language that specifically employs the strengths of the French language [37] and the role translation from French to English plays. *Carte* translates to English as a map and is indeed often a diagram of geographical space. However, depending on context *carte* also translates as menu. In both the map and menu translations of *carte*, it is clear that there is always both an engagement of past and future, documenting and planning. The map represents a terrain, what has already been discovered, yet also serves as a plan to explore that terrain anew. The menu documents the food that can be served in a restaurant and its manner of being served yet also serves as a means to plan the meal that will be eaten.

The integration of three-dimensional space and time in version three of the diagram also aligns with the intent discussed above that the artist and observer–user are *within* or immersed in a networked artwork. We behave through observation and interaction with other nodes of a networked artwork in a performative manner, ad-hoc and in real time. There is, as second-order cybernetics states, no observation from *without*, so we are actors in the system of a networked artwork. Not alone does this mean that we influence or change other nodes, and thereby the artwork as a whole for our experience and potentially other's experience, but it also means that other nodes can influence or change us within the context of the artwork. In this matter, all nodes are therefore considered equal.

The continued use of an image to visualise the diagram and its development discussed above is problematic. How can I, after all, state that the diagram is more than an image and map if it is merely visualised as a static image, if it is not a three-dimensional space and contains no time? Version three of the diagram (Fig. 3.9) with its lack of shade and its, while reduced in size, perfectly central artwork element node can still be seen as a flat two-dimensional diagram; three isosceles triangles in contact through their equal sides or legs instead of the tetrahedron it is intended to be. The diagram already maintains an abstraction, what was previously referred to as reductive and I will now term as simplified, from the networked artwork it represents. To continue to visualise it as an image, therefore, simply layers abstraction on abstraction and restricts it from being useful as a diagram and framework. To correctly facilitate the three-dimensional and time-based qualities the newly conceptualised diagram now intends to have it needs to become three-dimensional. This has been addressed by creating the diagram anew within a screen-based three-dimensional environment accessible online through a web browser (Fig. 3.11),[3] version four in the development of the framework.

The addition of three-dimensional space and time in the screen-based three-dimensional diagram, version four, offers several additional advantages over the use of an image to visualise the diagram. These additional advantages include the use of animation and, most importantly, interaction. Interaction allows an artist and observer–user to play and pause an animation that explains the diagram step by step; to interact with the diagram through a graphical user interface (GUI) that enables or disables the visibility of key points of the diagram, and to directly manipulate the diagram. Direct manipulation of the diagram is now of fundamental importance to the diagram and how an artist and observer–user experience it. Manipulation allows an artist and observer–user to rotate the diagram along its breadth, height and length or its three axes of X, Y and Z, and to move their eye, their vision, through it. This enables an artist and observer–user to position themselves at various points in and around the diagram allowing different points of view of the diagram, including the point of view seen from the nodes that represent artist and observer–user. As a result, version four of the diagram (Fig. 3.11) offers several advantages over version three of the diagram (Fig. 3.9). The diagram is now not just intended to be perceived and understood as being a three-dimensional space but to be experienced as such. By moving our eye inside the diagram and behaving with it more actively we become users inside of the space of the diagram and therefore have a sense of what it means to be *within* a networked artwork.

The positioning of the node artist, now at the bottom left corner of the tetrahedron configuration of the diagram (versions three and four), implies an ordering of nodes. This *first ordering* follows a left-to-right trajectory present in all triangular and tetrahedron configurations of the diagram. The ordering conforms to the idea that an artist's creation of a networked artwork, the typical creative processes of creating new nodes, such as an artwork element, as well as the techniques of positioning and juxtaposition of existing nodes [31–33] makes the artist the originating node or the point of origin of a networked artwork. The concept of the artist as an originating node of a networked artwork also conforms to the definition of a Peircean triadic model; a system of three subjects A, B and C, where subject A brings subject B into the same sort of correspondence or relationship with subject C that A has with subject C [38] (Fig. 3.6). As a result of the interdependence of the three subjects in a Peircean system it can only be diagrammed in a triangular configuration and cannot be reduced in number.

As mentioned above the node behaviour is layered separately to the plane containing the other three nodes. It is not intended to be on top or in front of them but instead simply removed or *away* from them in a non-hierarchical manner. The purpose of this is to highlight the distinctiveness of behaviour and, in particular, the idea of its emergence from other nodes. This implies a *second ordering* that follows

---

[3] An image of the screen-based three-dimensional diagram has been provided in this chapter. However, it is strongly advised to view this version online here: http://www.asquare.org/framew ork/.

a bottom-to-top trajectory in the tetrahedron configuration of the diagram (versions three and four). The tetrahedron continues to conform to a Peircean triadic model. Artist, artwork element and observer–user form the initial triadic model, now a part of the tetrahedron, while the tetrahedron's additional three sides are also triangular with behaviour forming the *away* most corner of each.[4] The emergence of behaviour is, therefore, equally the result of the nodal pair combinations of artist and artwork element, artist and observer–user, observer–user and artwork element as well as the combination of all three nodes artist, artwork element and observer–user. Similar to how in the novel Flatland the Sphere employs an analogy to describe parallel movements of forms where a point becomes a line, a line becomes a square and a square becomes a cube to the protagonist, A Square [34], the nodes of artist, artwork element and observer–user of the triadic model can be thought to be moved parallel to themselves into the third dimension.

Behaviour's non-hierarchical condition of being *away* from other nodes is, however, initially visualised as above or further up the Y axis when the screen-based three-dimensional diagram is first viewed. This is a consequence of the *second ordering* of nodes in a bottom-to-top trajectory, which itself builds on the *first ordering* of nodes in a left-to-right trajectory, and is what I will term as the *first view* positions of nodes in the diagram. The first view positions of the diagram's nodes correspond to the formation of a networked artwork. These positions and their implied hierarchy are, therefore, simply initial positions as the diagram can, thereafter, be rotated in any way conceivable. The rotation of the diagram allows the orientation of its nodes to subvert any understanding of up and down, front and back and left and right so any sense of hierarchy in the diagram is transient. As such, the diagram breaks with the traditional Peircean triadic model, which is delineated as A, B and C and is therefore perpetually hierarchical. Similar to Deleuze and Guattari's conception of the rhizome, itself diagrammatic, the diagram and networked art instead have no positions, no beginning or end [39].

---

[4] In the domain of networks nodes employing a triangular configuration provide direct links between all nodes and form the most optimised version of Baran's distributed network (Fig. 3.1). At larger scales this becomes more difficult to achieve and can be impossible to diagram.

3.5 "A Map *Is Not* the Territory It Represents" [40]

The latest version of the diagram, version five (Fig. 3.12), is very much an experiment undertaken at the time of writing this chapter. This version continues to incorporate three-dimensional space and time, however, is an attempt to increase how the diagram is experienced by also being immersive for an artist and observer–user. It achieves this by moving the diagram beyond version four's screen-based mode of use, ultimately an improvement over the diagram as image yet still very much a flattened experience, to instead employ a virtual reality (VR) headset and wireless hand-held controllers. The diagram is hand-drawn in VR. It is a return to the immediacy of version one of the diagram, itself a whiteboard drawing, and is perhaps, therefore, suggestive of version one's pedagogical context or application. An artist and observer–user can now move not just their eye but their whole body by walking through and around the diagram. They can manipulate their view of the diagram by tilting their head in various ways to observe it at different angles and, by moving their body around it, observe it from different sides and at different scales. The diagram can also be directly manipulated through natural gestures of their hands, such as lifting the diagram by raising a hand and rotating the diagram by rotating their wrist. Version five of the diagram, therefore, yields an even better experience of being *within* it and a networked artwork.

In the early stages of developing the framework, it was noted that the diagram was reductive. It was intended to be the simplest possible configuration of a networked artwork to communicate it yet be modular to allow scaling up towards more complex configurations. To continue to achieve this all subsequent versions of the diagram have remained reductive or what I now term as simplified. This quality of simplification, however, is not unique to my diagram. It is a quality that must exist in all diagrams for them to function and be useful as diagrams in line with the map-territory relation [40].

The quality of simplification is now more important than ever with version five of the diagram. Similar to version four of the diagram, its three-dimensional space and time are not the same three-dimensional space and time as the artists and observer–users. However, as a result of its immersive nature, the better experience of being *within* it, it is now closer than ever to that space and risks being confused with it or an illusion of it. Simplification allows the diagram to maintain its abstraction from a networked artwork. Simplification ensures that the diagram retains its representational qualities of reality. It is a map of existing networked artworks and yet simultaneously a plan for artworks to be created. In doing so, the diagram, as both map and plan, in a sense never becomes *as large as the territory* it maps and, therefore, does not become useless as a result of its ability to function as a diagram [41].

Transposing the diagram from two dimensions to three dimensions, incorporating time and subsequently becoming more immersive, effectively coincides with the diagram being developed through digital means. To be clear, the art practice that is simultaneously mapped and planned by the diagram is not a digital one. Networked

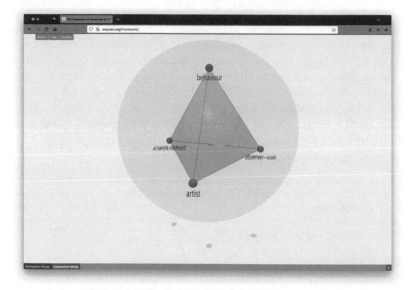

Fig. 3.11 The framework of networked art screen-based online drawing, version four, integrating three-dimensional space and time. Courtesy of Garrett Lynch IRL

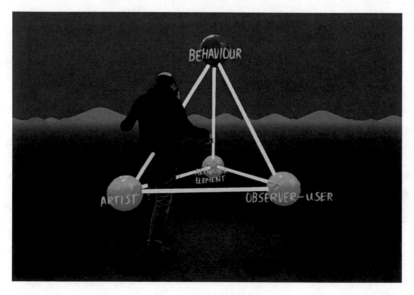

Fig. 3.12 The framework of networked art three-dimensional drawing, version five, as an immersive diagram. Courtesy of Garrett Lynch IRL

art's roots, to name just a few, in systems theory, cybernetics, systems art and conceptual art pre-date the digital era. With two of its four nodes, artist and observer–user, being specifically human and analogue while its remaining two nodes, artwork element and behaviour, are non-human and not specifically digital as they encompass both analogue and digital it has no claim on the specificity of being digital. However, the diagram as a framework of networked art has benefitted from what being digital can afford: three-dimensionality, time, animation, interaction and immersion.

The diagram, as both map and plan, in a sense never becomes *as large as the territory* it maps [41] but through being digital does it become its own territory? In Simulacra and Simulation, Baudrillard states that under contemporary conditions:

> abstraction is no longer that of the map, the double, the mirror, or the concept. Simulation is no longer that of a territory, a referential being or substance. It is the generation by models of a real without origin or reality: A hyperreal. The territory no longer precedes the map, nor does it survive it. It is nevertheless the map that precedes the territory—precession of simulacra—that engenders the territory [42].

One of these contemporary conditions can certainly be considered to be that of being digital. Maps, and therefore my diagram, can now be created 'virtually', preceding and independent of what is 'real'.[5] While the map may now no longer be visible within the territory creating an infinite regression, a Droste effect, through its representation of the territory it is nevertheless still there, it still exists, is still embedded within the territory just as the 'virtual' is embedded within the 'real'.

A space can already be understood to have emerged in version three of the diagram (Fig. 3.9) as a result of it changing from a two-dimensional triangle to a three-dimensional tetrahedron. This spatiality is subsequently increased in versions four and five and now that space is experienced similarly to an artwork's space. So unquestionably yes my diagram has become its own territory and its condition of being digital enables this. At the start of this chapter, I asked whether networked art could be uncoupled from technologies that rapidly changed; be wielded, worked, without the web and digitally disentangled. It seems that while networked art as a practice is by and large not specifically digital, its diagram can't be digitally disentangled. More importantly, the diagram is itself experienced as a territory. It is no longer simply a visualisation of the simplest possible configuration of a networked artwork, it is the simplest networked artwork. I am now no longer tracing a line along the path taken by others but drawing a map, a *hyperreal map*, of a new undiscovered territory.

---

[5] 'Virtual' and 'real' are considered problematised terms in all of my practice and writing. Often misunderstood as binary opposites the 'virtual' is in fact a part of what is 'real' according to Deleuze and Guattari's Ontological Quadrivium of real, possible and actual, virtual. To underline the problematic naming/positioning/opposition of the terms I employ them within scare quotes to suggest that what is considered 'real', physical/material, and what is 'virtual', non-physical/non-material, are arguably largely inaccurate descriptions.

References

1. Eubanks, V.: The Mythography of the "New" Frontier. https://web.mit.edu/comm-forum/legacy/papers/eubanks.html
2. Barlow, J.P.: A Declaration of the Independence of Cyberspace. https://www.eff.org/cyberspace-independence
3. Bookchin, N., Shulgin, A.: Introduction to net.art (1994–1999). http://www.easylife.org/netart/
4. McLuhan, M.. Understanding Media: The Extensions of Man. MIT Press, Cambridge, MA (1994)
5. Baran, P.: On Distributed Communications. I. Introduction to Distributed Communications Networks. The Rand Corporation, Santa Monica (1964)
6. Castells, M.: The Informational City: Information Technology, Economic Restructuring and the Urban-Regional Process. Blackwell, Oxford (1989)
7. ASC.: History of Cybernetics, Chapter 2: The Coalescence of Cybernetics. American Society for Cybernetics. https://asc-cybernetics.org/foundations/history2.htm
8. Buchloh, B.: Hans Haacke: memory and instrumental reason. Art in America, February, pp. 96–109 and pp. 157–159 (1988)
9. Skrebowski, L.: All systems go: recovering Hans Haacke's systems art. Grey Room **30**(Winter), 54–83 (2008). http://www.mitpressjournals.org/toc/grey/-/30
10. O'Doherty, B.: Inside the White Cube. University of California Press, Berkeley (1999)
11. Lippard, L.: Six Years: The Dematerialization of the Art Object from 1966 to 1972. University of California Press, Berkeley (1997)
12. Pierce, A.: Beuys, Hardt and Negri: One World—One Consciousness. No: ART2014-0992 (2014). https://doi.org/10.30958/ajha.2-2-1
13. Knighton, H.: Navigating the Waters with Micronesian Stick Charts. National Museum of Natural History, Smithsonian Institution, Washington, DC (2020). https://ocean.si.edu/human-connections/history-cultures/navigating-waters-micronesian-stick-charts
14. Klee, P.: Pedagogical Sketchbook. Praeger, New York (1972)
15. Rosen, M.: A Little-Known Story about a Movement, a Magazine, and the Computer's Arrival in Art: New Tendencies and Bit International, 1961–1973. The MIT Press, Cambridge, MA (2011)
16. Tate Gallery: The Tate Gallery 1984–86: Illustrated Catalogue of Acquisitions Including Supplement to Catalogue of Acquisitions 1982–84. pp. 489–494. Tate Gallery, London (1988). https://www.tate.org.uk/art/artworks/beuys-four-blackboards-t03594
17. Laperle, S.: Variations of a drawing: Sol LeWitt and his written instructions. RISD Gallery, Providence, RI (2013). https://risdmuseum.org/manual/45_variations_of_a_drawing_sol_lewitt_and_his_written_instructions
18. NSW Education Standards Authority (NESA): Visual Arts Years 7–10 Syllabus, New South Wales (2003)
19. Casemajor Loustau, N.: Mark Lombardi's topographies of power: the work in the map. Espace 103–104 (spring–summer) (2013). https://espaceartactuel.com/en/mark-lombardis-topographies-of-power/
20. Treister, S.: Suzanne Treister. https://www.suzannetreister.net/
21. Deller, J.: The History of the World (1997). http://www.jeremydeller.org/TheHistoryOfTheWorld/TheHistoryOfTheWorld.php
22. Lynch, G.: The Transformative Nature of Networks within Contemporary Art Practice (2018)
23. Watson, J.: Guattari's Diagrammatic Thought: Writing Between Lacan and Deleuze. Continuum, London (2009)
24. Umpleby, S.A.: What comes after second order cybernetics. Cybern. Human Knowing **8**(3), 87–89 (2001). https://www.academia.edu/65371025/What_comes_after_second_order_cybernetics
25. Bullough, E.: 'Psychical distance' as a factor in art and an aesthetic principle. Br. J. Psychol. **5**(2), 87–118 (1912). https://doi.org/10.1111/j.2044-8295.1912.tb00057.x

26. Ascott, R., Shanken, E.A.: Art and education in the telematic culture. In: Telematic Embrace: Visionary Theories of Art, Technology, and Consciousness, pp. 212–221. University of California Press, Berkeley (2007)
27. Braidotti, R.: Anthropos redux: a defence of monism in the anthropocene epoch. Frame **29**(2), 29–46 (2016). https://www.frameliteraryjournal.com/29-2-perspectives-on-the-anthropocene/1450/
28. Murdock, J.: Inhuman/nonhuman/human: actor-network theory and the prospects for a nondualistic and symmetrical perspective on nature and society. Environ. Plan. D Soc. Space **15**(6), 731–756 (1997). https://doi.org/10.1068/d150
29. Latour, B.: Politics of Nature How to Bring the Sciences into Democracy. Harvard University Press, Cambridge, MA (2004)
30. Rapp, R., De Lutz, C.: Introduction. In: Nonhuman Agents in Art, Culture and Theory. Art Laboratory Berlin, Berlin (2017). https://artlaboratory-berlin.org/events/non-human-agents-in-art-culture-and-theory/
31. Burnham, J.: Real time systems. Artforum **VIII**(1), 49–55 (1969)
32. Fineberg, J.: Appropriation. In: Art Since 1940 Strategies of Being, pp. 454–471. Laurence King Publishing, London (1995)
33. Bourriaud, N.: Relational Aesthetics. Les Presses du Réel, Dijon (2002)
34. Abbott, E.A.: Flatland: A Romance of Many Dimensions. Dover Publications, New York (1992)
35. Capra, F.: The Hidden Connections: Integrating the Biological, Cognitive, and Social Dimensions of Life Into a Science of Substainability. Doubleday, New York (2002)
36. O'Sullivan, S.: On the diagram (and a practice of diagrammatics). In: Situational Diagram, pp. 13–25. Dominique Lévy Gallery, New York (2016)
37. Morizot, B.: Penser le concept comme carte: Une pratique deleuzienne de la philosophie. In: La Géophilosophie de Gilles Deleuze. Mimesis, France (2017) https://hal.archives-ouvertes.fr/hal-01476141
38. Peirce, C.S.: The New Elements of Mathematics: Volume IV Mathematical Philosophy. Mouton Publishers, The Hague (1976)
39. Deleuze, G., Guattari, F.: A Thousand Plateaus: Capitalism and Schizophrenia. University of Minnesota Press, Minneapolis (1987)
40. Korzybski, A.: Science and Sanity: An Introduction to Non-Aristotelian Systems and General Semantics. Institute of General Semantics, New York (1994)
41. Borges, J.L.: On exactitude in science. In: Collected Fictions of Jorge Luis Borges, pp. 704–705. Penguin, London (1999)
42. Baudrillard, J.: Simulcra and simulations. In: Jean Baudrillard Selected Writings, pp. 166–184. Stanford University Press, Redwood City (2002)

Chapter 4
Exercising Digitally: A Multi-Perspective Analysis of Exergames for Physical Activity and Health Promotion

Lisa Röglin⬛, Anna Lisa Martin-Niedecken⬛, and Sascha Ketelhut⬛

Abstract Exergames are often considered an intriguing opportunity for promoting physical activity (PA) among various target groups. However, due to the large number of diverse products currently available under the term "exergames", it is not appropriate to label exergames per se as promising tools for addressing physical inactivity. In this chapter, the authors aim to analyze current exergaming products from different perspectives to identify their strengths, weaknesses, potentials, and risks for PA and health promotion. Furthermore, this chapter provides a stimulus for thought, inspiration, and suggestions for researchers, game designers, and publishers dedicated to exergaming. It also gives a broad overview of the history of exergames, focusing on the publishers' influence on product development and targeting ("gamification" versus "sportification"). Furthermore, the authors aim to provide a structured approach to clustering exergames based on the platforms or devices required to play them. Considering the current literature findings, the authors go on to summarize the physical, psychological, and cognitive effects of exergames and discuss the potential and limitations of current products regarding PA and health promotion. Subsequently, they provide research-based recommendations on what to consider when developing exergames by giving insight into a specific case. The work concludes with some possible future directions and an excursion into the metaverse.

Keywords Sportification · Gamification · Exergaming effects · Digital health promotion · Exercising in the Metaverse · Avatar-based exercising · Exergaming

L. Röglin (✉)
Institute of Sport Science, Martin-Luther-University Halle-Wittenberg, 06108 Halle (Saale), Germany
e-mail: Li.roeglin@gmail.com

A. L. Martin-Niedecken
Department of Design, Institute for Design Research, Zurich University of the Arts, 8031 Zürich, Switzerland
e-mail: anna.martin@zhdk.ch

S. Ketelhut
Department of Health Science, Institute of Sport Science, University of Bern, 3012 Bern, Switzerland
e-mail: sascha.ketelhut@unibe.ch

overview · Designing exergames · Physical activity promotion · Virtual reality exercising · Mixed-reality physical activity · Augmented reality fitness · Clustering exergames · Exergaming history · Exergaming ecosystem

4.1   A New Approach to Physical Activity and Health Promotion?

Regular physical activity (PA) is associated with lower all-cause mortality and a lower risk of cardiovascular disease, diabetes, and various cancers [121, 163]. Research indicates that PA can help prevent non-communicable diseases, accounting for 74% of deaths worldwide [160]. Moreover, PA can improve mood, reduce depressive symptoms, and enhance cognitive functions [41, 115]. Thus, regular PA is considered a foundation for a healthy lifestyle in all age groups. The risk reduction (mortality and morbidity) amounts to about 20–30% for the various diseases and is thus comparable to or even more effective than a drug-based monotherapy [81].

PA is generally defined as any movement generated by the skeletal muscles that raises energy consumption above the basal metabolic rate (energy requirement at rest) [22]. Accordingly, in addition to sports, activities performed during leisure time, at work, on the way to work or school, during household chores and gardening are considered PA. To maintain and promote health, a certain level of PA should be achieved. According to the World Health Organization (WHO), an adult should engage in at least 150–300 min of moderate to vigorous PA each week [161].

Although the positive effects of regular PA are widely recognized [98, 116], more than one quarter of adults worldwide do not meet the PA recommendation [53]. Various cultural developments and technical achievements in recent decades have fundamentally changed people's everyday lives and radically reduced PA [9, 104]. The widespread lack of PA in society has serious health consequences. Physical inactivity (not meeting PA recommendations) is ranked fourth as a risk factor in global mortality statistics [76] and is blamed for 5.5% of all deaths (approx. 3.2 million) worldwide and for 7.7% of deaths in so-called "high income countries", i.e. in the western industrialized countries. Moreover, physical inactivity results in an economic burden of INT$ 67.5 billion on the health-care system worldwide due to health care expenditures and loss of productivity [29]. Lack of PA has become one of the greatest health-related challenges of our time and the current trajectory of the physical inactivity trend is concerning.

In addition to physical inactivity, sedentary behavior is emerging as a novel risk factor for cardiovascular disease, diabetes, mental illness, and all-cause mortality [66, 148]. Sedentary behavior is defined as prolonged behaviors characterized by low levels of PA and thus low energy expenditure [150]. Contrary to general perceptions, sedentary behavior does not necessarily reflect a lack of PA. Individuals are considered to be active when they meet PA recommendations. This does not prevent them from also devoting a significant amount of time to sedentary behavior. PA and sedentary behavior are regarded as distinct but interrelated behavioral attributes with unique determinants and health consequences. The detrimental consequences of sedentary behavior are largely independent of the positive effect of PA on chronic disease risk [114]. Therefore, persons who engage in a high volume of sedentary behavior exhibit increased risks of morbidity and mortality irrespective of their level of moderate-to-vigorous PA [34]. Sitting time was responsible for 3.8% of all-cause mortality (approximately 433,000 deaths) in 54 countries worldwide [124]. According to [154] every hour spent sitting in front of the television reduces life expectancy by 21.8 min.

Increasing PA levels and reducing sedentary behavior have become a global public health challenge today. So if, on the one hand, the health benefits of regular PA are proven and generally known, but, on the other hand, more than one quarter of adults are unable to achieve the recommendations, there must be serious reasons that prevent them from engaging in regular PA. Indeed, research has uncovered numerous barriers (especially temporal, physical, and psychological). Changing and maintaining behavior to obtain health benefits, however, is often impeded by motivational issues, and lack of time [101]. Thus, there is a need to implement innovative lifestyle interventions that promote PA and decrease sedentary behavior.

Paradoxically, one promising way to encourage people to exercise more could be with the help of video games [91]. So-called exergames (exercise and gaming) combine physical and cognitive activities with immersive game experiences. In contrast to standard computer games, the player is physically active throughout the game, as he must perform a variety of movements to control the game. The player's movements are detected by different sensors (accelerometer, pressure sensor, image recognition methods and motion-capturing systems). According to research, exergames may be highly enjoyable, attention-grabbing, and intrinsically motivating [49].

Exergames, also known as motion-based games, exertion games, or active video games (Fig. 4.1), can usually be played in single-player and cooperative or competitive multi-player mode and are used in a wide variety of applications.

Typically, the player moves in front of a screen that display a virtual game scenario, which he controls by his physical input. In this way, commercially available exergame platforms such as the Nintendo Wii, Sony Move, or Microsoft Kinect and their associated games have provided entertaining PA experiences for the living room for over 10 years. However, there are also analog or hybrid exergames that often resemble traditional sports and can be used both indoors and outdoors, thus completely dispensing with the classic player-screen setting (e. g. Fitlights).

Manufacturers of game-based training and therapy applications such as virtual augmented climbing (ValoMotion), game-based robotic exercise therapy (Hocoma), or functional high-intensity interval exergaming (Sphery) are now becoming established in the fitness, rehabilitation, and esports markets [89]. Even though most commercial exergames were primarily developed for entertainment purposes [169], in recent years exergames have been considered valuable for encouraging participation in exercise, as well as for improving adherence to exercise and rehabilitation tasks [19, 137].

Fig. 4.1 Exergaming terms: exergames are active video games that combine physical and cognitive activities with immersive game experiences. Numerous terms are used synonymously in literature and everyday life to describe these kinds of games. Most of these refer to the same thing; others are generic or highly specific terms

Although most commercial exergames have been developed primarily for entertainment purposes [169], exergames are increasingly being appreciated for promoting PA and improving adherence to exercise and rehabilitation programs.

In general, depending on the software and hardware design, exergames can potentially be used to train a wide variety of physical (e.g., balance, speed, strength, endurance) and cognitive functions (e.g., attention, memory, executive functions) to achieve individual goals [96].

Numerous studies have investigated the potential effectiveness of both commercially available and specially developed exergames with various target groups (children, adolescents, seniors, or patients). The results provide evidence for promising cognitive, physical, and mental training effects.

In summary, there is a consensus that adequate PA leads to functional, morphological, and metabolic adaptations of various biological systems, thereby increasing physical and mental performance and promoting health. Unfortunately, a large portion of our society is disinclined to enjoy and participate in substantial amounts of PA [1].

In the last decade, there have been increased efforts to utilize digital games for serious purposes. Under the label "Exergames" serious games are developed, used and evaluated specifically for PA and health promotion.

In this chapter, we will give a broad overview of the history of exergames. We will further try to provide a structured overview of the wide range of products that are summarized under the term "exergames" and cluster the systems into different categories. Thereafter, we will summarize the current literature and discuss the possibilities and limitations of current exergames with a focus on PA and health promotion. Subsequently, we will discuss what to consider when designing an exergame and give an insight into a specific case. Finally, we will try to summarize this chapter and outline some future directions for researchers, game designers, and publishers.

4.2 History of Exergaming: A Brief Overview of Products and Developers

Within the last 10 years, exergames have become very popular and numerous exergaming products have been developed for the mainstream market. However, the history of exergames dates back much further than initially appears. This chapter aims to provide a brief overview of the turbulent history of exergaming, taking a closer look at selected products, especially those that have had a deeper impact on the market. In addition, the developers' influence on the respective exergames will be addressed.

4.2.1 Trial and Error

The earliest attempts to combine "movement" and "gaming" were made as early as 1982 (Fig. 4.2). Amiga's "Joyboard," for example, made it possible for the first time to control a video game by shifting body weight [17]. The "Joyboard" was a motion-controller designed like a conventional body scale, that was equipped with pressure sensors translating the body movements into the game. The only game specifically released for the "Joyboard" was the skiing game "Mogul Maniac" [16].

Also in 1982, Atari developed a prototype called "Atari's Puffer" [3, 44]. The "Puffer project" was used to develop a home entertainment system that connected the "Atari 5200 SuperSystem™ video game console" to an exercise bike. The game was controlled by the exercise bike, which was equipped with interactive pedals and handles. The virtual cyclist on the screen only moved when the player pedaled. Unfortunately, the project development stopped with the prototype and the final market launch was not realized for financial reasons [40].

Based on the "Puffer prototype," Atari developed a combination of an exercise bike and the Suncom Aerobics Joystick™ [45], providing an active ergometer-based video gaming experience. The joystick was connectable between most standard home trainers and an "Atari 2600 Video Computer System™". Various shoot-'em-ups, and driving games were available.

Interestingly, exergaming history is characterized by many experiments that did not succeed on the market. According to game designer Falstein quoted from Orland and Remo [113] and [54] this may be explained by the high pricing and low usability of many products of the 1980s and especially the 1990s. Furthermore, some early products may be considered ahead of their time, having failed to find a suitable target group when they were launched [54]. However, the early beginnings also saw some innovative products that continue to shape the exergaming market to this today. For example, the "Exus Foot Craz dance mat" developed in 1983 [46] is considered the forerunner of the dance game series "Dance Dance Revolution" (DDR) produced by the Konami Corporation's Bemani music games division. "DDR" was initially developed for arcades in Japan in 1998 and is now one of the most popular games worldwide with nearly 100 updated versions, including on video-game console systems [16].

The "Exus Foot Craz dance mat" was compatible with the Atari 2600 Video Computer System™ [16] and included, for example, the game "Video Reflex" [4], in which bugs displayed randomly on the screen had to be removed by touching a corresponding color area on the dance mat in time.

The principle of today's game "DDR" is based on a similar, quite simple but successful idea: while arrows are displayed on the screen in time with music, the player on the dance mat must step on the corresponding arrows at the right moment.

Notably, Nintendo (Namco Bandal Games) also adopted the concept of the "Foot Craz Dance mat" in the mid-1980s (released circa 1988) by developing a modified fitness mat, the "Power Pad" [40]. Various running and dancing games have been released for the "Power Pad" in the late 1980s and early 1990s including "Family Fun Fitness™," "Athletic World," or "Dance Aerobics" [16].

Another innovative and influential exergaming product released in the mid-1980s and designed by the RacerMate company is the "CompuTrainer". The "Compu-Trainer" connects to the rear wheel of a bicycle and provides electro-magnetic resistance and a connection to a computer. It is considered to be the first electronic indoor cycling trainer with an interactive course design software, which was found to have a major impact on the indoor training experience [120]. As the "CompuTrainer" has been continuously upgraded with innovations, a modified form of the product was still available until the 2010s. However, production finally ceased in 2017 due to competition from larger companies and technological progress [120].

4.2.2 "Gamification" Versus "Sportification"

Interestingly, exergaming history reveals that the initial impulses for the development of exergames came solely (with a few exceptions) from the games industry (e. g. Atari, Konami, Namco Bandal Games, Nintendo, Sony), which consequently had a strong influence on the focus of their products. Thus, the exergames of the early years were primarily aimed at the target group of "gamers," focusing on the fun factor and game design. The movement component was merely an extension of the game concept, not a serious "sportification" of the game or "gamification" of exercise programs. Some products were rather considered to provide "exercise by accident" because evaluated exercise or training programs were not usually integrated into the game. Therefore, from an exercise and health science perspective, most exergames were not intensive enough to promote health-related PA. Falstein cited from Orland and Remo [113] explained this development at a "Games for Health Conference" in 2008 as follows: "Nobody wants to go to an arcade to work up a sweat."

It is only in the last ten years that the focus of exergames has shifted. Fitness companies are increasingly using exergames and gaming elements to increase the attractiveness of training programs or PA interventions ("gamification"). These innovative approaches may help to increase motivation and to attract new target groups. Today, an increasing number of developers from the games industry (e.g. "Six to Start") provide exergames that include moderate-to-high-intensity exercise ("sportification")

The popularity of exergames has also led to growing scientific interest, as demonstrated by the increasing number of scientific publications (Fig. 4.3). In particular, sports science and many other fields such as neuroscience, computer science, and rehabilitation are investigating the various products from different perspectives. However, there are still large discrepancies in the consoles and games applied in research. The most commonly used games in exergaming research are the Wii-Fit Plus games [146]. Other exergames (especially non-console-based exergames, see Sec. 4.3) have received less attention in the scientific community.

Although the exergame industry is still in its infancy, and large-scale research studies on their efficacy are still sparse, exergames may present a new way of being active ("the new active"). For developing promising exergames in the future, interdisciplinary approaches are strongly recommended.

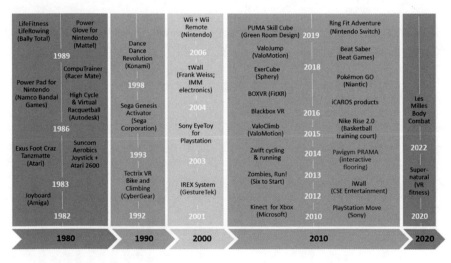

Fig. 4.2 Exergaming milestones (games and products): the figure illustrates the approximate release dates of various exergaming products. While in the 2000s the development of the "Nintendo Wii" console and especially the "Nintendo Wii Remote Controller" made exergaming popular on the mainstream market, the 2010s saw the development of numerous new products based on very different designs and technologies

4.3 Status Quo: A Clustering Approach to Current Exergaming Products

Today, numerous products are summarized under the term "exergames." However, these exergaming products could not be more different in structure, design, functionality, area of application, and target group. As mentioned earlier, exergames also differ significantly in their intensity and efficacy (see chap. 4 for more details). These differences should be taken into account when selecting an appropriate exergame for a respective application area or target group.

This chapter, therefore, attempts to provide a clustering of selected exergames by different categories to give an overview of the current exergaming market (Fig. 4.4). It is noteworthy that there are multiple ways of grouping because some exergames have several distinctive features. Previous scientific approaches that clustered and categorized exergames are no longer applicable today due to the rapidly growing and increasingly differentiated exergaming market [48].

Our clustering approach is based on the corresponding exergaming platforms or devices (Fig. 4.4) required to play the game. Nonetheless, there can be fundamental differences between exergames of the same category.

4.3.1 Console-Based Exergaming

Description The category "console-based exergaming" includes exergames that require a game console such as Nintendo Wii, XBOX Kinect (Microsoft, Remond,

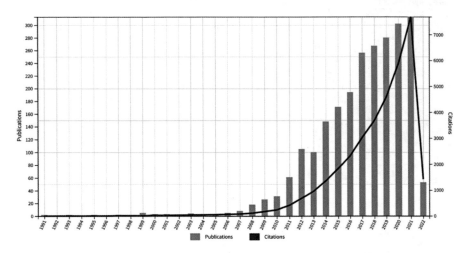

Fig. 4.3 Publications and citations in the field of exergaming from 1991–2022 (*source* webofscience.com (clarivate), search terms: "exergaming," "exergames," or "exertion games" accessed 20.02.2022)

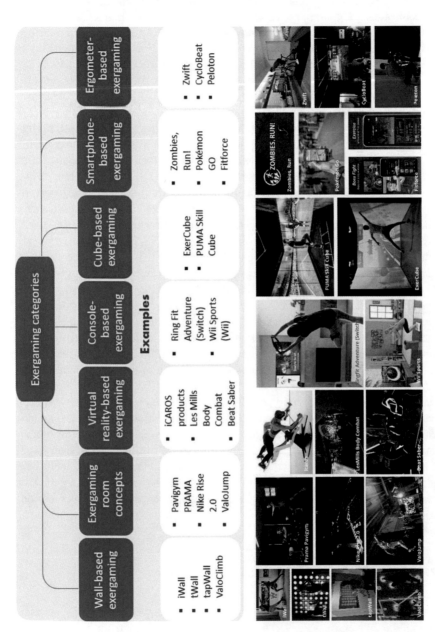

Fig. 4.4 Clustering of exergames with selected examples – a current suggestion

WA), Playstation Move, or Nintendo Switch, which capture movements and transmit them into the game/onto the screen, thus, providing a mixed-reality gaming experience. In some cases (depending on the exergaming title), further devices such as a game controller, a leg strap, a balancing board, or a dancing mat are required to play the exergame. Consequently, depending on the console, the translation of the movement into the game is based on different approaches. For example, Microsoft's Kinect is a motion-sensing input device that tracks the user's body movement through gestures and spoken words (voice recognition) [62]. Hence, the Kinect allows users to control and interact with the XBOX 360 games without requiring a handheld game controller or similar devices.

Console-based exergames are typically provided by the gaming industry. Big players are Microsoft (Redmond, WA) or Sony (Kyoto, Japan). Compared to conventional, sedentary video games, these exergames add an active component to the game. Current research shows that most exergames require light-to-moderate physical activity and are not based on individually designed training concepts. However, the amount of physical activity required, and calories burned can vary significantly from title to title [61]. It is worth noting that, for a long time, scientific studies focused primarily on Nintendo Wii-Fit games [146].

Setting and Target Group Console-based exergames are usually played at home (living room-based exergames) and are targeted at the whole family—even older people [88, 159]. Most of them have a single- and a multiplayer mode. No specific physical abilities are required to play these exergames. In some cases, console-based exergames have also been used in schools [135], care facilities, or similar institutions [112] as a means of promoting PA. For example, the Wii-Fit Plus games are the most commonly used games in the clinic [146]. In rehabilitation, exergaming has been used as a tool to improve balance and functional movement in a variety of populations [27, 52, 145, 147].

Gaming Titles Among the most popular console-based exergames are "Wii Fit" (Nintendo Wii), "Wii Sports" (Nintendo Wii), "Ring Fit Adventure" (Nintendo Switch) or "DDR". These games include various sports, such as dancing, yoga, bowling, and boxing. All in all, there is an exergaming title for almost every age group and preference. In 2012 alone, Microsoft (Redmond, WA) released about 100 kinect-enabled XBOX 360 titles [61]. However, some of the workout titles available today are merely exercise videos that have been enhanced with Kinect motion capturing.

4.3.2 Virtual Reality (VR)-Based Exergaming

Description This category includes exergames that are played using a VR headset immersing the player in VR. VR fitness experiences range from a full virtual gym displayed in the games (e.g. "FitXR experiences") to swinging virtual lightsabers and slicing blocks to music in "Beat Saber." In some cases, these exergames require additional devices, training equipment, or exclusive training platforms (e.g.

"ICAROS") for home use. In other cases, VR-based exergames are performed in a gym using the training equipment provided (e.g. "Blackbox VR"; location-based/ gym-based exergaming).

This category includes very different exergaming products, which are provided either by the games or the fitness industry. Consequently, these products address entirely different target groups. Three examples are described below.

ICAROS A leading brand for VR-based exergaming experiences is ICAROS (ICAROS GmbH, Germany). ICAROS provides various training equipment and devices that combine fitness, virtual reality, and mobile games for use at home, at boutique studios, rehabilitation centers, and group workouts at the gym. The products are designed for different age groups and fitness levels (https://www.icaros.com/ en/). For example, the ICAROS cloud (https://www.icaros.com/en/products/icaros-cloud) is an inflatable and digitally connected training device that resembles an over-sized balance board. This enables the player to fly or dive through virtual worlds while performing a full-body workout. The workout includes strengthening leg and upper body muscles and improves coordination and balance. The ICAROS Health, on the other hand, is designed for professional use in orthopedic and neurological therapy and rehabilitation. It offers effective core and balance training for people with physical limitations (https://www.icaros.com/en/products/icaros-health).

Blackbox VR A gym-based VR-exergaming approach is provided by Blackbox VR (Video: https://youtu.be/uZbiB90Kg4I). Blackbox VR enables the player to participate in gamified workout battles including artificial intelligence progression. The workout is based on resistance training and a high-intensity interval training workout program lasting 30 min. The better the performance during the workout, the more powerful are the in-game attacks (https://www.blackbox-vr.com/the-game/).

Beat Saber and its Modification Feed Saber Compared to the examples above, the two games Beat Saber and Feed Saber are not based on sophisticated training concepts. Rather, the movements in the game seem to be incidental ("exercise by accident"). The focus of both games is on listening to music. The music sets the beat, in which the player has to dodge glowing walls and slice boxes with a laser sword (Beat Saber, https://beatsaber.com/). In contrast, Feet Saber (a special modification of Beat Saber) is getting the player to play Beat Saber with the feet. Both games were designed by the Czech game developer Beat Games.

4.3.3   Cube-Based Exergaming

Description This category includes exergames constructed like a cube that frames the player with the help of (usually) 3 or more (video) walls and thus provides a mixed reality fitness experience. In contrast to wall-based exergames (see Sect. 4.3.6), the player is not only standing in front of a wall but is almost surrounded by it, increasing the immersion in the digital world. Sometimes the player can even interact with the game by touching the walls. In most games, motion trackers are necessary to control the game.

One example from this category and the exergaming room concept category is the ExerCube by Sphery Ltd (Zürich, Switzerland; https://sphery.ch/exercube/) which is described in detail in Chap. 5 and will not be further discussed in this section.

PUMA Skill Cube An example of a cube-based exergame is the so-called Skill Cube by PUMA (Herzogenaurach, Germany; video: https://www.youtube.com/watch?v=yawoGsmG4aY). The PUMA Skill Cube was created by the agency Green Room Design (Birmingham, UK; https://greenroomdesign.com/) and is only installed at a PUMA flagship store in New York's Fifth Avenue. This consists of several video walls, and the floor is covered with a multi-sport, synthetic turf.

The cube transports the player into a virtual football stadium or an abandoned warehouse to provide an authentic footwear-trial customer experience in the store. The cube experience consists of several trials (e.g., ladder and jump). It allows customers to play with sports stars (e.g. Antoine Griezmann, Lewis Hamilton, or Romelu Lukaku), who are presented as virtual training buddies. At the end of the experience, the players receive a score that is entered into a Leaderboard, allowing them to compete against other customers. Additionally, the players receive a personalized email for social sharing.

Unlike the other exergames, the Skill Cube does not focus exclusively on the exercise experience. Instead, it aims to fulfill the customer's emotional needs and the brand's commercial needs (increasing store visits and sales). Consequently, the Skill Cube is not a typical exergame used privately, in gyms, or in other health facilities. Nevertheless, it still has typical features of exergames.

4.3.4 Smartphone-Based or Mobile Device-Based Exergaming

Description In this category we have grouped exergaming apps designed for an augmented reality (AR) (fitness)-experience with the smartphone or other mobile devices. These mobile exergames can be used both at home and outdoors. In most cases, no additional devices or equipment are required to play the game.

This category includes games that could not be more different in their objectives and target audience. Two examples are presented in more detail below.

Zombies, Run! Zombies, Run! is an exergaming jogging app that aims to make the running experience of amateur or professional runners more exciting (https://zombiesrungame.com/). While walking, jogging, or running (anywhere in the world; GPS must be active), the player listens (through headphones) to a zombie story combined with music. The player's challenge is to complete a mission while being chased by zombies. The zombies adapt to the runner's speed, so the player can shake off the pursuers by accelerating. If caught by a zombie, the player loses collected items.

The app was developed by the British studio Six to Start (https://www.sixtostart.com/about/) and the English novelist and game writer Naomi Alderman. Zombies, Run! is funded by the UK Department of Health for the NHS.

Pokémon GO Pokémon GO, made by the computer game development studio Niantic (San Francisco, US, https://pokemongolive.com/), is a global indoor and outdoor gaming experience that requires movement to catch and fight virtual fantasy creatures. According to data from SurveyMonkey Intelligence, in 2016 Pokémon GO was officially the biggest mobile game in the United States, with 21 million active daily users [60].

Compared to "Zombies, Run!" Pokémon GO is not a fitness game aiming for exercise. Instead, Pokémon GO fits into the category "exercise by accident" as the required movements are not underlying training concepts and do not have to be executed precisely. Nevertheless, considerable distances can be covered while gaming, for which players are rewarded with medals. An announcement by Niantic revealed that as of December 7, 2016, players had reached a total of 8.7 billion kilometers walking outside in search of new Pokémon to catch [42]. However, movements can, unfortunately, be faked relatively easily while gaming: a fake GPS makes it possible to use all of Pokémon GO's features without having to move around.

4.3.5 Ergometer-Based Exergames

Description This category includes exergaming apps or computer simulations connected to a running or cycling ergometer/bike on a roller. These exergames aim to enhance the indoor ergometer workout experience by providing an immersive and joyful virtual training environment. This environment usually includes several different workout opportunities, different virtual worlds, and multi-player/ competition modes, allowing the player to compete with a global community while at home or in the gym. Usually, all necessary performance parameters, such as heart rate, cadence, and speed, can be tracked and recorded to adjust the training intensity.

The target group of ergometer-based exergames includes both non-professional and professional athletes. Interestingly, during the COVID-19-lockdown, professional competition preparations and some of the competitions were carried out virtually using ergometer-based exergaming apps.

Zwift Among the most popular multi-player online platforms for virtual cycling or running exercise is Zwift (Zwift Inc, California; https://us.zwift.com/). Zwift offers immersive workout experiences (e.g. exercising between volcanoes, jungles, or cyberpunk-inspired cities), training plans, social events with the global Zwift community, and races and competitions for every fitness level. The app aims to motivate non-professional and professional athletes as they complete their indoor workouts by merging exercise and adventure. To use Zwift, stationary compatible stationary running or cycling trainers are necessary. Additionally, a heart-rate monitor, a steering block, or a grade simulator can be added to the player's setup to achieve a realistic in-game experience. Within the Zwift experience, the player can choose between different worlds, routes, events, and workouts, which are created by professional coaches and consist of different levels. Furthermore, there is the possibility of customizing a personalized avatar. The required items (e.g., kit, bike, wheels, helmet, hair) are unlocked by leveling up, scoring achievements in-game, or completing special missions or group rides.

4.3.6 Wall-Based Exergames

Description Wall-based exergames usually consist of an interactive video wall, a touch screen or an augmented climbing wall in front of which or on which one or more players (multi-player mode) perform the movements to interact with the wall. In addition, motion sensor devices may be necessary to provide an immersive gaming experience.

Depending on the wall provider or brand (e.g., iWall, tWall, TapWall, SMARTfit, Pavigym Vertical 3.0), a variety of games is provided for different target groups (from young to old) and fitness levels (just-for-fun versus intense workouts). Usually, the games have different degrees of difficulty and duration, which can be chosen based on the player's condition and needs.

The application areas are just as diverse as the target group: schools, public places, fitness studios, hotels, or even airports. Furthermore, wall-based exergames are often used in the rehabilitation sector, and several games can also be played in a wheelchair.

iWall The iWall is a wall exergame created by CSE Entertainment, a developer of VR, interactive fitness games, and devices (Kajaani, Finland; video: https://youtu.be/JRmCOEF4rPk). It includes various game contents, and gaming experiences (e. g., Parkour, KayaKing, Ski Cross, StepUp!, JumpOn, or Flow Master) developed with the help of fitness and sports professionals. According to the players' condition and needs, the games can be adapted to allow different types of training (high-intensity cardio workouts versus family fun games) and target different muscle groups. The challenges help improve agility, strength, body control, balance, coordination, and reaction speed. For example, the game Parkour (video: https://youtu.be/DDNeKW 01dXs) offers a high-intensity workout experience requiring endurance, coordination, and cardio. The game is set on parkour routes above the rooftops of Manhattan. The player who finds the best route, moves in the most agile way and is the fastest wins the game. Faster running or knee lifting in front of the wall is translated into faster running in the game.

The iWall is already being used in various settings, such as schools, rehabilitation facilities, airports, and other public places.

tWall The principle of the tWall (IMM electronics GmbH, Mittweida, Germany; https://www.twall.de/) is based on LEDs that randomly light up and must be switched off by the player's touch. The reaction times and errors (e.g., wrongly activated fields) are recorded and are used to assess reaction speed and concentration.

There are several models of the tWall, and more than 100 games are available. Some games can be played with balls that interact with the wall instead of the player's hand. Players can compete against each other with the help of an event mode. A similar wall-based exergame is the SMARTfit Trainer (https://smartfitinc.com/).

ValoClimb ValoClimb (https://valomotion.com/valoclimb/) is an augmented climbing wall from the company ValoMotion, which focuses on exergaming solutions for entertainment centers, theme parks, and climbing gyms. ValoClimb technology is used to project interactive climbing games on to indoor climbing walls, augmenting the real-life climbing experience in a gamified way. The ValoClimb projection adapts to a variety of different locations.

The various games provided include different levels, modes, and themes for different types of player and occasion (from children's first climbing experiences to experienced climbers practicing their skills). For example, climbers can create their own routes and view their climbing videos to get instant feedback. Additionally, they can compare their climbing skills by overlaying two climbing videos. The competitive mode is promoted with the help of the ValoLeague.

4.3.7   Exergaming Room Concepts

Description In contrast to cube-based exergames, exergaming room concepts cover entire studios, gyms, indoor playgrounds, or trampoline arenas and/or digitally augment these settings. This creates a mixed- or AR training environment.

The room design usually includes several gamification elements or combines various exergaming products. In addition, special lighting effects, markings, and pressure-sensitive floors or walls provide an interactive and immersive indoor training experience that aims to inspire and motivate users.

Depending on the product, provider, design, and training concept, the target group ranges from children to ambitious athletes interested in playful alternatives to conventional training and exercise approaches. Some examples of different exergaming room concepts are presented below.

Pavigym PRAMA The Pavigym company specializes in the development of interactive flooring for gyms by integrating sensors, electronics, and LED lights into the floors. With PRAMA, an innovative training concept for gyms has been created (a software-controlled circuit training) that combines a selection of exercises with music and light (https://www.pavigym.com/de/prama, video: https://vimeo.com/210 943483), creating an interactive, immersive and gamified training environment. A PRAMA studio is open-plan, with no equipment or static components. Fitness experts regularly update the various exercises and programs with specially developed algorithms.

Within the PRAMA community, players can train together (group experiences) or compete against each other, gamifying the workout. Additionally, PRAMA CLOUD is a personal online result-tracking platform, helping users to follow their progress over time (e.g. heart-rate control) and be part of the digital community. With PRAMA HOME Pavigym has even created an online workout platform for home exercise.

The training method is aimed both at ambitious athletes and at those who usually avoid conventional gyms and/or training concepts.

ValoJump Another exergame from the company ValoMotion is ValoJump (https://valomotion.com/valojump/), which was designed especially for indoor trampoline halls. By using large screens installed next to the trampolines, the jumping experience is digitally enhanced and gamified. The players see themselves inside the gaming world on the screens and can interact with the game in real time. Notably, no wearables are required. The games are aimed at various age groups and offer different types of training. Single-player or multi-player modes are available.

For larger groups, ValoMotion has even launched a whole arena, the ValoArena (video: https://youtu.be/qFOOnOHe6yc), which is an unattended 6-player mixed-reality arena.

Nike Rise 2.0 Nike Rise 2.0 (video: https://youtu.be/nGuOA_EJ8qk) provides a digitized, interactive LED basketball court connected to tracking technologies and a coaching app to record, evaluate and improve the athletes' performance. The basketball players thus play in an entirely reactive environment that combines state-of-the-art technology and experiential design. Basketball coaches can use the coaching app to create drills and draw moves that are transferred directly to the court and are adjustable to specific skills, players, or teams. Consequently, the boundaries of traditional sports training are expanded. The tracking is used to collect data from each player to create detailed profiles, which help the coaching app to recognize players' strengths and weaknesses.

The basketball training system was developed by the creative agency AKQA (https://www.akqa.com/work/nike/rise/) in collaboration with Artisan (technical agency) and dandelion + burdock (live data visualizations for the court) for Nike (client) to inspire the emerging youth talents in China.

Similar digital training systems could be extended to other sports.

4.4   Potential Effects of Exergaming

By combining video games and their playful, motivational nature with physical exercise, exergames appear to be a suitable and appealing tool to facilitate PA. Exergames are often applied as innovative intervention tools in a range of therapeutic fields and populations. There is an increasing body of literature assessing the effects of exergaming on different health and performance outcomes. Next we will summarize current literature and discuss the possibilities and limitations of exergames for PA and health promotion.

4.4.1   Exercise Intensity and Energy Expenditure

As stated earlier, PA is defined as any movement that raises energy expenditure above the basal metabolic rate. Thereafter, exergaming must reach a specific intensity threshold to be considered PA.

A growing body of literature reports a significant increase in heart rate, oxygen consumption, and energy expenditure when playing exergames compared to sedentary behavior or playing normal video games [32, 65, 118, 143]. A systematic review of 28 laboratory studies measuring exercise intensity reported that intensity levels of exergames typically ranged from light to moderate intensity. Only two studies in this review reported that exergaming led to vigorous-intensity exercise [117]. Thus most commercially available exergames (if played correctly) achieve exercise intensities that can be considered light to moderate PA and thereafter form part of the daily recommended PA guidelines.

However, when analyzing exercise intensity during exergaming, there are major differences between the game consoles and the games played. As mentioned before, one of the most frequently examined exergame consoles is the Nintendo Wii. According to various studies, the mean heart rate reached during games on this console ranges from 44 to 77% of the maximum heart rate [18, 51, 57, 158]. Another commonly applied console is the XBOX 360 Kinect. For this console, current research reports exercise intensities between 60 and 77% of maximum heart rate [106, 149, 156], depending on the game played. For the well-known exergame "Dance Dance Revolution," a mean heart rate of 70% of the maximum heart rate was reported [144]. The highest exercise intensities have been reported for the Exer-Cube, reaching intensities of 86% of maximum heart rate and achieving a metabolic equivalent of task of 6.7 [69]. Thus, this exergame triggers exercises of vigorous intensity.

Previous studies have shown that exergames that involved predominantly lower-body or whole-body movement resulted in higher exercise intensity and energy expenditure when compared with isolated upper-body games [118]. Furthermore, playing exergames in multi-player mode results in significantly higher exercise intensities and energy expenditure than playing in single-player mode [110]. Besides the game design, the players' demographics seem to have an influence on exercise intensity and energy expenditure. Generally, younger players reported higher energy expenditure and reached higher exercise intensities than older players [111, 117]. Furthermore, gender and weight status seem to moderate exercise intensity and energy expenditure during exergaming [85, 111]. Although it is indisputable that some exercise is better than none and low-intensity exercise promotes positive health outcomes in inactive populations [99], it is widely accepted that higher intensity exercise leads to greater adaptations and more pronounced health benefits [129]. Thus, current exergames are often claimed to be too low in intensity to trigger relevant physical adjustments [14, 84]. Furthermore, most exergames lack individual tailoring of the exercise stimulus. To achieve optimal benefits and guarantee long-term performance development, it is essential to tailor the exercise stimulus, considering the person's preconditions (age, gender, risk profile, etc.) and performance progression (Ketelhut and [67].

4.4.2 Enjoyment

Enjoyment has been shown to be an important factor for engaging in PA [78]. According to the literature, exergames can increase PA adherence [139], as people find virtual versions of traditional exercises more enjoyable [37]. This is particularly true for less active individuals [63, 82]. Feedback, challenge, and rewards are especially under discussion as three mechanisms by which exergames may produce enjoyment [83]. Furthermore, it is not enough to sit and passively consume the game when playing exergames. The player must actively participate in the immersive, audio-visual, and narratively engaging game scenarios. Immersion represents the mental processes of having all the attention taken over with the sense of being surrounded by a completely different reality [108].

The immersive and engaging experience during exergaming can help shift the player's attention away from an intrinsic focus on their own body to an external and game-directed focus. This distraction from physiological cues during the exercise can help enhance enjoyment [157]. Therefore, exergaming is often perceived as a form of entertainment rather than exercise [73]. Even high-intensity interval training wrapped in an exergame is perceived as less demanding and more enjoyable [90].

Overall, exergames have been shown to increase enjoyment [51], training adherence [152], long-term motivation [86], engagement [83], immersion [82], and flow experience [92]. Even during high-intensity exergames, participants reported higher enjoyment, higher intrinsic motivation, and higher flow scores than in ordinary endurance exercise [68, 125], physical education classes [126], or a functional workout with a personal trainer [89]. Interestingly, overweight and obese individuals as well as young adults report higher enjoyment than both normal-weight and older individuals when playing exergames [111]. Since research implies that PA enjoyment is linked to higher levels of participation in PA [30], exergames may present a valuable tool to promote PA. Furthermore, well-designed exergames can help experience positive feedback from enjoyment and achievement. This is relevant as perceived physical competence supports prolonged engagement and can increase interest in new exercise experiences [7].

4.4.3 Health-Related Parameters

Anthropometrics Like traditional exercise, regular exergaming can be associated with positive anthropometric outcomes, including waist and hip circumference, body fat mass, body fat percentage, BMI, and body weight [102, 142, 151]. This has also been reported for obese and non-obese children and adolescents [87, 139, 153, 165]. Nonetheless, based on a recent umbrella review, the effects of exergaming on body composition, BMI, or other weight-related outcomes remain inconsistent [111]. Only in studies with higher exercise frequency (>3/week) and in more obese individuals is the effectiveness of exergaming on anthropometric parameters more robust [2, 111]

Physical Activity There is evidence that exergaming may be able to contribute to PA engagement on a short-term basis. Unfortunately, there is no clear evidence for the long-term effects of exergaming on PA [64]. A more recent review supports this notion [111]. The authors conclude that exergaming represents a promising approach to increasing weekly PA in the short term; however, without specific recommendations regarding frequency and duration, exergames failed to increase PA levels in the longer term.

Concerning sedentary behavior, research suggests that exergaming has the potential to replace some sedentary behavior [14, 75, 117, 131]

Aerobic Fitness and Cardiovascular Risk Literature on the effects of exergaming on health-related outcomes is still sparse. McBain et al. [100], Schürch et al., [131], and Yu et al. [167] report favorable changes in aerobic fitness in adults after exergaming interventions. In children, different studies report significant improvement in aerobic fitness after home- or school-based exergaming programs [28, 43, 70]. However, based on a recent review, current evidence does not support the long-term benefits of exergaming for aerobic fitness [111].

Concerning cardiovascular risk factors, exergaming showed similar effects to traditional aerobic exercise in improving endothelial function [56] and flow-mediated dilatation [103]. In overweight and obese children, a home-based exergaming intervention showed positive effects on systolic and diastolic blood pressure, total cholesterol, and LDL cholesterol [139]. In recent studies by Kircher et al. [71, 72], a single exergaming session positively affected blood pressure and other hemodynamic parameters. It can be assumed that exergames at higher intensities have the greatest potential to induce positive effects on health-related markers.

Motor Skills and Balance Especially in older demographics, exergames have been shown to improve balance control [15, 36, 74, 123], reduce fall risk [31, 130], and increase muscle strength [58]. Based on a current meta-analysis, exergaming-associated effects on balance, gait, muscle strength, upper limb function, and dexterity in stand-alone interventions are similar to traditional physiotherapy [122]. Thus, it is not surprising that the World Health Organization supports the use of exergames to promote PA in older adults as a preventive measure against disease and disability [162].

Regarding younger individuals, positive effects of exergames on different motor skills have been reported [70, 138, 155]. However, based on a current review, the results remain inconsistent and thus require further investigation [80].

Mental Health Acute and regular PA has been shown to have a positive effect on mental well-being [132], as PA can help reduce anxiety, low mood, and stress [21]. An active lifestyle and a high level of physical fitness are associated with a lower prevalence of depressive symptoms and illness, especially in old age [133, 168]. An increasing number of studies have shown that exercise has a comparable effect on the reduction of depressive symptoms to established psychological and pharmacological therapies [24]. Therefore, PA is used in treating affective disorders in the form of exercise therapy alongside pharmacotherapeutic and psychotherapeutic treatment as adjunctive therapy [24].

Positive effects of exergaming on social interaction, self-esteem, self-concept, motivation, and mood have been reported [20, 59, 77, 79, 94, 97, 125, 140]. According to a systematic review by [79], exergames may even positively alleviate depression. The effect size found in exergames was much greater than that found in aerobic exercise [8], similar to that in psychotherapy [25], but generally lesser than in other traditional forms of exercise [35, 136]. Nevertheless, the limited number of studies reflected that exergames are still a very new concept in depression research [79].

4.4.4 Exergaming and Cognitive Performance

There is mounting evidence to suggest that PA induces physiological and metabolic changes that, in turn, facilitate cognitive functions through structural and functional adaptations in the brain [12, 47, 55]. Similarly, studies shown that exergaming can positively affect executive functions, attention skills, and visuospatial abilities [10, 11, 107, 140, 141, 164]. Studies that compared exergaming to traditional types of exercise found similar or slightly superior effects of exergaming on executive functions. This might be an indicator that exergaming is a promising approach for preserving and facilitating cognitive and brain health [141]. A unique feature of exergames regarding training effects is the game-based combination of physical and cognitively challenging tasks—so-called dual-task training [93]. This offers more promising motor-cognitive effects than traditional, one-dimensional training approaches [5, 33, 127, 141]. Exergames thereby incorporate a cognitive task directly into a motor task. In summary, it can be concluded that exergaming might have the potential to maintain or facilitate cognitive and brain health and might be recommended in addition to traditional forms of exercise (Table 4.1).

4.4.5 Limitations of Current Literature

Despite these generally positive findings, numerous questions on the health-related effects of exergaming remain unanswered. In addition to several methodological deficits, a major problem is the extensive lack of long-term studies. Thus, it is not yet clear whether the potential that exists in principle can manifest itself in sustainable preventive effects. Additionally, current literature does not allow us to determine whether the often-reported enjoyment can last longer than the average study length of 6–8 weeks and thus results in long-term PA engagement. Furthermore, the dose–response relationship is poorly understood. Additionally, the studies show significant differences between the age of the subjects, the design, the measurement methods used, and the treatment conditions.

Table 4.1 Potential effects of acute and long-term exergaming

Dimensions	Potential positive effects
Physiology	– Energy expenditure – Heart rate – Cardiovascular risk factors – Physical activity – Motor skills – Mobility – Walking speed – Aerobic fitness – Strength – Coordination – Balance/fall risk
Cognition	– Visual-spatial abilities – Executive functions – Attention
Mental	– Social interaction – Self-confidence, self-concept, self-esteem – Motivation – Mood and enjoyment – Depression

4.4.6 Conclusion

Research investigating the use of exergaming for health and PA promotion is still in its infancy. Nonetheless, preliminary findings suggest that exergaming (more precisely: certain exergaming-products) is an option for engaging people in voluntary exercise and thus can contribute to positive health outcomes [142]. Furthermore, exergames may be used to support the development of regular exercise habits [39].

The empirical literature on exercise intensity has lent support to using exergames as a valid form of aerobic exercise [117, 118]. Exergaming can already be found as an option to increase PA in position statements [23]. Unfortunately, there is very little convincing evidence for the sustainable long-term effect of exergames. High-quality randomized experimental longitudinal studies with pre-post tests and strict conditional control are still pending in many areas. Furthermore, many questions remain as to design, dose–response relationships, and appropriate settings of use. To improve the efficacy of exergames, a clear direction should be provided concerning the intensity, duration, and frequency that are sufficient to produce health benefits while minimizing potential adverse events or loss of enjoyment [142]. It can be concluded that just owning an exergame does not automatically result in achieving PA recommendations and a consecutive risk reduction. It is important that exergames are played with an adequate duration and frequency to form a relevant part of the daily recommended PA guidelines. If positive health effects are to be achieved, exergames triggering higher exercise intensities and integrating sound training concepts should be applied.

4.5   Considerations for the Design of Exergames

To establish exergames as a real alternative or supplement to existing PA programs or training offers, multiple avenues have been identified–especially on the level of research and development (Fig. 4.5) and these were implemented in the design of the fitness game environment "the ExerCube" [91] and its ecosystem (Fig. 4.6).

To enable both entertaining and effective exergame training, it is highly recommended to involve an interdisciplinary team from all relevant disciplines (e.g., sports science, psychology, and game design), as well as the respective target group in the development of an exergame. A holistic, user-centered, symbiotic, and iterative design approach should be applied that considers all three levels of exergame design: the player's moving and feeling body, the mediating controller technology, and the audiovisual and narrative (virtual) game scenario.

Ideally, an exergame integrates a real-world, yet game-like training/movement concept with scalable intensity to interact and control the game, which is highly beneficial in its training effects and can be tracked accurately via sensors. In addition, the training/movement concept should be customizable and adaptable in real time to the player's abilities, requirements and goals, both in each session and over a longer training program.

The ExerCube (Sphery Ltd)—a physical immersive training hardware and its exergame experience "Sphery Racer" that takes the player on a virtual sci-fi underwater race—were designed following these guidelines. The virtual game scenario is projected on to three cube walls surrounding the player, who performs functional full-body workout exercises to control the game. Additionally, the player's physical (via heart-rate sensor) and cognitive performances (via reaction time) are tracked and used to continuously adjust the game's difficulty and complexity in real time.

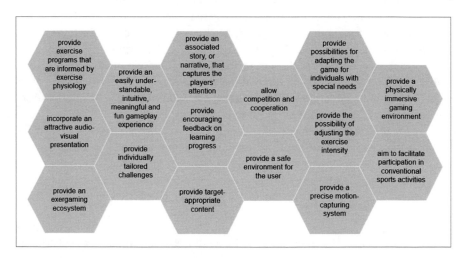

Fig. 4.5 Design recommendations for exergames

Fig. 4.6 The ExerCube and its "ecosystem" (Sphery Ltd)

Among others things, the Sphery Racer experience was proven to be an attractive and effective alternative to functional high-intensity interval training with a personal trainer [93, 95].

Other exergames mainly require arm movements, thus engaging less muscle mass. In particular, lower-body engagement during exergaming seems to increase exercise intensity. This was well demonstrated in a study by [105]. After adjusting a "Wii Sports" game and adding some more movement tasks, mainly engaging the lower limbs, the mean HR increased from 45.0 to 62.2% [105]. It has been shown that exergames that simultaneously recruit the upper and lower limbs to promote the greatest metabolic costs [50]. Furthermore, the precise motion-capturing system of the ExerCube may explain the vigorous exercise intensity. The ExerCube uses an HTC Vive infrared motion-tracking sensor that allows users to move freely while tracking both legs and arms. Thus, bodily movements are tracked more precisely than on consoles that use a hand-held motion sensor. Tracking bodily movements based on the position of a hand-held device can result in inaccurate movement execution or cheating [6].

Another great feature of an exergame is the potential choice of player mode. The Sphery Racer can be played by a single player, as well as by two players in the same cube (collaborative mode) and in different cubes (competitive mode). The cooperative multiplayer mode has been shown to evoke variously challenging physical and cognitive training experiences [97]. Social training experiences are very rich in social interactions and additional training stimuli, while single-player experiences tend to be more focused and intense.

Furthermore, exergames should be easily accessible and usable by players with different abilities. This includes players with certain physical limitations or disabilities. Therefore, the game should provide easily adaptable software and hardware (e. g., an adapted range of motion). Therefore, the Sphery Racer can for example also be played when sitting in a wheelchair. Currently, new exergame scenarios – e.g. for use in the geriatric and sports rehabilitation – are being developed for the cube hardware.

To increase long-term motivation, it could also be beneficial to build an entire ecosystem around the exergame. This could be done through elements familiar from gamification, such as leaderboards or competitions and the establishment of exergame league systems (e.g., ExerCube League: https://sphery.ch/exercube-league/) which help to build a story around the exergame. Furthermore, it helps to be able to use an exergame in different settings (e.g., at home or in the gym). Therefore, the ExerCube was adapted among other things for home use, featuring an on-body feedback system that replaces the physical cube hardware [119]. Finally, an app (https://devapp.sphery. ch/login) could serve as a connecting element of the different parts and, in addition to the exergame itself, also help to track progress, analyze training data and plan exergame training sessions to achieve an individual training goal.

Last but not least, a frequent further development of an exergame should always be accompanied by interdisciplinary research studies to ensure a certain quality and to derive potential avenues for new iterations.

4.6 Summary and Future Directions

4.6.1 In a Nutshell

Regular PA is considered a foundation for a healthy lifestyle in all age groups and is related to numerous positive health effects. Unfortunately, most people are not prone to enjoy and participate in substantial amounts of PA [1]. Various cultural developments and technical achievements in the last decades have created further barriers to engaging in PA, making it even more challenging to find ways to increase the amount of PA in our society. Given the strong appeal of gaming in general, exergames are often regarded as a promising way to encourage people to engage in more PA and thus promote health.

In this chapter, we gave a broad historical summary of the development of exergames. We also provided a structured overview of the wide range of products that are summarized under the term "exergames" by clustering the systems into different categories. Thereafter, we summarized the current literature and discussed the possibilities and limitations of current exergames for PA and health promotion. On the basis of the weaknesses of current consoles and games, we discussed how exergames should be designed to guarantee optimal effects. In this regard, we provided some best-practice examples from the ExerCube, an innovative exergaming setting.

It is clear that exergames are not an entirely new phenomenon. Exergames have been on the market since 1982; however, the scientific interest has mainly developed in the last fifteen years. Thus, it is not surprising that scientific evidence on the health effects of exergames is still unsatisfying. Nonetheless, current research shows that different exergames, if well designed and adequately used (frequency, duration, intensity, etc.), may exert numerous positive effects. Nonetheless, there are fundamental differences between the various consoles and games. Although, in their current form, exergames are unlikely to be the solution to the epidemic of physical inactivity, they appear to offer a potentially innovative strategy that can be used to reduce sedentary time, increase adherence to exercise programs, promote enjoyment of PA, and consecutively exert positive health effects. This may be especially effective for at-risk populations. Unfortunately, to date, most commercial exergames primarily target an appealing gaming experience but neglect major exercise principles, limiting their efficacy.

The ExerCube by Sphery presents a model of good practice, providing valuable information for game designers, researchers, and publishers. If current or future exergames fulfill these recommendations, they may be a valuable addition to existing forms of exercise approaches in prevention and therapy. In view of the ever-increasing use of technology in our society, exergames will become even more widespread, as they provide an inexpensive and easy-to-use way to exercise [91]. The trend to implement exergames in health promotion is steadily growing, as exergames bear the potential to gain access to new target groups (Ketelhut et al. 2021).

4.6.2 Future Directions

This chapter has identified and outlined several important issues for the field of exergaming. Current trends show that there is still much to be done to improve these games to make them a reliable tool to combat the epidemic of physical inactivity, and to promote health [91]. Therefore, we summarize the future directions to motivate researchers, game designers, and publishers.

More Long-term Interdisciplinary Research If exergames are to be used for PA and health promotion, it is imperative to gain insights into long-term effects and more detailed physical effects that go far beyond current research methods. This research will help to evaluate how prolonged participation in exergaming possibly alters enjoyment and if the reported positive effects persist in the long term. Furthermore, we need more data on which specific target groups benefit from which games and in which settings which exergames can be applied.

Putting More Science in the Game The findings from interdisciplinary, high-quality research must be considered in the process of game design. We envision that a close collaboration between exergame researchers, developers, and providers will pave the way for new, innovative, and effective exergames. Especially if the goal is to promote PA and health, the development process should be informed by exercise physiology.

Integrating Individualized and Adaptive Training Concepts Future exergames should consider general training principles to trigger exercise with adequate intensity, duration, and frequency that result in relevant physiological and psychological adaptations. To achieve maximal benefits, sophisticated training concepts must build the basis of the games. To avoid overloading users, pre-classification assessments (fitness test, cognitive abilities) should be included to determine the physical and cognitive game challenge. Furthermore, psychophysical assessments throughout the game are essential to allow reactive adjustments throughout the game.

Personalize for Specific Groups Exergames should be selected and used in an individualized and personalized manner. This refers both to the game consoles or game content and to the selected psychophysical challenge. Games should be customized to fit the player's characteristics (age, gender, level of cognitive and motor skills, interests, risk factors, goals). In particular, the target group's preferences regarding content and game context should be taken into account. Thus, we call for a user-centered game design and creating research-based overviews of which games are suitable for which target group.

Keep It Fun The level of enjoyment of an activity has been identified as one of the predictive factors for the effectiveness of an exercise program [128, 142]. Thus, exergames should provide an attractive and physically immersive gaming environment. The goal should be to achieve relevant health effects that are as sustainable as possible without diminishing the game experience.

Exergames for PA Promotion During COVID-19 Pandemic Home confinement restrictions in response to COVID-19 pandemic changed our everyday routines, modifying our PA (Cachón-Zagalaz et al. 2020). The possibilities for maintaining an active lifestyle decreased with an augmented risk of developing acute and long-term health impairments (Cuschieri et al. 2020). In this concern, new solutions are urgently needed to enable PA while adhering to social distancing and quarantine measures. Exergames can become a useful tool to promote PA even during the difficult conditions imposed by the pandemic.

4.6.3 Trends: Exercising in the Metaverse for a Healthy (Real)-Life?

4.6.3.1 Changing Fitness Habits

With the recent rebranding of Facebook as Meta, the development of the Metaverse (or the Meteverses) as envisioned by Marc Zuckerberg, and the related development of VR technology is proliferating. By combining aspects of digital gaming, social media, AR, VR, and cryptocurrencies, the Metaverse enables users to interact completely digitally. Moreover, the Metaverse transfers a large part of everyday life into the digital realm.

Notably, although the idea of the Metaverse is not yet entirely tangible and understandable to everyone, it appears that VR fitness is one of the segments of high interest in the Metaverse. Today's gym users may in the near future put on their VR headset, enter the Metaverse, and be immediately transported to the gym of their choice while (in fact) standing in their living room. Consequently, there is no need to travel or bring equipment usually required for conventional gym visits. This type of training in the Metaverse is not only a new experience for sports enthusiasts but also introduces entirely new target groups to the (virtual) gym. Companies such as Fit XR and Supernatural, which offer VR-based exergaming experiences (see Sect. 4.3.2), are popular and have a large customer base paying money for an online subscription (comparable to monthly gym fees). Although the Metaverse is still in development, gaming and exercising in the virtual world could evolve considerably within the next few years. Thus, both areas are driving the growth of VR headsets and playing a crucial role in Metaverse advancement.

4.6.3.2 Avatars as Motivators for (Real-Life) Healthy Behavior?

A significant component of the Metaverse is the development and customization of avatars representing users in the virtual ecosystem and giving them access to a unique community.

In exergaming, avatars play a major role in translating the player's personality into the digital training experience [26]. Thus, many exergames offer the possibility to customize an avatar before starting the game. In some cases, the avatar serves as a stand-in for the user and, in other cases, the player is customizing an in-game personal trainer. Furthermore, the avatar provides a visual feedback mechanism for the exercises performed in-game. For example, when performing squats, the avatar's glutes and legs light up [26]. Although the opportunities for avatar customization in exergames are limited in quality and quantity so far, focusing more on the customization of sports equipment than the avatar's body, avatars are considered essential to promoting the long-term adoption of healthy behaviors offline [26]. Interestingly, researchers studying the psychological effects of avatar appearance on user behavior in games and virtual worlds found that virtual behaviors can also persist outside the digital environment [166]. Thus, [26] concluded that not only do users affect their avatars, but avatars also have a significant effect on their real-life counterparts.

In the Metaverse, avatars (in "profile picture"-format or 3D) not only give users access to the virtual ecosystem but also allow them to feel that they can express their identity in the digital realm. Mainly, non-fungible tokens (NFT) avatars created by algorithms are an essential component for identifying the users of a community in a Metaverse, as every avatar is unique in its characteristics, attributes, and design. In short, NFTs are digital items with certificates of authenticity and ownership that are registered on blockchains (the technology behind cryptocurrencies), enabling transparent, decentralized, and immutable data storage [134]. Since each user can personalize the avatar as desired with digitally purchased accessories, clothing, hairstyles, and more, avatars are the best way to visualize each user's creativity. NFTs further allow them to monetize the results by using the avatar as their wallet.

Keeping in mind that avatars may have a significant effect on their real-life counterparts, they could potentially play an important role in future metaverse-based exergaming experiences, helping to present users at their best in the virtual world but also in the real world. Research suggests that humans can build social relationships with avatars and software-generated partners [13, 38, 109] and thus may even motivate players to live healthier and more conscious lives in the long term.

In this context, exergaming competitions or leagues (either on a small scale or inspired by major esport competitions) may serve as a motivational booster.

However, future research must investigate the relationships between avatars and their real-world counterparts, especially regarding metaverse-based fitness experiences and real-life health behavior. Moreover, further research may investigate the possible influence of the different types of avatars (e.g., "profile picture" format versus 3D or imaginative avatars such as animals, warriors, and science fiction characters versus avatars with identical human characteristics).

References

1. Allender, S., Cowburn, G., Foster, C.: Understanding participation in sport and physical activity among children and adults: a review of qualitative studies. Health Educ. Res. **21**, 826–835 (2006). https://doi.org/10.1093/her/cyl063
2. Ameryoun, A., Sanaeinasab, H., Saffari, M., Koenig, H.G.: Impact of game-based health promotion programs on body mass index in overweight/obese children and adolescents: a

systematic review and meta-analysis of randomized controlled trials. Child. Obes. **14**, 67–80 (2018). https://doi.org/10.1089/chi.2017.0250

3. Atari Compendium: ATARI VCS/2600 Unreleased/Prototype Games and Hardware: PUFFER (2022). http://www.ataricompendium.com/game_library/unreleased/unreleased.html#puffer. Accessed 21 July 2022

4. Atarimania: Video Reflex (2022). http://www.atarimania.com/game-atari-2600-vcs-video-ref lex_8414.html. Accessed 21 July 2022

5. Ballesteros, S., Voelcker-Rehage, C., Bherer, L.: Editorial: cognitive and brain plasticity induced by physical exercise, cognitive training, video games, and combined interventions. Front. Hum. Neurosci. **12**, 1–7 (2018). https://doi.org/10.3389/fnhum.2018.00169

6. Baranowski, T., Maddison, R., Maloney, A., Medina, E., Simons, M.: Building a better mouse-trap (exergame) to increase youth physical activity. Games Health J. **3**, 72–78 (2014). https://doi.org/10.1089/g4h.2014.0018

7. Barnett, L.M., Hinkley, T., Okely, A.D., Hesketh, K., Salmon, J.: Use of electronic games by young children and fundamental movement skills? Percept. Mot. Skills. **114**, 1023–1034 (2012). https://doi.org/10.2466/10.13.PMS.114.3.1023-1034

8. Bartley, C.A., Hay, M., Bloch, M.H.: Meta-analysis: aerobic exercise for the treatment of anxiety disorders. Prog. Neuro-Psychopharmacol. Biol. Psychiatry. **45**, 34–39 (2013). https://doi.org/10.1016/j.pnpbp.2013.04.016

9. Bell, A.C., Ge, K., Popkin, B.M.: The road to obesity or the path to prevention: Motorized transportation and obesity in China. Obes. Res. **10**, 277–283 (2002). https://doi.org/10.1038/oby.2002.38

10. Benzing, V., Heinks, T., Eggenberger, N., Schmidt, M.: Acute cognitively engaging exergame-based physical activity enhances executive functions in adolescents. PLoS ONE **11**, 1–15 (2016). https://doi.org/10.1371/journal.pone.0167501

11. Best, J.R., Nagamatsu, L.S., Liu-Ambrose, T.: Improvements to executive function during exercise training predict maintenance of physical activity over the following year. Front. Hum. Neurosci. **8**, 1–9 (2014). https://doi.org/10.3389/fnhum.2014.00353

12. Bherer, L., Erickson, K.I., Liu-Ambrose, T.: Physical exercise and brain functions in older adults. J. Aging Res. **2013**, (2013). https://doi.org/10.1155/2013/197326

13. Bickmore, T.W.: Establishing and maintaining long-term human-computer relationships. ACM Trans. Comput. Interact. **12**, 293–327 (2005)

14. Biddiss, E., Irwin, J.: Active video games to promote physical activity in children and youth: a systematic review. Arch. Pediatr. Adolesc. Med. **164**, 664–672 (2010). https://doi.org/10.1001/archpediatrics.2010.104

15. Bieryla, K.A., Dold, N.M.: Feasibility of Wii Fit training to improve clinical measures of balance in older adults. Clin. Interv. Aging. **8**, 775–781 (2013). https://doi.org/10.2147/CIA.S46164

16. Bogost, I.: The Rhetoric of Exergaming (2005). http://bogost.com/downloads/i.%20boogst%20the%20rhetoric%20of%20exergaming.pdf. Accessed 21 July 2022

17. Bolsø, E.I.: Big book of amiga hardware. Misk Hardware/Amiga Corp: Joyboard (2004). https://bigbookofamigahardware.com/bboah/product.aspx?id=716. Accessed 21 July 2022

18. Bosch, P.R., Poloni, J., Thornton, A., Lynskey, J.V.: The heart rate response to Nintendo Wii boxing in young adults. Cardiopulm. Phys. Ther. J. **23**, 13–18 (2012). https://doi.org/10.1097/01823246-201223020-00003

19. Burke, J.W., McNeill, M.D.J., Charles, D.K., Morrow, P.J., Crosbie, J.H., McDonough, S.M.: Optimising engagement for stroke rehabilitation using serious games. Vis. Comput. **25**, 1085–1099 (2009). https://doi.org/10.1007/s00371-009-0387-4

20. Byrne, A.M., Kim, M.: The exergame as a tool for mental health treatment. J. Creat. Ment. Heal. **14**, 465–477 (2019). https://doi.org/10.1080/15401383.2019.1627263

21. Callaghan, P.: Exercise: a neglected intervention in mental health care? J. Psychiatr. Ment. Health Nurs. **11**, 476–483 (2004). https://doi.org/10.1111/j.1365-2850.2004.00751.x

22. Caspersen, C.J., Powell, K.E., Christenson, G.M.: Physical activity, exercise, and physical fitness: definitions and distinctions for health-related research CARL. Public Health Rep. **100**, 126 (1985). https://doi.org/10.1093/nq/s9-IX.228.365-f

23. Chaput, J.P., LeBlanc, A.G., McFarlane, A., Colley, R.C., Thivel, D., Biddle, S.J.H., Maddison, R., Leatherdale, S.T., Tremblay, M.S.: Active healthy kids Canada's position on active video games for children and youth. Paediatr Child Heal. **18**, 529–532 (2013)

24. Cooney, G., Dwan, K., Mead, G.: Exercise for depression. JAMA - J. Am. Med. Assoc. **311**, 2432–2433 (2014). https://doi.org/10.1001/jama.2014.4930

25. Cuijpers, P., Muñoz, R.F., Clarke, G.N., Lewinsohn, P.M.: Psychoeducational treatment and prevention of depression: the "coping with depression" course thirty years later. Clin. Psychol. Rev. **29**, 449–458 (2009). https://doi.org/10.1016/j.cpr.2009.04.005

26. Czerwonka, S., Alvarez, A., McArthur, V.: One ring fit to rule them all? An analysis of avatar bodies and customization in exergames. Front. Psychol. **12**, (2021). https://doi.org/10.3389/fpsyg.2021.695258

27. Deutsch, J.E., Brettler, A., Smith, C., Welsh, J., John, R., Guarrera-Bowlby, P., Kafri, M.: Nintendo Wii sports and Wii fit game analysis, validation, and application to stroke rehabilitation. Top. Stroke Rehabil. **18**, 701–719 (2011). https://doi.org/10.1310/tsr1806-701

28. Dickinson, K., Place, M.: A randomised control trial of the impact of a computer-based activity programme upon the fitness of children with autism. Autism Res. Treat. **2014**, 1–9 (2014). https://doi.org/10.1155/2014/419653

29. Ding, D., Lawson, K.D., Kolbe-Alexander, T.L., Finkelstein, E.A., Katzmarzyk, P.T., van Mechelen, W., Pratt, M.: The economic burden of physical inactivity: a global analysis of major non-communicable diseases. Lancet **388**, 1311–1324 (2016). https://doi.org/10.1016/S0140-6736(16)30383-X

30. Dishman, R.K., Motl, R.W., Saunders, R., Felton, G., Ward, D.S., Dowda, M., Pate, R.R.: Enjoyment mediates effects of a school-based physical-activity intervention. Med. Sci. Sports Exerc. **37**, 478–487 (2005). https://doi.org/10.1249/01.MSS.0000155391.62733.A7

31. Duque, G., Boersma, D., Loza-Diaz, G., Hassan, S., Suarez, H., Geisinger, D., Suriyaarachchi, P., Sharma, A., Demontiero, O.: Effects of balance training using a virtual-reality system in older fallers. Clin. Interv. Aging. **8**, 257–263 (2013). https://doi.org/10.2147/CIA.S41453

32. Dutta, N., Pereira, M.A.: Effects of Active Video Games on Energy Expenditure in Adults: A Systematic Literature Review (2015). https://journals.humankinetics.com/view/journals/jpah/12/6/article-p890.xml

33. Egger, F., Benzing, V., Conzelmann, A., Schmidt, M.: Boost your brain, while having a break! The effects of long-term cognitively engaging physical activity breaks on children's executive functions and academic achievement. PLoS ONE **14**, 1–20 (2019). https://doi.org/10.1371/journal.pone.0212482

34. Ekelund, U., Steene-Johannessen, J., Brown, W.J., Fagerland, M.W., Owen, N., Powell, K.E., Bauman, A., Lee, I.M., Ding, D., Heath, G., Hallal, P.C., Kohl, H.W., Pratt, M., Reis, R., Sallis, J., Aadahl, M., Blot, W.J., Chey, T., Deka, A., Dunstan, D., Ford, E.S., Færch, K., Inoue, M., Katzmarzyk, P.T., Keadle, S.K., Matthews, C.E., Martinez, D., Patel, A.V., Pavey, T., Petersen, C.B., Van Der Ploeg, H., Rangul, V., Sethi, P., Sund, E.R., Westgate, K., Wijndaele, K., Yi-Park, S.: Does physical activity attenuate, or even eliminate, the detrimental association of sitting time with mortality? A harmonised meta-analysis of data from more than 1 million men and women. Lancet **388**, 1302–1310 (2016). https://doi.org/10.1016/S0140-6736(16)30370-1

35. Ensari, I., Motl, R.W., Pilutti, L.A.: Exercise training improves depressive symptoms in people with multiple sclerosis: results of a meta-analysis. J. Psychosom. Res. **76**, 465–471 (2014). https://doi.org/10.1016/j.jpsychores.2014.03.014

36. Fang, Q., Ghanouni, P., Anderson, S.E., Touchett, H., Shirley, R., Fang, F., Fang, C.: Effects of exergaming on balance of healthy older adults: a systematic review and meta-analysis of randomized controlled trials. Games Health J. **9**, 11–23 (2020). https://doi.org/10.1089/g4h.2019.0016

37. Farrow, M., Lutteroth, C., Rouse, P.C., Bilzon, J.L.J.: Virtual-reality exergaming improves performance during high-intensity interval training. Eur. J. Sport Sci. **19**, 719–727 (2019). https://doi.org/10.1080/17461391.2018.1542459

38. Feltz, D.L., Forlenza, S.T., Winn, B., Kerr, N.L.: Cyber buddy is better than no buddy: a test of the Köhler motivation effect in exergames. Games Health J. **3**, 98–105 (2014). https://doi.org/10.1089/g4h.2013.0088

39. Ferguson, B.: Games for wellness—impacting the lives of employees and the profits of employers. Games Health J. **1**, 177–179 (2012). https://doi.org/10.1089/g4h.2012.0023
40. Finco, M.D., Maass, R.W.: The history of exergames: promotion of exercise and active living through body interaction. In: SeGAH 2014—IEEE 3rd International Conference on Serious Games Appl. Heal. Books Proc. 1–6 (2014). https://doi.org/10.1109/SeGAH.2014.7067100
41. Fox, K.R.: The influence of physical activity on mental well-being. Public Health Nutr. **2**, 411–418 (1999). https://doi.org/10.1017/S1368980099000567
42. Frank, A.: Pokémon Go Players Have Walked Enough to Circle the Earth 200,000 Times (2016). https://www.polygon.com/2016/12/20/14027410/pokemon-go-player-stats-distance-traveled-pokemon-caught#:~:text=As%20of%20Dec.,200%2C000%20trips%20around%20the%20Earth. Accessed 21 July 2022
43. Gao, Z., Hannan, P., Xiang, P., Stodden, D.F., Valdez, V.E.: Video game-based exercise, Latino children's physical health, and academic achievement. Am. J. Prev. Med. **44**, S240–S246 (2013). https://doi.org/10.1016/j.amepre.2012.11.023
44. Goldberg, M., Caron, L., Lida, K.: Atari gaming headquarters. Project puffer page. Atari gaming headquarters—Atari Project Puffer Page (atarihq.com) (2012a). Accessed 21 July 2022
45. Goldberg, M., Caron, L., Lida, K.: Atari gaming headquarters. SUNCOM AEROBICS JOYSTICK. AGH Museum—Suncom Aerobics Joystick (atarihq.com) (2012b). Accessed 21 July 2022
46. Goldberg, M., Caron, L., Lida, K.: Atari gaming headquarters. In: FOOT CRAZ CONTROLLER BY EXUS. AGH Museum—Foot Craz Controller by Exus (atarihq.com) (2012c). Accessed 21 July 2022
47. Gomez-Pinilla, F., Hillman, C.: The influence of exercise on cognitive abilities. Compr. Physiol. **3**, 403–428 (2013). https://doi.org/10.1002/cphy.c110063
48. González, C.S.G., Adelantado, V.N.: A structural theoretical framework based on motor play to categorize and analyze active video games. Games Cult. **11**, 690–719 (2016). https://doi.org/10.1177/1555412015576613
49. Grasser, A., Chipman, P., Leeming, F., Biedenbach, S.: Deep learning and emotion in serious games. In: Ritterfeld, U., Cody, M., Vorderer, P. (eds.) Serious Games, Mechanisms and Effects. Routledge, New York and London (2009)
50. Graves, L.E.F., Ridgers, N.D., Stratton, G.: The contribution of upper limb and total body movement to adolescents' energy expenditure whilst playing Nintendo Wii. Eur. J. Appl. Physiol. **104**, 617–623 (2008). https://doi.org/10.1007/s00421-008-0813-8
51. Graves, L.E.F., Ridgers, N.D., Williams, K., Stratton, G., Atkinson, G., Cable, N.T.: The physiological cost and enjoyment of Wii fit in adolescents, young adults, and older adults. J. Phys. Act. Heal. **7**, 393–401 (2010). https://doi.org/10.1123/jpah.7.3.393
52. Griffin, M., McCormick, D., Taylor, M.: Using the Nintendo Wii as an intervention in a falls prevention group. J Am Geriatr Soc. **60**, 385–387 (2012)
53. Guthold, R., Stevens, G.A., Riley, L.M., Bull, F.C.: Worldwide trends in insufficient physical activity from 2001 to 2016: a pooled analysis of 358 population-based surveys with 1·9 million participants. Lancet Glob. Heal. **6**, e1077–e1086 (2018). https://doi.org/10.1016/S2214-109X(18)30357-7
54. Hansen, L.: The evolution of fitness: exergaming defined. The Evolution of Fitness: Exergaming Defined (clubsolutionsmagazine.com) (2007). Accessed 21 July 2022
55. Hillman, C.H., Erickson, K.I., Kramer, A.F.: Be smart, exercise your heart: exercise effects on brain and cognition. Nature **9**, 58–65 (2008)
56. Jo, E.A., Wu, S.S., Han, H.R., Park, J.J., Park, S., Cho, K.I.: Effects of exergaming in postmenopausal women with high cardiovascular risk: a randomized controlled trial. Clin. Cardiol. **43**, 363–370 (2020). https://doi.org/10.1002/clc.23324
57. Jordan, M., Donne, B., Fletcher, D.: Only lower limb controlled interactive computer gaming enables an effective increase in energy expenditure. Eur. J. Appl. Physiol. **111**, 1465–1472 (2011). https://doi.org/10.1007/s00421-010-1773-3

58. Jorgensen, M.G., Laessoe, U., Hendriksen, C., Nielsen, O.B.F., Aagaard, P.: Efficacy of nintendo wii training on mechanical leg muscle function and postural balance in community-dwelling older adults: a randomized controlled trial. J. Gerontol. Ser. A Biol. Sci. Med. Sci. **68**, 845–852 (2013). https://doi.org/10.1093/gerona/gls222

59. Joronen, K., Aikasalo, A., Suvitie, A.: Nonphysical effects of exergames on child and adolescent well-being: a comprehensive systematic review. Scand. J. Caring Sci. **31**, 449–461 (2017). https://doi.org/10.1111/scs.12393

60. Kain, E.: 'Pokémon GO' Is The Biggest Mobile Game In US History—And It's About To Top Snapchat (2016). https://www.forbes.com/sites/erikkain/2016/07/13/pokemon-go-is-the-biggest-mobile-game-in-us-history-and-its-about-to-top-snapchat/. Accessed 21 July 2022

61. Kamel Boulos, M.N.: Xbox 360 Kinect exergames for health. Games Health J. **1**, 326–330 (2012). https://doi.org/10.1089/g4h.2012.0041

62. Kamel Boulos, M.N., Viangteeravat, T., Anyanwu, M.N., Ra Nagisetty, V., Kuscu, E.: Web GIS in practice IX: a demonstration of geospatial visual analytics using Microsoft Live Labs Pivot technology and WHO mortality data. Int. J. Health Geogr. **10**, 1–14 (2011). https://doi.org/10.1186/1476-072X-10-19

63. Kappen, D.L., Mirza-Babaei, P., Nacke, L.E.: Older adults' physical activity and exergames: a systematic review. Int. J. Hum. Comput. Interact. **35**, 140–167 (2019). https://doi.org/10.1080/10447318.2018.1441253

64. Kari, T.: Can exergaming promote physical fitness and physical activity?: A systematic review of systematic reviews. Int. J. Gaming Comput. Simulations. **6**, 59–77 (2014)

65. Kari, T.: Promoting physical activity and fitness with exergames: Updated systematic review of systematic reviews. In: Dubbels, B. (ed.) Transforming Gaming and Computer Simulation Technologies across Industries, pp. 225–245. IGI Global (2016)

66. Katzmarzyk, P.T., Church, T.S., Craig, C.L., Bouchard, C.: Sitting time and mortality from all causes, cardiovascular disease, and cancer. Med. Sci. Sports Exerc. **41**, 998–1005 (2009). https://doi.org/10.1249/MSS.0b013e3181930355

67. Ketelhut, S., Ketelhut, R.G.: Type of Exercise Training and Training Methods (2020)

68. Ketelhut, S., Ketelhut, R.G., Kircher, E., Röglin, L., Hottenrott, K., Martin-Niedecken, A.L., Ketelhut, K.: Gaming instead of training? Exergaming Induces high-intensity exercise stimulus and reduces cardiovascular reactivity to cold pressor test. Front. Cardiovasc. Med. **9**, 1–10 (2022). https://doi.org/10.3389/fcvm.2022.798149

69. Ketelhut, S., Röglin, L., Kircher, E., Martin-Niedecken, A.L., Ketelhut, R., Hottenrott, K., Ketelhut, K.: The new way to exercise? Evaluating an innovative heart-rate-controlled exergame. Int. J. Sports Med. **43**, 77–82 (2022). https://doi.org/10.1055/a-1520-4742

70. Ketelhut, S., Röglin, L., Martin-Niedecken, A.L., Nigg, C.R., Ketelhut, K.: Integrating regular exergaming sessions in the exercube into a school setting increases physical fitness in elementary school children: a randomized controlled trial. J. Clin. Med. **11**, (2022). https://doi.org/10.3390/jcm11061570

71. Kircher, E., Ketelhut, S., Ketelhut, K., Röglin, L., Hottenrott, K., Martin-niedecken, A.L., Ketelhut, R.G.: A game-based approach to lower blood pressure ? Comparing acute hemodynamic responses to endurance exercise and exergaming : a randomized crossover trial. Int. J. Environ. Res. Public Health. **19**, (2022)

72. Kircher, E., Ketelhut, S., Ketelhut, K., Röglin, L., Martin-Niedecken, A.L., Hottenrott, K., Ketelhut, R.G.: Acute effects of heart rate-controlled exergaming on vascular function in young adults. Games Health J. **11**, 58–66 (2022). https://doi.org/10.1089/g4h.2021.0196

73. Klein, M.J., Simmers, C.S.: Exergaming: virtual inspiration, real perspiration. Young Consum. **10**, 35–45 (2009). https://doi.org/10.1108/17473610910940774

74. Laufer, Y., Dar, G., Kodesh, E.: Does a Wii-based exercise program enhance balance control of independently functioning older adults? A systematic review. Clin. Interv. Aging. **9**, 1803–1813 (2014). https://doi.org/10.2147/CIA.S69673

75. LeBlanc, A.G., Chaput, J.P., McFarlane, A., Colley, R.C., Thivel, D., Biddle, S.J.H., Maddison, R., Leatherdale, S.T., Tremblay, M.S.: Active video games and health indicators in children and youth: a systematic review. PLoS One. **8** (2013). https://doi.org/10.1371/journal.pone.0065351

76. Lee, I.M., Shiroma, E.J., Lobelo, F., Puska, P., Blair, S.N., Katzmarzyk, P.T., Alkandari, J.R., Andersen, L.B., Bauman, A.E., Brownson, R.C., Bull, F.C., Craig, C.L., Ekelund, U., Goenka, S., Guthold, R., Hallal, P.C., Haskell, W.L., Heath, G.W., Inoue, S., Kahlmeier, S., Kohl, H.W., Lambert, E.V., Leetongin, G., Loos, R.J.F., Marcus, B., Martin, B.W., Owen, N., Parra, D.C., Pratt, M., Ogilvie, D., Reis, R.S., Sallis, J.F., Sarmiento, O.L., Wells, J.C.: Effect of physical inactivity on major non-communicable diseases worldwide: an analysis of burden of disease and life expectancy. Lancet **380**, 219–229 (2012). https://doi.org/10.1016/S0140-6736(12)61031-9

77. Lee, S., Kim, W., Park, T., Peng, W.: The psychological effects of playing exergames: a systematic review. Cyberpsychol. Behav. Soc. Netw **20**, 513–532 (2017). https://doi.org/10.1089/cyber.2017.0183

78. Lewis, B.A., Williams, D.M., Frayeh, A., Marcus, B.H.: Self-efficacy versus perceived enjoyment as predictors of physical activity behaviour. Psychol. Heal. **31**, 456–469 (2016). https://doi.org/10.1080/08870446.2015.1111372

79. Li, J., Theng, Y.L., Foo, S.: Effect of exergames on depression: a systematic review and meta-analysis. Cyberpsychol. Behav. Soc. Netw. **19**, 34–42 (2016). https://doi.org/10.1089/cyber.2015.0366

80. Liu, W., Zeng, N., McDonough, D.J., Gao, Z.: Effect of active video games on healthy children's fundamental motor skills and physical fitness: a systematic review. Int. J. Environ. Res. Public Health. **17**, 1–17 (2020). https://doi.org/10.3390/ijerph17218264

81. Löllgen, H.: Gesundheit, bewegung und körperliche aktivität. Dtsch. Z. Sportmed. **66**, 139–140 (2015). https://doi.org/10.5960/dzsm.2015.184

82. Lu, A.S., Kharrazi, H., Gharghabi, F., Thompson, D.: A systematic review of health videogames on childhood obesity prevention and intervention. Games Health J. **2**, 131–141 (2013). https://doi.org/10.1089/g4h.2013.0025

83. Lyons, E.J.: Cultivating engagement and enjoyment in exergames using feedback, challenge, and rewards. Games Health J. **4**, 12–18 (2015). https://doi.org/10.1089/g4h.2014.0072

84. Lyons, E.J., Tate, D.F., Ward, D.S., Bowling, J.M., Ribisl, K.M., Kalyararaman, S.: Energy expenditure and enjoyment during video game play: differences by game type. Med. Sci. Sports Exerc. **43**, 1987–1993 (2011). https://doi.org/10.1249/MSS.0b013e318216ebf3

85. Mackintosh, K.A., Standage, M., Staiano, A.E., Lester, L., McNarry, M.A.: Investigating the physiological and psychosocial responses of single- and dual-player exergaming in young adults. Games Health J. **5**, 375–381 (2016). https://doi.org/10.1089/g4h.2016.0015

86. Macvean, A., Robertson, J.: Understanding exergame users' physical activity, motivation and behavior over time. Conf. Hum. Factors Comput. Syst. - Proc. 1251–1260 (2013). https://doi.org/10.1145/2470654.2466163

87. Maddison, R., Ni Mhurchu, C., Jull, A., Jiang, Y., Prapavessis, H., Rodgers, A.: Energy expended playing video console games: an opportunity to increase children's physical activity? Pediatr. Exerc. Sci. **19**, 334–343 (2007). https://doi.org/10.1123/pes.19.3.334

88. Marston, H.R., Freeman, S., Bishop, K.A., Beech, C.L.: A scoping review of digital gaming research involving older adults aged 85 and older. Games Health J. **5**, 157–174 (2016). https://doi.org/10.1089/g4h.2015.0087

89. Martin-Niedecken, A.L., Schättin, A.: Let the Body'n'Brain games begin: toward innovative training approaches in eSports athletes. Front. Psychol. **11**, 1–9 (2020). https://doi.org/10.3389/fpsyg.2020.00138

90. Martin-Niedecken, A.L., Rogers, K., Vidal, L.T., Mekler, E.D., Segura, E.M.: Exercube vs. Personal trainer: evaluating a holistic, immersive, and adaptive fitness game setup. In: Proceedings of the 2019 CHI Conference on Human Factors in Computing Systems. pp. 1–15. CHI, Glasgow, Scotland (2019)

91. Martin-Niedecken, A.L.: Towards balancing fun and exertion in exergames: exploring the impact of movement-based controller devices, exercise concepts, game adaptivity and player modes on player experience and training intensity in different exergame settings. Ph.D. Thesis. Technische Universität Darmstadt (2021). https://doi.org/10.26083/tuprints-00014186

92. Martin-Niedecken, A.L., Götz, U.: Go with the dual flow: evaluating the psychophysiological adaptive fitness game environment "Plunder Planet". In: Alcañiz, M., Göbel, S., Ma, M., Fradinho Oliveira, M., Baalsrud Hauge, J., Marsh, T. (eds) Serious Games. JCSG 2017. Lecture Notes in Computer Science, vol. 10622. Springer, Cham (2017). https://doi.org/10.1007/978-3-319-70111-0_4

93. Martin-Niedecken, A.L., Mahrer, A., Rogers, K., de Bruin, E.D., Schättin, A.: "HIIT" the Exer-Cube: comparing the effectiveness of functional high-intensity interval training in conventional vs. exergame-based training. Front. Comput. Sci. **2** (2020). https://doi.org/10.3389/fcomp.2020.00033

94. Martin-Niedecken, A.L., Mekler, E.D.: The ExerCube: participatory design of an immersive fitness game environment. In: Göbel, S., et al. (eds) Serious Games. JCSG 2018. Lecture Notes in Computer Science, vol. 11243. Springer, Cham (2018). https://doi.org/10.1007/978-3-030-02762-9_28

95. Martin-Niedecken, A.L., Rogers, K., Turmo Vidal, L., Mekler, E.D., Márquez Segura, E.: ExerCube vs. personal trainer: evaluating a holistic, immersive, and adaptive fitness game setup. In *Proceedings of the 2019 CHI Conference on Human Factors in Computing Systems (CHI '19)*. Association for Computing Machinery, New York, NY, USA, Paper 88, 1–15 (2019). https://doi.org/10.1145/3290605.3300318

96. Martin-Niedecken, A.L., Schwarz, T., Schättin, A.: Comparing the impact of heart rate-based in-game adaptations in an exergame-based functional high-intensity interval training on training intensity and experience in healthy young adults. Front. Psychol. **12** (2021). https://doi.org/10.3389/fpsyg.2021.572877

97. Martin-Niedecken, A.L., Segura, E.M., Rogers, K., Niedecken, S., Vidal, L.T.: Towards socially immersive fitness games: an exploratory evaluation through embodied sketching. CHI Play 2019 Ext. Abstr. Annu. Symp. Comput. Interact. Play. 525–534 (2019). https://doi.org/10.1145/3341215.3356293

98. Mathieu, R.A., Powell-Wiley, T.M., Ayers, C.R., McGuire, D.K., Khera, A., Das, S.R., Lakoski, S.G.: Physical activity participation, health perceptions, and CVD mortality in a multi-ethnic population: the Dallas heart study Reese. Am Hear. J. **163**, 1037–1040 (2012). https://doi.org/10.1016/j.ahj.2012.03.005.Physical

99. Matthews, C.E., Moore, S.C., Sampson, J., Blair, A., Xiao, Q., Keadle, S.K., Hollenbeck, A., Park, Y.: Mortality benefits for replacing sitting time with different physical activities. Med. Sci. Sports Exerc. **47**, 1833–1840 (2015). https://doi.org/10.1249/MSS.0000000000000621

100. McBain, T., Weston, M., Crawshaw, P., Haighton, C., Spears, I.: Development of an exergame to deliver a sustained dose of high-intensity training: Formative pilot randomized trial. J. Med. Internet Res. **20** (2018). https://doi.org/10.2196/games.7758

101. McGuire, A.M., Anderson, D.J., Fulbrook, P.: Perceived barriers to healthy lifestyle activities in midlife and older Australian women with type 2 diabetes. Collegian **21**, 301–310 (2014). https://doi.org/10.1016/j.colegn.2013.07.001

102. Mejia-Downs, A., Fruth, S.J., Clifford, A., Hine, S., Huckstep, J., Merkel, H., Wilkinson, H., Yoder, J.: A preliminary exploration of the effects of a 6-week interactive video dance exercise program in an adult population. Cardiopulm. Phys. Ther. J. **22**, 5–11 (2011). https://doi.org/10.1097/01823246-201122040-00002

103. Mills, A., Rosenberg, M., Stratton, G., Carter, H.H., Spence, A.L., Pugh, C.J.A., Green, D.J., Naylor, L.H.: The effect of exergaming on vascular function in children. J. Pediatr. **163**, 806–810 (2013). https://doi.org/10.1016/j.jpeds.2013.03.076

104. Monda, K.L., Adair, L.S., Zhai, F., Popkin, B.M.: Longitudinal relationships between occupational and domestic physical activity patterns and body weight in China. Eur. J. Clin. Nutr. **62**, 1318–1325 (2008). https://doi.org/10.1038/sj.ejcn.1602849

105. Monedero, J., McDonnell, A.C., Keoghan, M., O'Gorman, D.J.: Modified active videogame play results in moderate-intensity exercise. Games Health J. **3**, 234–240 (2014). https://doi.org/10.1089/g4h.2013.0096

106. Monedero, J., Murphy, E.E., O'Gorman, D.J.: Energy expenditure and affect responses to different types of active video game and exercise. PLoS ONE **12**, 1–13 (2017). https://doi.org/10.1371/journal.pone.0176213

107. Mura, G., Carta, M.G., Sancassiani, F., Machado, S., Prosperini, L.: Active exergames to improve cognitive functioning in neurological disabilities: a systematic review and meta-analysis. Eur. J. Phys. Rehabil. Med. **54**, 450–462 (2018). https://doi.org/10.23736/S1973-9087.17.04680-9

108. Murray, J.H.: Hamlet on the Holodeck: The Future of Narrative in Cyberspace. The Free Press, New York (1997)

109. Nass, C., Moon, Y., Carney, P.: Are people polite to computers? Responses to computer-based interviewing systems. J. Appl. Soc. Psychol. **29**, 1093–1109 (1999). https://doi.org/10.1111/j.1559-1816.1999.tb00142.x

110. O'Donovan, C., Hirsch, E., Holohan, E., McBride, I., McManus, R., Hussey, J.: Energy expended playing Xbox Kinect™ And Wii™ games: a preliminary study comparing single and multiplayer modes. Physiother. (United Kingdom) **98**, 224–229 (2012). https://doi.org/10.1016/j.physio.2012.05.010

111. O'Loughlin, E.K., Dutczak, H., Kakinami, L., Consalvo, M., McGrath, J.J., Barnett, T.A.: Exergaming in youth and young adults: a narrative overview. Games Health J. **9**, 314–338 (2020). https://doi.org/10.1089/g4h.2019.0008

112. Oesch, P., Kool, J., Fernandez-Luque, L., Brox, E., Evertsen, G., Civit, A., Hilfiker, R., Bachmann, S.: Exergames versus self-regulated exercises with instruction leaflets to improve adherence during geriatric rehabilitation: a randomized controlled trial. BMC Geriatr. **17**, 1–9 (2017). https://doi.org/10.1186/s12877-017-0467-7

113. Orland, K., Remo, C.: Games for health: Noah Falstein on exergaming history (2008). https://www.gamedeveloper.com/pc/games-for-health-noah-falstein-on-exergaming-history. Accessed 21 July 2022

114. Owen, N., Healy, G.N., Matthews, C.E., Dunstan, D.W.: Too much sitting: the population health science of sedentary behavior. Exerc. Sport Sci. Rev. **38**, 105–113 (2010). https://doi.org/10.1097/JES.0b013e3181e373a2

115. Pearce, M., Garcia, L., Abbas, A., Strain, T., Schuch, F.B., Golubic, R., Kelly, P., Khan, S., Utukuri, M., Laird, Y., Mok, A., Smith, A., Tainio, M., Brage, S., Woodcock, J.: Association between physical activity and risk of depression: a systematic review and meta-analysis. JAMA Psychiat. **79**, 550–559 (2022). https://doi.org/10.1001/jamapsychiatry.2022.0609

116. Pedersen, B.K., Saltin, B.: Exercise as medicine: evidence for prescribing exercise as therapy in 26 different chronic diseases. Scand. J. Med. Sci. Sport. **25**, 1–72 (2015). https://doi.org/10.1111/sms.12581

117. Peng, W., Crouse, J.C., Lin, J.H.: Using active video games for physical activity promotion: a systematic review of the current state of research. Heal. Educ. Behav. **40**, 171–192 (2013). https://doi.org/10.1177/1090198112444956

118. Peng, W., Lin, J.H., Crouse, J.: Is playing exergames really exercising? A meta-analysis of energy expenditure in active video games. Cyberpsychol. Behav. Soc. Netw. **14**, 681–688 (2011). https://doi.org/10.1089/cyber.2010.0578

119. Pickles, J., Schättin, A., Flagmeier, D., Schärer, B., Riederer, Y., Niedecken, S., Villiger, S., Jurt, R., Kind, N., Scott, S., Stettlerr, C., Martin-Niedecken, A.L.: Exergaming mit „On-Body" Feedbacksystem für den Heimgebrauch. In: dvs Hochschultag 2022, dvs Band 298. Edition Czwalina (2022)

120. Rainmaker, DC.: The End of An Era: CompuTrainer Ceases Production. The End of An Era: CompuTrainer Ceases Production | DC Rainmaker (2017). Accessed 21 July 2022

121. Reiner, M., Niermann, C., Jekauc, D., Woll, A.: Long-term health benefits of physical activity: a systematic review of longitudinal studies. BMC Public Health **13**, 813 (2013). https://doi.org/10.1186/1471-2458-13-813

122. Reis, E., Postolache, G., Teixeira, L., Arriaga, P., Lima, M.L., Postolache, O.: Exergames for motor rehabilitation in older adults: an umbrella review. Phys. Ther. Rev. **24**, 84–99 (2019). https://doi.org/10.1080/10833196.2019.1639012

123. Rendon, A.A., Lohman, E.B., Thorpe, D., Johnson, E.G., Medina, E., Bradley, B.: The effect of virtual reality gaming on dynamic balance in older adults. Age Ageing. **41**, 549–552 (2012). https://doi.org/10.1093/ageing/afs053

124. Rezende, L.F.M., Sá, T.H., Mielke, G.I., Viscondi, J.Y.K., Rey-López, J.P., Garcia, L.M.T.: All-cause mortality attributable to sitting time: analysis of 54 countries worldwide. Am. J. Prev. Med. **51**, 253–263 (2016). https://doi.org/10.1016/j.amepre.2016.01.022

125. Röglin, L., Ketelhut, S., Ketelhut, K., Kircher, E., Ketelhut, R.G., Martin-Niedecken, A.L., Hottenrott, K., Stoll, O.: Adaptive high-intensity exergaming: the more enjoyable alternative to conventional training approaches despite working Harder. Games Health J. **10**, 400–407 (2021). https://doi.org/10.1089/g4h.2021.0014

126. Röglin, L., Stoll, O., Ketelhut, K., Martin-Niedecken, A. L., & Ketelhut, S.: Evaluating Changes in Perceived Enjoyment throughout a 12-Week School-Based Exergaming Intervention. Children **10**, 144 (2023). https://doi.org/10.3390/children10010144

127. Schättin, A., Arner, R., Gennaro, F., de Bruin, E.D.: Adaptations of prefrontal brain activity, executive functions, and gait in healthy elderly following exergame and balance training: A randomized-controlled study. Front. Aging Neurosci. **8** (2016). https://doi.org/10.3389/fnagi.2016.00278

128. Schneider, M.: Intrinsic motivation mediates the association between exercise-associated affect and physical activity among adolescents. Front. Psychol. **9** (2018). https://doi.org/10.3389/fpsyg.2018.01151

129. Schnohr, P., Marott, J.L., Jensen, J.S., Jensen, G.B.: Intensity versus duration of cycling, impact on all-cause and coronary heart disease mortality: the Copenhagen City Heart Study. Eur. J. Prev. Cardiol. **19**, 73–80 (2012). https://doi.org/10.1177/1741826710393196

130. Schoene, D., Lord, S.R., Delbaere, K., Severino, C., Davies, T.A., Smith, S.T.: A randomized controlled pilot study of home-based step training in older people using videogame technology. PLoS One. **8** (2013). https://doi.org/10.1371/journal.pone.0057734

131. Schürch, Y., Burger, M., Amor, L., Zehnder, C., Benzing, V., Mieschler, M., Baur, H., Schmid, S., Bangerter, C., Nigg, C.R., Ketelhut, S.: Comparison of an exergame and a moderate-intensity endurance training intervention on physiological parameters. Curr. Issues Sport Sci. (CISS) **8**(2), 071 (2023). https://doi.org/10.36950/2023.2ciss071

132. Schuch, F., Vancampfort, D., Firth, J., Rosenbaum, S., Ward, P., Reichert, T., Bagatini, N.C., Bgeginski, R., Stubbs, B.: Physical activity and sedentary behavior in people with major depressive disorder: a systematic review and meta-analysis. J. Affect. Disord. **210**, 139–150 (2017). https://doi.org/10.1016/j.jad.2016.10.050

133. Schuch, F.B., Vancampfort, D., Firth, J., Rosenbaum, S., Ward, P.B., Silva, E.S., Hallgren, M., De Leon, A.P., Dunn, A.L., Deslandes, A.C., Fleck, M.P., Carvalho, A.F., Stubbs, B.: Physical activity and incident depression: a meta-analysis of prospective cohort studies. Am. J. Psychiatry. **175**, 631–648 (2018). https://doi.org/10.1176/appi.ajp.2018.17111194

134. Sestino, A., Guido, G., Peluso, A.M.: The Concept and Technicalities of NFTs. In: Non-Fungible Tokens (NFTs). Palgrave Macmillan, Cham (2022)

135. Sheehan, D.P., Katz, L.: The impact of a six week exergaming curriculum on balance with grade three school children using the wii FIT+TM. Int. J. Comput. Sci. Sport. **11**, 5–22 (2012)

136. Silveira, H., Moraes, H., Oliveira, N., Coutinho, E.S.F., Laks, J., Deslandes, A.: Physical exercise and clinically depressed patients: a systematic review and meta-analysis. Neuropsychobiology **67**, 61–68 (2013). https://doi.org/10.1159/000345160

137. Skjæret, N., Nawaz, A., Morat, T., Schoene, D., Helbostad, J.L., Vereijken, B.: Exercise and rehabilitation delivered through exergames in older adults: an integrative review of technologies, safety and efficacy. Int. J. Med. Inform. **85**, 1–16 (2016). https://doi.org/10.1016/j.ijmedinf.2015.10.008

138. Smits-Engelsman, B.C.M., Jelsma, L.D., Ferguson, G.D.: The effect of exergames on functional strength, anaerobic fitness, balance and agility in children with and without motor coordination difficulties living in low-income communities. Hum. Mov. Sci. **55**, 327–337 (2017). https://doi.org/10.1016/j.humov.2016.07.006

139. Staiano, A.E., Beyl, R.A., Guan, W., Hendrick, C.A., Hsia, D.S., Newton, R.L.: Home-based exergaming among children with overweight and obesity: a randomized clinical trial. Pediatr. Obes. **13**, 724–733 (2018). https://doi.org/10.1111/ijpo.12438

140. Staiano, A.E., Calvert, S.L.: Exergames for physical education courses: physical, social, and cognitive benefits. Child Dev. Perspect. **5**, 93–98 (2011). https://doi.org/10.1111/j.1750-8606. 2011.00162.x
141. Stojan, R., Voelcker-Rehage, C.: A systematic review on the cognitive benefits and neurophysiological correlates of exergaming in healthy older adults. J. Clin. Med. **8** (2019). https:// /doi.org/10.3390/jcm8050734
142. Street, T.D., Lacey, S.J., Langdon, R.R.: Gaming your way to health: a systematic review of exergaming programs to increase health and exercise behaviors in adults. Games Health J. **6**, 136–146 (2017). https://doi.org/10.1089/g4h.2016.0102
143. Sween, J., Wallington, S.F., Sheppard, V., Taylor, T., Llanos, A.A., Adams-Campbell, L.L.: The role of exergaming in improving physical activity: a review. **11**, 864–870 (2014). https:// /doi.org/10.1123/jpah.2011-0425
144. Tan, B., Aziz, A.R., Chua, K., Teh, K.C.: Aerobic demands of the dance simulation game. Int. J. Sports Med. **23**, 125–129 (2002). https://doi.org/10.1055/s-2002-20132
145. Taylor, M.J.D., Griffin, M.: The use of gaming technology for rehabilitation in people with multiple sclerosis. Mult. Scler. J. **21**, 355–371 (2015). https://doi.org/10.1177/135245851456 3593
146. Taylor, M.J.D., McCormick, D., Shawis, T., Impson, R., Griffin, M.: Activity-promoting gaming systems in exercise and rehabilitation. J. Rehabil. Res. Dev. **48**, 1171–1186 (2011). https://doi.org/10.1682/JRRD.2010.09.0171
147. Taylor, M., Shawis, T., Impson, R.: Nintendo Wii as a training tool in falls prevention rehabilitation: case studies. J Am Geriatr Soc. **60**, 1781–1783 (2012)
148. Thorp, A.A., Owen, N., Neuhaus, M., Dunstan, D.W.: Sedentary behaviors and subsequent health outcomes in adults: a systematic review of longitudinal studies, 19962011. Am. J. Prev. Med. **41**, 207–215 (2011). https://doi.org/10.1016/j.amepre.2011.05.004
149. Tietjen, A.M.J., Devereux, G.R.: Physical demands of exergaming in healthy young adults. J. Strength Cond. Res. **33**, 1978–1986 (2019). https://doi.org/10.1519/JSC.0000000000002235
150. Tremblay, M.S., Colley, R.C., Saunders, T.J., Healy, G.N., Owen, N.: Physiological and health implications of a sedentary lifestyle. Appl. Physiol. Nutr. Metab. **35**, 725–740 (2010). https:// /doi.org/10.1139/H10-079
151. Tripette, J., Murakami, H., Gando, Y., Kawakami, R., Sasaki, A., Hanawa, S., Hirosako, A., Miyachi, M.: Home-based active video games to promote weight loss during the postpartum period. Med. Sci. Sports Exerc. **46**, 472–478 (2014). https://doi.org/10.1249/MSS.000000000 0000136
152. Valenzuela, T., Okubo, Y., Woodbury, A., Lord, S.R., Delbaere, K.: Adherence to technology-based exercise programs in older adults: a systematic review. J. Geriatr. Phys. Ther. **41**, 49–61 (2018). https://doi.org/10.1519/JPT.0000000000000095
153. Valeriani, F., Protano, C., Marotta, D., Liguori, G., Spica, V.R., Valerio, G., Vitali, M., Gallè, F.: Exergames in childhood obesity treatment: A systematic review. Int. J. Environ. Res. Public Health. **18** (2021). https://doi.org/10.3390/ijerph18094938
154. Veerman, J.L., Healy, G.N., Cobiac, L.J., Vos, T., Winkler, E.A.H., Owen, N., Dunstan, D.W.: Television viewing time and reduced life expectancy: a life table analysis. Br. J. Sports Med. **46**, 927–930 (2012). https://doi.org/10.1136/bjsports-2011-085662
155. Vernadakis, N., Papastergiou, M., Zetou, E., Antoniou, P.: The impact of an exergame-based intervention on children's fundamental motor skills. Comput. Educ. **83**, 90–102 (2015). https:// /doi.org/10.1016/j.compedu.2015.01.001
156. Viana, R.B., Vancini, R.L., Vieira, C.A., Gentil, P., Campos, M.H., Andrade, M.S., de Lira, C.A.B.: Profiling exercise intensity during the exergame Hollywood Workout on XBOX 360 Kinect®. PeerJ **2018**, 1–16 (2018). https://doi.org/10.7717/peerj.5574
157. Warburton, D.E.R., Bredin, S.S.D., Horita, L.T.L., Zbogar, D., Scott, J.M., Esch, B.T.A., Rhodes, R.E.: The health benefits of interactive video game exercise. Appl. Physiol. Nutr. Metab. **32**, 655–663 (2007). https://doi.org/10.1139/H07-038
158. Willems, M., Bond, T.S.: Comparison of Physiological and Metabolic Responses to Playing Nintendo Wii Sports and Brisk Treadmill Walking (2009)

159. Wollersheim, D., Merkes, M., Shields, N., Liamputtong, P., Wallis, L., Reynolds, F., Koh, L.: Physical and psychosocial effects of Wii video game use among older women. Aust. J. Emerg. Technol. Soc. **8**, 85–98 (2010)
160. World Health Organisation: Global action plan on physical activity 2018–2030: more active people for a healthier world. World, Geneva (2018)
161. World Health Organisation: WHO Guidelines on Physical Activity and Sedentary Behaviour
162. World Health Organization: Global Recommendations on Physical Activity for Health. World Health Organization, Geneva (2010)
163. Wu, Y., Zhang, D., Kang, S.: Physical activity and risk of breast cancer: a meta-analysis of prospective studies. Breast Cancer Res. Treat. **137**, 869–882 (2013). https://doi.org/10.1007/s10549-012-2396-7
164. Xiong, S., Zhang, P., Gao, Z.: Effects of exergaming on preschoolers' executive functions and perceived competence: a pilot randomized trial. J. Clin. Med. **8**, 4–5 (2019). https://doi.org/10.3390/jcm8040469
165. Ye, S., Lee, J., Stodden, D., Gao, Z.: Impact of exergaming on children's motor skill competence and health-related fitness: a quasi-experimental study. J. Clin. Med. **7**, 261 (2018). https://doi.org/10.3390/jcm7090261
166. Yee, N., Bailenson, J.: The proteus effect: the effect of transformed self-representation on behavior. Hum. Commun. Res. **33**, 271–290 (2007). https://doi.org/10.1111/j.1468-2958.2007.00299.x
167. Yu, T.C., Chiang, C.H., Wu, P.T., Wu, W.L., Chu, I.H.: Effects of exergames on physical fitness in middle- aged and older adults in Taiwan. Int. J. Environ. Res. Public Health. **17** (2020). https://doi.org/10.3390/ijerph17072565
168. Zahl, T., Steinsbekk, S., Wichstrøm, L.: Physical activity, sedentary behavior, and symptoms of major depression in middle childhood. Pediatrics. **139** (2017). https://doi.org/10.1542/peds.2016-1711
169. Zyda, M.: From visual to virtual reality to games. IEEE Comput. Soc. **1**, 25–32 (2005)

Chapter 5
Close Encounters of the Immersive Kind: Embodied Fundamentals and Future Directions of Affective Virtual Reality (VR) Design

Kate Gwynne

Abstract As the virtual reality (VR) landscape has evolved, so too has our understanding of the technology's greater purposes. Of present interest to researchers and practitioners is not just the compelling experience of *presence*, but increasingly, its affective possibilities (Riva et al. 2007; Banacou et al. in Front Hum Neurosci, 2016; Milk in https://www.ted.com/talks/chris_milk_how_virtual_reality_can_create_the_ultimate_empathy_machine?language=en, 2015; ; Abraham in Contemp Theatr Rev 30:474–489, 2020; Herrera et al. in PLoS One 13, 2018; Flavian et al. in J Hosp Market Manag 30:1–20, 2020; Martins in https://www.latimes.com/entertainment-arts/story/2020-04-26/coronavirus-vr-virtual-reality-theater-tender-claws-live-actors, 2021; Marcolin et al. in IEEE Comput Graph Appl 41:171–718, 2021). For example, participants have referred to the potency of emotions such as awe (Abraham in Contemp Theatr Rev 30:474–489, 2020), social connection (Martins in https://www.latimes.com/entertainment-arts/story/2020-04-26/coronavirus-vr-virtual-reality-theater-tender-claws-live-actors, 2021) and empathy (Herrera et al. in PLoS One 13:e0204494, 2018) that can arise in response to VR environments. Researchers have therefore suggested that "affective VR" (Marcolin et al. 2021, p. 172) will play a crucial role in the future of VR design, with the potential to improve lives and behaviours (p. 172). In this chapter, I propose that the exploration of VR Narrative design practices, in highlighting embodied engagement techniques which extrapolate the platform's affective possibilities, can contribute best practices which may have implications in a range of emerging VR design contexts, such as those which may one day comprise an "embodied internet" (Zuckerberg 2021) known as the Metaverse. In particular, the findings from my study draw attention

The research informing this chapter was undertaken with the support of an Australian Government Research Training Program Scholarship.

K. Gwynne (✉)
Sydney, Australia
e-mail: kate.gwynne@gmail.com

to the choreography of encounters as a pivotal co-creational design strategy and precursor to affective potential.

5.1   Introduction

5.1.1   Affect and the Future of Virtual Reality Design

Exploration of the affective potential of VR technologies may be traced back to the mid-late 90s with works such as *Osmose* (1995) and *Ephémère* (1998) by immersive artist Char Davies. Visitors to Davies' abstracted landscapes, traversed through breath and body movement, often reported being profoundly moved [15]. Today, given the lowered-cost and public availability of virtual head-mounted display (HMD) systems, research and practice-led interest into the affective possibilities of VR environments has since blossomed. Of particular interest is how the technology's embodied and sensory dimensions are implicit in such potential [1, 5, 18, 22, 33, 41].

Recent VR technology developments focused on enhancements to *immersion*, often espoused as the ultimate goal of VR design [9] may often be motivated by affect, given the correlation between emotional states and engagement ([18], pp. 8–9). Critic Todd Martins' experience of immersive theatre production *The Under Presents* [3] during the USA's first lockdown is a recent example [43]. Using motion-capture technology, which works by transferring motion data from an actor wearing a special suit with sensors to an avatar [4, p. 1], artistic director Jacqueline Lyanga was interested in exploring the kind of emotional interplay that might arise as a result of this dynamic [30]. Martins shared how this artistic intention translated into his experience of a virtual hug with an actor and a group of other participants:

> This was my first hug in months, and even though there was no one actually there — *there* being my kitchen—it felt pretty dang good. Everyone snapped in celebration. Moments later I removed the virtual reality headset, but not simply because work beckoned. No, this odd yet innocent moment in an absurdist work of tech-focused improvisational theatre had not just put a smile on my face, it brought me to tears.

As Martin's experience highlights, enhanced immersion isn't the only motivation driving virtual reality innovation and artistic experimentation, but increasingly the affective potential that arises through these more natural, in the sense of approximating everyday corporeal experience, interactions [14]. Even beyond entertainment, immersive experiences such as those that might one day form part of a future Metaverse, have the potential to offer a different kind of affective experience than today's dominant entertainment platforms—transportive, enacted through a range of sensory experiences and felt on a deeply visceral and personal level.

There is considerable commercial interest, driven by Meta and large commitments by companies such as Sony and Lego to Epic Games' Metaverse plans [21] in the development of an "embodied internet" [56]. Described by computer scientist Louis Rosenberg as, *"a persistent and immersive simulated world that is experienced in* the *first person* by large groups of *simultaneous users* who share a strong sense of *mutual presence*," [42] the Metaverse will facilitate and enrich many of the things we currently do in the real world [, p. 23], such as learning, work, healthcare, exercise and entertainment. Rosenberg argues that the Metaverse will be enabled via two distinct technological experiences: virtual, avatar-based VR worlds, and augmented, which combine real environments overlayed with virtual elements [42]. While the applications of these technologies will likely vary, their success will not doubt be based on the degree to which they can engage, and entice people to return.

Design principles, such as those highlighted through the study of VR narratives, an artform with motivations and techniques tailored towards creating affective possibilities [33] may therefore have applications in the creation of experiences that one day make up the Metaverse. In particular, insights raised through the embodied states framework which I have developed draws attention to the centrality of encounters— between participants, participants and virtual agents, and participants and responsive environments—in extrapolating the deeper possibilities of such environments. I define key terms *affect* and *emotion* and their relationship to the core dimensions of VR environments, *presence*, *immersion* and *embodiment*, before proceeding into a more detailed analysis of the insights raised by the study.

5.1.1.1   Affect and Emotion

Affect and *emotion* are often used synonymously [46]. As Brian Massumi argues in *Parables for the Virtual: Movement, Affect, Sensation* (2002) however, the terms refer to different logics and orders [31, p. 17]. Affects are pre-personal or pre-conscious while emotions are social [46].

The concept of *affect* originates from Baruch Spinoza's *Ethics* (1677). Spinoza defined affect in terms of "affections of the body," referring to how the power dynamics of the body shift and as a result the mind forms an idea of this shift [29, p. 10]. This early conceptualisation drew attention to affect as a corporeal experience. Massumi conceptualises affect as the experience of *intensity*: an emotional state, sometimes a "shock" [31, p. 37] which is charged with motion but is not yet activity [31, p. 26]. Given that affect is pre-conscious and a state of "inbetweenness", it remains slippery and abstract. It needs to be qualified, narrativised and given subjective meaning. This mental process is understood through the term emotion [, p. iv], and explains its more popular usage [32], especially as a way to categorise and market different storytelling genres to audiences.

The abstract qualities of affect can be better understood through real world examples such as in the case of music therapy, as discussed by Eric Shouse [46]. When an elderly patient lost leg mobility following a hip fracture, physiotherapists used music to regain her movement. While nothing else had worked, when exposed to sound, her leg tapped on its own accord. This example highlights affect's significance as a precursor to will and consciousness [46] and also its importance as an embodied mechanism in its own right.

The aspect of affect which makes it of particular interest to this study is the way, as Massumi draws attention to, that it "accumulates" through "relatedness" as "swells of intensities" that pass between people (2010, p. 2). Affect comes about in response to the dynamics of human interaction in the world, what Massumi refers to as participation (Massumi 2016 in Seighworth 2017). It is the transmutability of affect which makes it, as Shouse describes it, such a powerful social force [46].

5.1.1.2   Affect, Emotion and Art

In *What is Philosophy* (1991) Deleuze and Guattari, drawing on Spinoza's theory, propose that artistic work can be understood as a series of sensations made up of "percepts" and "affects" (1991, p. 64). Percepts refer to the materiality of the artwork, while affects refer to the "block of sensations" that are drawn out from the affections that the viewing of the artwork prompts (1991, p. 167). In *Affect as Method* Anna Hickey-Moody describes this act of reception thus: "New realities imagined in art are communicated through kinaesthetic economies of affect, relays of sensation between an artwork and consumer" [24, p. 86]. As these earlier and more recent theorists therefore emphasise, affect is crucial to the process of art reception essentially due to its propulsion through relatedness. Affect passes from the artwork to the viewer in the exchange that is fundamental to art viewing.

In the context of discussions around storytelling and audience impact, affect is often overlooked or encapsulated in the more commonly used term emotion. In his meditations on the art of modern-day screenwriting, for example Robert McKee defines story not in terms of plot, though of course it is an obvious feature of the artform, but in terms of its capacity to induce what he calls "meaningful emotion" (p. 111). Like Aristotle, it is the ability of story to provide meaning where life does not, that in his mind defines a "good story" (p. 111). McKee explains how in film, emotional resonance emerges as characters are submitted to pressures that force them into dilemmas. Beats, scenes, sequences and acts, he explains, are structured to intensify towards the story's climactic resolution (p. 42). The visceral audience responses to these techniques, such as laughing, crying, and sitting on the edges of their seats, is the affective assemblage, following Deleuze (Buchanan 2013, p. 5), of this emotional viewing experience. While affect and emotion refer to different processes, they are intrinsic and interrelated corporeal and mental aspects of art reception.

5.1.1.3   Affect and Co-creational Media

Conceptualisations of affect shift in the context of co-creational media [12, 26]. While the physiological process involved in affect is unchanging, the way this state is arrived at in co-creational media involves more complex dimensions: the expression of *agency* and the experience of sensory *immersion*.

Brenda Laurel defines agency as the "power to take action" [26, p. 117]. Even simple actions, she argues, can have a significant impact on the participant's experience (p. 117). Director and interactive content creator Alon Benari explained to me in an interview how as a filmmaker, every technique, such as music, acting and close-ups, is used to increase engagement with the story and characters [66]. In co-creational media, he argued, there is another tool for this purpose: choice. He elaborated:

> I sit in this imaginary date, and my date says, "I'll just take salad " Then, the waiter says, "Okay. What will you have, sir?" I have three options. Just the fact that there is a choice of three options teaches me a lot about this character. If the options are, she's taking salad, and the options are, "I'll take salad as well," "I think I'll stick to something light," or, "You know what? I'm not sure yet." That's one thing, but if it would be like, "Steak." I don't know. Even the choices that you present tell the story… It's supposed to be a tool for evoking emotion, and not just a novelty. (2017)

Here Benari alludes to how the choices that the participant makes, in effect the way they express agency, helps to illuminate aspects of the character which in turn contribute to an emotional connection and therefore, an emotional payoff. This, he argues, creates a higher level of immersion and engagement because the outcome depends on the participant. According to Benari, agency and immersion are integral to the way emotion is experienced in co-creational media. Whereas the type of emotion is what is highlighted in the experience of more passively viewing a painting or watching a film, in co-creational media the experience shifts. Brenda Laurel takes the emotional payoff associated with the drama as a given and instead draws attention to the "feeling of completion" and an "apprehension of the shape of a whole" that a participant experiences then they act as a co-creator [26, p. 122]. Co-creational media therefore nurses possibilities for complex and multi-layered experiences of affect due to the way it involves the participant as a collaborator.

5.1.1.4 The Core Dimensions of Virtual Reality Environments

VR environments, experienced through an ego-centric first-person point of view, reframe choice and agency through a focus on the participant's body and senses. Interactions in immersive environments shift into increasingly embodied modes, through close spatial proximity with other participants, virtual characters or responsive environments. The way we understand affect therefore in turn shifts as emphasis moves to new experiential dimensions: *presence, immersion* and *embodiment*. This in turn reinforces the way affect is experienced as intensity that passes between bodies [31, p. 2]. The social and participatory origins of affect theory therefore make it a fitting term to apply in the understanding of the kind of states experienced while engaging in VR works. The embodied origins of affect are foregrounded through the experience of VR.

5.1.1.5   Immersion, Presence and Embodiment and Their Interrelationship with Affect

In the context of VR experience, prominent VR researcher Mel Slater outlines immersion as "what the technology delivers from an objective point of view" while presence is a "human reaction to immersion" [48, p. 1]. The objective properties are made up of "sensory motor contingencies" (SCs) by which he is referring to the sensory dimensions of immersive environments [49, p. 3550]. As Schwartz and Steptoe outline, virtual reality systems are often differentiated from other digital media technologies through tracking of body movements, surround sound, wide field of view, egocentric visual and sensory stimuli through which stereoscopic imagery may be perceived [45, p. 3]. This allows for more natural sensorimotor actions to be performed, unlike non-immersive systems, such as a keyboard and mouse [45, p. 3]. While Schwartz and Steptoe qualify VR systems as "immersive" and two-dimensional interfaces as "non-immersive" that is not to say that preceding interfaces do not create the experiential phenomenon of immersion. The main difference between immersion in the context of virtual reality systems and other media forms is an increasing shift to sensory conceptualisations of the term, from immersion as "the willing suspension of disbelief" [, p. 270] and general involvement with media, to the sense of "being there" in another location.

According to Slater, we feel a compelling sense of "being there", what he calls "place illusion" (PI) [49, p. 3551], in a virtual world due to its complex sensory dimensions. He argues however that there is also another important dimension to VR presence, the perception that, "what is apparently happening is really happening" [49, p. 3553]. He calls this "plausibility illusion" (Psi) [49, p. 3553], using the example of the direct address where a woman looks at the participant directly in the eye (identified on page 11 as the *encounter state*) as a technique that evokes Psi.

The term embodiment, refers, as Simon Penny details, to the physical characteristics of the body of an agent and the causal role of these characteristics in cognitive functioning [38, p. 182]. In VR environments, the sense of embodiment has been associated with (1) sense of self-location, (2) sense of agency and (3) a sense of body ownership [20, 50]. Slater explains that, "SoE toward a body B is the sense that emerges when B's properties are processed as if they were the properties of one's own biological body [51]." The participant may feel embodied in a virtual environment when they can interact in that environment through an ego-centric point of view and perform a range of actions from that perspective. Embodiment in virtual reality environments, as Biocca argues, is therefore the pinnacle of the progressive merger of the body and technology through the pursuit of what he calls "total immersion" [9]. The coupling of sensors and the display to the participant's body, he elaborated, has allowed for this experience to become a reality [9].

In *Elements of a Cyberspace Playhouse* [55] computer programmer Randy Walser anticipated that the embodied dimensions of virtual reality environments, by giving "a virtual body" and a "role" to the audience would be integral to its future significance. In his predictions he drew attention to the advent of a virtual body or avatar as a crucial element of this experience. Slater echoes this emphasis, arguing that the manipulation of body representation can be a large contributor to embodied experience and also presence [50, p. 382], by, as he explains, acting as the "focal point where PI and Psi are fused" [49, p. 3554].

Embodiment is therefore a key dimension of virtual reality's enhanced immersive properties, but it is also, as is the interest of this study, of particular significance in eliciting emotions. Recent studies have sought to understand the relationship between these core experiential dimensions of virtual environments in affective outcomes. For example, researchers have found, using a variety of different experiment approaches,[1,2] that the degree of presence and immersion impact the intensity with which affect is experienced (Marcolin et al. [28], p. 1), and vice versa (Riva et al. [41], p. 1). Marcolin et al. and Riva et al. also found that less presence inducing experiences were correlated with weaker affective responses ([28, p. 173; 2017, p. 55]). In another study exploring VR for a pre-experience of a hotel room, an immersive 360 experience was not only more effective at eliciting emotion, because it was more embodied, but they also found that emotion in turn influenced engagement ([18], pp. 8–9). Katarina Pavic et al. meanwhile found that the degree of immersion, interactivity, content, sensory modalities and participants' characteristics were all key factors for eliciting positive emotional states [37, p. 1]. These studies confirm the important relationship between the core dimensions of VR environments and affective possibilities as well as highlighting the increasing significance of affective VR.

5.2 Embodied Fundamentals of VR Narrative Design

The exploration of a group of ten[3] VR narrative works produced between 2016 and 2018 reveals insights, illuminated through artistic experimentation, about the embodied fundamentals of virtual environments. Through my research process I identified eight embodied states which I propose are integral to the affective possibilities of the artform. The states include *1. Dialogic, 2. Somatic (See, Feel, Move, Smell, Taste, Hear and Breathe), 3. Encounter, 4. Performative, 5. Ludic, 6. Expanded-self, 7. Embodied Archetype and 8. Climactic.* Of these states, the encounter state emerges as a pivotal design strategy, through the alchemy of co-presence and dramatic conflict, in framing co-creational, and in turn, affective possibilities. I outline my research methodology below before proceeding into a closer analysis of the study's findings.

---

[1] Marcolin et al. [28] developed a database of VR environments designed to elicit frustration (anger-based emotion), disgust, fear, happiness, and sadness.

[2] Riva et al. [41] designed three virtual parks designed to elicit anxious, relaxing and neutral emotional states to compare the relationship between emotion and presence.

[3] VR Noir (2016), The Abbot's Book (2016), Carne y Arena (2017), Broken Night (2017), Inanimate Alice: perpetual nomads (2018), Caliban Below (2018), Awavena (2018), Awake VR (2018), Spheres (2018) and Wolves in the Walls (2018).

5.2.1 Research Methodology: Tuning into My Intuitive First-Person Perspective

The research methodology involved the analysis of a group of VR works to identify design processes, interviews with creators to gain a closer understanding of approaches to generating narrative virtual reality works, and a review of relevant literature.

I viewed some of the works in exhibition spaces,[4] while the majority were downloaded from the Viveport, Steam, Oculus app stores or made available directly from the creators and experienced in the VR Lab at the School of Arts and Media, UNSW.[5] While the works in the study were diverse in genre—from installation art, journalistic, young adult e-literature, gothic noir to crime drama—they were unified by offering some sensory engagement and interaction, to varying degrees, such that a diversity of approaches to spatial narrative design could be explored. Given that insights were to be interrogated in the development of an original VR script, my own subjective, first-person experience of the works was used as a key piece of research data.

In her book *Designing with the Body: Somaesthetic Interaction Design* (2018),[6] Kristina Höök proposes that the first-person perspective is not a simplification of, but instead a necessity of movement-based interaction design [25, p. 3]. She argues for the tuning into the "soma,"—referring to the self in its combined dual facets of mind and body, and which emphasises the embodied dimensions of thought and the emotional precursors of physical responses in order to reach design insights [25, p. 3]. The first-person perspective of the designer, she argues, is what allows them to train their, "designerly judgments and somaesthetic sensibilities" (p. 169). She elaborates (p. 169):

> It is not "unclear" to you when you engage with an interaction. You know when it touches your soma. It is a reality, but a reality you can only probe through your own somaesthetic sensibility.

The approach that Höök outlines, a process of tuning into my minute-by-minute experience of the body and its interrelationship with thought and feeling, was a process also echoed by interviewees [59, 66–68]. This was not an automatic state for me, given that I had never before engaged with a technology or a story in such an immersive, embodied manner. However, as I learned to tune into the subtle corporeal states involved in these experiences, it became crucial to accessing the deeper complexities of the works. The affective responses, often intensely visceral and performative[7] that I refer to in my findings are therefore personal and do not

---

[4] VR Noir (2016), Carne y Arena (2017), Awavena (2018).

[5] The Abbot's Book (2016), Broken Night (2017), Inanimate Alice: perpetual nomads (2018), Caliban Below (2018), Awavena (2018), Awake VR (2018), Spheres (2018) and Wolves in the Walls (2018).

[6] Höök's work on somaesthetic theory draws on the original work of American philosopher Richard Shusterman (2012) [47].

[7] Jeremy Bailenson categorises some participants of virtual reality experiences, given the intensity of their emotional or embodied responses, as "high presence." (2018, p. 21) My own affective and

speak to a collective. They are representative of my unique relationship to the content and experiential dimensions of the works, and are no doubt influenced by my race, gender, age, interests, cultural background and experiences. That said, the opportunity to carry out interviews allowed me to counterbalance personal experience with creator experiences and motivations. Through this process commonalities emerged in how I and the interviewees understood the experiential dimensions of the works.

As is common given the lived-experience interest of qualitative research, a semi-structured interview approach was followed [6, p. 1]. This approach fit with the aims of the interviews to gain a "behind the scenes" understanding of the conceptual thinking, processes and approaches that led to design decisions in the works analysed. Interviews were carried out in-person, via video conference or telephone depending on interviewee location and communication preference. The duration of the interviews was approximately 1-h, with the potential for a follow-up discussion if required.

The final aspect of the methodology involved a literature review of the field which included reference to theoretical, media and creative practice viewpoints on the subject of virtual reality design and associated artforms. From this process, techniques emerged across the works which were categorised in an excel sheet and from this sheet grouped into the eight aforementioned states, detailed in the below framework. The resulting embodied states framework which I have developed proposes a grammar through which VR narrative design can be conceptualised.

5.2.2   Findings: Proposing an Embodiment Grammar

The states outlined in the embodied states framework refer to moments where important shifts have occurred, over the duration of the VR narrative work. They highlight what I am doing, seeing, touching, thinking and feeling, as if I were able to press pause on the simulation and examine the complex dimensions involved at that particular moment in time. In order to conceptualise the states I have separated them. However, just as it is impossible to separate my mind from my corporeal experience, the states cannot exist alone but through the complex interrelationships with other states. Most often emphasis shifts to one or a combination of states to produce a specific outcome (Fig. 5.1).

These eight embodied states highlight design set-up (dialogic[8]), sensory modes of engagement (somatic) and co-creational strategies (encounter, performative and ludic) which I propose act as precursors for deeper affective possibilities experienced in VR narrative environments. The dialogic state relates to early framing decisions

---

performative responses seem to fit this description, and may be explained by my extensive childhood background in ballet and dance performance.

[8] VR narratives can be viewed as an evolution of Phaedra Bell's theory of dialogic productions (2000, p. 44). In her exploration of multimedia theatre, Bell applied the term "dialogic" (2000, p. 44) to describe the dynamic and tensions between live and pre-recorded elements. The synchronicity of these elements is enhanced when they appear to be in dialogue, even though they are not.

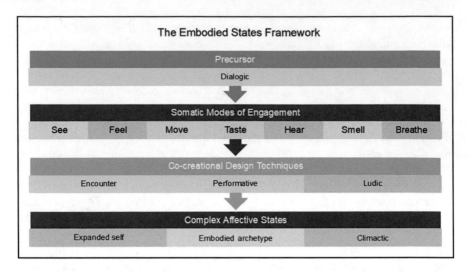

Fig. 5.1 The Embodied States Framework: VR Narrative design can be conceptualised through eight embodied states (seven key sensory modes most common to VR narrative works are counted under the one overarching *somatic* state)

involving live and pre-recorded elements which set the sensory guiderails for how I experience the work.

A central challenge of VR narrative design involves the reframing of interaction through the participant's senses. Seven possible modes, representing seven different sensory channels (those most often observed in VR work), are therefore possible. Design techniques analysed often highlighted the affective and co-creational value of limiting rather than aiming to create a wide range of sensory experiences. Through selective sensory engagements, emerging from the environmental design such as a swinging hammock (move) in *Awavena* (2018) or sand crunching under bare feet (move and feel) in *Carne y Arena* (2017), creators emphasise key experiential states through which the participant can inhabit a perspective or bring their own interpretation of events, without being overtly directed to do so. By structuring sensory experience through *patterns, rhythm* and *repetitions* these focused actions not only lead to embodied expressions of agency, but crucially, into an embodied experience of the character's perspective. The *somatic* state is often defined by constraints rather than range. By selecting sensory modes of engagement which are naturally revealed through the environment, and the character's relationship to it, the creator in turn opens up possibilities for meaningful co-creational involvement in the work.

Three complex co-creational states, dependent on the aforementioned dialogic possibilities and sensory elements illuminated through the environmental design, are a feature of the collaborative aspects of VR narrative environments. The three co-creational techniques identified include encounters, performative and ludic techniques. The encounter state impacts the participant by bringing together, often in moments of intense visceral exchange, the narrative and co-creational possibilities of co-presence situations. As Robert McKee argues, conflict between characters is often a main driver of story ([32], p. 101). These situations also lead, through the actions that force characters to make, into deeper revelations of self. In a narrative space, such encounters can be deeply impactful, prompting me into an embodied action as a result.[9] The choreography of encounters stands out as a central design strategy in virtual content creation, its affective potential augmented when combined with other co-creational design strategies.

As Kate Nash, Associate Professor of Media and Communication at the University of Leeds outlines, VR can be conceptualised as a performative medium because of the way it addresses the participant from a first-person point of view [34, p. 124]. By drawing on performative constructs from artistic practices such as immersive theatre and dance, designers bring to life my dramatic purpose as I traverse narrative spaces and embody other points of view.[10] Often, a transition into a performative state comes about through an encounter with a virtual character in the way they prompt me into an embodied response or shift my sense of self (See Figs. 5.2, 5.4, 5.5, 5.6, and 5.10). Performative states can act as a potent co-creational technique, often corresponding to moments of embodied revelation and climactic resolution.

VR narrative design techniques identified in this study, shifting my sense of self through co-creational design techniques outlined in *The Embodied States Framework* on page 11, do so often without any change in virtual representation, and instead through co-creational techniques which play on my own propensity and desire to imagine. Like actor Elizabeth Hess's experience of her expanding sense-of-self in her embodied character work [23, p. 2], I move through shifting states as I engage with objects, the environment around me and other virtual characters. In these experiences, structured through performative and ludic principles, I gain insight into other ways of seeing and being, such as the dehumanising experience of an illegal border crossing in *Carne y Arena* (2017) or the mind-altering sensory experience of Hushahu's drug induced vision in the forest in *Awavena* (2018) (See Fig. 5.3). My sense of self expands not just through the embodied perspective taking of another, but, through this corporeal and mental reconfiguration at the intersection of, as Hess calls it "self and otherness," (2016, p. 2) to question my own sense of identity, preconceptions and my relationship with the world around me in the process.

The flow through states, as outlined in my framework, may culminate in the affective experience of inhabiting another point of view. The embodied archetype can

---

[9] My response to the border patrol officer swinging a gun in my face in *Carne y Arena* (2017), where I found myself crouching on the ground, my hands pressing into cold sand, is an example of this. The possibilities for a response from this encounter situation are made possible by the enhanced dialogic set-up that Iñárritu establishes, where the untethered backpack and the expansive feel of the large gallery space, allows me to embody the scene through a range of movements.

[10] For example, in *Wolves in the Walls* (2018) interactions with Lucy were conceived of through the principles of dance choreography, where her body was in constant movement, swirling and colliding in a pull and push of dramatic energy with mine. Similarly, environmental design is used to shift me into different performative states and perspectives, such as the cold air-conditioned waiting room in *Carne y Arena* (2017) where I must wait, in a long-drawn-out sequence, for the alarm to sound, and the door to open.

Fig. 5.2 Participant's view inside the immersive HMD, looking towards Hushahu as if travelling on a boat with her

Fig. 5.3 Participant's view inside the immersive HMD, experiencing the forest as Hushahu does during her vision state

Fig. 5.4 Participant's view inside the immersive HMD, looking at Lucy as she draws virtual hands to map onto their real ones

Fig. 5.5 Participant's view inside the immersive HMD, interacting with Lucy

Fig. 5.6 Participant's view inside the immersive HMD, embodying the perspective of a woman being interrogated by a detective. © Eko

be viewed as the emotional pinnacle of this journey as I come to a deeper experience of and identification with the character. It is a state most compelling when brought into being through active work I have done—such as catching pig puppet as Lucy's imaginary friend (*Wolves in the Walls* (2017)), discovering the hideous rendering of Caliban staring back at me from the mirror *Caliban Below* (2017) (See Fig. 5.10), and sweeping to the ground to escape a gun to my face in *Carne y Arena* (2018)— and through these actions bringing my own personal interpretation to events. The embodied archetype allows for a different kind of visceral empathy, not experienced as a distant observer but through my own work, as I reach, stride, crouch and peer into dark corners where surprising revelations can be found.

The climactic state is the culmination of all preceding co-creational modes. It is rendered through the dialogic set-up, the somatic engagements, the complex co-creational states that, when structured with affective impetus, lead to a moment of intense visceral and emotional complexity. Participants of virtual environments report the experience of awe and wonder in these moments [1], of the surprising revelations suddenly uncovered through the pleasure of their own embodied actions and the thrill of kinesthetic resonance. The climactic state highlights the theatrical magic of simulated worlds, of bodies and environments which defy the limits of physics, and where the participant may even forget, as I often did in these fleeting moments, the technological illusion that makes it all possible.

While I have outlined a linear flow which emerged as common, to varying degrees of complexity, to this group of works, I am in no means suggesting that all VR narrative experiences must follow such a configuration, or that all preceding states must be employed to achieve a certain affective state. As many artists have illustrated, established grammars and conventions are open to reinterpretation and innovation often happens through a radical reconceptualisation of the status quo.[11] No framework or grammar should be considered the final answer to any creative problem, but instead may act as an open invitation to interrogation and reinterpretation. What is perhaps most relevant about it, in the flow of co-creational states, their interrelationships and affective dimensions is the way the framework positions VR narrative design within the larger philosophical concepts of art as a process. As Deleuze's concept of the assemblage invites creators to do, rather than seeking out the answers, to instead frame the possibilities ([27], pp. 41–42). The framework offers some structural suggestions through the flow and interrelationship between states for how to generate such potential.

5.3 The Encounter State as Pivotal to VR Design

The experience of co-presence isn't inherently affective, instead, as the analysis of the works in the study highlights, it relies on the way interactions, what I call encounters with other participants, virtual agents or responsive environments are choreographed. Below I outline three design principles illuminated through the study which helps to shed light on how encounters can act as a pivotal co-creational technique. The first principle involves an understanding of how co-creational possibilities are dependent on the dialogic set-up of the work. I explore some example dialogic framing techniques and their co-creational impact through the analysis of *Carne y Arena* (2017) and *Awavena* (2018), particularly good examples of this relationship for the way they utilise environmental design, sensory elements and responsive simulations in design solutions. The second principle focuses on how creators use encounters to establish a compelling sense of character embodiment, even when sensory experience, or representational cues are limited. The third principle refers to how encounters can be structured through performative and ludic techniques which act to augment the work's co-creational and affective possibilities. Across the techniques and experiential possibilities highlighted, the significance of techniques which work with absence, juxtapositions and the in-between is highlighted, emphasising the value of nuanced

---

[11] An example of this is a work outside of the study titled *I, Philip* (2014). In the work, I inhabited the robot recreation of Philip K Dick. An experience that is considered 360-degree film and offered no material means of engagement, when applied in the right context, became vested with embodied meaning.

conceptualisations of the core experiential dimensions of VR environments and their relationship with affect in design approaches.[12]

5.3.1 Framing Encounters Through Dialogic Possibilities

Encounters, their co-creational and affective possibilities, depend on framing decisions. Framing techniques specific to VR design can be understood through the concept, per Phaedra Bell, of "intermedia exchange" [7, p. 44]. Drawing on Bell's theory of "dialogic media" [7, p. 44] I propose that VR narrative framing techniques can be understood through this conceptualisation of media for the way techniques increase the synchronicity between the live and simulated elements. I examine a range of dialogic framing techniques employed by creators in the works *Carne y Arena* (2017) and *Awavena* (2018) in order to understand this concept and its experiential influence.

I experienced *Carne y Arena* (2017), a VR art installation, at LACMA in 2017. It began by walking through a heavy red curtain and down a hall to a door. Through the door was an air-conditioned room lined with old shoes (worn during attempted illegal border crossings into the USA) where, after removing my own shoes, I had to wait, chill seeping into my bones, until an alarm sounded. I proceeded through another door into a dimly lit cavernous gallery space where cold sand crunched underneath my bare feet and a cool draft blew from an unseen location. Moving through rooms and into a state of increasing anxiety, I experienced a sense of immersion into another place and frame of mind long before I put on a headset.

Director of *Carne y Arena* (2017), Alejandro Iñárritu, refers to the importance of sensory elements for the way they draw the participant's attention to their body, and therefore, their emotional intelligence. He says, "Your feet touching the sand has a huge impact; your body's telling you this is true" (Iñárritu in [53]). The significance of these sensory elements becomes apparent once I put on the headset and backpack (so that I can experience the simulated world untethered) and fix the headset over my glasses. As I find myself surrounded by 360-degree imagery of the Sonoran Desert, the environmental design creates a sensory reinforcement of what I see, thus increasing the synchronicity between the live elements and prerecorded 360-degree film.

In her exploration of multimedia theatre as a "dialogic media production" (2000, p. 44) Bell seeks to understand the dynamic and tensions between live and prerecorded action. She explains (2000, p. 44) the phenomenon in multimedia theatre thus,

---

[12] In order to explore design techniques in detail, some of the descriptions of the works contain spoilers. To view the works prior to reading my descriptions, see the links to works section listed at the end of the chapter.

Inter-media exchange is the mutual acknowledgement of images produced by separate media and their accompanying interchange of dialogue, glance, attribute, equipment or other currency such that the images cohere and appear to coincide in the same time and space.

As Bell details, various techniques can be employed by creators to generate synchronicity between what would otherwise be disjointed media forms. An example of the kind of interchange Bell is referring to can be observed in the adaptation of the Czech play *The Eleventh Commandment* (1950) [p 40]. Filmed sequences of characters' "offstage or past actions" in the play were projected onto the wall at the back of the stage to create a "lightly ironic or downright farcical comment" on that which was coming to pass in the live action (2000, p. 40). The play concluded with a live character firing a pistol at a character onscreen, who immediately fell to the ground. The synergistic use of these mixed media elements therefore augmented the dramatic impact of the work, in this case through irony and surprise. In the case of a multimedia production such as this, Bell also argues however that the synergy relies on the "mind of the perceiver" (2000, p. 44) to deduce meaning from the separate elements. There remains an inherent dissonance, where the live is in contrast to the recorded image's lack of intelligence (2005, p. 45). This dissonance is a crucial element of multimedia theatre and continues even with more immersive media forms such as VR.

The cinematic representation and associated limits which Bell discusses in the prerecorded image, have now been superseded by a technology that is better described, as Kate Nash points out, as "a medium of simulation" [34, p. 123]. A simulation not only allows for the sense of immersion into another place, positioning the audience at the centre of the action, but also creates opportunities for greater synchronicity between the filmed or CGI and live action dimensions.[13] VR technologies are still nascent however, and the tensions that Bell described as inherent to dialogic media persist, with implications for how we experience immersive stories. As Miriam Ross explains, the ideal of full immersion cannot be realised, instead, when we experience a VR narrative world, we are engaged in a constant negotiation across both real and virtual spaces [43]. Virtual reality is therefore the experience of being in two places at once [43, p. 302]. Any perceived synchronicity still relies to a large degree on the participant's willing suspension of disbelief and players own desire for immersion [, p. 33] in response to the framing techniques of the creator.

In her exploration of the experience of her body as "missing" while experiencing a VR work, Sita Popat, Lecturer in Dance at the University of Leeds, is interested in how the "proprioceptive senses" come into greater focus [39, p. 360] as the participant contends with the disharmony between body schema and body image [39, p. 371] that come about due to this disjunction, in particular when there is no avatar

---

[13] Evolving technologies will continue to enhance this synchronicity, such as the through the advent of motion capture (outlined on page 2) [30], as well as AI advancements, which are being applied in the development of VR games such as the forthcoming *Peaky Blinders: The King's Ransom* (2022) see [14].

representation of the participant's body in space.[14] In her reflections on the experience of *White Island* (2014), a performance installation involving a virtual hot-air balloon ride over an illustrated Arctic landscape, Popat details the strange experience of an embodied visceral reaction of the illusion of hurling towards a virtual representation of an immense mountainside. She raises her arms to her face, yet without the visual reflection of her body doing this action [39, p. 361]. She argues therefore that, without an avatar creating a connective membrane between the physical and virtual, "the body in VR is experienced as blurred, being both virtual and physical, absent and present, compounded and indivisible, even though body and environment have different materialities" [39, p. 360]. She explains how she grips onto the thick hanging rope that is used to guide the height and direction of the balloon in order to feel stabilised and comforted during the experience. The rope becomes the membrane that enhances the synchronicity between her physical body and the 360-degree imagery.

Iñárritu's use of set-design, sand under foot, wind and cavernous dimly lit gallery space work like the rope in *White Island* (2014) creating a bridge between media forms. The synergy that such dialogic elements generate act as a framing strategy which impacts my experience in several ways. Firstly, I feel more immersed in the desert simulation. I also feel more grounded in my body and in the simulation, making me feel a sense of embodied presence in the narrative space. This establishes the experiential conditions which are vital precursors to the way I respond to subsequent dramatic encounters.

In the following passage (in italics) I detail my experience of *Carne y Arena* (2017) during a visit to the Los Angeles County Museum of Art (LACMA) in California. I employ this italicised technique throughout the chapter ahead of the analysis of the works. The aim of the approach is to convey the immediacy of embodied experience and affective impact of design techniques given that viewing the works may be difficult for readers who do not have access to an immersive HMD system.

Carne y Arena (2017)

Soon I hear voices, or more specifically, a woman wailing. A group approaches. They are tired, one woman limps, being helped by another. As they near me, I approach them and immediately seeing their predicament, I want to reach out. I cannot touch them, but when I come closer, following the group, from inside one of the immigrant's chests a red beating heart illuminates. My embodied actions have effected a change in the simulation. All of a sudden jarring lights and the whirr of propellers cut through the darkness. A patrol vehicle sweeps in. A blinding light. Someone shouts at me to get down. A dog is barking. The patrol officer approaches and, in a surprise development, his gun swings on me. I freeze in fear. He shouts his orders again. I crouch down. My heart is pounding in my chest. The sand cuts into my palms.

Through these encounters, first coming across the group, then the confrontation with the patrol officers, I am prompted into a co-creational response. Being in close physical proximity to the immigrants causes me to draw nearer, and in doing so, I

---

[14] Research has shown that a virtual body mapped to the participants' in one way to generate presence and cohesion between physical and simulated spaces [5, 49].

generate a change in the environment, illuminating a beating heart. Next, when the patrol sweeps in, I react as if the situation were real—a measure, as Slater points out, of immersion [49, p. 3554]—crouching to the floor when the officer shouts at me to do so. These encounters therefore rely on a compelling sense of psychological presence and sensory immersion which, as I detailed earlier, are established through dialogic framing techniques.

The encounters create co-creational possibilities when they are framed through a dialogic set-up that offers the space, sensory elements and synchronicity that allow me to engage with full use of my body. This in turn creates possibilities for deeper affective responses as a result. As Iñárritu himself details, his aesthetic aspirations were not simply immersion, but more importantly, to connect with "human emotion" (2017). He described his aims with using VR thus (Iñárritu in [35]):

> My motivation was to use the latest technology to explore the ancient theme of migration and subordinate it to serve and connect to the human emotion. Not just as an observer in a passive way or as a new way of escapism, but actually to blend an abstract narrative space with a physical and sensorial experience in an active way.

Here he draws attention to the way he conceptualises the senses and co-creation as a channel into emotion. My affective experience reflects this. By reacting I feel it viscerally, through my own unique embodied actions, through the discomfort of cold sand cutting into my palms, tears welling up uncomfortably in the display. By reacting to the drama as a co-creator, deeper, more layered affective possibilities therefore emerge. I shift from a visitor to an actor, implicit in the drama, not looking on at the immigrants, but for a few moments, feeling the world as they do. As Kriston Capps reflected in a review of the piece [13]:

> I was always mindful that I was on a stage, watching a drama play out, until that drama turned on me. The symbolic value of Border Patrol agents turning their flashlight beams on the viewer won't be lost on anyone, even if they see it coming.

Thus, Capps echoes my own experience of the work, in how the encounter situation impacts the participant when the drama "turns on" them. The affective experience which motivates work like this is made possible by framing techniques which comprehend the participant's status across two spaces, the physical and the simulated, as integral to its embodied possibilities. This dialogic framing, which either constrains or enables co-creational possibilities, was identified across the works studied, with varying design approaches and experiential outcomes.

In *Awavena* (2018), creator Lynette Wallworth uses environmental design and sensory elements like in *Carne y Arena* (2017), such as a swinging chair and a baton held in my hand, but in this case co-creational possibilities are also dependent on the responsiveness of the simulation. These differing emphases highlight how framing techniques are linked to the creators' aesthetic aims, budget and also the technical abilities of the production team.

Like *Carne y Arena* (2017), I experienced *Awavena* (2018) in a dimly lit gallery space at Carriageworks in Sydney. The experience was marked by two distinct sections, and for the purpose of this discussion I will summarise the beginning then focus mostly on the second part. In the first part I had the experience of travelling to the Amazon to meet Hushahu, the first female Sharman of the Yawanawa people.

Before commencing the second part of the experience, I was given a cylindrical shaped black baton to hold in my hand. Once inside the simulation, Hushahu begins to detail, in a voice over, how a drug-induced vision is integral to her initiation as a Shaman. She explains the deep connection that she experiences with the forest during this year-long journey into a state of altered consciousness, and as she does, it is then me standing inside the forest. In the following passage I reflect on my experience of *Awavena* (2018) shown at Carriageworks, Sydney, in 2018.

Awavena (2018)

Particles of light hover in the air all around me. As I move my hand holding the baton through the particles, they react to my movements, scuttling away as I sweep my hands through them. Through this kinaesthetic action I feel that I have become Hushahu, experiencing what she saw, during her year-long forest visions. Through my embodied actions I also feel her connection with the forest all around me, just as she did. I am drawn up inside a tree, up and up until I feel I have crossed over into another dimension, a realm usually inaccessible, entirely invisible to me. In doing so, for moments I have not just embodied Hushahu, but transcended myself.

The encounter in this work involves both my interactions with the forest, and my experience embodying Hushahu's perspective during a vision state. In this scene Wallworth uses the technology's capabilities to simulate a way of experiencing the natural world as inherently connected to the human corporeal and sensory experience, through movement and breath. This was, as Tashka the chief of the Yawanawa explained, a crucial reason for telling the story of the Yawanawa using immersive technology. After watching Wallworth's previous VR work, *Collisions* (2016), he saw immediately the connection. "When I put on the sunglasses (headset) it opened a window like medicine does—," he reflected, "Like a vision. It made a portal" [54]. VR enables viewers of the work to experience the world as Hushahu did, which was integral to the message the tribe wanted to convey. Importantly, through Hushahu's visions, women's stories that had become lost to the tribe were returned to her. When she brought these stories back to the tribe's chief, they began a conversation of how they could change their practices. To convey this message, it was vital that the work brought the participant into the world of the Yawanawa so that they could experience their knowledge in a deeply sensory way [54].

While the gallery setting, where I can stand while experiencing this work, and the baton which I hold in my hand, create the basic synergies necessary for the forest encounter, the synergy between the simulation and my embodied corporality takes on greater importance than in *Carne y Arena* (2017) in order to share this message. Wallworth elaborated on her technical and artistic goals in a presentation she gave in 2019 at ACMI Melbourne:

It has to show that the forest is aware of you. It has to show you that everything is alive. In VR, you can do something like that. I can use headtracking in the headsets so that it knows where you are looking. And those points from the point cloud data in the forest can move towards you and can suggest to you that they are aware of you looking at them. We can use breath to disturb those particles and give an understanding of the connectedness of my breath to a huge forest.

Wallworth thus draws attention to the dialogic elements which translated into this simulated encounter—the participant's gaze, the technology's capacity to track where their head moves, point cloud data that responds to the participant's interests and particles of light that react to their every breath. In this case sensory channels are limited to breath and movement, but there are many other possibilities, as highlighted in *The Embodied States Framework*, depending on the sensory possibilities inherent in the narrative space, and the aesthetic motivations of the work.

While in the previous section I mentioned the affective possibilities which arise through encounters with other participants [30], Wallworth's work also highlights the affective potential of encounters with the natural world, even when the representation is abstracted (See Fig. 5.3). Another example of this is the work of Char Davies such as *Ephémère* (1998), where participants traversed a simulated natural environment through breathing and body balance. The artist detailed how the experience of floating through the landscape, disembodied from their regular human form, where participants could witness natural events such as a seed pod opening, was profoundly moving. It was described by some participants as "euphoric", while causing others to weep [15]. Davies reflected how, "...the artist's goal is to connect the immersant not to others but to the depths of his or her own self," with one participant reflecting it was, "the most evocative exploration of the perception of consciousness that I have experienced since I can't remember when" [15]. Such experiences highlight how the embodied and sensory dimensions of VR can lead to new ways of engaging with not just the environment, but also the self. Wallworth, in reflecting on responsive works such as *Awavena* (2018) calls this an experience of intimacy, just like the creators of *The Under Presents* (2020). *Awavena* (2018) draws attention to how encounters of many kinds, interpersonal, environmental, or even more abstract and imaginative, can lead to deeply affective experiences.

Exploration of *Carne y Arena* (2017) and *Awavena* (2018) highlights how dialogic strategies create the necessary precursors for such possibilities. They also draw attention to how, rather than the ideal of full immersion, VR experience can be understood as a constant negotiation across physical, simulated and imaginative spaces. Different framing approaches such as sensory elements or responsive environments may be employed depending on the aesthetic and affective motivations of the work. For example, across both works, sensory elements in the physical environment are used to draw attention to the participants' corporeal intelligence and reinforce what is depicted in the display, thus generating a sense of immersion that is an important precursor for a range of embodied responses. By framing encounters through a dialogic set-up, these works highlight how deeper, more layered affective possibilities can emerge.

5.3.2 Structuring Encounters Through Performative and Ludic Techniques

As I have mentioned, co-presence alone does not lead to the affective possibilities of VR environments. Often, and certainly the case in the VR narrative works studied, it is the way these encounters are framed and then choreographed through co-creational techniques that generates such potential. In this section I will explore co-creational design strategies and how they are structured alongside encounters. The analysis not only examines how the affective possibilities of virtual environments are determined through the interrelationship of these techniques, but also how such techniques are translated from other media forms into immersive spaces, emphasising my embodied corporality as the central element through which co-creation is experienced.

5.3.2.1   Performative Techniques

The relevancy of performance as an experiential precursor for VR design is well documented. In this section, I examine the affective and co-creational significance of techniques inspired by immersive theatre and the grammar of dance and how they were applied in the development of *Wolves in the Walls* (2018).

Reflecting on the creative process in an article in 'No Proscenium', writer and director of *Wolves in the Walls* (2018) Pete Billington describes how performative constructs offered solutions to dramatic questions, such as the mechanics of spatial engagement and the role of the audience, which they found couldn't be answered through the grammars of gaming and feature film animation alone [8]. Collaborating with *Third Rail Projects,* a New York based group creating site specific, immersive and experiential performance works, they began to explore other artforms for solutions (2019). Immersive theatre added the additional dimension needed in order to bring the work to its full dramatic potential. In particular, reference to the production *Then She Fell* (2012) illuminated "Intimacy, specificity and catered experience," (2019) such as through quiet interactions between characters and audience members, which could be applied in their work. In 'No Proscenium', Billington described how they saw this intimacy translating into VR [8]:

> The nature of the one-on-one interaction. *Why it is so powerful and connective?* These ideas emerge inside *Wolves* as something we call "haiku moments." Little quiet interactions with just Lucy and the audience. They give us time to take each other in, bond, and try to understand one another. In *Then She Fell*, these moments are so successful, it feels like you are falling in love.

Like Martins experience of intimacy when experiencing a virtual hug, in *Wolves in the Walls* (2018) the techniques are applied to give the encounters with Lucy more dramatic, co-creational and in turn affective potential. The way the participant inhabits the space with Lucy is also important, because drama and interaction happen spatially rather than in two-dimensions. To solve issues of volume and distance the creators drew on dance choreography which strongly influences immersive theatre

(2019. *Wolves in the Walls* (2018) can therefore be understood as a dance in the way Lucy is in constant motion as she "weaves" (2019) through rooms. The experience of intimacy and intensity of co-presence experienced through the encounter with Lucy is highlighted from the very first scene (See Figs. 5.4 and 5.5). In this passage I reflect on my experience of the beginning of the work, downloaded from the Oculus Store (see page 48, list of works) and viewed it on an Oculus Rift headset.

Wolves in the Walls (2018)

In the opening scene of Wolves in the Walls (2018), I am standing in the attic of a rickety old house. It is there that I meet a young girl named Lucy.

"Oh, I drew you kind of tall," she says, appraising her work, after sketching an outline of my 'body' in chalk.

"Don't move... now you're like me," she says. "Hello."

Lucy has a predicament. There are wolves in the walls, she tells me.

"I've been finding clues. But now we can do it together... Oh Sorry...That's one" she draws me a hand with her chalk, "...hold still silly... And two."

Now when I move the handheld controller and press the buttons, the chalk-drawn hands move as if they are my own. I am a blank slate onto which Lucy has painted on an identity: her imaginary friend. My task is obvious from the outset, to be her sidekick, like Watson is to Holmes, on a mission to find clues of the wolves living in the walls.

During this exchange, Lucy walks fluidly and elegantly around me. Through this quiet moment as she draws my hands, she pulls me into her make-believe world. As I respond to Lucy, moving my new virtual hands to test them out in space, I enter a kind of techno-mediated spontaneous enactment, one which isn't just focused on doing things but, as the story progresses, on nudging me into a performative role. Performer Ruth Zaporah similarly alludes to the significance of interactions between herself and the other actors (Zaporah in [10, p. 128]):

I sense the body as no different than the space it is moving in and the sound it is moving to. If I am improvising with a partner, each of our bodies becomes an extension of the other. I perceive her body as no other than my own; her voice, my voice; her story, mine.

Thus, Zaporah highlights the way the dynamics of spatial presence with another performer lead her, almost hypnotically, into a performative state. This is also my experience while inhabiting the space with Lucy, a co-creational dance as she guides me through a series of rooms in the house and I help her to uncover clues, right through until the final scene, when the wolves do emerge from the walls, and we must fight them together in a wolf battle. Dance choreography and immersive theatre set the dramatic guidelines, through embodied action in the space, and intimate exchanges, for the affective possibilities of this final scene, when I recover Lucy's beloved pig puppet, thus fulfilling my role as Lucy's beloved side kick.

When combined with encounter situations, performative techniques enhance the co-creational possibilities of a work through the alchemy of inhabiting a physical space with another character. As the progression of events in *Wolves in the Walls*

(2018) highlights, techniques and principles inspired by performance art, such as those established in immersive theatre or dance, can be integral to solving the dramatic questions of how participants inhabit virtual spaces with characters, experience a dramatic role and generate affective possibilities.

5.3.2.2 Ludic Techniques

There is an established link, as many theorists have argued, between agency and affect. Howard Rheingold for example suggests that an interaction's quality should be judged, among other elements, on its ability to "move people" ([40], p. 17). In this section I explore how ludic techniques taken from games inform the interactions in *Wolves in the Walls* (2018) and crucially, how the structuring of these ludic techniques leads to affective possibilities. I continue with the close analysis of *Wolves in the Walls* (2018) in this section then explore the way encounters, ludic and performative constructs work together to result in particularly complex and layered affective experiences. I downloaded the work from the Oculus Store (see page 48, list of works) and viewed it on the Oculus Rift at the VR LAB in the School of Arts and Media at UNSW.

Wolves in the Walls (2018)

In the second scene of "Wolves in the Walls" (2018), I visit the kitchen with Lucy. Her mother is making jam, and so Lucy instructs me to label the jars and stack them on the shelf behind me. Using my handheld controllers, I pull the lever on the machine, then watch the jar tumble through it, before picking one up, again by pressing the button on the handheld controller, and placing it on the sideboard behind me. I do this a few times, in a repetitive task involving actions which recall for me a vintage game like Tetris or Pac Man. When the contents of the jars suddenly disappear from the shelves, and both Lucy and I witness it, we have, through these collaborative actions, uncovered another vital clue that something strange is indeed going on in the house.

This scene sets up a recurring motif throughout *Wolves in the Walls* (2018)—the ability to flex my sensory involvement through micro-tasks which are not merely gratuitous expressions of agency but lead to revelations of story and character. In each scene in *Wolves in the Walls* (2018) I am given a task to complete, which in some way progresses the story or sets events up for a later pay-off. An example is the scene with Lucy's brother where I play a virtual reality videogame called "Wolf slayer". Wielding a lightsaber-esque sword, I must defeat a relentless stream of wolves as they bound towards me.

The ludic experience is again simple, with the satisfaction I experience linked to increasing kinaesthetic command of the controllers as I successfully eradicate the wolves. While this is enjoyable, the real reason for completing this task only becomes evident later, when the wolves finally do emerge from the walls, and I must fight them in a real wolf battle to try to take back the house and rescue Lucy's beloved pig puppet. The first mock wolf battle acts like a game level which must be completed before progressing to the next one, with increasing stakes adding to the tension.

While stacking jars and slaying wolves with a lightsaber at first appear to be gratuitous actions, when they are experienced in virtual reality, they generate novel embodied experience. Never before have I completed virtual tasks through the movement of an imaginary friend's avatars hands mapped perfectly to my own, felt the tingle of satisfaction of placing a jar down on the machine through the synchronised action of my virtual hand and my real one holding a handheld controller, so that it can be filled with jam and stacked on the shelf. The action itself, as a mimetic construct, is not new however. Instead, it draws on ludic engagement techniques long established in video games.

In the book, *In game: From Immersion to Incorporation* (2011), videogame researcher Gordon Calleja, outlines ludic involvement as one of the six dimensions which contribute to the sensation of presence [12, p. 165]. He defines ludic involvement as the choices players make within a game and the associated repercussions [12, p. 165], while also listing rules, goals, missions and rewards as some of the key properties of game design. How such properties are conceptualised of course differs with different types of games. The ludic structures in *Wolves in the Walls* (2018) share commonalities with narrative, role-playing and adventure games such as *The Witcher* (2007) and *Grand Theft Auto IV* (2008). Calleja elaborates [12, p. 51] on how this kind of structure works in a game like *Grand Theft Auto* IV (2008):

> ... a virtual environment which has a series of mini-games embedded in it and whose scripted narrative has a linear structure of progression with strict ludic properties. To advance the scripted narrative, players must complete missions that usually involve a number of linked goals which can be achieved only by adhering to a set of relatively strict parameters.

The structure that Calleja outlines is very similar to what I experience while playing *Wolves in the Walls* (2018), where the missions, stacking jars and slaying wolves with a lightsaber act like micro missions subserving the story [12, p. 151]. These missions allow me to express agency in the progression of the narrative. Some goals, as Calleja outlines, act as prerequisites for the completion of other goals (p. 151).

A notable commonality between *VR Noir* (2016), *Wolves in the Walls* (2018) and *Caliban Below* (2018) is that each of these works employ micro and macro ludic structuring techniques in service of dramatic motivations. They also draw on detective and gothic horror storytelling tropes commonly used in narrative and role-play games. Unlike works like *Awavena* (2018) and *Carne y Arena* (2017) which incorporate interactions that are more meandering and exploratory, with revelations of meaning happening along the way, works like *Wolves in the Walls* (2018) are framed with an overt goal from the outset which is the perfect framing for the introduction of tasks.

As Lucy's imaginary friend, I am her detective sidekick, on a mission to find out if there are wolves in the walls. Each mission we complete together progresses us towards that goal. It is only by helping with filling and stacking jars on the sideboard that I experience the jars, right beside me, being emptied by a sly and unseen interloper. Meanwhile, by playing "Wolf Slayer," I learn important skills, the purpose of which is only later revealed with the final mission, to defeat the wolves with lightsabers, and recover Lucy's pig puppet. The goal to find pig puppet comes to supersede the original goal and is a clever ludic strategy with dramatic impetus and ultimately affective dimensions. Through the missions, I am not just a bystander to Lucy's plight, but an active participant in it. Thus, when she loses pig puppet, I feel not just ludic impetus to recover it, but emotional, because of the connection I have formed through inhabiting a collaborative space with Lucy. The final mission proves an opportunity not just to defeat the wolves but to recover pig puppet, appearing tossed into the air, during the wolf battle.

Another notable ludic aspect of the experience is the repetitive nature of it. It was only in returning to the experience for a second time that I was able to catch pig puppet for Lucy, after a failed first attempt. The sense of achievement that I felt at this moment was, as interaction designer Brenda Laurel draws attention to, palpable [26, p. 122]. As Laurel argues, such moments can be quite impactful due to, "the feeling of completion that is implicit in the final apprehension of the shape of a whole of which one has been a co-creator" [26, p. 122]. Experiencing the outcomes of actions which I have taken lends these outcomes greater weight.

In reflecting on the experience of *Carne y Arena* (2018) in an interview with *Variety*, Iñárritu also argues the significance of the second viewing. He elaborates [61]:

> In Milan there was an art critic who saw it two times. He told me something I truly believe. He said the second viewing is essential. There are so many things going on, and then your feet are touching the sand and the breeze — all of that combined with the illusion in your brain, you back up a little bit. The second time, it's not as overwhelming. It's a very different experience. And you'll discover a lot of detail. There are a lot of secrets that I hide there, that nobody has seen. But I think in time people will discover a lot of things that are there.

Here Iñárritu explores how more details and experiences within the environment are revealed, like hidden easter eggs, on a second viewing. It was only in the second viewing, for example, that Iñárritu describes how some viewers notice the beating heart in the centre of the immigrant's chests—a poignant message of the deep human cost of the immigration crisis. This evocative detail echoes the flowers in Char Davies *Ephémère* (1999), entering into blooming when a curious participant comes close enough [16]. In these moments, such as discovering the immigrants beating heart, it is as if Iñárritu has leaned into me and whispered a secret. "*Oh, you've found it. This was it—the thing I really wanted to show you, if you were willing to seek it out.*" In these moments, of discovering beating hearts and catching beloved toy pigs, where I move beyond the superficial, and find myself inhabiting the deeper layers of the work, the full affective depth and meaning is revealed.

Works such as *Wolves in the Walls* (2018) employ micro and macro ludic structuring techniques common to adventure games and subordinate them to story goals. Micro-missions are framed through repetitive tasks involving fluid embodied actions. This not only leads into novel expressions of agency in an immersive environment, to build kinaesthetic skills that will be later put to the test, but, in combination with performative constructs and structuring techniques, to cultivate a stronger sense of embodied presence as Lucy's imaginary friend. Similarly, creators draw on game design techniques like hiding easter eggs and allowing them to be revealed only through the embodied curiosity of the participant or making tasks or missions so difficult they need to be repeated, in order to enhance the affective experience of the work. Crucially, these ludic techniques are translated into immersive worlds by centring the actions and revelations on my embodied actions and my senses.

Wolves in the Walls (2018) draws attention to how the three core co-creational strategies relevant to VR narrative design, encounter, ludic and performative states, create the possibilities for more layered affective experiences. By reframing these techniques through embodied actions, such as stacking jars by reaching out with my handheld controllers or sweeping my arms through the air as if holding a lightsaber, they do not just become novel expressions of agency, but lead to deeper, more visceral experiences of character.

5.3.3 Choreographing Encounters to Generate Character Embodiment

In this section I continue my discussion through particular focus on character embodiment. I begin by conceptualising embodiment as it relates to immersive narrative creation before outlining some overarching dramatic approaches. I explore three works which highlight different techniques used to create compelling experiences of seeing through the perspective and acting through the bodies of a range of perspectives.

5.3.3.1 Dramatic Approaches to Character Embodiment

The interdisciplinary origins of embodiment [38, p. 196] are especially relevant when it comes to understanding the core dimensions of VR, given the complex experiential modes which make up the artform. The three aspects of embodied experience which have already been noted, include, (1) sense of self-location, (2) sense of agency and (3) a sense of body ownership [20, 50]. These can be understood as technical foundations. To evoke a sense of character embodiment, however, these technical aspects must also be considered in balance with artistic techniques which act on the participant's suspension of disbelief.

As I have established, the performative dimensions of experiencing a character from a first-person perspective draw on embodied performance concepts such as role-play, improvisation and immersive theatre. The interactive dimensions of character embodiment draw on ludic principles. These constructs often accompany encounter situations, which I will expand upon in the subsequent section. Importantly, these performative and ludic strategies come into greater focus in VR narrative design solutions to embodiment because of representational absence. This is not necessarily a limitation which needs to be overcome but can also be seen as a design strategy with affective potential. Before I proceed into an examination of some of these more specific techniques however, it is important to establish two different overarching dramatic approaches to character embodiment which are linked to this lack of body representation.

In an interview with me, writer Mike Jones distinguishes between "pre-defined" or "tabula rasa" (blank slate) for how to frame the participant's sense of character [68]. In a pre-defined scenario, the participant is told who they are from the outset (*VR Noir* 2016; *Spheres* 2018). In a "tabula rasa" scenario they discover it through their involvement in the story (*Carne y Arena* 2017). As Jones elaborates [68]:

> If you're doing tabula rasa, then the decisions you force the character to make, the dilemmas you force them into, will compel them to shape a character in a certain way... to become a character you wanted them to be if you give them the right set of impetus.

Here Jones draws attention to how character embodiment need not be a fixed concept but instead an experience that evolves, through time and space, as a co-creational act and where unique possibilities emerge through spontaneous interactions between actors and participants. Interestingly, this is also how he conceptualises presence in a VR environment, not just the idea that the participant feels a sense of transportation, but that, through being given choices and expressing agency in those choices, their reason for being there really matters [68].

To add to Jones' theory for how character is defined, a third interesting possibility exists. The participant is allocated a clear role from the beginning, but this role shifts or evolves through an unexpected revelation (*Caliban Below* 2018, *Wolves in the Walls* 2018). In "tabula rasa" and this third possibility, the creator sets up a series of events which, as the participant reacts to them, do not just lead them to the end of the story, but to gaining a deeper experience of the character. In the following examples I explore the way VR narrative creators establish a sense of character embodiment through a variety of techniques which place an emphasis on the co-creational and affective possibilities of walking in the shoes of another.

5.3.3.2   *Broken Night* (2016), Sense-of-Self and the First-Person Address

The opening scene of *Broken Night* (2016) plunges me into the uncomfortable first-person perspective of a woman being interrogated by a detective (see Fig. 5.6). Following this establishing scene, there is a juxtaposition as the POV switches, and I watch this woman reliving her memory of the event, influencing the story through the way I direct my gaze. In an interview, the director of the project, Alon Benari explained how and why they came to this solution for managing point of view, and what obstacles they needed to overcome. As the team approached VR to work on *Broken Night (2017)* they knew, as Benari outlined, that they would need to determine what the right participant experience for the medium was (2017). They soon decided that an interface which placed the viewer in a removed third person position, which had worked in their other interactive projects, wouldn't in this instance. He described why in an interview with me (2017):

> As we were working on this, it became clear that the expectation of the viewer when they go into VR is to have some kind of sense of self within this world. If I'm just floating in this

world, it's not as powerful. That's where we had this idea, "Okay. Let's say, I am her, but the majority of the story is her memory where she is third person."

In this reflection, Benari echoes other theorists who have posited that 360-degree simulations are best experienced by an ego-centric first-person perspective [26, 40, 51], and also reinforces Slater's theory on the elements which contribute to embodied presence.

In *Broken Night* (2017), in the scenes where I am inhabiting the woman's perspective in the interrogation room, the experience of "sitting" across from the detective creates an embodied experience of the character through the confluence of two factors. First, there is a dramatic intensity that comes from an actor looking directly at me which prompts me into a heightened performative state. Kate Nash reflects on this experience of "direct address" [34, p. 127], in the film *Ground Beneath Her* (2016), an observational piece where the viewer watches Sabita as she rebuilds following the Nepal earthquake of 2015. In one moment in the film, Sabita looks at the camera and reaches out, as if she were offering the viewer a flower [34, p. 127]. Nash argues that this use of direct address not only impacts the participant's experience of presence, but positions them as a benefactor, creating a "remediation of NGO communication in VR" [34, p. 127]. While in *Ground Beneath Her* (2016) the participant plays the role of the benefactor, in *Broken Night* (2017) this technique is used to prompt me into embodying the role of the woman being interrogated. This performative construct is reinforced by tracking capabilities in the display, which allow me to move my head left and right, to take in the room as if I am sitting right there. Importantly, I enter this embodied state through limited sensory experience. As Slater found, the experience of presence, in this case character embodiment, can be evoked through "minimal elements" (2005, 337), the rest being filled in by "cortical processing" (2005, 337).

Like *Broken Night* (2017) other works in the study reflect the way dramatic encounters can lead me to an experience of character embodiment even when sensory engagement is limited. In fact, the moments when character embodiment is experienced, through an expanded sense-of-self or even a deeper affective experience of an archetype, my embodied actions were very focused.

In her book *Computers as Theatre* [26], Brenda Laurel draws attention to the significance of constraints and selectivity in co-creational media design (p. 100, 118). She explains that constraints are an essential part of any human–computer interaction, not only because of the need to work within hardware and software limitations (p. 100) but also because of the relationship between constraints and creativity (p. 101). "Limitations," she explains "—" constraints that focus creative efforts—paradoxically increase our imaginative power by reducing the number of possibilities open to us" (p. 101). Laurel explains how constraints are recognised as a technique to aid creativity in the art of theatrical improvisation (p. 106). She refers to *commedia dell'arte*, where actors are constrained in choice and action (p. 106). Such techniques were no doubt vital in producing comedic or dramatic possibilities. This idea is reinforced through the works included in the study, where a few selective sensory engagements were often chosen by creators[15] as the way through which I express agency and make decisions, but more importantly, for how they construct a sense of embodiment and role-play.

Broken Night (2017) is an interesting example of character embodiment because it establishes the minimal sensory and embodied conditions through which it may be experienced. In this case, the experience of a direct address from the detective across the table, combined with head-tracking and the claustrophobic mood of the room, prompts me into this performative state, allowing subsequent interactions to occur through this embodied construct.

5.3.3.3   Monsters in Mirrors and Shadow Men on the Telephone (Caliban Below (2018) and Awake VR (2018))

In this section I will explore the overarching dramatic approach to character embodiment, "tabula rasa" and "predefined" as raised by writer Mike Jones, through analysis of encounter techniques in *Awake VR* (2018) and *Caliban Below* (2018). In *Awake VR* (2018) I extend on the discussion previously introduced to highlight how a moment of focused sensory experience can lead, through juxtaposition, to an impactful experience of embodied revelation. In the case of *Caliban Below* (2018) I examine how a "predefined" approach shifts with the process of progressive character revelation, and how a surprise twist leads to deeper affective experiences. I should note that the proper analysis of techniques has necessitated spoilers. Both works are available online to view (listed at the end of the chapter) ahead of this textual analysis.

The telephone call in *Awake VR* (2018) is an example of how a moment of focused sensory experience leads to a shift in my sense-of-self and my relationship to the environment around me. I begin through the experience of "tabula rasa" where I discover Harry at home alone, teetering on madness, wheelchair bound and grieving for his lost wife, Rose. This first scene is mostly carried by Harry's monologue and some minimal interactions with the environment. I can "blink" to a new location by pointing my handheld controllers at a globe, or by pointing at a "ticker" to stop it, thus waking Harry. However, my engagement is largely upheld by my intrigue with Harry and his predicament as he launches into a dialogue about the disappearance of his wife and the mystery surrounding his experience of lucid dream states. The below passage details my embodied experience of key scenes in the work, which I downloaded from the Steam Store (see page 48, list of works) and viewed on an Oculus Rift (Figs. 5.7 and 5.8).

Awake VR (2018)

As Harry returns to his lucid dream state to replay what happened to Rose, I witness Harry giving Rose a necklace. The scene switches and I find myself standing in a living room,

---

[15] Examples include repeatedly turning the silver wheels to light the passageway in *Caliban Below* (2018), wielding a light saber in a game environment and later putting my skills to the test in a real wolf battle at the conclusion of *Wolves in the Walls* (2018).

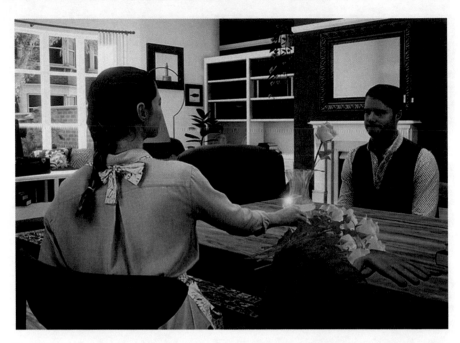

Fig. 5.7 Participant's view inside the immersive HMD witnessing Harry give a necklace to Rose. © New Canvas

alone. A phone rings. I approach and pick it up with the handheld controllers. "You're a hard man to reach, Harry'" says the "Shadow Man" on the telephone.

Through the encounter with the Shadow Man on the telephone my sense of embodied presence shifts, recalling the gun swinging on me in the pivotal moment in *Carne y Arena* (2016), where I become implicated in the action. In this scene in *Awake VR* (2018) I cross the room to go to the telephone and pick it up, all through my intuitive reaction to the drama, to create a spontaneous transition from a viewer to embodying Harry. Thus, by choreographing this encounter through sensory and co-creational modes of engagement, what Jones calls "decisions you force the character to make," [68] at a moment of information revelation, I am prompted into a state of embodied role-play. This example again highlights Laurel's point about the value of constraints in co-creational media for the way they focus efforts and increase imaginative potential by reducing possibilities. She elaborates [26, p. 117].

> But we mustn't fall prey to the notion that more is always better, or that our task is the seemingly impossible one of emulating the sensory and experiential bandwidth of the real world. Artistic selectivity is the countervailing force – capturing what is essential in the most effective and economic way.

Just as Laurel outlines, by focusing my sensory experience around the act of picking up a telephone, rather than a range of actions, the embodied impact increases. Furthermore, by structuring the sensory actions through an encounter, I am nudged

Fig. 5.8 Participant's view inside the immersive HMD embodying the perspective of Harry. © New Canvas

into a role. Elizabeth Hess in her embodied character work, sheds light on how this process occurs, highlighting how, through interpersonal interactions, she comes to a deeper experience of character. She explains, "As a spontaneous narrative unfolds, the actor organically discovers an archetype through a pattern of sustained interactions with others" (2017, p. 70). The process of character identification that happens through encounters with others that Hess details is certainly true for me in the scene in *Awake VR* (2017). When picking up the telephone as a result of the Shadow Man's actions, his words crystalise the revelation that in this scene, I am Harry.

The Shadow Man on the telephone draws together the three states of being which Hess furthermore outlines are key to embodied character experience: behavioural, physiological and psychological. In this scene I express instinct (behavioural) through the decision that, when the telephone rings, I must answer it. I cross the room by walking and reaching out to pick up the telephone, which fulfills the physiological state. Lastly, through my mental assembly when the Shadow Man refers to me as Harry, I have employed the psychological state. Like in embodied acting, the emphasis in VR narrative design is drawn to these states as the way to access the imaginative and affective centre of the character. Beginning with the experience of "tabula rasa," the Shadow Man on the telephone highlights how character embodiment as a co-creational experience arises through white space, juxtaposition, and the careful structuring decisions of creators.

5.3.3.4   Caliban in the Mirror—*Caliban Below* (2018)

Caliban Below (2018) uses similar techniques as *Awake VR* (2018), however instead of developing a sense of character from "tabula rasa", the character is "predefined", and encounters (See Fig. 5.9) are used to progressively develop the character before subverting expectations at the end (See Fig. 5.10). I should note that while the descriptive passage details the concluding section of the experience, in the beginning, prior to my journey down into the catacombs I am told by a voice-over narrator that I am playing Caliban, the scion of a broken lineage and the mystery lying hidden beneath the ruins of a decrepit Italian abbey. The subsequent passage details my experience of the final climactic scene in *Caliban Below* (2018). I downloaded the work from the Steam Store (see page 48, list of works) and experienced it on the HTC Vive.

Caliban Below (2018)

I ascend some stairs and explore the alcove, ignoring a door to the outside to check there isn't anything else in here to discover. As I return from one side, I sense something moving in the corner of my eye, a dark shadowy outline as I pass the mirrored door. I stop and approach. As I walk towards my reflection I pass into the light. A hideous creature with sharp curving teeth, sunken grey skin and long sinewy arms stares back at me. As I move back and forth transfixed by my reflection, his body moves in perfect sync as I do, a visceral and embodied trick which makes me feel that the horrible reflection is who I am, even though I know it is just a computer-generated illusion. A voiceover commences of a shocked Caliban as he finally remembers he has been stuck down here, as if in a dream, as this hideous creature. He was once just a boy but no longer. The image fades out and the credits role.

There are two dimensions to the choreography of this surprise encounter with Caliban in the mirror that shed light on its affective possibilities. Firstly, the careful structuring of embodied actions and encounters through which progressive information is communicated prior to this moment is important. Prior to this event, I have lit the passageway by turning silver wheels, and encountered the ghost of a young woman, a forgotten love interest of Caliban's who I discover (through Caliban's voice over) has died down here by some fault of his own though he can't remember how. I have also come upon the ghost of Avando (See Fig. 5.9) who comments how Caliban has become lost down in the catacombs. Memory and the passing of time are emphasised while the environment and story is steeped in the haze of mystery. When I get to this final surprise event, of revealing Caliban's true self "no longer a boy," the previous encounters suddenly make sense. In these techniques, Conelly translates storytelling approaches to character development, into immersive space. As author and story consultant Robert McKee argues ([32], p. 107),

The function of structure is to provide progressively building pressures that force characters into more and more difficult dilemmas where they must make more and more difficult risk-taking choices and actions, gradually revealing their true natures, even down to the unconscious self.

Thus, McKee highlights the way effective story structuring leads to character revelation. In *Caliban Below* (2018) the question of Caliban's character lies in its representational absence, a forgotten self, waiting to be discovered. McKee elaborates

Fig. 5.9 Participant's view inside the immersive HMD embodying the perspective of Caliban during an encounter with the ghost of Avando

on the affective revelation that this technique creates. "The storyteller leads us into expectation, makes us think we understand, then cracks open reality, creating surprise and curiosity, sending us back through the story, again and again" ([32], p. 237). The revelation works here as a result of the interactive and dramatic structuring that led to this event. I realise that the silver wheels I turned to light the passageway were done so not as a visitor, but as a creature who has long become a part of this subterranean world, lost to his memories. The revelation gives further meaning to the embodied actions that went before, and therefore to the affective resonance of embodying Caliban.

The second element impacting embodied character presence is the co-creational nature of the revelation. As the narrator reveals who I am, I as co-creator make this so. I move back and forth in front of the mirror, fascinated by the vision in front of me, in turn projecting onto the character my own movement, feelings and expressions. The visual representation here is not a constant but takes on meaning through juxtaposition between absence and representational presence. This experience of looking upon a virtual mirror and seeing the reflection, synchronised with my embodied actions, is entirely new, filling me with awe due to a "visceral sense of personal involvement" as Rheingold explains, where I feel the experience not just see it ([40], p. 55). Thus, the affective dimensions of the character are made more complex through the visceral experience of embodied actions, the revelation which

Fig. 5.10 Participant's view inside the immersive HMD embodying the perspective of Caliban during an encounter with Caliban in a mirror

happens through personal discovery, and the awe induced through the novelty of the embodied surprise.

Caliban Below (2018) is an example of how, beginning with a "predefined" set-up, creators can evoke a deeper experience of character embodiment by taking advantage of representational absence, or by using juxtapositions of presence and absence to create affective impact. Crucially, structuring techniques which evoke a sense of character through embodied interaction and encounters which communicate information at key dramatic moments lead up to these climactic moments and shed light on character when understood in retrospect.

5.3.3.5 Watch Out for Those 'OilyMen'—Embodied Suspense in *Inanimate Alice: Perpetual Nomads* (2018)

I conclude this section by exploring a final work from the framework: *Inanimate Alice: Perpetual Nomads* (2018). Of particular interest is how the embodied experience of the character, Alice, is impacted through embodied suspense. Encounters are central here but as a work of e-literature the creator, XR artist Mez Breeze, takes a different approach, through text-based messaging and what she calls, in an interview with me, a form of "co-opted agency" (2018) to generate a sense of anxiety which is reinforced through sound design and interactive mechanics (See Figs. 5.11 and 5.13).

As the first VR instalment of the *Animate Alice* e-lit educational series for children and young adults, the work needed to provide continuity in terms of narrative arc and characterisation, while also creating, as creator Mez Breeze describes it, a "seamless progression via motion mechanics specific to VR" (2018). The experience follows the story of teenage Alice who, after her bus breaks down, needs to find a charger for her phone. In the following passage, I detail my experience of *Inanimate Alice: Perpetual Nomads* (2018) as it approaches a climactic moment towards the end. I downloaded the work from the website (see page 48, list of works) and experienced it on an Oculus Rift.

Inanimate Alice: Perpetual Nomads (2018)

I receive another alert from "Whispurring Nomads," a group messaging service on my phone. The participants I have been chatting to, "PLAYA" and "Unicrony," tell me that the 'OilyMen' are coming for me (Alice doesn't want to believe them but doubt creeps in). As I continue to look for a charging station inside the Infodome, the music tempo increases. I have a creeping anxiety that maybe these 'OilyMen' they've been talking about really are coming to get me. In response to the increasing tempo of music I seem to move more quickly, even though I am not. I finally find a charging station, and exhale in relief, but then everything shuts down. The music reaches a crescendo. It's a make-believe world, a young adult experience, but even so I found myself on edge, my heart rate increasing in time with the frenzied music.

Echoing the theme of juxtapositions as raised through analysis of *Awake VR* (2018) and *Caliban Below* (2018), this work highlights how encounters with *PLAYA*

Fig. 5.11 Participant's perspective embodying Alice during an encounter with Playa using the Whispurring Nomads Chat application

and *Unicrony*, and their affective impact in generating suspense, are framed through contrast between stasis and embodied movement all experienced through a first-person experience of character (See Figs. 5.12 and 5.13). In an interview with me (2018), creator Mez Breeze explained the reason for this dynamic:

> Having the player progress through the work in the first-person role of Alice seemed intuitive. The live editing in the chat app scenes allowed the player sustained immersion levels through a type of co-opted agency, similar to how a reader or viewer is guided to identify with main protagonists when reading a novel or watching a movie, but with the added aspect of an overlaying of twinned embodied (as opposed to disembodied) immersion and enhanced engagement.

Breeze therefore highlights how the messaging exchange generates immersion through a sense of "co-opted agency" (2018) and embodiment of Alice. Co-opted agency, where the participant expresses agency but within guardrails, is of course, following Brenda Laurel [26, p. 101], a common feature of interactive works to generate enhanced engagement. It is important to note that the embodied experience is sustained in these text exchanges because of the ability to carry out embodied actions earlier, such as picking up objects by staring at them and teleporting through the "footsteps" icon, which establishes a kinaesthetic dynamic when navigating through the environment.

Inhabiting Alice's perspective, I load the *Whispurring Nomads* app when prompted, then witness her crafting messages to PLAYA and Unicrony, before being

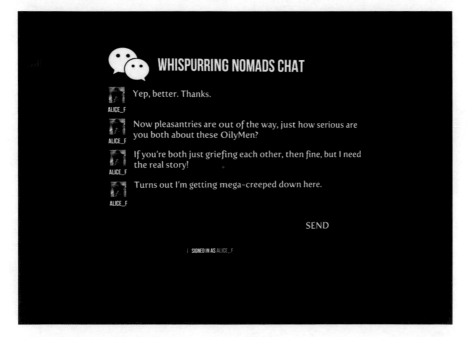

Fig. 5.12 Participant's perspective embodying Alice while using the Whispurring Nomads. Alice's rising anxiety about the OilyMen is conveyed in the tone of the text

Fig. 5.13 Participant's perspective moving through the Infodome to find a phone charger. Alice's increasing anxiety about the OilyMen is conveyed in the tone of the text and the increased music tempo

prompted to choose whether to send them. While I don't have total agency, these decision moments increase my investment in the story and embodied presence as a co-creator (See Figs. 5.11, 5.12 and 5.13). Despite there being no co-presence experience in the sense of inhabiting close spatial proximity with these characters, the *Whispurring Nomads* chat exchange works on a dramatic and embodied level in a similar way to encounters through structuring techniques, though in this case emphasising suspense. As Breeze explained in our interview, she wanted to create a "build-up of tension" and "crescendo-ish progression," the sense of being "bombarded" to the final climax scene (2018). The chat exchange is crucial to this suspense by the way it introduces a potential threat (when the participants claim the 'OilyMen' are coming to get Alice), and conveys Alice's increasing anxiety through the way she responds in her messages back to PLAYA and Unicrony. While these interactions are differentiated by stasis, the search for the charging station through the "footprints" teleportation and increasingly feverish music, increases the tension through deferring the attainment of the goal of reaching the charging station. As Breeze explained, "high tempo" music that climaxed with "an electronic scream sound" was crucial to this "swelling approach" (2018). The feeling through teleportation, juxtaposed with the rising tension through the text exchanges, increases my visceral experience of suspense.

In *The Cambridge Companion to Narrative Theory* (2018) Matthew Garrett, author and Associate Professor of English at Wesleyan University, explores the importance of such delays and juxtapositions in narrative structuring (2018, 26):

> … the whole gamut of dilations, digressions, and shortcuts that literary and cinematic narrators employ to represent story events – all the means of "generalized distortion" by which a narrative discourse, in its "pseudo-time," fails straightforwardly to repeat the linear chronology of a story are the very essence of narrative.

Thus, Garrett highlights how playing with time, and holding off information, is a pivotal technique in narrative structuring [19]. In the case of *Inanimate Alice: Perpetual Nomads* the background music, increasing in tempo, the feeling of hurrying through the Infodome (by gazing towards the footprints and blinking to that location) reinforces for me the feeling that time is running out. Changes in rhythm, such as the text-based messages, and sections of movement through the Infodome, become a key strategy to defer information, which in turn generates suspense. These techniques all act on my embodied experience of Alice through a visceral, kinaesthetic experience which juxtaposes encounters with PLAYA and Unichrony with my own movements through the space. It also highlights the importance of sound design not just in generating engagement but also its impact on embodied presence.

The analysis of *Broken Night* (2017), *Awake VR* (2018), *Caliban Below* (2018) and *Inanimate Alice: Perpetual Nomads* (2018) highlight a range of co-creational structuring techniques which can act as a precursor for the experience of character embodiment and its potential affective dimensions. The direct address, as the interrogation room experience in *Broken Night* (2017) highlights, prompts me into a performative state due to the intensity of co-presence, as well as tracking capabilities in the display, where I can move my head as if I am the woman. This technique also highlights how minimal sensory elements can have a notable impact on the experience of presence, focusing the participant's imaginative faculties. *Awake VR* (2018) and *Caliban Below* (2018) highlight how, progressing from "tabula rasa" or "predefined" approaches to character, the participant gains insight into the character through embodied actions and the mental assembly that occurs through encounter situations with other virtual agents. The sensory experience is focused, in turn increasing the embodied experience through the impact that constraints have on my capacity to imagine. Affective impact happens through the juxtapositions of representational absence and, presence, as well as revelations that occur through my co-creational choices. In moments such as picking up the telephone or moving back and forth in front of the mirror at a key revelation and twist in *Caliban Below* (2018) the complex affective dimensions of the experience of character emerge, felt viscerally, kinesthetically and through the intensity of the performative dimensions of simulated environments. Lastly, in the case of *Inanimate Alice: Perpetual Nomad* (2018) the work highlights how sound design and juxtaposition between exchanges with other characters and sections of embodied momentum through the simulation, act on the affective experience of character by generating suspense. Importantly, these techniques again highlight the centrality of encounters in virtual space, of the way these co-presence situations prompt me into an embodied response, in generating

affective outcomes. They also highlight how representational absence is not necessarily a hindrance but instead an affective design approach, where character secrets
lie waiting to be discovered through participant involvement.

The embodied states framework draws attention to the centrality of encounters in
elucidating the co-creational and therefore affective possibilities of virtual environments. Through the three design approaches outlined on page 11, framing encounters
through dialogic possibilities, structuring co-creation through ludic and performative
techniques and generating character embodiment through co-presence situations, I
have sought to identify the key structuring techniques which generate such possibilities. In this analysis, the value of representational absence, juxtapositions and the
in-between in affective design solutions is highlighted. I examine the implications of
these insights in how we may conceptualise presence, immersion and embodiment
in comparison to perspectives outlined in the introduction, before considering the
implications of these findings in the evolving field and future of VR design.

5.3.4 VR Narrative Design Contributions to Conceptualisations of Presence, Immersion and Embodiment

While VR technologies have been the subject of decades of research, the recent
proliferation of more affordable, publicly accessible headsets, have enabled specific
enquiry across a broader scope of applications. VR narrative design practices
contribute to this growing body of knowledge of what defines the experiential dimensions of virtual environments. With design considerations prioritising affect, conceptions of presence, immersion and embodiment are in turn impacted. The goal of
full-immersion, common to earlier literature and still a major driver of innovation,
especially in gaming, becomes complicated. To feel present in a VR narrative is not
just the result of sensory immersion, but also the experience of dramatic action, a state
of willing suspension of disbelief, and, as writer Mike Jones eloquently describes it,
an act of "forgetting" [68]. It may rely on framing strategies that begin long before
a headset is put on, through environmental design such as a swinging chair, that
draws my attention to my body as the focal site of meaning. Such strategies highlight
how presence isn't a *moving into* a digital space but constantly *moving between*,
a sensation of being in two places at once, or a shifting from absence into presence, to achieve a desired effect at a particular dramatic juncture. Through absence,
such as the lack of body representation, VR narrative design practices in turn create
possibilities. In the experience of "tabula rasa" I have the opportunity, through carefully structured encounters and co-creational engagements, to create the character
through my own embodied actions and decisions, to experience, therefore, the affective resonance of self and otherness. Crucially, by focusing the sensory experience
during key scenes and narrative moments, such co-creation is enabled by allowing my
imagination to be focused, as XR artist Mez Breeze details, through the experience

of co-opted agency. Embodiment therefore becomes a co-creational act, generated through a collaboration between the participant and the simulated world, in dramatic possibilities.

5.4   Conclusion: The Future of VR Design

5.4.1   The Future of VR Design

As the VR industry evolves, encounters in virtual space, such as between participants, participants and virtual agents, and participants and responsive environments, will be a central dimension of VR design, not just in entertainment contexts but across a variety of application where affect is integral to design motivations.

One example is in healthcare. In the project *Wonder VR: Interactive Storytelling through VR 360 Video with NHS Patients Living with Dementia (2020)*, researchers used immersive technology to treat dementia patients during long hospital stays, and found that responses such as awe, nostalgia and joy were key to patients' improvements on subjective wellbeing (2020, 474). The researchers developed narratives based on locations with special significance to the patients, such as a walk-through Epping Forest, a beach-based experience, and a 360-degree film of another patient's favourite animals. Patients experienced a set of timed interactions and movements, conducted with a facilitator. Interactions including inviting the patients to reach out and feel leaves from the woodland or placing feet in a tray of sand to reinforce, through sensory dimensions, the beach scene in front of them. Researchers noted how these real-world tactile interactions, echoing the sensory experiences and dialogic set-up (on a much more basic level) used in works like *Carne y Arena* (2017), increased the sense of immersion that the patient felt while visiting these simulated locations. This highlights how structuring techniques identified in VR narrative design, in particular the use of sensory elements to enhance immersion, also apply in a therapeutic context. For the researchers, there was a connection between positive outcomes, such as reduced agitation and restlessness of patients (2020, 481) and the importance of multisensory stimuli. Immersion, they reflected, by providing an escape from the environment, in turn improved wellbeing (2020, 481).

Encounters act as a co-creational strategy to create the potential for such outcomes. The experience of the forest, and the engagement with the animals, was made interactive through the sensory experience of touching leaves, or, in the case of the woman with the affinity with animals, in describing what animals she saw to a nurse who was present with her. This highlights how encounter choreography as identified in VR narrative design can be applied in other settings, even if in a more simplistic way, with implications for affective possibilities. Projects like *Wonder VR* (2020) represent just the inception of what might be possible for future health applications of VR, such as the scenario depicted in the Black Mirror episode *San Junipero* (2016).[16]

The study outlined in *Building long-term empathy: A large-scale comparison of traditional and virtual reality perspective-taking* [22] is another example of how principles of encounter design illuminated through the embodied states framework may apply in tangential spaces, in this case in potentially influencing pro-social behaviour with advocacy applications (2018, p. 28). In the study researchers sought to ascertain the effectiveness of a perspective-taking experience of becoming homeless in a virtual reality environment versus a traditional perspective-taking exercise, as well as against other less immersive forms of media. The results (2018, p. 1) found that,

... participants who became homeless in VR had more positive, longer-lasting attitudes toward the homeless and signed a petition supporting the homeless at a significantly higher rate than participants who performed a traditional perspective-taking task.

The study also found (2018, p. 1) that,

... a significantly higher number of participants in the VR condition signed a petition supporting affordable housing for the homeless compared to the traditional and less immersive conditions.

The researchers thus highlighted a strong link between embodied perspective taking, empathy and even pro-social behaviour. They also noted that those who experienced the virtual reality perspective taking exercise felt more connected and empathetic toward the homeless than the less immersive versions (2018, 1). The narrative VR experience was adapted from a documentary titled *Hotel 22*, and interviews with people who have experience being homeless, and follow a series of events that depict the real-life experiences of people who become homeless due to such circumstances such as medical bills, domestic violence or drug addiction (2018, 7). In the experience, participants move through a series of scenes with interactive tasks and choices. It begins in an apartment where the participant has lost their job and must sell items to try to make the rent. While living in their car, they are approached by a police officer, leading to the vehicle becoming impounded. While on the bus, they are approached by two men who threaten them. On the bus the participant also meets with other homeless people through which they learn about their stories (2018,7). The researchers noted that the higher levels of interactivity and immersion experienced in VR led to more empathy and positive attitudes (2018, 4). While they did not specifically test impact of interactions and responsiveness of the experience on presence and empathy, they speculated a strong correlation between enhancements in these features and pro-social behaviour (2018, 32). These results echo the insights raised in the embodied states framework. Given that encounters formed a key part of interactive structuring in the scenes in the *Becoming Homeless* simulation, we

---

[16] In the episode, a fictional holiday town set in the 1980s is revealed to be an "immersive nostalgia therapy", allowing patients, in this case both terminally ill, to access a world of memories for therapeutic reasons.

can also hypothesise the important role of encounters as a co-creational strategy with affective impact, such as the feeling of dehumanisation (one of the affective measures of the study) which may have occurred during encounters with the police and men on the bus, as well as the feeling *that* being present with other homeless people had on the experience of empathy.

These two examples highlight how encounter design, emerging from the principles of conflict driven storytelling and translated into immersive space, may be employed as co-creational strategies in a variety of contexts, especially where emotional experiences such as awe, connection or empathy are integral to design motivations.

These developments point to a future where entire sections of hospitals and floors of universities may become spaces dedicated to immersive experiences—where immersive media designers fulfill important roles, even as changemakers, such as has been exemplified through the immersive work of Australian filmmaker Lynette Wallworth [54]. The findings from this study in turn highlight how encounters in virtual space will be central to design techniques which bring this future world into being.

5.4.2 Conclusion

The choreography of encounters, as highlighted through design approaches that prioritise dialogic framing, focused sensory engagement, and co-creational engagement strategies, including ludic and performative techniques, as a way to generate affective possibilities, may have applications in a range of design contexts. Such design techniques may be increasingly relevant as VR and other immersive technologies become more integral, through the development of the Metaverse, to how we work, communicate and spend our leisure time. Rather than reinforcing the desire and quest for full immersion, however, VR narrative design approaches instead invite designers to embrace the in-between: the absences, white spaces and juxtapositions. By revealing the centrality of encounters in affective design, the insights raised from this emerging artform instead focuses on framing, through the alchemy of collaboration in cyberspace, for new possibilities.

In memory of Michael (Mike) Conelly and his pioneering work in VR

1970—2022

References

1. Abraham, N.: Wonder VR: interactive storytelling through VR 360 video with NHS patients living with Dementia. Contemp. Theatr. Rev. **30**, 474–489 (2020). https://doi.org/10.1080/104 86801.2020.1812591

2. Adams, W.: Conducting semi-structured interviews. In: Newcomer, K., Hatry, H., Wholey, J. (eds.) Handbook of Practical Program Evaluation, 4th edn, pp. 492–505. Jossey-Bass, a Wiley imprint, San Francisco, California (2015)
3. Ahmadinejad, T.: Tender Claws: The Under Presents. https://tenderclaws.com/theunderpres ents (2019)
4. Andreadis, A., Hemery, A., Antonakakis, A., Gourdoglou, G., Mauridis, P., Christopoulos, D., Karigiannis, J.N.: Real-time motion capture technology on a live theatrical performance with computer generated scenery. In: 2010 14th Panhellenic Conference on Informatics, pp. 148–152 (2010)
5. Banacou, D., Hanumanthu, P., Slater, M.: Virtual embodiment of white people in a black virtual body leads to a sustained reduction in their implicit racial bias. Front Hum Neurosci (2016). https://doi.org/10.3389/fnhum.2016.00601
6. Bearman, M.: Eliciting rich data: a practical approach to writing semi-structured interview schedules. Focus Health Profess Educ A Multi-Profess J **20**(3), 1–11 (2019). https://doi.org/10.3316/informit.002757698372666
7. Bell, P.: Dialogic media productions and inter-media exchange. In: Kan, L. (eds.) J. Dram. Theory Crit. **14**(2), 41–56 (2000)
8. Billington, P.: Behind the Scenes: How We Made 'Wolves in the Walls' VR with Third Rail Projects. https://noproscenium.com/behind-the-scenes-how-we-made-wolves-in-the-walls-vr-with-third-rail-projects-dd50eeb14bf6 (2019)
9. Biocca, F.: The Cyborg's dilemma: progressive embodiment in virtual environments. J. Comput. Mediat. Commun. **3**(2) (1997). https://doi.org/10.1111/j.1083-6101.1997.tb00070.x
10. Broadhurst, S., Machon, J.: Identity, performance and technology: practices of empowerment, embodiment and technicity. Palgrave Macmillan, Palgrave Macmillan UK, Basingstoke, Hampshire, New York, London (2012)
11. Buchanan, I.: Assemblage theory and its discontents. Deleuze Stud. **9**(3), 382–392 (2015). https://doi.org/10.3366/dls.2015.0193
12. Calleja, G.: In-Game: From Immersion to Incorporation. MIT Press, Cambridge, MA, USA (2011)
13. Capps, K.: Alejandro Iñárritu's VR Film 'Carne y Arena' Is Pretty Real. Bloomberg. https://www.bloomberg.com/news/articles/2018-06-07/alejandro-i-rritu-s-vr-film-carne-y-arena-is-pretty-real (2018)
14. Cox, S.: AI Brings Peaky Blinders Characters to Life. Goldsmiths, University of London. https://www.gold.ac.uk/news/peaky-blinders-vr/ (2019)
15. Davies, C.: Virtual space. In: Penz, F., Radick, G., Howell, R.: Space: In Science, Art and Society, pp. 69–104. Cambridge University Press, Cambridge, England (2004)
16. Davies, C.: Ephémère: landscape, earth, body, and time in immersive virtual space. In: Reframing Consciousness. Intellect Books, Portland, OR (1999)
17. Deleuze, G., Guattari, F., Tomlinson, H., Burchell, G., Burchell, G.: What Is Philosophy? Columbia University Press, New York, United States (1991)
18. Flavian, C., Ibanez-Sanchez, S., Orus, C.: Impacts of technological embodiment through virtual reality on potential guests' emotions and engagement. J. Hosp. Market. Manag. **30**, 1–20 (2020). https://doi.org/10.1080/19368623.2020.1770146
19. Garrett, M. (ed.): The Cambridge Companion to Narrative Theory. Cambridge University Press, Cambridge (2018)
20. Gorisse, G., Christmann, O., Amato, E.A., Richir, S.: First- and third-person perspectives in immersive virtual environments: presence and performance analysis of embodied participants. Front. Robot. AI **4** (2017)
21. Harris, O.: Black Mirror: San Junipero. Season 3. Episode 4. Netflix. https://www.netflix.com/au/title/70264888 (2016)
22. Herrera, F., Bailenson, J., Weisz, E., Ogle, E., Zaki, J.: Building long-term empathy: a large-scale comparison of traditional and virtual reality perspective-taking. PLoS One **13**, e0204494 (2018). https://doi.org/10.1371/journal.pone.0204494
23. Hess, E.: (2016) Acting and Being. Palgrave, Macmillan, UK

24. Hickey-Moody, A.: Affect as method: feelings, aesthetics and affective pedagogy. In: Coleman, R., Ringrose, J. (eds.) Deleuze and Research Methodologies. Edinburgh University Press, Edinburgh, United Kingdom (2013)
25. Höök, K.: Designing with the Body: Somaesthetic Interaction Design. The MIT Press, Cambridge, Massachusetts (2018)
26. Laurel, B.: (1991) Computers as Theatre. Addison-Wesley Pub, Reading, Mass.
27. Law, J.: After Method: Mess in Social Science Research. Routledge, New York, NY (2004)
28. Marcolin, F., Wally Scurati, G., Ulrich, L., Nonis, F., Vezzetti, E., Dozio, N., Ferrise, F., Stork, A., Dasule, R.C.: Affective Virtual Reality. How to Design Artificial Experiences Impacting Human Emotions. IEEE Comput. Graph. Appl. **41**, 171–178 (2021). https://doi.org/10.1109/MCG.2021.3115015
29. Marshall, E.: Spinoza's cognitive affects and their feel. Br. J. Hist. Philos. 16, 1–23 (2008). https://doi.org/10.1080/09608780701789251
30. Martins, T.: Coronavirus effect: Actors at home inhabit live VR theater, Los Angeles Times. https://www.latimes.com/entertainment-arts/story/2020-04-26/coronavirus-vr-virtual-reality-theater-tender-claws-live-actors (2021)
31. Massumi, B.: Parables for the Virtual: Movement, Affect, Sensation. Duke University Press, Durham, NC (2002)
32. McKee, R.: Story: Substance, Structure, Style, and the Principles of Screenwriting. Methuen, London (1997)
33. Milk, C.: How Virtual Reality Can Create the Ultimate Empathy Machine. TED. https://www.ted.com/talks/chris_milk_how_virtual_reality_can_create_the_ultimate_empathy_machine?language=en (2015)
34. Nash, K.: Virtual reality witness: exploring the ethics of mediated presence. Stud. Doc. Film **12**, 119–131 (2018). https://doi.org/10.1080/17503280.2017.1340796
35. Obrist, H.: Filmmaker Alejandro G. Iñárritu on Breaking the Fourth Wall. Another Magazine. https://www.anothermag.com/art-photography/10124/filmmaker-alejandro-g-inarritu-on-bre aking-the-fourth-wall (2017)
36. Oculus, V.R.: VR Visionaries: Fable Studio. https://www.oculus.com/blog/vr-visionaries-fable-studio/ (2018)
37. Pavic, K., Vergilino-Perez, D., Gricourt, T., Chaby, L.: (2022) Because I'm Happy—An Overview on Fostering Positive Emotions Through Virtual Reality. Front. Virtual Reality **3**
38. Penny, S.: Making Sense: Cognition, Computing, Art, and Embodiment. MIT Press, Cambridge, MA, USA (2017)
39. Popat, S.: Missing in action: embodied experience and virtual reality. Theatr. J. **68**, 357–378 (2016). https://doi.org/10.1353/tj.2016.0071
40. Rheingold, H.: Virtual Reality. Simon & Schuster, New York, Sydney (1991)
41. Riva, G., Mantovani, F., Capideville, C., Preziosa, A., Morganti, F., Villani, D., Gaggioli, A., Botella, C., Alcañiz Raya, M.: Affective interactions using virtual reality: the link between presence and emotions. Cycberpsychol. Behav. **10**, 45–56 (2007). https://doi.org/10.1089/cpb.2006.9993
42. Rosenberg, L.: There are Two Kinds of Metaverse. Only One will Inherit the Earth. https://big think.com/the-future/metaverse-augmented-virtual-reality/ (2022)
43. Ross, M.: Virtual reality's new synesthetic possibilities. Televis. New Media **21**, 297–314 (2020)
44. Sanchez-Vives, M.V., Slater, M.: From presence to consciousness through virtual reality. Nat. Rev. Neurosci. **6**, 332–339 (2005). https://doi.org/10.1038/nrn1651
45. Schwartz, R., Steptoe, W.: (2018) The immersive VR self: performance, embodiment and presence in immersive virtual reality environments. In: Papacharissi, Z. (ed.) A Networked Self and Human Augmentics, Artificial Intelligence, Sentience. Routledge
46. Shouse, E.: Feeling, emotion, affect. M/C J. **8**(6) (2005). https://doi.org/10.5204/mcj.2443
47. Shusterman, R.: Body Consciousness: A Philosophy of Mindfulness and Somaesthetics. Cambridge University Press, Cambridge (2008)
48. Slater, M.: A note on presence terminology. Presence Connect **3** (2003)

49. Slater, M.: Place illusion and plausibility can lead to realistic behaviour in immersive virtual environments. Philos. Trans. R. Soc. B Biol. Sci. **364**(3549) (2009). https://doi.org/10.1098/rstb.2009.0138
50. Slater, M.: The Sense of Embodiment in Virtual Reality. PRESENCE: Virtual and Augmented Reality. MIT Press (2012)
51. Slater, M., Spanlang, B., Sanchez-Vives, M.V., Blanke, O.: First person experience of body transfer in virtual reality. PLoS One **5**, e10564 (2010). https://doi.org/10.1371/journal.pone.0010564
52. Third Rail Projects: Then She Fell. https://thenshefell.com (2012)
53. Tapley, K. Alejandro G. Iñárritu on 'Carne y Arena' and the Academy's Movie Museum. Variety, November 14. https://variety.com/2017/film/in-contention/alejandro-inarritu-carne-y-arena-academy-museum-interview-1202610586/ (2017)
54. Wallworth, L.: Shamanic Visions in VR, ACMI Melbourne Presentation. https://lynettewallworth.com/lectures (2019)
55. Walser, R.: Elements of a cyberspace playhouse. In: Proceedings of National Computer Graphics Association '90, pp. 19–22 (1990)
56. Zuckerberg, M.: Founder's Letter, 2021. https://about.fb.com/news/2021/10/founders-letter/ (2021)

The Embodied States Framework: List of Works

57. Benari, A., Zubalsky, T.: Broken Night. Eko, NYC, U.S.A. (2017)
58. Billington, P.: Wolves in the Walls, Fable Studio, London, U.K. https://www.oculus.com/experiences/rift/2272579216119318/ (2018). Accessed 18/03/22
59. Conelly, M.: The Abbot's Book. Blackthorn Media, L.A., U.S.A. https://store.steampowered.com/app/434430/Abbots_Book_Demo/ (2016). Accessed 18/03/22
60. Conelly, M.: Caliban Below. Blackthorn Media, L.A., U.S.A. https://store.steampowered.com/app/649890/Caliban_Below/ (2018). Accessed 06/04/19
61. Iñárritu, A.: Carne y Arena. Los Angeles County Museum of Art. Touring website: https://carne-y-arena.com/ (2017). Accessed 18/03/22
62. McNitt, E.: Spheres. Oculus Facebook Technologies, LLC. https://www.oclus.com/experiences/rift/1859625197439973/?locale=en_US (2018). Accessed 06/04/19
63. Mez Breeze: Design and Bradfield Narrative Designs. Inanimate Alice: Perpetual No Mads. https://mezbreeze.itch.io/perpetual-nomads (2018). Accessed 06/04/19
64. Start, V.R.: Awake VR. https://store.steampoered.com/app/845900/Awake_Episode_One/ (2018). Accessed 01/10/19
65. Wallworth, L.: Awavena. Carriageworks, Sydney. https://www.viveport.com/6792ef3d-0775-4ab4-b3d3-3d9c15b64d47 (2018). Accessed 18/03/22

Interviews

66. Benari, A.: Interview with Alon Benari, Director, Eko. https://eko.com/ (2017). SAM, UNSW
67. Breeze, M.: Interview with Mez Breeze, Founder of Mez Breeze Design. http://mezbreezedesign.com (2018). SAM, UNSW
68. Jones, M.: Interview with Mike Jones, Writer, Script Producer and Story Editor. http://mikejones.tv (2017). SAM, UNSW

Chapter 6
Creating Digitally: Computer Films/ Screenlife/Zoom

Gregory Dolgopolov

Abstract What the pandemic taught us is that an audience will spend more time in front of a range of screens for their work and entertainment. In order to engage with these experiences, filmmakers need to employ a new language of representation. Screenlife and Zoom films are two intersecting approaches to remote screen-based production that have proven successful with audiences. Both Screenlife and Zoom films employ a new language and format of visual content production where everything that the viewer sees happens on a computer, tablet or smartphone screen. The computer screen is now the camera lens and the mise en scene, the apps are the locations, and the desktop is the set and the site of exhibition. Pre-production becomes an integral part of post-production. The computer is simultaneously the screen, the camera, the narrator and competing voices in various constantly moving and layering windows. We get to know so much about the protagonist—whether it is by how they use their cursor or how they interact with others online. These films are potentially cheaper to produce than live action films and allow productions to engage big name stars and save on locations and often on crews—as during lockdown many actors became their own camera operators and lighting crew. This chapter examines computer screen films *Searching* (2018), *#SidIAdoma* (2020) and *Host* (2020) to consider the nuanced difference between Screenlife and Zoom films.

Keywords Screenlife · Zoom · Timur Bekmambetov · Searching · #SidIAdoma · Host · Profile · Pandemic

6.1 Introduction

Screenlife works because, frankly, we deal with computer screens all the damn time. We instinctively understand the meaning of everything on screen, and not just in terms of icons and terminology. A computer screen represents a character's subjective point of view, as well as their inner thoughts. The manner in which people move a mouse, or type, or arrange

G. Dolgopolov (✉)
University of New South Wales, Sydney, Australia
e-mail: gregd@unsw.edu.au

163

their desktop, or use apps, speaks volumes to a person's psyche. When somebody's typing, we're literally seeing them form thoughts in real-time – including editing or even censoring themselves. As a result, Screenlife films can be extraordinarily intimate experiences (Timur Bekmambetov quoted in an interview with Andrew Todd [17]).

In order to create, grow and thrive in a post-pandemic world, filmmakers need to develop the tools, skills and new language competencies to transform their audio-visual storytelling into a pandemic-proof format. One highly promising solution that has emerged over recent years has been Screenlife and Zoom films. Telling stories on computers or mobile phones using accessible domestic technologies with the potential to work remotely, asynchronously, sustainably utilising existing digital artefacts and budget consciously is a promising strand of future filmmaking. However, because everything now happens online, filmmakers are faced with a significant problem of representation—how to depict and engage with online ecosystems, digital logics and multiple points of view in constructing intimate and contemporary character representations and compelling narratives in engaging modern formats? We no longer accept unrealistic depictions of the virtual reality of working on computers—we are so familiar with the tools of contemporary digital communications, productive apps and entertainment formats that we expect to find nuances in how a particular character uses their devices and how their desktop appears and the details, pauses and style of how they type a text message and what emojis they may use. A key issue now is how do you tell a story that takes place on a computer, mobile or other screen formats that range from dashcams, reverse parking screens, tablets to surveillance cameras and multiple messaging apps operating simultaneously? One of the key advocates for a new language is the director and producer Timur Bekmambetov who came to fame with the Russian fantasy thriller *Night Watch* (2004) before directing the breakout action thriller *Wanted* (2008) in the USA. Bekmambetov claims that Screenlife is a new screen language. In an interview he explains the concept (Image 6.1):

> *Screenlife is a new format of visual content that has grown from independent projects to full-length, world-renowned films, documentaries, and TV shows. Its main idea is that everything that the viewer sees happens on the computer, tablet, or smartphone screen. All the events unfold directly on the screen of your device. Instead of a film set — there's a desktop, instead of the protagonist's actions — a cursor. If you are involved in video production, cinema, or even video games, Screenlife is a new expressive environment for you, the potential of which is yet to be discovered* [5].

Part of the tools available for desktop or computer films is the use of videoconferencing as a communication tool—whether it is Zoom or Skype or FaceTime. In this chapter I argue that it is time to sketch some distinctions between Zoom films and Screenlife as now we have sufficient cinematic examples and variations of the broader concept of computer screen films to be able to examine greater nuance in these different, but at times, convergent production methods. Zoom films, or films made in zoom or similar video conferencing software are a subset of Screenlife or computer screen films. One of the key differences between Screenlife and Zoom films is about point-of-view (POV). In Zoom films the audience is positioned as a member of the community—while we cannot see ourselves in the nestled window

Image 6.1 A collage of Screenlife and Zoom serials

line up, we feel that we could be there. We have an equally privileged position to the other participants, and after a few years of working from home and regularly using zoom, many audiences have developed the competencies of engaging with the format productively and creatively. However, in the Screenlife world the audience adapt to the key protagonist's POV; their cursor movement dictates what we look at and our attention is manipulated by the editors as some layers are foregrounded, some apps appear bigger and the narrative is managed through the dynamic interplay of multiple computer tools being used by the protagonist. We often only see them and their reactions through video conferencing or messenger apps.

Here we occupy their idealised position that is well positioned to search for clues, to work collaboratively, to uncover material from others and to share intimate emotions. Some computer screen films that have been dubbed Screenlife by their producer Bekmambetov, such as the *Unfriended* [28], use the Zoom film format of the online group chat, but then transform this by sharing videos or other online content to provide greater sophistication. Whereas many pandemic era Zoom format series retain the multi-windowed chat format without ever exiting the ecosystem and without adding much extra computer usage and breaking the audience's voyeuristic point of view.

6.2   Screenlife Origins

In defining desktop or computer screen films or what is now better known as Screenlife, we need to make a number of distinctions and while there are substantial crossovers between the formats, employing a working taxonomy is useful.

A screenshot is to photography what screencasting is to videography. Screencasting is the technique of digitally recording a sequence from a dynamic computer screen. Screencasts are typically recordings of what is happening on a screen, which might be a web browsing session or a software demonstration with voice-over narration. The term "screencast" was first used by the writer Jon Udell in an article published on *InfoWorld* in 2005, describing the benefits of using this technique to show how computer applications worked [18]. Screencasting is an important technique at the backend of the production of all computer/desktop/zoom and Screenlife film and there are various software applications that do this to various extents of control. The emphasis of the screencast is "reported in real-time by a guide showing a particular task" [19]. But Screenlife films, especially sophisticated productions such as *Searching* [25] and *Profile* (Bekmambetov 2018) are asynchronous and are edited in sections of what could be called 'found footage'—constructed aspects of a character's desktop or mobile interface—component parts that can be used to cleverly construct character. These types of films are created more like animation than screencasting as the process offers the editors far more control over what the audience are looking at and when and how to move their attention from one area of the screen to another. In screencasting that level of control is not there as it is recording a 'what you see is what you get'. Furthermore, the usage of Adobe Illustrator allowed the *Searching* editorial team access to vector graphics that retained the sharpness of digital artefacts such as calendar reminders or message bubbles no matter how much the objects were zoomed in or moved around the character's desktop. For the casual observer, many computer films may appear to be screencast, but in order to achieve far more control, filmmakers may need the more sophisticated approach of Screenlife post-production. Screenlife takes that extra step of guiding audience attention in screen control and shaping the point of view and subjectivity to create the experience of greater subjective engagement.

The Zoom or the Chatroom film is a variation of the screencast film as all that the audience sees is what is occurring on the desktop or on the computer or mobile phone screen with its recognizable interface and multiple windows. The audience is positioned more objectively. In contrast, a Screenlife film ranges across multiple interfaces, the desktop, various messenger apps, web cams, mobile interfaces, and the 'off-desktop' world—with other cameras picking up the character's action—sometimes revealing the protagonist looking and interacting with their computer screen such as in the film *Cyberbully* [22]. The protagonist Casey Jacobs (Maisie Williams) is a typical British teenage girl who lives her life mainly online. The action takes place entirely in Casey's bedroom and in multiple online environments. We see Casey largely from the POV of the computer in video chats with her friends as well as her POV of the computer screen—so far—so Screenlife—but the director also chose to vary the computer POV to a variety of side-angles and in-situ shots of Casey in her bedroom interacting with her computer—this is a more objective perspective than the very highly first person subjective POV offered through the mediation of the screen whether looking at Casey from the POV of her friend Megan or seeing and operating her screen as Casey and by extension, the audience. This variation on the screen language is closer to classical cinema's representation of a character using a computer, but with the hybrid Screenlife aspect used to create a deeper engagement. That the film steps out of a solely desktop ecosystem such as the one created in *Searching* is an interesting variation on the emergent screen language.

The desktop film is no longer dependent on the film medium. As Catherine Grant suggests, it is a filmmaking practice "which uses screen capture technology to treat the computer screen as both a camera lens and a canvas" [8]. Many desktop films do not use a camera. The camera appears to be the computer or mobile phone screen. The pre-production becomes an integral part of the post-production. The computer is simultaneously the screen, the camera, the narrator and competing voices in various constantly moving and layering windows. Interestingly these films pre-suppose a cinema audience. Surprisingly, it is the experience of watching a desktop film in a cinema that is so compelling as the encounter feels to be one of dislocation. Watching a film such as *Searching* (2018) that opens with the 2001 released Windows XP desktop landscape creates a sense of confusion as it appears to replicate the computer user experience, but without the user interactivity and strong feelings of nostalgia for the time when XP was the dominant and unifying desktop experience. Indeed, desktop films create the semblance of audience interaction, as though they can have an impact on the narrative outcome perhaps far more so than the active audience engagement generated by classical cinema.

Desktop films are hybrid constructions. They utilise pre-existing materials such as software tools, audio messages, video clips, messenger apps, photographs, calendar reminders and other digital devices in meaningful narratives. They appropriate and redefine and blend together these digital tools and artefacts. Some of the films like *Dniukha!* [23] moves between various screen interactions and seemingly live-action sequences that are then presented on other screens. It is largely a post-production genre—gathering, combining and re-appropriating existing material into new contexts. In the Screenlife films such as *Searching*, the point of view of the father character and the viewer are essentially the same, making the format unusually close to shooting from the first-person perspective. At the same time, the viewers find themselves in an extremely advantageous position; they peep through the various activities of the protagonist on the screen, penetrating their inner world—they feel their emotions through their tactile use of the digital world.

In contrast, Zoom films such as *Host* (2020) or Chat Group films or series such as *#SidYaDoma* (2020), the main characters are no longer implied, but are shown directly, often as participants in a conference or group call, as well as characters in various media files. The audience here does not have the same embedded, voyeuristic position because the accepted conventions of a Zoom conference imply that the characters of the events unfolding on the screen are aware of the presence of a third-party observer. The narrative of this type of presence among other performers and simultaneous voyeurs is justified by the wide accessibility of the video conference format. Anyone can join a group call or video conference at any time, breaking the intimacy of what is happening, as, for example, in the movie *Unfriended* [28], when the explanation of the two lovers left alone in the conference call is interrupted by the sudden appearance of an unidentified third participant in the call. In contrast, the characters in a film such as *Searching* [25], *Cyberbully* [22] or *Profile* (Bekmambetov 2018) often do not know who is watching them as they have multiple interactions utilising various software and messenger apps and the continuous editing and movement across manifold interactions places the audience in the position of voyeurs—watching their character's actions and decisions unfold.

6.3 From Desktop to #Screenlife

The first entirely "desktop film" was *Thomas in Love* [27], a Belgian black comedy set in the benevolent near future about an agoraphobic young man who has spent the past eight years locked up in his sealed apartment communicating with the outside world entirely through his computer. His therapist suggests that Thomas should use an online dating service paid for by his insurance company in the hope that he will meet a real woman and fall in love thereby aiding his recovery. Thomas tries a range of virtual partners, animated cyber dolls and an online brothel in his search for love and connection in a virtual world, but for Thomas there is love and the desire to connect in real life. We never see his face, but we hear his voice. He operates the computer interfaces. The viewer's experience is entirely from the point of view of the actor playing Thomas (Benoit Verhaert). The audience sees what Thomas sees on his computer screen, unmediated by a computer frame. We inhabit Thomas—from interrupting a cybersex session to take his mother's regular video calls, to speaking to the insurance people, the vacuum cleaner tech support, his therapist and several virtual lovers and cybersex dolls. They see Thomas. In a sense, they see us. There are lots of close-ups and direct to camera addresses.

The first "desktop" feature film to enjoy box office success and effectively utilise the format is the teen-slasher *Unfriended* [28]. The film was made on a very modest budget of less than $1 million, but grossed more than $64 million theatrically worldwide and more than $4 million in the US home entertainment market [16] bringing it to the attention of studios and launching the 'Screenlife' concept popularised by the film's producer, the Kazakh producer-director Timur Bekmambetov. This pulpy, teen horror film was Bekmambetov's proof-of-concept for his grand vision of Screenlife. "It was important for me to make it as a movie, not as a short or some YouTube video, because it was a statement that Screenlife is not just a gimmicky, one-time trick. It's a new film language" [14] (Image 6.2).

In the desktop or the computer screen film the range of communicative possibilities is vast. The action can move rapidly from browser windows and search engines to messenger screens, pop up ads, video calls, saved videos, text messages, open web cams and multi-windowed video chats. The screen real estate appears as a standard desktop with all the multi-tasking options available to build a complex and sophisticated narrative. Viewers assume a highly subjective first-person point-of-view by appearing to be the protagonist as they surf between a Skype or Zoom conversation before anxiously deleting a reminder that their rent is due and moving quickly between Google image search and Facebook messenger and then onto stored image

Image 6.2 *Unfriended* [28]. Image Courtesy Bazelevs Productions

files all the while hearing the protagonist breathing and observing their hesitation in their cursor movement or typing and re-typing a specific search term. This reticence would often give greater insight into their character, even more so than the reveal of their Spotify playlist or email notifications from their psychiatrist.

Bekmambetov enthuses, "Screenlife films are universally relatable to people around the world, and audiences intuitively understand this storytelling language, because it's how we live. I believe our devices can reveal more about a person than they would ever say out loud, because we never lie to our screens," [9]. Desktop films are not a new subgenre, rather they are a new and innovative screen language that exploits audiences' digital competencies. Screenlife films present the protagonists' action, telephone or messenger conversations and photo archives on a desktop computer screen. The screen language tracks the narrative from the protagonist's first-person perspective with the cursor directing the viewer's gaze. Based on ubiquitous digital technologies and shared new media languages and competencies, desktop films restage a users' mediated communication experiences that are at once dynamic, intimate and authentic in that so many of the tools, apps and competencies are well known from domestic and daily use and provide audiences with an emotional connection.

Initially calling the format a "screenmovie", Bekmambetov wrote a manifesto for conceiving the virtual space and the production methodology that identified three main pillars that appear to be applicable to most "Screenlife" productions [2]:

1. The setting is virtual reality in general and one specific computer screen, belonging to one character. The action never moves outside of the screen, unravelling on the display of one character's gadget. The size of the screen (i.e. the frame boundaries) remain a constant. The appearance of new visual elements has a rational explanation and corresponds with the formats of life in a virtual space: The viewer must constantly be aware of where

exactly the action occurring at any given moment originated. The camerawork is stylized to resemble the behaviour of a digital gadget's camera.

2. Unity of Time – all the action takes place in real time—here and now, while the film is put together by means of in-frame montage without any visible transitions (as if shot in one continuous take). All the action takes place in real time—here and now, while the film is put together by means of in-frame montage without any visible transitions (as if shot in one continuous take). The editing stage thus becomes the paramount stage of production, replacing the screenwriting stage.

3. All the sounds in the film originate from the computer. Their origin can always be rationally explained; the viewer has to understand at all times where the music track is coming from.

From 2015 to 2018 Bekmambetov developed what he called "Screenlife" software to allow for the seamless production of these films and made it freely available. The software allows filmmakers to record sessions in Google Chrome along with links and videos that record webcam footage and microphone sounds using the cursor for direction. Bekmambetov moved predominately into Screenlife production as producer, but also as a director on *Profile* (2018). Therefore, Screenlife became both a presentation and production format where the diegesis takes place entirely on the screens of various devices from desktops to mobiles using all forms of video messenger apps to engage viewers familiar with their aesthetics and applications.

The format will no doubt experience further growth and public exposure following the significant 2020 deal between Universal Pictures and Timur Bekmambetov's Bazelevs Studio to produce and distribute five features made in Screenlife. The financial success of *Unfriended* (2015), made for $1 million and grossing $65 m worldwide and *Searching* (2018) that grossed $75.5 m on an $880,000 budget were clearly influential factors behind the decision. Bekmambetov plans to apply the format to several genres from horror and thrillers to romantic comedies and sci-fi. He was quoted as saying, "Screenlife is a great way to tell stories about modern reality, which is happening predominantly online and on our screens. And now, given restraints on traditional workflows during the pandemic, there's a strong surge of interest in Screenlife as it's one of the few available ways for filmmakers to keep shooting movies" [15]. Bekmambetov claimed in his manifesto that the chief innovation of Screenlife was its capacity to create a sense of stream of consciousness that the viewer experiences as a special, privileged insight into their character. He explains that "it enables the author to explore the psyche of a character in a new way via their interaction with virtual reality. The user writes a message to a conversation partner, the cursor freezes, as if hesitant; the viewer is able to observe the emotional transitions of the character and his actions through the way the character's cursor moves, to understand his background and motivation" [2]. The horror or investigative thriller in the Screenlife format are highly conducive to the genre.

Not all Screenlife projects are created equally. Films such as *Profile* and *Searching* feature considerable innovations in screen language and extensive post-production. They were both carefully constructed, technically sophisticated and narratively tight feature films that performed well at film festivals globally. *Unfriended* won the "Most Innovative Film" at the Fantasia Film Festival.

In 2018 *Searching* won the Sundance Film Festival Audience Award and *Profile* won the Audience Awards at the Berlin Film Festival and the SXSW Film Festival.

6.4   Zoom Films

Films made in the Zoom videotelephony software (Zoom Meetings commonly shortened to Zoom developed by Zoom Video Communications) or replicating the group video chat (Skype or Facetime are often used) are variations on the computer screen film. These films are more limited and confined to a discrete ecosystem than other Screenlife films such as *Dashcam* (Savage 2021) or *Dniukha!* or *Searching*—largely by narrative design. The Zoom films became very popular during the global lockdowns allowing for remote production methodologies to emerge. The ubiquitous and often utilised Zoom software became both the pre-production communication program used by the actors and the crew and in many instances the actual program used to record the narrative action in real-time.

The quarantine-era Zoom films or simplified Screenlife series with multiple characters, but little post-production process that appeared during the lockdown of 2020 were far more focused on clever writing, strong performances and contemporary comedy situations. Unlike the Screenlife preference for horror, the Zoom films were about community-oriented situational comedy or humorous sketches. A dozen or more series that were released on the streaming services in Russia exploited the benefits of Screenlife for covid-safe production processes so that content could be made remotely, with either very small crews, minimal set decorations and props and with the actors acting as cinematographers and designers – picking the best backgrounds and lighting and ensuring that their sound recording was optimised.

Even though the Russian quarantine-era serials were universally dubbed as "Screenlife series" by the Russian mainstream press [1], it is surprising just how different these series are from the Bekmambetov manifesto and the films that preceded them. Surprisingly, the protagonist's point of view disappears in these series. In *Searching, Profile, Unfriended* and *Dniukha!* (2018), there is a clear impression of the protagonist leading an investigation, with viewers following the diegesis at the same time as the protagonist does. There is a clear character perspective. By contrast, the Zoom film series do away with a singular protagonist, and instead the screen real estate is shared with multiple characters and multiple, unprivileged points of view. There is no key protagonist and no singular perspective. An all-in-it-together ethos permeates these series. The reasons for using the Screenlife format are clearly more pragmatic than stylistic—there is no moment where the protagonist's cursor hovers indecisively, no typing and then retyping a search query, no deep character insights. This is replaced by a synchronous, unified connection of time and place between multiple characters (Image 6.3).

Perhaps the most well-known example of a Zoom film in the Anglophone world was *Host* (2020) the British 56-min supernatural horror film. The film grew out of a short film zombie-prank idea by filmmaker Rob Savage that went viral and prompted a

Image 6.3 *Host* (Savage 2020). Image courtesy Vertigo Films

larger Zoom call length production. The film was released on the horror/supernatural focused premium streaming service Shudder (www.shudder.com). *Host* became a runaway hit, scoring 100% on Rotten Tomatoes and attracting glowing reviews from hard-boiled horror fans. Brian Eggert from *Deep Focus Review* wrote "it preys on our fear of isolation in a way that feels incomparable to any other horror movie right now. *Host* may go down as a rare anthropological document of what life was like for a relatively brief (and also excruciatingly long) period of time" [4]. In *Host* the action takes place in a video group chat on Zoom during the pandemic lockdown when a group of friends (Haley Bishop, Jemma Moore, Emma Louise Webb, Radina Drandova, Caroline Ward and Edward Linard—the actors all use their real names) decide to hold a virtual séance. Of course, things go awry. The story takes place entirely on Haley Bishop's computer with her friends on Zoom along with Seylan, a Scottish medium that they bring in to summon the dead as a fun shared activity during lockdown. Not everyone takes the séance seriously, there is a sceptic ready to undermine the activity but instead this creates a virtual vessel allowing the spirits to enter the mortal realm and wreak havoc as one by one the friends begin to experience increasingly frightening supernatural activities in their houses. The friends go to explore disturbances in their lockdown houses—bumps in the attic, shadows in the hallway, creepy sounds in darkened seemingly empty rooms. The characters take their laptops with them so that their mates can watch on in terror as they are thrashed around by the invisible, demonic spirit. They cower, they plot, they scheme, they attempt to flee, but eventually all the friends, bar Jemma and Hayley, are horrifically destroyed by the summoned demon with only the timing out of the Zoom call ending the action.

Host was made during the quarantine lockdown and Savage directed the actors remotely. According to writer Dino-Ray Ramos [13] "All the actors operated their own cameras, pulled off their own practical effects and lit their own scenes." *Telegraph* reviewer Tom Fordy noted, "found footage is a sub-genre that upgrades with the technology, retooling itself with new creative means and scares. The *Blair Witch Project* used shaky cam to make horror a first-person experience [...] *Host* does all that [...] But it uses Zoom's familiar features—filters, custom backgrounds, profile pics, even the free call time limit—to literal haunting effect. [...] Savage said that the process of making *Host* was like "building the train track in front of ourselves as we were speeding into production. It wouldn't have worked if we'd tried to apply a normal framework" [6]. The Zoom structure was alternating between close ups of the friends sitting in the nestled windows virtually holding hands to enable the séance to having wide angle, full frame action sequences where the characters film on their own laptops, negotiate floating filters and jump scares leading into savage attacks and their eventual gory deaths. *Host* was a clever, well considered, and fast turn-around film completed in 12 weeks from conception to distribution. *Host* adapted a new technology and cultural practices for film production and used Zoom both narratively, thematically and technically to stunning and relevant effect.

6.5 *#Sidiadoma* [7]

In contrast to the found footage supernatural horror Zoom format of *Host* or *Unfriended*, a number of comedy series appeared especially in Russia during 2020 quarantine on the new streaming services. They were made using Zoom quickly, remotely and employing a serialized narrative with large casts of well-known actors attracting considerable media and audience attention. *#SidYaDoma* that translates as 'Staying at Home' was a weekly 8-part 20 min comedy series that launched on the streaming platform PREMIER in Russia in April 2020. Each episode was produced and filmed using Zoom by some of the best-known Russian A-list actors. The series attracted a high rating on the Russian film search portal, *Kinopoisk* with 6.8/10 based on 2880 votes. Audience size was unavailable (Image 6.4).

The plot revolves around the employees of a provincial cleaning supplies business, KlinXoz and the complex private life of their boss, Gennadii Tsvetkov (Aleksandr Robak) during the quarantine. Tsvetkov is trying to save his business and staff from going under during the enforced lockdown and is encouraging his staff to continue working normally, but remotely. Meanwhile his new wife (Iuliia Aleksandrova) is stuck in Bali and he speaks to her regularly on messenger while she takes part in all forms of spiritual rituals while he is trapped at home with his mother-in-law. Tsvetkov's ex-wife (Anna Mikhalkova) is not wasting any time during quarantine and is planning on marrying Tsvetkov's new wife's yoga teacher, while also managing their son's affairs and keeping up a warm relationship with all the KlinXoz employees. Meanwhile the company's secretary, Iuliia, is upset because her secret lover has chosen to self-isolate with his family rather than with her. He takes all his Zoom calls from the apartment's closet in order to achieve a modicum of privacy. Head of logistics and security Sergei (Gosha Kutsenko), a former special forces officer and

Image 6.4 *#SidIaDoma* [7]. Image courtesy Premiere

now a heavy drinking single man creates as many problems for the company as he manages to solve while sitting in an empty indoor swimming pool where aside from the multiple bottles of strategically hidden vodka, he is storing the company's entire toilet paper stockpile. He is supported by the company's accountant who dreams of having a relationship with him, but is stuck at home trying to work and home-school two young children who turn her life into a living hell. The series explored the company's employees' complex inter-relationships as they try to survive amidst the deprivations of the lockdown with all the communications shared on Zoom.

The success of the series was its representational authenticity both narratively and through the zoom/mobile phone interface. It captured the everyday lives of working people trying to make ends meet, look after one another, and continue having some semblance of private life while working from home, often with hysterical consequences. The director used Zoom to record the actors through various takes working remotely. The action takes place entirely within the frame of the computer screen, which captures their video work meetings, their private messenger calls, mobile chats and all other forms of communications. However, there is no privileged view from the protagonist's screen. Point of view is shared equally among the workers. The actors filmed the series working from home. Director Olga Frenkel claimed that the enforced remote working protocol created numerous new opportunities:

> *I enjoy working through the screenplay methodically without having to think about the limitations imposed on hiring equipment and sound stages. I like the opportunities of quickly reshooting a scene. I like being able to change the script without being dependent on economic or organisational factors. Normally there is at least one year from the moment you come up with an idea to when you get a chance to film the pilot, but we wrote and filmed the pilot in just 10 days! And I really like the intimate set up, just me, Zoom and my favourite actors* (Frenkel [7], my translation).

The remarkable thing about this series and part of the reason it attracted such positive feedback from viewers—"Well done! Very funny series. Cheered me up!" [11]—was that the series not only captured the conversations and issues and concerns that people were going through in their everyday lives during the pandemic, but it also represented people's daily life while working remotely and their heavy reliance on screens for all forms of communication, whether private or work related as well as everything in between. The series' aesthetic aims were to represent a contemporary authenticity with genuinely amateurish moments of frozen screens and software foibles.

#SidIAdoma effectively employed the Zoom dialect of the Screenlife language adapting to the fast turn-around series. The diegesis moved efficiently from the group Zoom meetings screen, where all key members were visible at the same time in the widely recognised multi-split screen format of non-hierarchically nested windows, to duologues that were interrupted by mobile messenger calls and even some inevitable faux pas with the character's accidentally sending "private" photographs to their work colleagues. *#SidIAdoma* exhibited some effective screen language innovations that developed the Zoom film iteration of the Screenlife format. The series used close ups with a high proportion of actors making a direct, intimate address to the webcam with few long distance or unmotivated camera set ups. Especially at times of high emotional drama, the actors moved into extreme close ups with their faces right up against the camera, hair over the lens, poorly framed but authentic moments creating an unusual sense of intimacy even for such a cosy format. There were no second cameras, no alternatives for coverage. In contrast to Timur Bekmambetov's feature film produced Screenlife projects there was no protagonist and no structured, consistent point of view. Emily Wei argues that as "immersive as Screenlife films could be, there is an omnipresent undertone of voyeurism concomitant with the awareness that we are looking at someone else's personal computer, a lingering feeling not quite easy to dismiss in spite of the identification with the protagonist" [20]. In this case the shared voyeurism was the intent to demonstrate the ethos of **we are all in it together**. Actor's eyelines were cleverly employed to focus just off centre, towards the bottom left of the screen so as not to create that impression that the actors were looking directly at the lens, directly at the viewer whether they were watching on a big screen TV or on their tablet or mobile phone. This created an experience of intimacy in that viewers were positioned as being part of the action, maintaining a clear feeling of connection, but also with a partial experience of voyeurism as if "zoom bombing" a meeting. Dynamism was produced through editing and movement between the actors' screens with frequent interruptions creating constant changes on the 'home' screen. Dialogue based sound bridges ensured continuity between scene fragments and, as with the standard Zoom interface, there were audio-privileged rapid transitions on the main screen when one of the participants spoke, otherwise they were all visible in small squares at the bottom of the main screen. The regular use of screen glitches and image stuttering was used to reinforce a feel of the authenticity of shaky internet connection and a haunting sense of realism—a technique to reinforce viewers' connection with comfortable and familiar referents.

The beauty of the Zoom film for this domestic comedy genre is that the format is character-centred, prioritising the script and relationships, and limiting the action outside of the performance in the frame of the computer screen. The focus is on characters' responses to one another, the attention to small details in character development, the capacity to watch people listening and responding to one another with no special effects or zany interiors. The format has focused attention on the way people speak and react and listen to what is around them and how they respond to their own loneliness and sensitivities to all forms of domestic anxieties.

The Zoom metaphor of bringing everyone together on one screen created a form of unity through a shared audio-visual narrative experience. Confined to the immobile, unmoving Zoom cells of existence—unable to go beyond their shared frames into the limitless realms of the online world—the characters are forced into a form of equivalence with their would-be viewers. *#SidIAdoma* created a symbolic community, while temporarily ignoring the rampant virus and its attack on intimacy.

6.6   *Searching* [25]

Similarly, *Searching* (2018) is a quest for intimacy and activating community, but the presentation format and structure of Screenlife in comparison to Zoom creates a different audience and protagonist point of view. Here everything is seen from the perspective of the computer screen—the audience sees what the protagonist brings up on his computer screen. The infiltration of the group Zoom séance in *Host* and the skype chat in *Unfriended* by a demon is not nearly as sophisticated as the complex story of a father searching for his missing daughter by tracking her online activities and searching her laptop for clues (Image 6.5).

Unlike the Zoom films such as *Host* or series such as *#SidIaDoma,* the action takes place in real time and from the shared perspective of the characters without hierarchy other than that dictated by audio queues. *Searching* is fixed on to the father's point of view of his computer screen and takes place over a week with temporal gaps in between the action. The audience sees the father, David Kim (John Cho) in photos, shared video calls and FaceTime. We see his reactions to various situations and an archive of family photos featuring him with the family, but the action is mainly from his perspective. We see the computer screens along with David. We do not have any privileged information. In the Zoom films, the character's audio levels (when they speak) determine which of the equally nested array of characters is shuffled into a highlight in the video feed or enlarged into a single shot or two characters appear larger on the screen. This shuffle of prominence based on who is speaking or leading the action works similarly to a shot/reverse shot technique—holding the viewer's attention on what is important. In *Searching*, director Aneesh Chaganty, sought a more cinematic experience by focusing on montage, multiple split screens, layering and dynamic changes between shot framings. The editorial team employed an animation editing technique by panning and scanning the desktop, layering it with multiple sources of information and then zooming in on an important message so as

Image 6.5 Image *Searching*
[25]. Bazelevs Production

to feature them and allow the audience to read alongside David. The audience are positioned as digital sleuth undertaking the search for David's daughter along with him. The various information leads that appear and are laid out on his desktop are multi-layered and the action alternates between various apps gaining greater screen real estate as their narrative prominence progresses. As Andy Young notes, "the *Searching* editorial team built their own workflow, using motion graphics techniques from After Effects to craft computer-screen 'performances' that accompanied the powerful live-action performances of John Cho and Debra Messing. The result is a film where even the movements of the screens and cursors are evocative" [21]. The movement of the cursor acted in a way to guide the audience's gaze and attention in a sense replicating David's gaze and subsequent (cursor) action. Similarly, there are several powerful moments of characterisation and emotion that are associated with the cursor and typing and retyping messages. Before David realises that his daughter Margot (Michelle La) is missing, we see him typing her a message about how proud her mother would be… Rather than just allow for the pragmatic sending of information, we see David thinking, then deleting and then retyping with a different tone. In those brief yet wonderfully poignant moments, we get an insight into what David is thinking, then censoring himself in trying to find the appropriate way to communicate effectively and compassionately with his teenage daughter.

In contrast to the Zoom films, Screenlife films like *Searching* and *Profile* (2018) intimately privilege one character's very personal point of view. In considering how to make their initial short film idea work as a feature film first-time feature director, Aneesh Chaganty, said "so we knew that if this medium that we were using is full of information, let's make information both the obstacle and the objective of the story" [12]. The story in this thriller becomes one of untangling this digital web, navigating the clues, negotiating the red-herrings, and making sense of the emotional labyrinths. The father, David, an IT expert, in his desperate search for his missing daughter uses all the contemporary tools of communication, surfing seamlessly between all the standard social media platforms, all the finder windows, Gmail inboxes, calendars and file systems. When David navigates his own disparate digital memories there seems to be a strong connection between the real world and his digital footprints, but when he opens Margot's computer in the search for clues, he realises that he is searching for his daughter in more ways than just her physical presence. It becomes clear that he lost her years earlier as there is a massive gap between what he thought he knew of Margot and what he finds out from contacting her friends and people captured in her photos and videos and the reality of her connection with others is far more complex and troubled than he could have imagined. There is an urgency in his search and a desperation to find his daughter alive. This is a clever and surprisingly emotional capacity of Screenlife films to provide character definition through a navigation of their desktop, the speed at which they click through and think of the digital connections in their urgent search. For example, *Unfriended* starts as a skype video chat between the key protagonist Blaire (Shelley Henning) and her boyfriend Mitch (Moses Jacob Storm) from the perspective of Blaire's Macintosh laptop. We may notice that she has a chess game open on her desktop and a torrent site downloading in the background and she plays 'Lost Cities' on her Spotify. How she types, how she moves her cursor and what things she is interested in can tell us as much about her as what she wears, how she speaks and what she says. Perhaps even more. We get an insight into how she thinks. In *Searching* the filmmakers created two different looks for each of the main character's computers as we get to discover them through their digital presence and artefacts.

The opening sequence is the most powerful part of the film. The first image is a very bright vista of green rolling hills of the Windows XP desktop. Watching this in the cinema on a very large screen was disconcerting as it seemed so strangely out of place, so big and so uncannily nostalgic. The scene unfolds to reveal some 16 years of complex family memories and life events through the prism of David going through these digital memories compressed into edited moments—all the relationship courting rituals, photographs, birth and piano lesson videos, emails and doctors' appointments—this sequence masterfully captures the highpoints in David's life with his family—his wife, Pamela (Sara Sohn) and their daughter Margot (Michelle La) and all the way up to Pamela's slow demise from cancer. Like a silent film this information is presented visually, dynamically and in an incredibly moving and compelling way. In commentary on the film, the filmmakers explained that they realised that they couldn't make a film set entirely on a computer screen cinematic, so they immediately dug in with the emotional character beats in the opening 10 min.

"We knew we had to like beat the concept right off the bat and kind of show the audience the full potential of what we could do." The opening details the beginning of Margot's digital life on through her childhood and mother's eventual death, and they refer to the sequence as *"Up* meets a Google commercial" [10]. The opening sequence worked very effectively to provide a sophisticated backstory, a kind of digital anthropology of the ups and downs of a family over two decades without any voice over narration and making that a highly emotional journey all confined to a computer screen and its databases. In order to find clues of his daughter David must not only seamlessly navigate the multiple windows and layers of Margot's contradictory social media history, but delve deep into the nefarious maze of the internet and the psychology of teenagers leading multiple lives.

The Screenlife world of this thriller is totally appropriate given how compelling it is to watch David in close-up googling clues on his computer. However, surprisingly not the entire world of the narrative exists in the Screenlife ecosystem. The audio tracks capture the mouse clicks and the birds chirping outside David's window and his breathing and the loading of the computer websites, but these sounds are unnervingly supplanted by non-diegetic music that is designed to amplify the thriller aspects of the film—high pitched strings and bass rumble at moments of realisations or discoveries. This broke Bekmambetov's third pillar of Screenlife—the Unity of Sound. Going outside of the Screenlife world was unsettling in a different way to the giant Windows XP vista at the beginning as the audience anticipated that the entire world of this narrative exists within the computer screen universe rather than have the traditional music score come in artificially from outside. In this sense the commitment to the Screenlife format appeared to waver for the filmmakers. However, this did not diminish the film's favour with festival audiences as *Searching* won the prestigious Sundance Audience Award in 2018.

6.7 Conclusion

Abating Bekmambetov's techno-boosterism somewhat, there is much merit in his claim "that Screenlife is a new film language". It does have its own rules, conventions and production methodologies. Some of them borrow from traditional filmmaking while some such as the cursor hover, the dynamic multiple window navigation, the typing and retyping and the subjective point of view are among many unique innovations. During the lockdown of 2020, Zoom films and Screenlife content really took off—not only examining topical issues such as the horrors of quotidian life under quarantine, but as effective production technologies for working remotely. This new generation of screen entertainment could be characterised by remote production processes that incorporated variations on the "Screenlife" format, mobile communications and videoconferencing technologies. The appearance of A-list actors in scripted comedies often becoming their own cinematographers in their own homes were further novelties that could potentially become inscribed as standard future production methodologies. The Zoom film format gave rise to a new "dialect" of the established screen language. Gone was the emphasis on the protagonist's point of view seen in the original Screenlife films. This was now replaced by a far more equitable communication format. The Zoom film series altered Bekmambetov's manifesto format to present material that was more accessible, intimate and authentic and implied a shared and equal screen space between actors and viewers that balanced their pleasure and pedagogy in learning how to adapt to the complexities of communication under lockdown. The future of this new cinematic language is no doubt hybrid as there will be incorporations of existing techniques and approaches alongside adaptations of new logics and methodologies where the diegesis may no longer remain confined exclusively to the space of the desktop to go beyond it outside the computer ecosystem. The next few years will no doubt demonstrate the efficacy and opportunity and engagement of these films with audiences and how that new language is accepted and modified.

References

1. Al'perina, S.: 'Sem' noveishikh serialov pro samoizoliatsiiu' (2020). Rossiiskaia Gazeta, May 12. https://rg.ru/2020/05/12/sem-novejshih-serialov-pro-samoizoliaciiu.html. Accessed 20 July 2022
2. Bekmambetov, T.: Rules of the Screenmovie: The *Unfriended* Manifesto for the Digital Age (2015). Movie Maker, 22 April. https://www.moviemaker.com/unfriended-rules-of-the-screen movie-a-manifesto-for-the-digital-age/. Accessed 7 May 2022
3. Bekmambetov, T.: Screenlife-Serial Vzaperti (2020). Afisha, 18 June. https://cetre.ru/category/afisha/Screenlife-serial-vzaperti-/. Accessed 21 July 2022
4. Eggert, B.: Host (2020). Deep Focus Review, 7 October 2020. https://deepfocusreview.com/reviews/host/. Accessed 15 July 2022
5. Ferrari, A.: IFH 468: The New Film Language of "Screenlife" with *Wanted* Director Timur Bekmambetov (2021). Indie Film Hustle, 13 May 2021. https://indiefilmhustle.com/timur-bek mambetov-Screenlife/. Accessed 23 Feb 2022
6. Fordy, T.: Zoom, The Horror Movie: How the Brits Behind Host Made a Chilling Lockdown Masterpiece (2020). The Telegraph, 16 August. https://www.telegraph.co.uk/films/0/zoom-hor ror-movie-brits-behind-host-made-chilling-lockdown-masterpiece/. Accessed 20 July 2022
7. Frenkel, O.: #SidIAdoma (2020). Vokrug TV. https://www.vokrug.tv/product/show/158679 42221. Accessed 20 July 2022
8. Grant, C.: On Desktop Documentary (or, Kevin B. Lee Goes Meta!) (2015). Film Studies for Free, 6 April 2015. https://filmstudiesforfree.blogspot.com/2015/04/on-desktop-documentary-or-kevin-b-lee.html. Accessed 23 July 2022
9. Grater, T.: Screenlife Creator Timur Bekmambetov Plans Indian Film Slate, Partners with Graphic India & Reliance Entertainment (2020). Deadline, 21 December 2020. https://dea dline.com/2020/12/Screenlife-timur-bekmambetov-indian-film-slate-graphic-india-reliance-entertainment-1234660003/. Accessed 23 July 2022
10. Hunter, R.: 31 Things We Learned from the 'Searching' Commentary (2018). Film School Rejects. https://filmschoolrejects.com/31-things-we-learned-from-the-searching-com mentary/. Accessed 23 July 2022
11. *Kinobase*: Audience Comments for #SidIaDoma (2020). https://kinobase.org/serial/117259-sidyadoma. Accessed 25 July 2022
12. Murphy, M.: How 'Searching' Uses Tech Devices as Narrative Devices (2018). The New York Times, 24 August. https://www.nytimes.com/2018/08/24/movies/searching-movie-tec hnology.html. Accessed 20 July 2022

13. Ramos, D.: Shudder Invokes Quarantine Spirits with Remotely Filmed Horror Film *Host* (2020). Deadline, 7 July. https://deadline.com/2020/07/shudder-host-rob-savage-quarantine-zoom-remotely-filmed-1202978631/. Accessed 27 July 2022
14. Rinder, G.: The History of Screenlife Films: 10 Key Movies in An Exciting New Genre (2021). GQ, 25 June 2021. https://www.gq.com/story/history-of-screenfilms-searching-host-unfriended. Accessed 28 July 2022
15. Rosser, M.: Universal Signs Five-Film Deal for Timur Bekmambetov's Screenlife Projects (2020). Screen Daily, 10 June. https://www.screendaily.com/news/universal-signs-five-film-deal-for-timur-bekmambetovs-Screenlife-projects/5150487.article. Accessed 21 July 2022
16. The Numbers (2014). Unfriended. www.the-numbers.com/movie/Unfriended. Accessed 25 July 2022
17. Todd, A.: Why the Time Is Right for "Screenlife" Movies Like 'Searching' and 'Unfriended' (2018). Slash Films, 28 August. https://www.slashfilm.com/560609/Screenlife-movies-and-the-future/?utm_campaign=clip. Accessed 23 July 2022
18. Udell, J.: Let's Hear It for Screencasting (2005). InfoWorld, 11 February 2005. https://www.infoworld.com/article/2670005/let-s-hear-it-for-screencasting.html. Accessed 24 July 2022
19. Udell, J.: What Is Screencasting? An Interview with Jon Udell (2006). MasterNewMedia, 2 May 2. https://www.masternewmedia.org//#ixzz7aQ7vaCjA. Accessed 26 July 2022
20. Wei, E.: Screenlife Films and Immersion Cinema (2018). Medium, 21 December 2018. https://medium.com/emergent-concepts-in-new-media-art/Screenlife-films-and-immersion-cinema-ae784f2a6ea5.
21. Young, A.: Made in Frame: Cutting the Thriller "Searching" in Adobe Premiere Pro (2018). Frame.Io Blog. https://blog.frame.io/2018/09/10/made-in-frame-searching/. Accessed 26 July 2022

Filmography

22. *Cyberbully* (2015) Directed by Ben Chanan [Film]. UK, Channel 4
23. *Dnyukha!* (2018) Directed by Roman Karimov [Film]. BestFilm. Eu Distribution
24. *Host* (2020) Directed by Rob Savage [Film]. Shudder Distribution
25. *Searching* (2018) Directed by Aneesh Chaganty [Film]. Bazelevs Production
26. *#SidIaDoma* (*#SittingAtHome* (Russian: *#СидЯдома*) (2020) Directed by Olga Frenkel [Series] Premiere
27. *Thomas in Love* (2001) Directed by Pierre-Paul Renders [Film]. IFC Films
28. *Unfriended* (2014) Directed by Levan Gabriadze, [Film]. Bazelevs Production and Blumhouse Productions

Chapter 7
Designing Interrogative Robot Theater: A Robot Who Won't Take No for an Answer

Sahar Sajadieh and Hannen Wolfe

Abstract "You are the hottest thing in the room! I couldn't help but come over to introduce myself." That's how the conversation began between the human-sized female robot and an audience member in a corner of the room during *Come Hither to Me!* In this robot theater the robotic agent charms the audience with her seductive humor and subtly enters them into a provocative dialogue that surfaces their stereotypical biases in gendered social interactions. *Come Hither to Me!* exemplifies "Interrogative Robot Theater," our performative and critical method for social robotics research with an objective of designing robotic embodiment and interactivity for theatrical performances and public interventions. We apply various design and theater-making methods to develop a socially engaging, fun, and playful interactive experience for the audience. Using humorous conversation and embodied interaction design, our feminist robot theater makes a satirical performative commentary on misogynist dating culture and stereotypical gender roles. Inspired by the male-centered pickup artist community guidelines, we designed a chatbot decision tree for our female-gendered robot actor that flirts and provokes conversation with participants of all genders, subverting the imbalanced power dynamics of sexist social interactions. This interventionist theater-making methodology builds upon social justice-oriented interaction design, interrogative design, and Theater of the Oppressed. Through the application of this approach, *Come Hither to Me!* interrogates and problematizes gendered intimacy and agency in social interactions.

Sahar Sajadieh and Hannen Wolfe are equally contributing authors.

S. Sajadieh (✉)
University of California, San Diego, CA, USA
e-mail: sahar@saharsajadieh.com

H. Wolfe
Colby College, Waterville, ME, USA
e-mail: hewolfe@colby.edu

© The Author(s) 2023
A. L. Brooks (ed.), *Creating Digitally*, Intelligent Systems Reference Library 241,
https://doi.org/10.1007/978-3-031-31360-8_7

7.1   A Theoretical Framework

Interrogative robot theater is a method for developing socially-engaged robotic art and activism, as well as an analytical framework for performance studies, media theory, and social science research. To design a thought-provoking interaction with a robot in a performative context, we apply theater-making techniques and design methodologies. We use robots as a deformed and rather humorous representation of the human with a purpose of asking questions, evoking dialogue, and analyzing particular societal paradigms. In our method, interrogative robot theater, we propose using play and humor to lighten the discomfort caused by examining underlying sociopolitical issues within our communities. Our robot theater, *Come Hither to Me!*, exemplifies the application of this method for art practice and activism. *Come Hither to Me!* implements our interrogative design methodology to approach, engage, and provoke the audience using satirical flirtatious conversations with a female robot [52, 63]. This feminist robotic performance explores gender roles and agency in inequitable social interactions. Using Augusto Boal's interventionist method, Theater of the Oppressed, our artwork explores the dramaturgical politics of the space between human and robot.

Interrogative robot theater builds upon interrogative design, social justice-oriented design and critical computing. According to Krzysztof Wodiczko, interrogative design is a type of design methodology that "takes a risk, explores, articulates, and responds to the questionable conditions of life in today's world" [61]. While interrogative design focuses on examining current problems and providing solutions, social justice-oriented design incorporates political responsibility into proposed actions [16]. On the other hand, critical computing uses the medium of computing to question the ways that people interact with computational technology [22]. Despite using robot theater as a medium, we are not questioning the way people interact with robots, but instead, the way they interact with each other. The robot is a comical, exaggerated, and satirical representation of the human in society and its performance brings to the surface underlying social expectations and biases. In order to critique problematic interpersonal dynamics, we designed and developed the robot's interactions by codifying patterns of behavior and communication.

Interrogative robot theater also draws from the devised political theater of Augusto Boal, Theatre of the Oppressed, specifically his theatrical methods of Forum and Invisible Theater [7]. We borrowed his term, "spect-actors," (spectator + actors) which emphasized the important active role of the audience as an agent of change. For Boal, the audience members are both spectators and actors, since they watch the performance while making meaningful modifications and taking a crucial part in dramatic actions as the performance unfolds. In our proposed method, by applying ideas from Theater of the Oppressed, we playfully question viewers' beliefs and simultaneously make them active players in these theatrical acts.

In addition to incorporating activist approaches in our method, interrogative robot theater examines and re-envisions conventional roles of the traditional theater practice. Building a character, which was mainly the responsibility of the actor (and to

Table 7.1 Elements of creating an interrogative robot theater

Design modality	Equivalent role in theater	Function in theater production
Dialogue design	Playwright's role	Playwriting
Interaction design	Director's role	Blocking/staging a play
Character design	*Actor's role	Character development

*This can be partially the responsibility of director and casting director

some extent the director and casting director) is re-imagined as robot character design in our method. Staging a play and playwriting, which were the roles of the director and the playwright respectively, were adapted to robot interaction design and robot dialogue design in interrogative robot theater (see Table 7.1). This approach explores and redefines the process of making a theatrical production for robot theater and is applied in our performance, *Come Hither to Me!*

Interrogative robot theater is a theatrical performance in which the audience, over the course of interacting with one or more robot actor/s through a series of dramatic actions, is provoked to question or take action toward societal systems of oppression. While we are establishing interrogative robot theater as a novel creative practice in robotic arts, some prior works may fall within its definition. To be considered interrogative robot theater, the artwork needs to fulfill two distinct requirements. First, it needs to be interrogative, meaning that it intentionally raises questions about social systems of oppression. Second, the work needs to involve spect-actors, audience members who are involved in the performance, interacting with a robot performer.

7.2 Robot Theater: A New or Old Practice?

Robot theater is a performance showcasing a robot actor: an embodied interactive agent with a physical inorganic body and the ability to move in the performance space, which is perceived to have agency by the audience. Heather Knight, a Human Robot Interaction (HRI) researcher, considers robot theater to be an interdisciplinary field, encompassing robotics research, acting theory, cognitive neuroscience, behavioral psychology and dramaturgy [37]. When Knight describes robot theater, she focuses on narrative performances that include robots. She clarifies that "theatrical 'performance' can range from literal stage with audience to pre-meditated collisions with human environments, as in guerrilla theater or street performance. The term 'narrative' refers to a scripted or improvised sequence of coherent actions (and often speech) by one or more characters resulting in the construction of a timeline-based interaction arc (storyline)" [37].

With elements of theatricality and performativity entering into the field of social robotics, the research in robot theater and human robot interaction have become more intertwined. There are a few design methodologies for HRI that have been derived from the study of robot theater. Knight describes eight lessons learned about robot theater [37]. Nishiguchi et al. discuss a theatrical approach to human robot interaction, in which the contemporary colloquial theater theory is applied to HRI which when analyzed created a list of ten rules of how a robot should move and gesture when interacting with a robot [42]. There is a similarity between these rules and the rules defined by Kahn et al. when applying the concept of design patterns to designing social behavior in human robot interaction [35].

7.2.1 Robot Theater and Human Robot Interaction

Actor training techniques, particularly Stanislavski's Method acting, have been incorporated into HRI research. Hoffman's research has brought multiple theatrical concepts to HRI design including continuity and responsiveness from Method acting [25, 26], a movement-centric design approach [27], and a hybrid control system for puppeteering robotic stage actors live [28]. Knight's research with a robot barista used Method acting by having the participant act out a scene with the robot and then rate the interaction [24].

Similar to interrogative robot theater, some human robot interaction research projects focus on the performative acts of robots and the participants' interactions with them. However, only a few of these examples address social issues. Vilk and Fitter designed a robot that could perform stand-up comedy and studied the audiences' response. They used techniques from comedic performance (e.g., set-up and punchline) and had stand-up comedians write the jokes. Using the amount of the audiences' laughter as an evaluation method, Vilk et al. found that a robot with good timing was funnier, and adapting the robot's dialogue to the audiences' cues improved their reaction to individual jokes [57]. Knight also used audience tracking to train a robot stand-up comic [38]. While these works incorporate audience response in the adaptive performance, they are not usually politically charged. One example that explored social issues was Knight's robot barista that interjected into its clients' conversations while serving them. The study questioned participants' beliefs on data privacy and companies' ability to sell and use the users' data [24].

7.2.2 Robotic Performances and Performers

In robot theater a robot becomes a theatrical or social performer in a performative context or social scenario [24–26, 37, 39]. In contemporary theater history, there has been a connection between robots and theater since the term robot was introduced into the English language through Karel Čapek's play, R.U.R., in 1921 [10]. In theater, robots have used slapstick techniques in Commedia dell'arte performances [65], which was possible due to their physical presence and ability to move on stage. More recently in the play, I, Worker (2008), Hataraku Watashi explored human and robot gendered relationships and labor [9]. Heddatron explored the robotic nature of a woman's life devoid of freedom, by having robots force the lead to play *Hedda Gabler* repeatedly [43]. While *I, Worker* and *Heddatron* used robots in performances to highlight social issues, there was no direct interaction between the audience and the robots.

The emphasis of new media art practice on interactivity created a fertile ground for robotic performance to incorporate audience interaction with robots. Robots from early cybernetic art, *Rosa Bosom (R.0.S.A. Radio Operated Simulated Actress)* and *Robot K-456*, were originally designed for theatrical performances. Rosa was created in 1965 to perform as the Queen of France and won the Alternative Miss Universe competition in 1985 [31]. She was a human-sized robot that was remote controlled by her creator, Bruce Lacey, to go around and kiss people [3]. Rosa's act of kissing bystanders and making cat-calling whistles can be viewed as a performance of street harassment. If Lacey's intention in this performance was to question this form of interpersonal behavior in public space, the work could be considered as interrogative robot theater.

Nam Jun Paik's *Robot K-456* first performed in *Robot Opera* in 1964 [32]. During an exhibition of *Robot K-456* at the Whitney Museum of American Art in 1982, the robot was removed from a pedestal, guided down a street in New York City to the intersection of 75th Street and Madison Avenue and was hit by a car. In an interview Paik ironically called this work "the first catastrophe of the 21st century." While the collision was organized by Paik, the audience did not know that the driver was an actor (Bill Anastasi). Paik described the work as interrogating the "potential problems that arise when technologies collide out of human control" [32, 45]. *The First Catastrophe of the 21st Century* performance can be viewed as an example of Boal's Invisible Theater. Invisible Theater is a theatrical method proposed by Augusto Boal as part of his Theater of the Oppressed, in which a politically-charged theatrical act is reenacted in public space without the viewers being aware that it is a performance. While in *The First Catastrophe of the 21st Century* the robot and driver were performers, the audience surrounding the work did not know about it.

The Helpless Robot (1987) by Norman White [34], *Stupid Robot* (1985) by Simon Penny [48], and *Nose Wazoo* (1990) by Jim Pallas with Jim Zalewski [46] interacted with spectators through proxemics in different ways. *Nose Wazoo* played with the line between personal and social space, attempting to touch visitors with its nose when approached. *Stupid Robot* was designed to be reminiscent of a legless beggar and it shook a can of metal scraps when visitors were within its close proximity [48]. In *The Helpless Robot*, the robot would ask for help to draw people near, but if people came close, it would start to insult them. Its verbal interaction with the audience alternates between begging the audience to help it move its metal axis and bossing them around, followed by insults and accusations in a louder and louder voice in an abusive loop. While all three of these works were robots that responded to the spectators, *The Helpless Robot* bordered on interrogative robot theater.

In *Female Figure* (2014) by Jordan Wolfson, a female robot in a white corset, transparent skirt, and knee-high boots gave monologues using a male voice and danced to pop songs. To gain agency over the spectators and make them feel objectified, the robot used facial recognition software to "look at" them, returning their (male) gaze [6, 41]. In this way, the robotic performance both interacted with spectators (making eye-contact and responding to participants' presence) and interrogated the objectification of women and the male gaze.

7.3 Interrogative Robot Theater: A Practice-Based Approach from Character Design to Staging

We apply various interventionist art-making techniques to our performative human robot interaction design in order to turn robot theater into a medium for critiquing sociopolitical issues. Interrogative robot theater applies Wodiczko's interrogative design approach to the interaction design of robot theater. It incorporates critical computing and Dembrowski et al.'s social-justice oriented interaction design. Our methodology is also inspired by Augusto Boal's Theater of the Oppressed as a socially-engaged performative platform.

According to Wodiczko, interrogative design can function as a critical mirror questioning the user's preconceptions and assumptions about the self and others. Its objectives should be "to articulate and inspire communication of real," to make invisible issues visible, and to become "an opening through which a complexity of the lived experience can be recalled, memorized, translated, transmitted, perceived, and exchanged in a discursive and performative manner [61]." Interrogative robot theater uses Wodiczko's design approach to challenge the audience to probe into their unconscious biases and prejudices.

Dombrowski et al. describe six different design strategies for social justice-oriented interaction design: designing for transformation, recognition, reciprocity, enablement, distribution, and accountability [16]. Among these design strategies, the one that is most relevant to interrogative robot theater is "designing for recognition" [16]. In this way, we are using robot theater to identify unjust societal expectations and to question them. Similar to social justice-oriented interaction design, interrogative robot theater is committed to conflict, reflectivity, and personal politics. The methodology causes conflict through dialogue by questioning one's personal ethics in an entertaining yet critical way, and hence, striving to make viewers reflective without getting defensive.

Interrogative design and social justice-oriented interaction design methods do not specify the material or medium in which they are represented. Critical computing, on the other hand, explores the application of computation to "disempower hegemonic norms and socio-technical conditions" [22]. While critical computing is less focused on design and more focused on the medium, it has goals similar to interrogative design [23, 36]. Interrogative robot theater is a medium-specific approach akin to critical computing.

The interaction design of interrogative robot theater is influenced by the structure of Forum and Invisible Theater, political theater-making techniques pioneered by Augusto Boal [7]. While Invisible Theater makes a political intervention into pre-existing public spaces through performative acts that blend into social life, Forum Theater creates a new space in order to find potential solutions to an audience member's interpersonal, yet political, conflict. Even though the presence of robots in public spaces immediately introduces a performative quality into the robot's actions, our interrogative robot theater's provocative, interventionist mission can be hidden from the general audience who think they are only playfully interacting with a robot. As robots move into public spaces, there will be more opportunities for creating invisible interrogative robot theater.

7.3.1 Staging/Blocking: Interaction Design (Director's Role)

One of the main roles of the director is designing the blocking, the choreography of actors' movements in interaction with each other, the stage, and the audience. In our robot theater design approach, blocking is part of the process of the robot's embodied interaction design in various encounters with the audience, the space, and objects. Edward Hall's concept of proxemics is defined as "the interrelated observations and theories of man's use of space as a specialized elaboration of culture" [19]. It can be used to explore the line between comfort and discomfort in designing the blocking of interrogative robot theater. By getting too close or too far from the audience's social and personal spaces, the robot pushes their physical and metaphorical boundaries.

7.3.1.1 Forum Theater as a Model for Interaction Design

In Forum Theater a social scenario is presented in a performance and spect-actors can propose and enact new physical actions to collectively respond to the issue. Forum Theater includes the protagonist, the oppressed person defeated or frustrated by the oppressors who reinforce the problem. Unlike the protagonist, the oppressor may be multiple entities. It also includes the Joker that acts as the facilitator, the enabler, and the mediator for the group [8]. Forum Theater involves the replacement of any actor by the spect-actors who believe that they can offer alternative resolutions and are encouraged to interject and try out other solutions.

In interrogative robot theater spect-actors participate in a performative oppressive scenario. In this method, the robot plays the role of the oppressor, which can help create an alienation effect and make the audience observe the conflict from an outside perspective. We refer to this kind of interrogative robot theater as "Robot Forum Theater." By taking an active role in the performance, the spect-actor is allowed to objectively engage in this process of collective drama therapy to address a sociopolitical subject matter. Similar to Boal's approach, robot forum theater does not give the audience the satisfaction of solving the issue through a dramatic catharsis, but instead leaves them charged to make changes in society.

7.3.1.2   The Director-Joker

In Forum Theater, the role of the director is comparable to the Joker who wears many hats as the moderator, master of ceremonies, and workshop leader. The Joker works with the spect-actors to design a scene that contains a potential solution. The first iteration of the scene is performed only by actors. Afterward, the Joker introduces the idea that spect-actors can stop the performance and take over the role of any actor in the next reenactment, if they disagree with the proposed resolution. This process continues until the Joker closes the forum and ends the performance.

Commonly the theater director is also the casting director who chooses the actors, which in some ways the Joker may do as well. A traditional theater director imagines the vision of the dramatic text and theatrical performance, and implements it in the process of staging the work. In Forum Theater, the Joker obtains the vision of the performance from discussions with participants in the forum, and creates a theatrical scene based on that. The Joker together with the team of actors designs the constraints of the space and blocking of the initial scene. In the iterations of the scene, once the spect-actors swap in for actors, the Joker's role becomes much less significant. In other words, the Joker and actors slowly take a step out and let the spect-actors take charge of the dramatic direction of the work, and metaphorically take control of the actions needed to be taken in their social lives.

The role of the director in interrogative robot theater is closer to a video game director, interaction designer, or interactive art designer. In these interactive works, the director designs multiple paths, considering potential options that the participant may choose by their speech and movement. In *Digital Performance*, Steve Dixon discusses various levels of interactivity [15] based on the openness of the system and the level and depth of user interaction: navigation, participation, conversation, collaboration. Based on these categories, the participants in interrogative robot theater experience the performative interaction as a conversation: a dialogue between the user/participant and the interactive computational system. Technically, the interaction may also fall into the categories of participation ("users helping to bring to life the environment's sensory features") or collaboration ("the user and computer creating art together"). The category depends on the implementation of the robot's interactivity, complexity of the system, and the impact of spect-actors' input on the creation of the performance. In a more scripted version of the robot dialogue, as part of a less complex interactive performance, the interactivity category would be participation. The category would be considered collaboration in the case of a more complex conversational agent and elaborate interaction between the robot and spect-actors who construct of the performance together.

7.3.1.3   Choreographing the Physical and Emotional Spaces
                 of Interaction

Boal's Image Theater and Forum Theater explore the physical and emotional spaces of interaction between the audience and the actors. Forum Theater is constructed on top of Boal's Image Theater and to some extent can be considered as its animated form [7]. Image Theater tells a story and evokes dialogue by creating "tableau vivants". Tableau vivant (French for "living sculpture") is a static scene that can be used to express a sociopolitical issue through the choreography of still bodies and exploring their spatial relationship in the space. The specific form and orientation of the actor's body, as a sculptural object and in relation with other bodies—with its particular cultural meanings and metaphors—are quite expressive as a theatrical means. Developing and applying robots with extensive physical capability and gestural flcxibility, such as Hanson Robotics Sophia or Boston Dynamics Atlas humanoid robots, can extensively improve the nature of performance interactions in this type of robot theater.

7.3.2 Building a Character: Character Design (Actor's Role)

We propose a new classification and approach to robot theater and a new method for robot character design, inspired by the process of building a character in theatrical performances and Osada et al.'s method for defining social robot characters [44]. Similar to the approach proposed in their paper, we consider two main categories in the process of character design: "Robot Embodiment Design," which they refer to as surface design, and "Robot Behavior Design." We chose the word embodiment over surface to be inclusive of the physical and interactive aspects of the design of the robot's body. In this categorization, which we call the Descartes' mind-body dualism of robot character design [14], we differentiate between the process of designing the body and the mind of the robot.

Robot embodiment design, the design of the body, includes designing the robot's exoskeleton, costume, and makeup as well as designing the robot's movements, gestures, and voice. Robot behavior design is the process of identifying and implementing the personality and behavior of the robot. When designing for interrogative robot Theater, the robot's behavior and embodiment inform each other. We propose, however, that defining the robot's mind should precede the design of its body, in robot theater and other performative scenarios for social robots or any social interactive agents. In other words, their physical and interactive embodiment design should be created based on their personality and identity.

7.3.2.1 "Casting" the Robot

Traditionally in a theatrical production, the director and/or the casting director casts actors to play the roles of the characters in a play. They have to balance giving the actor enough agency to perform his/her role while ensuring their vision is portrayed. In robot forum theater, even though the casting process is different, the director "casts" a robot as the oppressor. The director/artist would identify and design the hardware and software requirements for the robot to perform the tasks and actions that are needed for the role. In this sense, the role of the director in robot forum theater becomes similar to the role of a puppet theater director. The robot actor itself can be considered as a technological embodiment of Edward Gordon Craig's über-marionette which he discussed in 1907 [12].

By choosing to cast a robot as an actor, the director is engaging with the audiences' preconceptions about robots, and the historical/sociopolitical context of the application of robots in art, literature, and society. The term "Robot" was introduced into the English language from the Czech word "Robota," which is translated to "forced labor" [1]. The robot has been used in media to both illustrate the repetitive nature of contemporary human life and the inflexible system that forces the masses to work like machines. When casting for an oppressive scenario, the decision to have a robot actor introduces different layers of meaning into the work. The robot can be used as a shorthand to question the validity of the system it represents.

7.3.2.2   The Robot Actor

While character design for theater involves the actor exploring and creating the character during the rehearsal and workshopping stages of the performance production, character design for robot theater is an integral part of developing the project as a whole. Human actors in theater use various techniques such as searching for inner motives of the character [54, 55] to justify their actions and portray their emotions properly. According to Richard Schechner, the theater actor creates a character on stage that is neither the character in the play nor their own character—living a double negative, somewhere between not me and not not me: "while performing, actors are not themselves, nor are they the characters" [53].

In Robot theater, a robot actor becomes a robotic embodiment of the character. Its personality is mostly identified and conceived through the process of embodiment, interaction, and optionally dialogue design with various project goals in mind. In some cases, the behavior of adaptive robots using various machine learning technologies may change over time for better or worse, based on the new knowledge they obtain from the environment and their interactions with humans. An example of this is Tay, Microsoft's chatbot, which became vulgarly racist after twenty-four hours of learning from the discriminatory comments of users online [33].

7.3.3   Script Writing: Dialogue Design (Playwright's Role)

The process of playwriting and dramaturgy in our interrogative robot theater design is an integrated component of the interaction and character design of the robot. While in some robotic performances the dialogue can be scripted, in others the robot can learn over time from verbal interaction with the audience using natural language processing (NLP) techniques, or it can also be a combination of both. Regardless of using scripted dialogue or NLP, gathered data/information about the system is used to create the algorithm that constructs the oppressive situation. Through this approach the dialogue design draws from interrogative design, with the dialogue expressing the oppressive system in Forum Theater.

Inspired by interactive filmmaking, screenwriting, and video game design, one approach to dialogue design is to use a dialogue decision tree. In this approach, having written dialogues as possible responses to a potential set of options alludes to a system that expects people to fall into predefined boxes. This type of system frequently has default responses when the visitor's response doesn't fit into any of the predetermined options. In a Kafkaesque way, the participant is placed in the middle of a system with the robot's dialogue representing the bureaucracy of an established structure [60]. This may be restrictive and frustrating for the audience, provoking a sense of alienation in the participant. If applied properly, it can be used as a performative means to emphasize that the participant cannot break out of the oppressive situation because all acceptable paths are accounted for.

As an alternative approach, natural language processing models trained on large human conversation data sets could be used to generate the oppressor's dialogue. Models trained on data from the Internet can introduce exciting, unexpected materials into the performance or create unwanted biases in the robot's dialogue, which may require more human supervision. It is important to consider these potential issues when using NLP for the dialogue design of the robot. However, if carefully and properly applied, these mistakes could be repeated consciously in interrogative robot theater to further create dialogue representing an oppressive system. By intentionally training the model on pre-existing biased text (e.g., sexist or racist comments), the robot can represent the collective biases of a group of people.

7.4 Research Methodology Implemented: *Come Hither to Me!*

Come Hither to Me! is a feminist robotic performance, inspired by the Boal's Theater of the Oppressed and Brecht's Epic Theatre. We applied our interrogative robot theater design methodology in our performance, *Come Hither to Me!* to bring attention to the gamification of dating, as a microcosm of social interactions, and to interrogate how societal stereotypes affect interpersonal relationships. This robotic artwork was exhibited in May 2018 at the Media Arts and Technology digital arts symposium at the California NanoSystems Institute - UC Santa Barbara and the Santa Barbara Center for Arts, Science, and Technology. Another version of this performance was presented at the ACM CHI (Human Factors in Computing Systems Conference) Interactivity Exhibition in Glasgow, Scotland, in May 2019. By incorporating ideas from Forum Theater, the artwork problematizes misogynistic behavior.

Come Hither to Me! showcases our custom-built female robot, ROVERita. The character design of ROVERita, comprised of embodiment and behavior design, was inspired by Stanislavski's theatrical method of building a character and other actor training approaches, in addition to the character design method proposed by Osada et al. [44, 54, 55]. We also incorporated ideas from Boal's Forum Theater into the interaction design of *Come Hither to Me!* We consider the participants to be the spect-actors of our performance, who would respond to and test different solutions for the performative, oppressive scenario during their interaction with the robot.

Through our particular verbal and embodied interaction design for this performance, our robot becomes the antagonist of this theatrical work. ROVERita plays the role of the oppressor by drawing on pickup artist media and guide books. As defined by Cosma et al., these guidelines are "men's sex advice, centered on cultivating masculinity markers by obtaining sex from multiple women" [11]. The interaction design of the robot for this performance draws from theater blocking and proxemics research. It also applies various critical design/theater-making techniques. The dialogue design of the robot was inspired by Futurist Theater and interactive narrative scriptwriting, using pickup artist strategies as an algorithmic model for the conversational flowchart.

7.4.1 Character Design: ROVERita

ROVERita is our human-sized female-gendered robot that we developed as a platform for human robot interaction and robot theater research (see Fig. 7.1). In our robot theater, *Come Hither to Me!*, ROVERita interacted with the audience by approaching them and initiating a flirtatious conversation, which over time became more intense and humorous [52]. Our robot identified the participants in the room, made her way toward one, and started the interaction with a pickup line (Fig. 7.2). Using the ROVER platform [62], ROVERita could detect the "hottest" person in the room with a heat sensor, move toward them by avoiding obstacles, stop three feet away, and begin a conversation. The participant could choose to respond and engage in a conversation with the robot, or walk away, in which case the robot would turn around and choose another participant. The robot's goal was to ask the participants out and obtain their phone numbers [52, 63]. Our goal, as the creators and designers, was to create an entertaining, socially-engaged interaction between the audience and the robot; to place the audience in the middle of a societal issue and leave them charged so they would reflect and act in related situations in their daily lives.

7.4.1.1 ROVERita, the Oppressor in the Forum Theater

Come Hither to Me! can be considered an implementation of robot forum theater, with ROVERita as the oppressor, the participant as the spect-actors, and the artist/s as the Joker facilitating the interactions in the performance. If chosen by ROVERita, the spect-actors (our protagonists) can interact with ROVERita for the whole duration of the performance, or walk away any time they no longer wish to engage. The

Fig. 7.1 In *Come Hither to Me!* ROVERita interacts with a female participant

objectifying behaviors and proposed microaggressions in pickup artist guidelines are used as a model for ROVERita's interaction design with the audience.

In our political performance, ROVERita is simply acting out societal forces when she is actively pursuing bystanders because she is a subject-object. The oppressive force is that of the patriarchal society. The robot reenacts the oppressive sociopolitical situation in a vicious cycle. By having ROVERita perform the same scenario repeatedly, it allows the audience to test different solutions. Pickup artist strategies represent the sexist patterns of behavior and accepted gender roles in society. ROVERita's reenactment of a pickup artist in an exaggerated and satirical way brings the audience subtly and playfully inside this problematic situation, making them feel insulted enough to leave with a sense of unease. The conflict between personal agency and societal structures impels the dramatic action and through that knowledge drives the spectator to action.

Fig. 7.2 ROVERita is a
seductive female interactive
robot

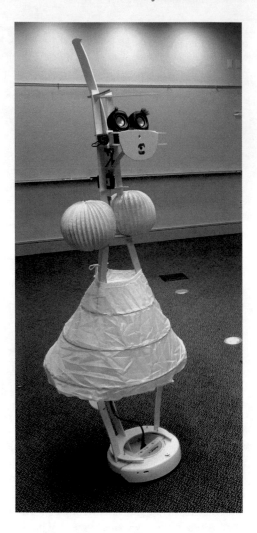

7.4.1.2   ROVERita as a Brechtian Character-Object

Bertolt Brecht proposes the idea of the character-object in which "the character is not absolute subject but the object of economic or social forces to which [they] respond and in virtue of which [they] act" [7]. ROVERita being a robot emphasizes that the character is an object (or a character-object). The robotic nature of the actor emphasizes her lack of agency and illustrates her being simultaneously an oppressor and an object/victim of the socio-economic pressure. ROVERita in that way becomes a representation of social forces engaging with the public.

7.4.1.3 Not ROVERita, Not Not ROVERita

In explaining the transformative quality of performances and rituals, Richard Schechner describes an actor's liminial state of temporarily leaving one's body (the actor's body), but not quite embodying another's (the character's body): a state between not me and not not me [53]. In the case of autonomous robots, including ROVERita, since they do not have an inherently embedded personality, they become (at least initially) whom the designers envision them to be.

As an interactive agent playing the role of the pickup artist, ROVERita has a different conversation with every participant. Similar to an improv artist, in her role as the antagonist of the Forum Theater, ROVERita uses different "strips of behavior" (in her case, different series of dialogues, movements, and gestures) but in various orders and rhythms in different interactions [53].

ROVERita's dialogues are transformed based on new conversation data from the audience after her interactions in different performances. She gradually becomes a character different from the archetypal oppressor as envisioned by the creators, and over time is adapted to specific responses of the audience. In this way, ROVERita's act puts her in a playful and performative state between not ROVERita and not not ROVERita.

Moreover, the character of ROVERita, having been designed based on the pickup artist guidelines, is an affectatious persona, fluctuating depending on whom the target is, in order to better manipulate the subject. Therefore, in interacting with her artificial character, the audience experiences a facade in flux, further highlighting the liminal state of her character. In the audience's perception, even though they don't witness a liminal persona between an actor and a character (such as in a film or a play), they envision struggles of a robot to be perceived as a believable humanoid character, or a machine in the state of becoming human (i.e., not machine…not not machine).

7.4.1.4 Robot Behavior Design

We designed a character for the robot with a predefined set of interactive and behavioral patterns. With an objective of designing a theatrical feminist critique of sexism in dating culture, we employed pickup artist strategies as a self-critical tool. Pickup artist strategies, tactics mainly used by men to approach and seduce women, were used in the interaction and character design of our female robot, as well as her dialogue design and dramaturgy. The assemblage of these different elements of design and the juxtaposition between them together define who ROVERita is and how she is perceived by the audience.

We first identified the characteristics of a pickup artist persona, using popular pickup artist books and online resources, and applied them in designing the character of our robot [50, 56]. While these guidelines are under the guise of self improvement material, they extensively objectify women and promote troubling ideas about consent [64]. Pickup artists follow certain patterns in their behavioral approach and body language as well as composed lines and scenarios as seduction strategies. They try to make their target feel simultaneously special and desperate for their attention. These are some of the pickup artist techniques that we extracted from the guidelines and applied in our design process: 1. backhanded compliments (referred to as negs), 2. pickup lines, 3. excessive confidence, 4. intensive persistence, 5. push and pull approach, 6. the rapid pace in dating approach and asking targets out, 7. false time constraint, 8. pretence of being laid back, 9. proudly peacocking [4, 21, 56].

Additionally, we used Stanislavski's system of Method acting to identify and apply the central inner motives of the character [54]. In this case, our pickup artist robot's main inner motive is to engage the audience in a conversation and to get as many phone numbers as possible for potential dates. Even though the audience members don't expect to go on a real date with the robot and the interactions are rather humorous and satirical, the objective is for the robot to engage them long enough in the conversation to ask them out and eventually get their phone number. To have a straightforward guideline for our character and interaction design process, we intentionally set a simple inner motive for the robot. This seems to be a common strategy in designing how robots interact with humans [24, 29, 30, 57].

7.4.1.5   Robot Embodiment Design

The character embodiment design of ROVERita consists of the identification of her sensory, cognitive, and interactive skills, in addition to the design of her embodied form, motions, and gestures. The design of her body entails making decisions about her skeletal structure, shell, skin, facial and bodily features, and alterable exterior elements such as makeup and costume. We analyzed our performance requirements and ROVERita's expected behavioral patterns using theater dramaturgical approaches and interrogative methods, and then determined the necessary capabilities for ROVERita. Based on those requirements and other elements of her envisioned character, we came up with the design of her body's shape and gestures.

As the main antagonist of our theatrical act, our robot needed to identity objects and people in the room, choose one person at a time, and be able to move in the space, go toward them, and stop at a proper distance. She also needed to be able to move back and forth in order to explore the participant's social and personal space (not intimate space, in this version), as defined by Hall [20], and turn around and leave, when dramaturgically needed. Thus, an iRobot CREATE base (equipped with a bump sensor) was installed as feet at the bottom of ROVERita's bodily structure so she could navigate the performance space.

She also has the ability to receive and interpret information from her surroundings using various sensors and cameras. ROVERita uses bump sensors for collision avoidance and proximity sensors for distance measurement, specifically for waist-high objects. ROVERita has a low-resolution heat camera to find the hottest area in her field of vision, ironically "the hottest person in the room." She also has a video camera and applies a face detection algorithm in real time to determine if the hot object is a person, so she would avoid hitting on objects such as a radiator, by mistake. For detecting faces, we used the OpenCV library's implementation of Haar feature-based cascade classifiers [58].

7.4.2 Interaction Design (with Space and Audience): Staging and Designing Interactivity

In *Come Hither to Me!* we applied our interrogative design methodology for creating robot theater. Our performance surfaces and interrogates normative sexist prejudices within our daily interactions. As noted by Wodiczko, design is considered interrogative when increases the level of ethical alertness by asking and provoking questions [61]. *Come Hither to Me!* problematizes societal gender biases and elicits questions about the expected adequate amount of female agency in approaching a potential partner through a satirical commentary on pickup artists.

Historically, women had restrictions on the ways they could publicly express their interests when choosing their romantic partners. For instance in the Victorian era, women were not allowed to publicly and explicitly flirt, they had to use non-verbal cues such as dropping a handkerchief or specific ways of holding their fan to send a signal to their potential desired partner [2, 5]. *Come Hither to Me!* asks to what extent these cultural norms and restrictions have remained in our society, both in the context of dating and other types of gendered encounters. Through performative interactions between the audience and the robot, the artwork touches upon and subtly ridicules societal restrictions and cultural norms which set invisible boundaries to the shape and amount of accepted female agency within our social structures.

As mentioned earlier, in the process of creating *Come Hither to Me!* as an example of interrogative robot theater, we also used social justice-oriented interaction design and Theater of the Oppressed. In their second proposed design strategy for social justice, Dombrowski et al. describe designing for recognition as an approach for "identifying unjust practices, policies, laws, and other phenomena," followed by articulation and framing of the problem through the process of design [16]. Our artwork uses robot theater as a platform for recognition and further expression of the hidden systemic sexism interwoven into our communication paradigms. Using the female robot as a pickup artist provoked an immediate reaction in the audience as it felt subconsciously wrong, challenging the audience to reflect on their gender preconceptions.

7.4.2.1 Theater of the Oppressed and Blocking of *Come Hither to Me!*

Inspired by Boal's Theater of the Oppressed, the blocking of *Come Hither to Me!* is designed to provoke dialogues and thoughts about subtle oppressive microaggressions in society. Our interrogative design approach in creating this artwork borrowed various creative and critical techniques from Forum and Invisible Theater. In our

Table 7.2 Embodied gestural vocabulary designed for ROVERita's interactivity and related examples in the performance

Gestural vocabulary	Examples of the robot's interactions
Moving forward	When the robot gets verbal confirmation from the audience to start/continue the conversation
Turning right/left	When the robot locate a hot point on her right/left, she turns toward that direction
Turning around and leaving	When the participant in conversation with the robot leaves or doesn't response for more than a certain period of time
Backing up	When the robot hit a physical barrier
Wiggling	When the robot want to show her excitement toward the participant's response
Wobbling/meandering	When the robot is searching for a new conversation partner

performance, the Forum Theater's process of workshopping potential resolutions for a social issue in the performance was replaced with audience participation in ROVERita's interactive act and informal group discussions. As opposed to Forum Theater, in *Come Hither to Me!* the original scenario is not created based on a sociopolitical issue raised by an audience member. Instead, the performance brings into attention an oppressive issue that we, the artists, identified and symbolically turned into a flirtatious conversation with a female robot. This theatrical robotic artwork playfully exposes audience's sexist normative beliefs in society, while they respond spontaneously to the humorous yet oppressive acts of our pushy flirtatious robot.

While in Forum Theater any audience member at any time during the performance can get out of their seat to participate in the act, in *Come Hither to Me!* the robot is the one who chooses with whom to interact. Moreover, in Forum Theater any participant can replace and play the role of any of the characters, but in our robot forum theater, spect-actors can only play the role of the protagonist and interact with the oppressor robot. In *Come Hither to Me!* the spect-actors are unaware that they are addressing a social issue in their interactions, and therefore their reactions to the situation becomes more spontaneous and impulsive. This playfully puts them in the middle of a problematic real-life scenario and has them experience the situation as it unfolds. Those viewers who witness ROVERita's conversation with a spect-actor and decide to interact with ROVERita afterward are comparable to the spect-actors of Boal's Forum Theater who replace a character and try a new solution.

7.4.2.2   Gestural Interaction Design

In *Come Hither to Me!*, ROVERita leads the conversation and asserts her agency as the one who chooses her partner through her direct flirtatious dialogues, persistent attitude, and her embodied movements and gestures. Designing the blocking and interactive motions for robots are the main components of interaction design in a robotic theatrical performance. Considering the movement capabilities of our robot, we defined gestural vocabulary for ROVERita's interactions with the audience and the performance space (see Table 7.2). In the process of staging this interactive performance, we applied this vocabulary in an expressive and performative way to create the illusion of agency in her actions and responses, and to evoke emotive reactions in the audience.

ROVERita's interaction design echos the Futurists' humorous, yet aggressive, interactive performances, as well as some later robotic artworks, such as Bruce Lacey's *Rosa Bosom (R.0.S.A. Radio Operated Simulated Actress)* (1965) and Norman White's robot in *The Helpless Robot* (1987). Rosa Bosom, similar to ROVERita, was a human-sized robot that actively pursued bystanders. Unlike ROVERita, Rosa Bosom was not autonomous, instead being remote-controlled by Lacey [49]. At the beginning of *Come Hither to Me!*, our robot locates the hottest body (with a face) in the field of view of her heat sensor, goes toward them, and initiates the conversation. This action performatively illustrates the agency of the robot and makes the approached audience member feel chosen. Either being the person selected by the robot in a group of people or being approached by her as the only person in the room can make the audience member feel special and evoke the curiosity of the chosen person and the witnessing audience.

ROVERita uses a low-resolution heat sensor to locate a person in the room, and bump and proximity sensors to avoid obstacles in her way. Once ROVERita finds a participant to interact with, she stops three feet away from her target and initiates a conversation by playing an audio file. When her communication partner is speaking, she listens while transcribing what they say and recording video of their facial response. The transcription is used to determine what she would say next. While ROVERita can see a face, she continues to talk to the participant. The final interaction is when either they move away or the conversation ends. Then, she starts moving toward the nearest hot area that she locates in the room.

ROVERita uses exaggerated, theatrical non-verbal cues to show her interest. These embodied gestures, such as wiggling, show her excitement when she receives positive responses from the spect-actor or when she tries to hold the their attention (see Table 7.2). For instance, after she asks the participant out on a date or asks them whether they would prefer wine or ice-cream and they confirm their interest, she moves forward a little and wiggles her body as a sign of excitement. She then follows through with a compliment and asks for their number: "I love how passionate you are! ... Give me your number and I'll keep you posted" (see Fig. 7.3).

Table 7.3 Examples of ROVERita's introductory, flirtatious, and negative comments

Introduction	Hi, you looked so beautiful over here, I had to come up and say hello. Hey you! You look gorgeous! Can I talk to you for a minute?
Flirtatious comments	Your eyes are breathtaking. You smell really good. Wow! You're just so. . . Wow! I was feeling a bit off today, until you turned me on.
Negative comments	Ew! You spit on me! I like the way you give up fashion for comfort. Hey, nice elbows. Really, I mean it! Your hair is so beautiful! But wait, is that a wig?

7.4.2.3 Choreographing Spatial Relationship: Physical and Emotional Distances of Oppression

While in Boal's Forum Theater, spatial relationships were defined through full-bodied interactions with human actors, in *Come Hither to Me!* ROVERita's interactions were less expressive because of the limitations of her physical motions and embodied gestures. We designed different movements, gestures, and blocking as potential responses of the robot. Like a directorial approach in designing theater blocking, these choices were based on beat changes in dialogues and different anticipated spect-actor reactions.

To convey her interest in a person, ROVERita approaches the participant by moving into their personal space. Moving closer to the participant and intruding into their territory can be viewed as an act of dominance [47]. With ROVERita representing the oppressor in robot forum theater, our blocking attempts to place her in a more dominant position over the course of the interaction, as she gets closer over time to the spect-actor. When she loses interest in the participant, after several attempts and having failed to obtain their phone number, she turns around and leaves, while making a rather offensive exit comment (e.g., "You are not that hot anyway! What a waste of time!").

In our version of robotic Theater of the Oppressed, we explored these spaces by arranging the spacial orientation between ROVERita and the spect-actor. We used Edward Hall's concept of proxemics and interpersonal distances for allocating an appropriate distance between them in different situations [19]. For instance, at the beginning when the robot finds an audience member to interact with, she stops three-feet away from them, which falls into the far phase of their personal space (2.5–4 ft) [19]. She starts the conversation with a compliment and a follow-up question: "Hey you! You look gorgeous! Can I talk to you for a minute?" (see Table 7.3). If the audience member shows interest in interacting with ROVERita, she will take another step forward to get a little closer to them and carry on with another affirmative flirtatious line (e.g., "Oh! You smell really good!" or "Your eyes are breathtaking!"). By moving forward when she receives bodily or verbal confirmation from the audience showing their interest in continuing the conversation (see Table 7.2), ROVERita gradually moves from the far phase (2.5–4 ft) to the close phase (1.5–2.5 ft) of the audience's personal space, but never enters the intimate space of the audience (less than 1.5 ft) [19].

ROVERita will leave if she is ignored. If at a point during the conversation ROVERita can no longer see the participant's face for a certain period of time (e.g., they looked away from her or left her field of view) she responds aggressively to grasp their attention: "I wasn't done talking to you!" If ROVERita expects the participant to answer a question and they don't, she responds humorously: "What happened? Cat got your tongue?" In either of these cases if the participant continues to be silent or does not reappear in ROVERita's field of view, she turns around and looks for a new conversation partner.

7.4.3 Dialogue Design: Playwriting + Dramaturgy

The dialogue design of ROVERita in *Come Hither to Me!* exemplifies the application of our interrogative robot theater creative methodology. Her dialogue structure is designed as a series of questions with flirtatious comments and lightly insulting compliments (negs) interspersed throughout the conversation, based on pickup artist strategies with a humorous and yet critical approach [56].

We identified the main beat changes in the dialogue design of ROVERita, using modern theater directorial approaches for play analysis and blocking design [13, 26]. Each beat change within dramatic dialogues in a play shows the character's different state of mind or tactic used in interactions with others to achieve their inner motives. In our dialogue design, each section with a different beat was constructed based on a proposed tactic in pickup artist strategies with an inner motive of getting a date (e.g., compliment, negs, push/pull, etc.).

We developed a flowchart to identify common categories of questions and answers and to predict potential responses and follow-up questions. The conversation starts with a pickup line, then follows with small talk questions to identify commonalities with the participant. These questions become more sensual and tempting over time. After establishing a rapport, the robot states that she has time constraints and asks about interests in common activities. Once a common activity interest is confirmed, the robot asks for the participant's phone number to go on a date. Throughout the conversation, ROVERita listens for keywords (or different combinations of keywords and phrases) in the participant's responses, so that she can answer appropriately. In situations where the robot doesn't find any known keywords, she responds with "neutral" statements to divert the conversation in a different direction without sounding like she did not understand the participant (see examples of neutral statements in Fig. 7.3).

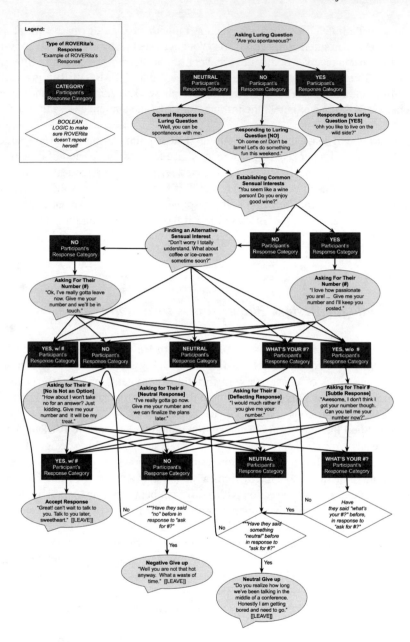

Fig. 7.3 This diagram shows an example of the dialogue flowchart in the conversation endgame, when ROVERita won't take NO for an answer and persists in getting the participant's number

7.4.3.1 Pickup Artist Strategies and Parodic Dialogue Design

The idea of lightly insulting compliments is referred to as negs [40]. These negative compliments are used by pickup artists as a tactic for lowering their target's self-esteem, and potentially making them respond positively to requests. The concept is supported by studies which show a correlation between lower self-esteem and more compliance with the requests of others [18, 59].

A research study focusing specifically on pickup artist strategies in heterosexual courting categorized three main measurable traits used in these strategies: competition, isolation, and teasing [21]. We applied mainly competition and teasing to the interactions of the robot. In regard to teasing, ROVERita alternates between compliments and negs. She applies the pickup artist pull/push tactic: she first shows interest in the participant by giving positive comments and focusing her attention on them, and then she pulls away her positive attention in a playful way by turning around the compliments or even making slightly insulting comments. She also pushes the participant's boundaries by making inappropriate personal comments that are either overly flirtatious or insulting, and then makes a joke and acts with an excessive niceness and politeness.

With respect to competition, ROVERita does not easily give up when she is rejected and keeps persisting by changing strategy and suggesting alternative options (see the flowchart in Fig. 7.3). If the participant says that they have a significant other, ROVERita responds with a specific strategy that is referred to in the pickup artist community as the "boyfriend destroyer," in which she begins undermining the significant other by making fun of them while simultaneously admiring the participant (e.g., "What a moron! I would never do that." or "I can't imagine who can deserve you!"). This strategy in real-life can potentially result in the isolation effect, but in the context of this performance and short span of the conversations, it mainly has a comical ice-breaking effect. Before inviting the participant to a date, ROVERita also attempts to incite their competitiveness by asking provocative questions, such as "Are you spontaneous?" followed by suggesting going out for some fun activities. If the participant states that they aren't spontaneous she teases them for being boring.

In the end, ROVERita does not easily give up when she is rejected and keeps changing strategy and suggesting alternative options. For each of the potential participant reactions, ROVERita has a set of strategic response subcategories with the objective of getting their phone number. For example, a participant could 1. say yes to coffee but would not give their phone number, 2. say no to giving a phone number, 3. ask for ROVERita's number instead, 4. ask when they would get coffee, and 5. finally give ROVERita their number. Figure 7.3 shows ROVERita's responses in each of those situations and provides an overview of the dialogue design of the conversation endgame.

7.4.3.2   Aggressive Dialogue of the Oppressor

Using Forum Theater model, the dialogue design of *Come Hither to Me!* applies data produced by an oppressive system (e.g., pickup artist guidelines representing systemic patriarchal paradigms) to create an algorithm for verbal interaction. The dialogues and intense interactions with the audience are also inspired by the Futurist performance tradition where the audience members were being provoked by insult, humor, and misogyny. The interaction and dialogue design of ROVERita echo both the tradition of replacing a human actor with a machine and direct physical and verbal interaction with the audience. Not only did the audience become active agents in Futurist artworks, but in some performances they were the subject of direct aggression on stage such as being shouted at and chased as part of the dramatic actions of the scenes [15]. Some later interactive installations, such as Norman White's *The Helpless Robot*, are other instances of the application of abusive humorous dialogue and audience manipulation in digital performance. Like ROVERita, the robot in *The Helpless Robot* is aggressively insistent [15].

In *Come Hither to Me!*, ROVERita's dialogue makes the spect-actor annoyed and uncomfortable but entertained enough to still want to continue the conversation. This allows for ROVERita to persistently objectify and ridicule them. The artwork strives to leave the participant impacted through humor and insults, either insulted or entertained (or both), regardless of when ROVERita or the participant leaves. There are multiple ways the scenario can end, without any of them offering any kind of cathartic satisfaction. The lack of resolution leaves the spect-actor charged to reflect and hopefully make change.

The dialectical tension between different components of ROVERita's design— embodiment, interaction, and dialogue design—simultaneously places her in a liminal state between various identities, a state of becoming [17]. This tension contributes to breaking down the preexisting boundaries between male and female, heterosexual and homosexual, and human and machine. In this way, through the application of humor, aggression, and play, the artwork provokes reactions in the audience to one's stereotypical expectations in real-life scenarios.

7.5   An Epilogue

Using pickup artist techniques, in our theatrical robotic artwork we bring to the surface underlying significant cultural issues in interpersonal relationships. We codified pickup artist behaviors into a dialogue with the robot as a microcosm of the patriarchal dynamics of society. Our robot represents an archetypal human in a position of power. We use various contrasting elements of design to create a performative assemblage that both theatrically entertains and confuses the audience's senses, triggering the audience to reflect and act. Inspired by existing models of political art practices, such as Boal's Theater of the Oppressed and Wodiczko's interrogative design, we propose a new socially-engaged performance practice: interrogative robot theater.

Through the social dance of a humorous robot in desperate need of finding a date, *Come Hither to Me!* illustrates the application of this method in the context of a flirtatious conversation with the audience.

7.5.1   Interrogative Design and Robot Forum Theater

Interrogative robot theater explores using robots as agents of change and mediums for social intervention. Having the robot role-play the oppressor not only makes the audience more engaged and entertained but also creates a sense of Brechtian alienation on stage, allowing them to view the issues more objectively. The current version of robot forum theater that we discussed brings to the audience's attention a societal issue chosen by the artists. Using this approach, we, the artists, are crowd-sourcing solutions to scenarios that we find pressing and relevant. If the same issue was presented to spect-actors, they would create entirely different scenarios. By incorporating the spect-actors into the design process, the work could be more relevant to them and their communities. This would democratize the design process.

In traditional Forum Theater, there are three primary roles: the protagonist, the antagonist (the oppressor), and the Joker. In our version of the Theater of the Oppressed, we focused on ROVERita as the oppressor and the sole antagonist. In other variations of robot forum theater one or more actors or robots could represent the oppressor. An instance of this application could contain multiple robots with related roles as the oppressor, for which a similar algorithm and training data set could be used for each robot. Two robots could also play opposite one another as the protagonist and antagonist. The robot protagonist's dialogue potentially could be generated from the recordings of spect-actors' reactions to the robot antagonist. When the robot antagonist and robot protagonist perform together, they would be reenacting the solutions proposed by the spect-actors.

Robot forum theater could be performed in various kinds of public spaces, using different modes of liveness and modalities of interaction [51]. For instance, the spect-actors could be online participants who would respond live to the issues brought up by the sociopolitically-charged performance. The robot actor could then apply and reenact the proposed ideas live online. Having the performative interaction happen online in the form of a text-based online drama by a disembodied robot (i.e., a chatbot) could be another possibility for the application of robot forum theater.

In *Come Hither to Me!* we mainly applied Edward Hall's politics of space. We could further extend the politics of gesture and the application of tableau vivants in robot theater. In Image Theater, Boal applies tableau vivant by creating provocative and politically charged sculptural arrangement of still bodies in space. We could further develop and apply his approach in interrogative robot theater, by designing robots with more malleable and expressive facial and bodily gestures. In these scenarios the robots would interact with the participants with their full bodies and explore the physical and emotional spaces in between, through using their embodied gestures and movements. Moreover, by workshopping the performance we could create forum-like environments and take into consideration the social issues and personal experiences of the audience in the process of creating the performative oppressive scenario to be reenacted by robots and spect-actors.

7.5.2   A Double Negative Identity of a Robotic Performer

As the creators of *Come Hither to Me!* we performed all the roles of the actor, casting director, director, and playwright. In robot theaters that mainly use natural language processing for verbal interaction design, the traditional roles of director and playwright may be eliminated. In that case instead of writing a set of lines, the conversation designers will most likely be machine learning programmers who will define a set of rules and train a neural network. This shift changes the notion of authorship, since the lines are not written by a playwright but generated by the robot's neural network. The participants will also become the co-authors of the dialogues and co-creators of the character because the data from their interactions will be used to train the neural network. In the process of building and embodying a character in traditional theater, actors have to do more than pretend, but live a double negative: not me…not not me. "The actress is not Ophelia, but she is not not Ophelia" [53]. In a machine learning based robotic character and dialogue design, we propose that the robotic actor lives in a different liminal double negative space. It is, simultaneously, in a constant state of becoming not not me and not not you, since it doesn't have a static character and its personality, words, acts, and identity exist in a state of flux in a liminal space between itself and others.

References

1. Robot. Oxford reference. https://www.oxfordreference.com/view/10.1093/oi/authority. 20110803100425123 Accessed 1 Sept 2022
2. Cassell's Family Magazine. v. 18. Cassell (1891). https://books.google.com/books? id=uKPQAAAAMAAJ
3. Nightmare robots. British pathé ltd. (1968). https://britishpathe.com/video/nightmare-robots/ query/nightmare+robots. Accessed 1 Sept 2022
4. Albright, D.: How to Meet and Pick Up Women. New Tradition Books (2010)
5. Aldrich, E.: From the Ballroom to Hell: Grace and Folly in Nineteenth-Century Dance. Northwestern University Press (1991)
6. Berger, J.: Ways of Seeing. Penguin, UK (2008)
7. Boal, A.: Theater of the Oppressed. Pluto classics, Pluto (2000)
8. Boal, A.: Games for Actors and Non-Actors. Routledge (2005)
9. Borggreen, G.: "Robots cannot lie": performative parasites of robot-human theatre. In: Robophilosophy, pp. 157–163 (2014)
10. Capek, K.: R.U.R. (Rossum's Universal Robots). Penguin (2004)
11. Cosma, S., Gurevich, M.: Securing sex: embattled masculinity and the pressured pursuit of women's bodies in men's online sex advice. Feminism Psychol. **30**(1), 42–62 (2020)
12. Craig, E.G.: The Actor and the Über-Marionette. The Mask (1907)

13. DeKoven, L.: Changing Direction: A Practical Approach to Directing Actors in Film and Theatre: Foreword by Ang Lee. Routledge (2006)
14. Descartes, R., Haldane, E.S., Ross, G.R.T.: Meditations on First Philosophy in Focus. Psychology Press (1993)
15. Dixon, S.: Digital Performance: A History of New Media in Theater, Dance, Performance Art, and Installation. MIT Press (2007)
16. Dombrowski, L., Harmon, E., Fox, S.: Social justice-oriented interaction design: outlining key design strategies and commitments. In: Proceedings of the 2016 ACM Conference on Designing Interactive Systems, pp. 656–671 (2016)
17. Felix, G., Guattari, D.: A thousand Plateaus· Capitalism and Schizophrenia (Trans. by Massumi, B.). University of Minnesota, Minneapolis (1987)
18. Gudjonsson, G.H., Sigurdsson, J.F.: The relationship of compliance with coping strategies and self-esteem. Eur. J. Psychol. Assess. **19**(2), 117 (2003)
19. Hall, E.T.: The Hidden Dimension. Doubleday & Co (1966)
20. Hall, E.T., Birdwhistell, R.L., Bock, B., Bohannan, P., Diebold, A.R., Jr., Durbin, M., Edmonson, M.S., Fischer, J., Hymes, D., Kimball, S.T., et al.: Proxemics [and comments and replies]. Curr. Anthropol. **9**(2/3), 83–108 (1968)
21. Hall, J.A., Canterberry, M.: Sexism and assertive courtship strategies. Sex Roles **65**(11–12), 840–853 (2011)
22. Harrell, D.F.: Toward a theory of critical computing: The case of social identity representation in digital media applications. Code Drift: Essays in Critical Digital Studies (2009)
23. Harrell, D.F.: Designing empowering and critical identities in social computing and gaming. CoDesign **6**(4), 187–206 (2010)
24. Hedaoo, S., Williams, A., Wadgaonkar, C., Knight, H.: A robot barista comments on its clients: social attitudes toward robot data use. In: 2019 14th ACM/IEEE International Conference on Human-Robot Interaction (HRI), pp. 66–74. IEEE (2019)
25. Hoffman, G.: Hri: four lessons from acting method. Citeseer, Tech. Rep. (2005)
26. Hoffman, G.: On stage: robots as performers. In: RSS 2011 Workshop on Human-Robot Interaction: Perspectives and Contributions to Robotics from the Human Sciences. Los Angeles, CA, vol. 1, p. 21 (2011)
27. Hoffman, G., Ju, W.: Designing robots with movement in mind. J. Human-Robot Interact. **3**(1), 91–122 (2014)
28. Hoffman, G., Kubat, R., Breazeal, C.: A hybrid control system for puppeteering a live robotic stage actor. In: RO-MAN 2008-The 17th IEEE International Symposium on Robot and Human Interactive Communication, pp. 354–359. IEEE (2008)
29. Hoffman, G., Vanunu, K.: Effects of robotic companionship on music enjoyment and agent perception. In: 2013 8th ACM/IEEE International Conference on Human-Robot Interaction (HRI), pp. 317–324. IEEE (2013)
30. Hoffman, G., Weinberg, G.: Interactive improvisation with a robotic marimba player. Auton. Robots **31**(2), 133–153 (2011)
31. Hoggett, R.: 1965—rosa bosom—bruce lacey (British). Cybernetic Zoo (2010). http://cyberneticzoo.com/robots/1965-rosa-bosom-bruce-lacey-british/. Accessed 1 Sept 2022
32. Hoggett, R.: 1964—robot k-456—nam june paik (Korean) & shuya abe (Japanese). Cybernetic zoo (2015). http://cyberneticzoo.com/robots-in-art/1964-robot-k-456-nam-june-paik-korean-shuya-abe-japanese/. Accessed 1 Sept 2022
33. Horton, H.: Microsoft deletes 'teen girl' AI after it became a Hitler-loving sex robot within 24 hours. The Telegraph **24** (2016)
34. Kac, E.: Towards a chronology of robotic art. Convergence **7**(1), 87–111 (2001)
35. Kahn, P.H., Freier, N.G., Kanda, T., Ishiguro, H., Ruckert, J.H., Severson, R.L., Kane, S.K.: Design patterns for sociality in human-robot interaction. In: Proceedings of the 3rd ACM/IEEE International Conference on Human Robot Interaction, pp. 97–104 (2008)
36. Kao, D., Harrell, D.F.: Mazestar: a platform for studying virtual identity and computer science education. In: Proceedings of the 12th International Conference on the Foundations of Digital Games, pp. 1–6 (2017)

37. Knight, H.: Eight lessons learned about non-verbal interactions through robot theater. In: International Conference on Social Robotics, pp. 42–51. Springer (2011)
38. Knight, H., Satkin, S., Ramakrishna, V., Divvala, S.: A savvy robot standup comic: online learning through audience tracking. In: Workshop Paper (TEI'10) (2011)
39. Lin, C.Y., Tseng, C.K., Teng, W.C., Lee, W.C., Kuo, C.H., Gu, H.Y., Chung, K.L., Fahn, C.S.: The realization of robot theater: humanoid robots and theatric performance. In: 2009 International Conference on Advanced Robotics, pp. 1–6. IEEE (2009)
40. Markovik, E.V.: The Mystery Method: How to Get Beautiful Women into Bed. St. Martin's Press (2007)
41. Mulvey, L.: Visual pleasure and narrative cinema. In: Visual and Other Pleasures, pp. 14–26. Springer (1989)
42. Nishiguchi, S., Ogawa, K., Yoshikawa, Y., Chikaraishi, T., Hirata, O., Ishiguro, H.: Theatrical approach: designing human-like behaviour in humanoid robots. Robot. Auton. Syst. **89**, 158–166 (2017)
43. Nylund, M.: With a technical vision and dedicated performances, 'Heddatron' is a thought-provoking triumph. The Lafayette (2020). https://www.lafayettestudentnews.com/blog/2020/02/28/pen-with-a-technical-vision-and-dedicated-performances-heddatron-is-a-thought-provoking-triumph/ Accessed 1 Sept 2022
44. Osada, J., Suzuki, K., Matsubara, H.: Role and value of character design of social robots. In: International Conference on Entertainment Computing, pp. 338–350. Springer (2020)
45. Paik, N.J., Stooss, T., Kellein, T.: Nam June Paik: Video Time, Video Space. H.N. Abrams, New York (1993)
46. Pallas, J.: Nose wazoo. Leonardo **24**(1), 83 (1991)
47. Patterson, M.L., Sechrest, L.B.: Interpersonal distance and impression formation. J. Pers. **38**(2), 161–166 (1970)
48. Penny, S.: Simon Penny (2011). http://simonpenny.net/ Accessed 17 Mar 2016
49. Reichardt, J.: Cybernetic Serendipity: The Computer and the Arts. Praeger (1969)
50. Ribeiro, M.H., Blackburn, J., Bradlyn, B., De Cristofaro, E., Stringhini, G., Long, S., Greenberg, S., Zannettou, S.: From pick-up artists to incels: a data-driven sketch of the manosphere. arXiv:2001.07600 (2020)
51. Sajadieh, S.: Cute or creepy, that is the question of liveness: Can artificial actors perform live? Artnodes **32**, 1–9 (2023)
52. Sajadieh, S., Wolfe, H.: Come hither to me: performance of a seductive robot. In: Extended Abstracts of the 2019 CHI Conference on Human Factors in Computing Systems, pp. 1–4 (2019)
53. Schechner, R., Lucie, S.: Performance Studies: An Introduction. Routledge (2020)
54. Stanislavski, C.: An Actor Prepares. A&C Black (2013)
55. Stanislavski, C.: Building a Character. A&C Black (2013)
56. Strauss, N.: The Game. Canongate Books (2006)
57. Vilk, J., Fitter, N.T.: Comedians in cafes getting data: evaluating timing and adaptivity in real-world robot comedy performance. In: Proceedings of the 2020 ACM/IEEE International Conference on Human-Robot Interaction, pp. 223–231 (2020)
58. Viola, P., Jones, M.J.: Robust real-time face detection. Int. J. Comput. Vision **57**(2), 137–154 (2004)
59. Walster, E.: The effect of self-esteem on romantic liking. J. Exp. Soc. Psychol. **1**(2), 184–197 (1965)
60. Warner, M.: Kafka, weber and organization theory. Human Relat. **60**(7), 1019–1038 (2007)
61. Wodiczko, K.: Critical Vehicles: Writings, Projects. MIT Press, Interviews (1999)
62. Wolfe, H., Peljhan, M., Visell, Y.: Singing robots: how embodiment affects emotional responses to non-linguistic utterances. IEEE Trans. Affect. Comput. (2017)
63. Wolfe, H.E., Sajadieh, S.: A robo-pickup artist breaking gender norms. In: Proceedings of the 2022 ACM/IEEE International Conference on Human-Robot Interaction, HRI '22, pp. 1210–1211. IEEE Press (2022)

64. Wright, D.: The discursive construction of resistance to sex in an online community. Discourse Context Media **36**, 100402 (2020)
65. Wurst, K.R.: I comici roboti: performing the lazzo of the statue from the commedia dell'arte. In: AAAI Mobile Robot Competition, pp. 124–128 (2002)

Chapter 8
Actor-Flower-Mesh-Work: Making Environments Together

Rocio von Jungenfeld ⓘ and Dave Murray-Rust ⓘ

Abstract Things simply push back. We had to grapple with this fact while working on our project *Lichtsuchende* and realising how things (which sometimes may feel inanimate) and their digital aspects continuously participate in the environments that they are part of and co-create, as well as in their own making. As researchers, we inevitably started to think about how we could better understand what was happening in this creative process and about how the combination of physical and digital agents and the entangled aspects of these and their complexities could be unpacked. How are we making these environments together? Through this process of inquiry a particular framework started to emerge which allowed us to explore and discuss the challenges and highlights of creating digitally (in our case an interactive installation) using a Latournian and an Ingoldian approach. Using Latour's *actor network* and Ingold's *meshwork* as theories helped us recognise the interrelations between the things in, within and outside the installation and their interrelations. We seek to understand the ways in which these two viewpoints can be applied as methodologies to unpack the factors involved in the creation of an artificial society and the emergence of a shared environment made of non-human and human things.

Keywords Kinetic sculpture · More than human · Actor network theory · Non-human actors · Meshworks · Object oriented ontology

8.1 Introduction

In this paper we use an Ingoldian 'meshwork' and a Latournian 'actor network' approach to unpack the complexities of devising and constructing complex situations that combine creativity and technology. To do this, we draw on our experience of

R. von Jungenfeld (✉)
University of Kent, Canterbury, United Kingdom
e-mail: r.von-jungenfeld@kent.ac.uk

D. Murray-Rust
Delft University of Technology, Delft, Netherlands
e-mail: d.s.murray-rust@tudelft.nl

creating and exhibiting the interactive installation *Lichtsuchende*, a society of robotic light-seeking flowers. We are interested in how ideas relate to *things,* through the selection and manipulation of materials and the crafting and redesign *of* (and *with*) these *things*. In particular, we draw on our observations of the interactions, relations and correspondences between the installation, its elements and the people that visited the space.

When discussing the making process, the exhibiting of the artwork, the potential of materials and the vitality of *things*, we use Tim Ingold's idea that making is an open process that is composed of a *mesh* of continuous interrelations where people, processes and *things* intertwine. It is at these intersections, and when looking at them from within the *mesh-*, that we can see how things *-work* and where *things* (tangible and digital) dissolve disciplinary boundaries and become *enmeshed*, participating in and co-creating each other's environment.

Using Bruno Latour's theory of actor-networks we analyse certain instances of these relations and investigate how materials, things, and people participate in multiple networks simultaneously. For this, we examine a selection of actors that participate and play active roles in the artwork, and the relationships these actors establish as they engage with one another. To assist unpacking Latour's Actor-Network-Theory (ANT) we draw on object oriented ontology (OOO) via Graham Harman and their take on post phenomenology and potentiality, the hidden or unexplored qualities that things possess.

There are tensions between these two approaches, with Ingold's view being that of an embedded and embodied experience of things within environments, while Latour's being (at least in relation to early work, 1988) that of an analytical description of constituent parts and the relations that parts establish at different points in time and depths within networks within networks. However, these two positions are useful to help us analyse creative practices, their technological aspects and how *things* relate with each other. The two approaches enable us to investigate from within and from outside. With Ingold's perspective we are inside, embedded in a *mesh* of interrelations, while with Latour's we are outside, observing how *things* work together and are interrelated from afar.

We develop a discussion of these theories somewhat auto-ethnographically around the artwork *Lichtsuchende,* a group of robots with social behaviours enacted through light. On a basic level, they sleep, search for light, recharge, get excited and go back to sleep. However, their trajectories of becoming are not that straightforward, and by looking at the moments of *becoming* we can delve into understanding the artwork better using two distinctive approaches: Ingold's *meshwork* and Latour's *actor-network*. These viewpoints can be applied as methodologies to unpack the factors involved in the creation of an artificial society and the emergence of a shared environment made of human and non-human *things*, but we do not attempt to formally divide or reconcile the perspectives. Rather, we use the Ingoldian and Latournian frameworks to inquire into the process of making, the becoming of *things* and the making of the environments in which they perform.

8.1.1 The *Lichtsuchende*

Lichtsuchende is an interactive installation, consisting of a society of biologically inspired, cybernetic creatures loosely resembling sunflowers [33]. Each robot can swivel its head on two axes, tracking bright light with its sensors, and can emit beams of light through a cluster of powerful LEDs (see Fig. 8.1). The tracking of light is their most distinctive behaviour, as they attempt to focus their gaze on the brightest light source nearby. This is the basis of their socialisation, they quietly look around the space, sending out beams of light in the hope of making contact with others. They follow a Maslowian psychology—our version of the hierarchy of needs [31]—through internal states, and when they find an interactional partner, their mutual gazes lead to a stroboscopic outpouring of joy, followed by exhaustion. The society works autonomously as the robots interact with each other, moving through states of dormancy, exploration, communication and repose, but is open to interacting with humans using torches or other light sources. Visitors can join in with the behaviour of the society through their physical interactions, learning to read the states and responses of the robots through patient and curious interaction.

Technically, each robot is made of a combination of (i) a custom circuit board (PCB) that holds a microcontroller with a cluster of LEDs and supports several light sensors (see Fig. 8.2); (ii) two servo motors that allow the 'head' of the robot to twist horizontally and vertically; (iii) a supporting skeleton made of transparent acrylic. They have been presented in several installations, where a variable number of robots are set up in a darkened space, with some locally sourced material (rocks, bricks) used to ballast the lightweight bodies. The processes involved in their making

Fig. 8.1 *Lichtsuchende*, details of a robot head with lights on and another robot in the background

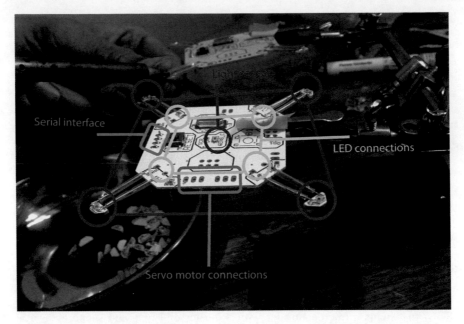

Fig. 8.2 Printed Circuit Boards (PCBs) being assembled; image features two PCBs, a hand, soldering iron and components. The graphics feature the layout and position of components

included (among others) designing circuit diagrams, fabricating PCBs, soldering and assembling components, laser cutting acrylic sheets, coding and uploading. We needed to continually adapt our ideas to the possibilities of the processes and the animacy of the things we were designing with and often encountered the vibrancy of their materials, things and their qualities pushing back, showing us other ways forward (or perhaps sideways), which highlighted to us that *making* is a *mesh* of continuous relations where processes and *things* intertwine and share in the making of their environments.

8.2   Starting to Make *Things*

Interactive works such as *Lichtsuchende* serve to discuss how making is a process through which ideas influence the development of *things* and *things* influence the development of ideas. Here the emphasis is not on whether it is the idea or *thing* that comes first, but on the things that carry from one to the other, and the dynamics established between them in the process of making. Movement between the two is a fluid process, where one is interpreted into the other through materials, senses and algorithms, and this interpretation is revised continuously in the process of making. We refer to making as the process of design and redesign where *things* are made, but also to the process of making as *being*, of participating in the environment and establishing connections within it and with other *things*.

As we make things with our hands, write variables with our fingertips, or build theoretical arguments for our papers, we develop a richer understanding of our relations with the *things* we make things with, the environments where these *things* are and the technologies which enable us to make them. This idea resonates with Lambros Malafouris' theory of material engagement [30]: making may be conceived as a process of mutually shaping. The materials we manipulate with our hands, whatever it is, for example dough with its lively yeasty fungus, shapes the bread as much as our kneading along with the humidity and temperature of the room. We have certain control over the bread we want to make, or the code we want to write, for we select the flour, yeast and recipe, or the language, variables and conditions, but in the process of making or writing the materials and the *things* themselves also participate in their own making. The qualities of the materials invite the maker to play, and in this play the making is negotiated. It is not a one-way action, where the maker imposes its will on the materials to shape the *thing* (i.e. hylomorphism, which Ingold, amongst others, criticises as contrary to process; [21], 20–21), but the materials and the *things* themselves push back during the making process and drive the maker to reflect and adjust, in what Donald Schön coined as *reflection-in-action* [37] although only in the context of human experience, not taking other *things* into account.

8.2.1 Becoming a Thing

There is a moment, in the development of a piece like *Lichtsuchende*, when the combined materials become a *thing*. When exactly this *thinging* happens is hard to discern; we have reflected on this elsewhere [34]. *Thinging* is a denominative verb defining the process of becoming some-*thing*. The verb also describes how a thing gets tweaked or redesigned over and over in its process of becoming, and then *thinging* applies to the process in which the initial *thing* continues transforming and turns into its thingyness. This idea of converting nouns into verbs reminds us of Bill Watterson's Calvin and Hobbes 25-Jan-1993 vignette: "[Calvin]: verbing weirds language; [Hobbes]: maybe we can eventually make language a complete impediment to understanding" [41]. This is not exactly our intention, but we are aware of the challenge to reach understanding through language. *Thinging* as a verb has been previously used to throw light, ontologically speaking, on the issue of what things *are* [13, 21]. Martin Heidegger in *The Thing* (*Das Ding* 1950) says "The thing things. Thinging gathers" (172) and "The thing things world" (178) which taken out of context may seem bold statements [15]. In the essay, Heidegger uses the example of a jug to discuss the complexity of its *thinging*. The jug is a thing made of clay, but its *thinging* (its potential of containing liquid or other substances) is made of the void where this containing happens. Thus the thing, here the jug, has the capacity of *thinging* (of becoming a thing) and gathering other materials in relation to the world where it *things* (verb).

So, what is this *thing* we are talking about? Is it an object, a material, a program, a thought? Is it physical, can we touch it? Is it visual? Is it responsive? From an object oriented ontology (OOO) point of view [5] everything is an object, no matter how small or what they are. If we take one of the light-seeking robots as an example, a *thing* is something that has the potential to be active or become activated, or something that has already acquired a place in an environment, enacts its potentialities and participates in and in relation to the environment and those other things around it. Different scholars approach this idea of *thinginess* or *thinghood* differently. For instance, in *Being and Time,* Heidegger states that *things* cannot reach out to the world around them, and objects are mere caricatures of things. A thing is a *Zeug,* "*etwas um zu...*" [15], something for humans to use, to do something with, which is an early take on the question of what things are, and which holds no stance in relation to the *Lichtsuchende.*

Although in German language being-there (*da-sein*) is applicable to people and other things, Heidegger reserves *Dasein,* the possibility of *being there* only to humans [13]. This is a point which is contested by developments in entanglement theories for instance in Ian Bogost's tiny ontology (2012) or Karen Barad's agential realism amongst others [1]. In a Latournian sense, things—whether material or immaterial— have agency and any actor has the potential to act. But is giving agency not already an imposed anthropomorphic way of looking at and conceiving of things anyway? Maybe, as Ingold brings up, thinking about the potential of things is more valuable than their agency, an approach which also aligns with Bennett's vitality of materials and Barad's agential realism. Some *things* may not be able to act by themselves, while they have the possibility of being acted upon and continuing their trajectories of becoming. In our installation, the only action a rock or a brick could perform by itself was balancing on other rocks, bricks or the *Lichtsuchende* bases and continuing with its gravitational pull towards the ground. However, rocks or bricks had the potential of becoming while they pulled their mass towards the ground, no matter where they were, they had potential and participated in the *meshwork.* When the foot of a person kicked the rock or brick, they could suddenly knock other rocks and bricks in their vicinity and the bases of other light-seeking flowers. Drawing on Ingold and also on Bennett, we could call this potential to become the *vitality* of the rock or brick rather than its agency.

In the context of the *Lichtsuchende, things* may be made of materials and computation. Their materials and algorithmic structures are shaped in the process of making. There are some aspects of them which are perhaps temporarily fixed, such as their atomic composition or computational language. Some materials have qualities which are obvious and immediately apprehended, such as the stone is heavy and its surface porous, other qualities are more difficult to grasp or recognize: their trajectories in the environment, their vitality, potentiality, composition. *Things* have the potential to become other things or be part of other *things*, participating in assemblages. So, the heaviness of the rock has the potential of making the rock become a weight that grounds the base of the robot flower firmly to the ground. Its hardness and sharp edges have the potential to become part of another *thing* such as an axe. The malleability of copper gives the material the potential to become a thin thread, its conductivity the potential of becoming a wire or cable. What the material qualities of things are, shape what things are prior to being manipulated by the artist or craftsperson. Their qualities also limit what these materials can be converted into, and whether they can become part of other things, and if so, of their relations. Manuel DeLanda coined these as *relations of exteriority* [6], where things are in relation rather than isolated, and these relations modify the thing, yet the thing is more than its relations, which also change over time.

The creation of digital things raises questions about *thingness*, not least because they can be fluid and evasive in ways that matter finds harder. Physically identical hardware running different programs looks the same from the outside, but becomes a very different object. Redström and Wiltse's *multi-intentionality* [36] is a useful concept here, to make sense of the idea that the aspects of a digital object that we (think we) understand can vary wildly; the things we encounter can have different possibilities for meaning, that change over time, and between people.

8.3 Material Qualities, Performances, Flows and Mattering

As we transform selected and collated materials into *things,* we engage in the process of making, which is about working with materials and contributing to their trans-formation, instead of shaping materials into things or imposing ideas onto materials or code (i.e. hylomorphic model). Our work is oriented towards engaging with the potentialities of *things*, bringing to the fore some of their qualities, their *vibrant matter* [4]. We conceive of making as a process that requires flexibility from both makers and materials. It is a process in which all participants (people, materials, ideas, environment) are woven, entangled into a fabric which is enmeshed and constantly changing; which is shaped by and shapes others, and resonates with current research in the field of interaction design [7]. It is unhelpful to consider elements in isolation, we have to consider them in relation to other things. Here, we could talk of making as a textile or fabric, the processes and elements being the threads that compose this fabric, expanding in all directions. This fabric however, is not made of straight lines or vector-like movements, they are tangled, much like sweetpea tendrils, or particles in Brownian motion (see Fig. 8.3, Ingold and Schrödinger). Borrowing the term from Henri Lefebvre, Ingold calls it a *meshwork* [19, 20]: "the trails along which life is lived" [20]. We propose considering the whole process of making the *Lichtsuchende* and them making their way through the environment, as a *meshwork* where each robotic flower and each of the elements that compose it or have contributed to its making are represented as squiggly lines that are tangled.

Ingold's approach to making draws among others on Paul Klee's *Pedagogical Sketchbook* [25]. There, Klee describes finished forms as dead ends, while he asso-ciates the action of giving form as a life-giving exercise [19]. This idea of shaping and forming as an action which contributes to the process of life is strongly aligned with

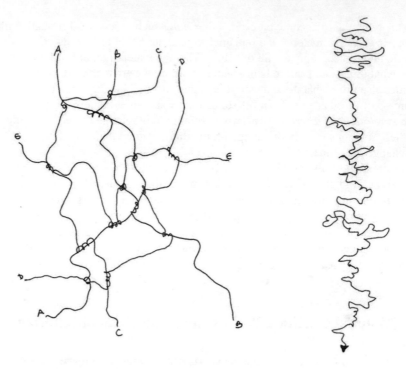

Fig. 8.3 Drawings based on Ingold [20] and Schrödinger ([38], first published 1944)

our experience of making the *Lichtsuchende*, and therefore we are reluctant to accept that making is a hylomorphic process where matter is transformed and shaped into preconceived, defined forms. There is a bit of preconceiving in making, there is no way of escaping that completely [34], but it is part of the idea-development process which changes when *making* takes place. In the case of the *Lichtsuchende*, *making* cannot be defined as manufacturing, but as a continuous process of exploring and finding ways of bringing things into life. From an industrial design perspective where a prototype has passed all the necessary tests and is sent finally to production, we could say the synthetic polymer (methyl methacrylate) is shaped into different size sheets and transformed into a clearly defined shape. Even in the process when acrylic sheets are industrially manufactured, where quantities are precisely measured and temperature regulated to accurate degrees, each sheet turns out slightly different. In the process of becoming an acrylic sheet the melted methyl methacrylate exposes its melting qualities, its atoms arrange and tangle up in a unique way. When extruded into the mould, the material does not always flow in the same way, its areas not exposed to the room temperature evenly. Similarly, when cutting the acrylic parts using CAD (our templates being similar to those featured in *Oregon Reef*, Beesley [2] 49), each individually cut shape will have undergone a slightly different shaping process, with the laser cutting through it more or less evenly. Even when the manufacturing process is strictly controlled, the state of the material is in flux.

This idea of materials and processes contributing to the outcome of an artwork is clearly present in Ken Goldberg and Joseph Santarromana's *Telegarden* [10]. In this work, online users instruct a robotic arm to insert seeds, water plants or monitor specific specimens in the garden. The artwork, the garden, is the result of a collaborative making process. The artists had to come up with the idea, work around design issues such as the physical structure of the garden, the mechanics and programming of the robotic arm and the internet platform where people could input actions and interact. This making process involved people, materials and technology, each of which contributed in its own way to the *Telegarden* as a *thing* in its own right. The artists may have contributed to a higher degree, because of their drive to materialise the idea of the piece, but the materials and technology available also played their role in that they shaped what the artists could do with them, which refers back to Malafouris' *How Things Shape the Mind* [30]; the things we create and have around shape the way in which engagement with them is possible, which brings us back to the notion of potentiality. In the *Telegarden*, the robotic arm inserts seeds and nurtures the plants, the plants grow in relation to the input of users which in turn triggers the actions of the robotic arm. Users contribute to the garden and work around what was there: as it forms, dries out or suffers from over watering, vitamin and mineral overdose or deficiencies, plagues. The artwork, now only available online in the form of archived audiovisual documentation, was in a continuous making process for almost a decade, in what Karen Barad calls *mattering* [1]; practised materiality, performed matter.

With *Lichtsuchende*, once the first prototype with glue and toothpicks was developed, a series of more sturdy robots were created. They were made following a template (same materials & program). In principle, all were the same, but they ended up being only similar. Each robot was in a sense unique, displaying the threads that had participated in their process of becoming: robot heads showed signs of the soldering process as marks or burns from the iron and flux. Despite the expectation that the surface mounted components on PCBs were going to speed up our assembly process, that was not always the case. The actions that had happened days before and in another country had an effect on the evening we spent scratching our heads wondering why the PCBs were not working. After much troubleshooting, we realised the LEDs had been originally soldered the wrong way round, and they had to be taken out and replaced. This shows that traces of actions apply even at a distance and across time. The threads that converged that evening (to name a few) involved a bunch of PCBs, a table, soldering irons, a multimeter, hot drinks, tired eyes, pliers, helping-hand stands, zip bags full of components.

The making process of *Lichtsuchende* (2015–2019), its *mattering,* may also be considered as a continuous process which has not finished yet. Despite having spent the last few years in boxes rather than installed, we have plans for its decommissioning, but we have yet to find a clear way of doing this. Surely, this *being-in-the-making* process has not been as lengthy as that of the *Telegarden*, at least not yet. However, the underlying idea that making is an ongoing process is clearly present in both works. When the artwork was installed for the first time in a vault (*Lichtsuchende*) or ready to accept requests from online users (*Telegarden*), its making was not over. The pieces may have needed to be turned on and off, elements within them needed to be taken down, refilled or rearranged, parts repaired, programs fixed. When these types of art-science projects take off in the public domain they are unfinished and continue *mattering* as the projects run their programs, their materials degrade and *things* break (and are sometimes fixed or replaced). The process of making and learning about the materials that compose the *things* we make and the workings of these *things* is endless, because *things* change along with our understanding and the relationships that other *things* initiate towards them. "Despite all claims to the contrary, crafts hold the key to knowledge" [26], *things* learn about making and materials in the processes of making, there is no shortcut to this process of growth [21].

It is interesting to think about how creating digitally engages differently with the flows and vitality of materials. Many aspects are similar: senses of *being-in-the-making*, going through processes of deformation, shaping, structuring, aggregation are just as present with digital materials. Some senses of vitality are enhanced: the replicability of digital materials supports re-use and appropriation of the artefacts created, giving a potential for greater reach but also unintended consequences. Animacy can be more strongly present, as the sensing and reacting capabilities of digital systems can be more vivid and direct than those of the traditional materials of making. However, there is less distinction between the two than might be imagined; Joler's *"Anatomy of an AI"* [22] shows how far the sociomaterial tendrils of a relatively disembodied digital experience can stretch, through the human labour that supports the creation and processing of data, the rare earth metals embedded in the construction, out to the recycling and landfill centres that are where their *mattering* takes a different form.

8.3.1 Different Things

"To exist is to differ; difference, in one sense, is the substantial side of things, what they have most in common and what makes them most different (Gabriel Tarde 1895/1999: 73)" [27].

Although the robots were designed using the same materials, processes and building strategies, each acquired a unique 'robot-al' (instead of 'person-al') quality. Building the robots mostly by hand resulted in individual creatures, with their own graciousness and flaws, similar to Jonnet Middleton's Unity Panda project where components for each panda were knitted separately following a 1946 pattern and later assembled [32]; each knitted fabric having its peculiarities: tighter or missing stitches, and each panda having different finishes, seams (alignment), stuffing, eyes, facial expression (see Fig. 8.4). In theory the robots were identical (following a template), but in practice each creature had its own history and physical singularities which affected its movements, reactions and apparent behaviour. When we observed at the beginning that the robots' movements were too fast and jerky, meaning they could not find each other, we slowed the *Lichtsuchende* down, modifying or adding new parameters to their behaviour. In a sense we bioengineered (in a robot-artificial sense of course) their behaviours, shaping their personality, as well as their physical state, replacing parts (the equivalent of limbs or organs in humans) or improving parts before reinserting them into the robots' live system. The modifications were part of a companionship design process [34].

Needing repair meant some *Lichtsuchende* had to enter what we called the 'robot clinic' and be seen by us or a technician (as during GLOBALE: Exo-Evolution exhibition at ZKM [9]) who after a first assessment would determine the level of injury and set up a recovery plan (e.g. soldering, laser cutting new parts, swapping old servos for new ones, re-programming). This relationship with the robots involved a level of compassion for the creatures, in an ontic sense, it involved caring for them. Having to leave a broken robot in a box labelled 'for repair' was sad, because we had built, looked after, tested, observed, fixed, or reprogrammed it, yet it was suddenly inert, comatose.

When trying to bring them back to life from a vegetative state we could either replace faulty or broken parts (physical) or re-boot their system by reprogramming them completely (psychic). Replacing broken components or acrylic parts was a routine procedure, but we felt reprogramming the creatures was like wiping out their memory with electroshock therapy, as in Michel Gondry's *The Eternal Sunshine of the Spotless mind* [11]. Although the reprogramming therapy involved reinstating a programme which was close to their previous one, it involved overwriting their previous program and therefore deleting any behaviour patterns they may have had with a new one. In the process of their *becoming* and us being enmeshed with them, we learned and incorporated that learning into new physical setups or algorithmic rules which the robots would perform.

Again, creating digitally has an interesting relationship with individuality, with the possibility for both more and less variation between pieces. The digital aspects of a piece can be reproduced perfectly, transmitted, shared, copied—all the affordances of digital technology that support idealised multiplication. Indeed, it is relatively impossible to engage with anything digital without making copies, as files are transmitted, stored, cached, displayed, rendered, at each moment occupying a different substrate while to some extent containing the same information. In the artwork here, the digital aspect of the robots was, barring error, identical, and their differences came mainly from variation of their physical characteristics and histories. However, digital works also provide the seed for individuated pieces, where a single work can have many instantiations for instance: an infinitely repeating series that a participant will experience a different moment of at each time; or a generative work that includes some level of randomness, so it may react differently each time; or a work that allows for a collection of related outputs to be made using different random seeds. However, for us in the *Lichtsuchende*, it was the variability that came from identical programs meeting subtly divergent physical matter that held interest, the constant diffractive tendency of matter to push back, and for objects in the world to acquire their own histories and differences.

Fig. 8.4 Jonnet Middleton's *Unity Panda* project. Image credit: chris + keir https://www.keir.xyz/work/unity-panda/

8.4   Things and Environments

For any making to take place, environments in which to make things have to be available. *Things* do not exist in a vacuum, they inhabit the worlds of which they are part of and which they make with their actions or inactions. Moving past the phenomenological dwelling and *Dasein* [14], *things* are there and their *being* there contributes to the shared environments they participate in. Since the drive to make things is inherent in humans, and humans live in environments, we could say that living in the environment is also a form of making. This is something that Ingold has extensively discussed in a series of articles about environments [18]. As we grow and dwell in the environment we make *things*, and in our interactions with *things* we make the *things* in which and with which we live (a pot, a CAD design, a fire, a machine, a program, a tent) and in turn these *things* make us; some of these *things* are material, some are not.

Environments may have different qualities and be tangible or immaterial. We can talk about the environments where our imagination plays (our dreamlands), the environments where data flow (digital architectures), the environments where micro-biota live (skin tissue). If we accept the premise that environments exist in relation to the *beings* that live in them [8] but also in relation to the *things* that are in them, then artificial lives made of atoms and energy such as the *Licthsuchende*, a group of light-seeking robot flowers, are certainly also in environments. These environments are composed of both non-human and human things, and each is enveloped by what Jacob von Üxküll calls their *Umwelt* [39], their unique way of understanding the world around them. Although their means to relate to the environment and the world around them are limited to a single physical sense, their ability to change the orientation of their sensory apparatus means the *Lichtsuchende* are able to establish active relationships with the environments where they are installed. Environments here are not simply passive, but contain the constellations of other surrounding things. While on a basic level, changes in light levels lead to changes in the robots' behaviours, on a more experiential level, creating a social fabric required a delicate tuning of the speeds of response and reaction for the robots. In order to create a *correspondence* between individuals [19] they needed their movements and actions to resonate with the rest of the group. If this was tuned too fast, as in early trials (e.g. at Inspace, Fig. 8.5), they would appear hyperactive and unable to relate to each other, creating a rather stressed sense of individual isolation. When their sensing and response were working in concert, their *Umwelt* led them to engage fruitfully with others around them.

In this sense then, the fine-tuning of a single parameter (e.g. speed of response) had a significant impact on how the environment was constituted or manifested for the *Lichtsuchende*.

The environment where artworks such as Rafael Lozano-Hemmer's *Vectorial Elevation (Relational Architecture #4)* [29] develop is considerably different. In *Vectorial Elevation*, the environment is two-fold, part of it is directly installed in Mexico City on the roofs of buildings and is projected into the air above the city, while

Fig. 8.5 *Lichtsuchende* waiting to be installed and tested as a large group for the first time at Inspace, Edinburgh

the other part is online (users select vectorial elevations) and on servers (processing of the settings selected through the user interface). The *Lichtsuchende* are closer to Philip Beesley's *Hylozoicground* [3], living architectures that respond to the movements of people in the environment, sensing their proximity and moving their articulated parts accordingly. In Beesley's artwork, there is no obvious action for people to perform, *Hylozoicground* senses presence, while our robotic creatures can be activated when people point light directly at them using a torch, or reflecting light using their bodies.

In terms of the making and creation involved in the *Lichtsuchende*, as with many other artworks, several different environments were involved, allowing different engagements between the various components, and different interactions. Here we will look at two—the environments where the robots were created, and the ones in which they were let loose to do their roboting. However, the environment where the robots were packed in boxes and shipped around, and stored for long periods is only brushed upon, and would merit a long discussion about the slow process of decay of its material components (electronics, acrylic) and perhaps even permanent death which is beyond the scope of this chapter.

The robots were constructed in what we could term the 'environment of creation' (Fig. 8.6). Here, as LEDs and PCBs flow through the global supply chain and into our workshops, we bring them into relation with shaping and modifying forces—soldering irons, heat guns, lasers, and they begin to take physical form. The animacy and vitality here is generally a physical, mineral, chemical kind—the apparent ways that solder 'knows' where to flow on a well prepared board, the transitions from liquid to solid as cyanoacrylate meets free air or the moment when heated acrylic suddenly becomes fluidly deformable and open to the possibilities of movement. Alongside the physical shaping, a more digital connection to the world starts to take place: the initial *Umwelt* of a microcontroller concerns clock signals from a crystal oscillating and information pushing in on the serial connections, activating receptive structures that allow for the patterning of its memory with programs. A series of gradually more complex test programs build up first initial possibilities for sensing and action, in terms of voltages that correspond to light levels, angles that indicate particular positions to be taken, and the possibility for structured communication with a connected device—sharing information and receiving instruction. As the testing and building progressed, for each device this initial state was built upon to create more refined and actively created parts of the senseworld: as light sensing was combined with movement, the *Umwelt* of the *Lichtsuchende* expands to have a sense of 'brightest' with the attendant possibility of orientation, and a slightly meaningful relation with the world starts to form. However, in this environment, stimulus was degenerate and minimal—moments of input would be interspersed with disassembly; interactions were technical, codified, programmatic, invasive, with little opportunity to relate internal states to environmental happenings, or to connect with others. The environment was oriented towards technical production rather than being supportive of robot activity: there were many bright lights that would draw their attention, physical arrangements were geared towards ergonomics for humans rather than robot socialisation, objects present were designed for testing, probing and evaluating rather than open ended interaction, and minimal thought was given to experience and aesthetics.

After a period in the environment of creation, the robots transitioned to a performance space, where there were more open possibilities. The lighting was carefully controlled to provide the optimum ground for them to communicate without distraction; they were in a space with many other robots, allowing the more social connective aspects of their *Umwelt* to come to the fore, as they formed alliances, corresponded,

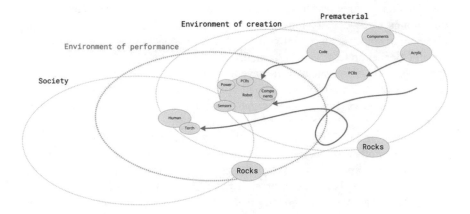

Fig. 8.6 Material flows through the environment of creation into the environment of performance

interacted. There was often a sense of group dynamics, the creation of a social environment within the larger environment of the room, that shaped their actions and behaviours through their sense world. It was this environment that the public were invited into, from the wider context of their daily lives, into a partitioned space, set up more for these light-seeking robots than people, and invited to engage with the objects on their own terms.

8.5   Actors at Work: ANT Briefly Unpacked

Bruno: "Oh, I love it. I am a serial redescriber. Now I know who I am. [LAUGHTER]" [28].

We use Harman's study of ANT as an entry point to help unpack some aspects of Latour's Actor Network Theory which are useful in our attempt to better understand the *Lichtsuchende* and their process of becoming. ANT is widely credited as the first comprehensive theory which assigns agency to non-human actors, allowing anything (including the *Lichtsuchende*) to be conceived as actors in the network. Harman (a key OOO scholar) prefers Latour, because he attributes ontic capacities (the ability to care for) to things, unlike Heidegger who attributes it only to people [28]. Do Latour's actors, in our case a bunch of robotic creatures, care for the actors around them? That is difficult to discern, but we can look into whether they relate to other things around them; that is more feasible.

At the core of ANT are actors and their relations, and how when combined, they forge a net of actions. According to Harman, ANT is based on irreduction, actors, alliances and translation:

(1) **Irreduction** refers to the need to avoid simplification. Things and their relations are complex, irreducible, yet when describing them, we have to reduce them to some degree. The *Lichtsuchende* are made of materials and parts, as well as lines of code and electric current. If we start unpacking these we go deeper (copper, plastic films, functions, solder, resistors, electric charge, to name a few) but end up having to reduce things, else we hit the infinite regression paradox where "each actor is a black box containing other actors ad infinitum" [28]. When considering the *Lichtsuchende* in a particular set up (e.g. Vault 13, see Fig. 8.7), other things and complex relationships become apparent, yet we still have to reduce them when describing a humid stone-based vault, stones serving as ballast to acrylic bases, robotic heads turning around searching for light, power supplies feeding current to the robots, visitors moving and carrying torches that point at things.

(2) **Actors** are anything that contribute to a *network*, this includes things and relations. As briefly described in relation to irreductions, actors are of varying sizes, shapes, they come and go, are real or fictional, material or immaterial, made of flesh or other stuff, while also made of other actors, and may be part of larger actors. As Harman states: "all actors are equally real [although not equally strong]" [28], and when analysing and describing, we have to consider them in terms of their actions, and the effects that these actions have on other actors or the relationships between

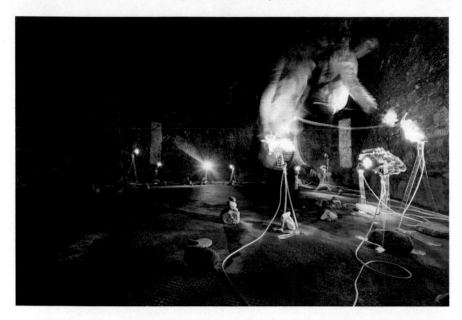

Fig. 8.7 *Lichtsuchende*, Hidden Door, Edinburgh. Image credit: Chris Scott (@chrisdonia)

actors. When looking at an instance as captured in Fig. 8.7 (long exposure: 1/30 shutter speed, f/8 aperture, EFL 16 focal length, ISO 100), we notice some obvious actors frozen at work: stonewalls, concrete floor, white power cables, light beams, acrylic structures, visitor crouching down/standing/reaching over with their hand, robotic creatures' heads turning and moving, stones acting as ballast. Actors are more than their relationships and as Latour states "*anything* that does modify a state of affairs by making a difference is an actor" [27]. There are actors at play that are not visible or captured in the image but have left a trace: the torch the visitor holds, the curtain shielding the space from daylight and keeping it relatively dark, the signage installed at the entrance giving guidance to visitors, or the sounds that different actors produced and played in the space.

(3) **Alliances** are connections between actors that can be strong or weak. These alliances are defined by where and to which other actors any actor turns to in order to forge connections. Latour's theory is aligned with secular occasionalism; things interact with each other at a local level without the intervention of a top-down figure or idea. As argued elsewhere, horizontality is a more even ground to discuss relations between things [24]. All relations require a mediator which, as mentioned, Harman criticises since if each actor requires a mediator, and every mediator is an actor, then what mediates between mediator and actor? (see earlier discussion of infinite regression paradox) [28]) This is an issue from a metaphysical standpoint, but for Latour each actor is a mediator, what is between them is simply another mediator [28]. For example, two robot flowers are linked to the power socket through a cable, this cable is placed between the flowers and the socket by another actor (technician)

who links them and is a tangible ally that acts when plugging things together. Their alliance and connection to other actors remains even after having left the room. Immaterial elements are linked, the program that runs the robot flower is allied to it through a circuit; this circuit has components that are linked by the solder that another ally added to them. The solder is made of the alliances between its metal molecules, and so on and so forth. The program is linked to the exhibition through many actors, all those involved in the set-up of the infrastructure and the organisation of the actors that are at play, along with those unpredictable actors (e.g. visitors) that appear over time (see Fig. 8.8).

(4) Translation explains how a thing and its description at any point can never be exactly the same because processes are applied in the translation and therefore the initial thing turns to be another thing [28]. Any description or analysis, as we are trying to do here in explaining the *Lichtsuchende*, is a translation. We acknowledge that in attempting to explain things, even when simply describing elements within a *network*, we are translating and reducing (see Fig. 8.9). It is this simple, we cannot unpack every actor, as each relationship of actions happens as an unrepeatable instance and each action changes the actor in the following action (translation). Actors change with these translations: "everything is in a state of perpetual perishing" [28]. Figure 8.9 captures the state at a particular moment, in that instance, the arrangement of actors (relations and things) was as pictured, yet after 1/30s passed, *things* would have changed: the visitor would have moved, robot heads turned, the point in their algorithm at which each robots was would have passed, the configuration of the lights projected into the space by different actors. Some things may seem permanent, i.e. the stones or wall, yet they are also in flux, only that their timescales differ, stones are slower than electric current.

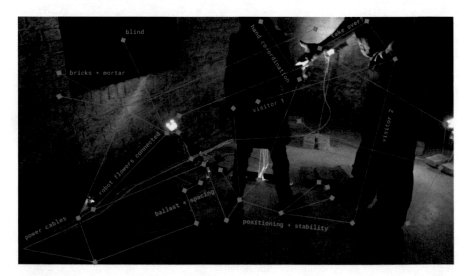

Fig. 8.8 Still from video showing some key actors in the installation and their relations at Update_ 5, Zebrastraat, Gent

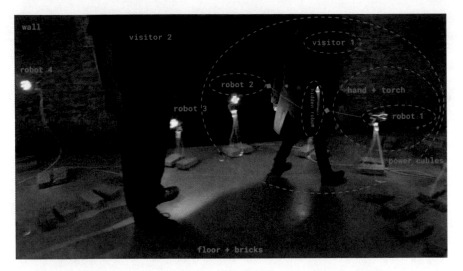

Fig. 8.9 Still from video showing key actors, groups and some subgroups and alliances at Update_ 5, Zebrastraat, Gent

For digital works, this broader sense of what can be an actor, and how far to follow relations is important to understanding the socio-cultural embeddings of works and their production. Recognising the agency of programming languages, libraries, datasets, discussion fora, repositories feels increasingly crucial to understanding the acts of creation. In a more local sense, the *Lichtsuchende* highlight the agencies of the various *things* brought together, as torches and spaces shape human behaviour, and interactions with the robots establish various physical and conceptual relationships. The materiality of the digital is easy to read in an agential way, as responses can be coded, interactions scripted, behaviours carefully shaped to enact various forms of 'liveness'.

8.6 The Active *Mesh* at Work

Here we bring *meshworks* and actors together to shine light onto things and the threads that help us understand what is at play in the environments that robots and other actors co-produced and shared. For this, we discuss how things and their relations changed over time as the *Lichtsuchende* are part of what Ingold (following Deleuze's work) refers to as *lines of becoming* (2011, 83–4) within the *meshwork*.

The environment is unique to each actor; a robot's world (its *Umwelt*) is different to the environment of the torch—that is hitching a lift with the visitor, nonetheless, their environments sometimes overlap (see Fig. 8.10), particularly when they are performing their *being* in the world in relation to each other. The environment changes depending on the setup and configuration, and how *things* are distributed across space and over time. In relation to the *Licthuschende* every exhibition space offered a relatively different environment where they could be active actors and relate to others, non-humans and humans. There are aspects of the environment which different actors may share, for instance, robot 1 and robot 2 may have shared the power supply, program and floor; or visitors, torch and robots may have shared black-out room and air temperature.

For some of these actors (i.e. torch) to be able to share the environment and participate in the *mesh* of relations, another crucial actor needed to be at work. Only when visitors hold the torches in their hands, carrying and performing gestures with them, the torches are able to modify the state of the robots and activate something in these other actors, leaving a trace. The non-human and human things, e.g. the torch and hand, merge in *cyborgian intentionality* (based on [17, 40]), where the robot-torch relationship is mediated by other agents, in this case the human, hand and moving body. Hence, the participation of the torch in the *meshwork* of relations is contingent on them being carried by and *ready-at-hand* (like Heidegger's hammer) with the visitor. The torch cannot activate itself (turn on) nor can it enter a performative state on its own. The torch becomes an active actor when embodied, and only falls back to Heidegger's *present-at-hand* state when the visitor places it back on the table on their way out, or the batteries run out. In relation to the latter, in *Being and Time* (1977 (1927)) Heidegger calls this the "disturbance of reference" which happens when something does not work properly or fails to work at all [12, 28]. At such a point, the relationship of correspondence between torch, hand, robot and

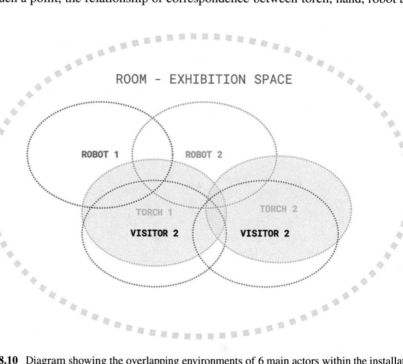

Fig. 8.10 Diagram showing the overlapping environments of 6 main actors within the installation

visitor, which Peter-Paul Verbeek coins as having *composite intentionality*—where technological things also have intentionality [40]—breaks. The gestures that were performed together, their entangled *lines of becoming,* diverge. As Ingold describes in 'The Textility of Making' and in relation to flying kites [19], the wind, kite, string, and person are "trajectories of movement, responding to one another" (215). Once the correspondence between them is disrupted, the kite loses its ability to fly, the dance of relations falls apart.

A reminder of the importance of potentialities seems appropriate here, since elements in the environment (things, which Harman calls objects) have qualities that are not always included in the network of relations [28]. However, as Harman points out for Latour "alliances are more important than hidden individual essences and potentialities" [28]. In Latour's case, potentialities are secondary, while for Ingold and *vital material* scholars such as Bennett, these are crucial to the *things* themselves because hidden potentialities, although dormant, participate in the *mesh* of relations that develop over time rather than having an effect on a particular instance.

In the *Lichtsuchende* installation, we could refer to the potentiality of robots to shine light at other robots, at the room and torch. When looking at a snapshot of the installation (see Fig. 8.11) some robots are asleep or taking a break, and their capacity to shine is dormant. Yet this potentiality is vital since without it robots would struggle to find each other and make *alliances*, or enter the *mesh* of relations where correspondence with other actors is possible. Also, the stones of the installation (which were borrowed from the beach and later returned) have the hidden potentiality of rolling in the shore and corresponding with the waves, but this capability disappears in the installation where the relations between the stones and other things are based on weight and size. Their rolling capacity is dormant, and only occasionally activated by other actors that may cause them to fall, roll over, knock down other things in their trajectory. In any instance where actors are at work, there are more potential relationships than those appearing to have been established. In Fig. 8.11 we see a robot, torch and child closely engaged and active, while other robots are facing away and have their headlights off. What is happening does not preclude the apparently dormant robots from establishing a connection with other actors, and corresponding with them in a fraction of a second from now. Associations are circumstantial, and increase or decrease depending on the number of actors that are active and at work at any moment, but *things* and their relations exist as threads over time, they are evolving actions within the *mesh,* they converge, correspond and push each other, they are *trajectories of becoming.*

To illustrate this we present two ways of looking at what is happening between the light-seeking robot flowers and other actors: (1) Ingold's lines in the *meshwork,* and (2) Latour's actors in the *network.* For this, we review visual documentation of the installation and apply visual analysis methodologies discussed elsewhere [23].

Fig. 8.11 Robot, torch, hand and child enmeshed in correspondence at Summerhall, Edinburgh

8.6.1   Ingold's Lines in the Meshwork

In Fig. 8.12, we have a sequence of images extracted from the video documentation showing an area of the installation where a number of actions are being performed. The lines of two visitors, a torch and a flower are overlaid all together onto the sequences to highlight movement and positioning, as well as points where and when these *things* may have converged. This mesh of lines is simplified in Figs. 8.13 and 8.14 so that particular *things* can be discussed in detail.

In Fig. 8.13: visitor 1 and the torch are drawn together in a form of *cyborgian intentionality* [40], as the technological object mediates their relation to the world. These two actors are enmeshed, needing each other to have an effect on other actors, particularly on the other technological actors (i.e. robot flowers) that are somewhat fixed to the floor. When following those two threads of action we see how they somehow move along together, crossing over, converging, getting entangled and activating other actors. The visitor is occluding our view of how things are being entangled—we cannot see the robots or torch in detail, however even from the back we can see the arm (that holds the torch) moving sideways and rotating, the lights and shadows in the space changing (reflections on walls, floor, fabric, skin). The two threads are at play, together over time.

In Fig. 8.14: visitor 2 and the robot are drawn as moving along, yet staying somehow apart, their paths not actually converging in the *mesh*. If we had overlaid the threads of visitor 1 (with the torch) and the robot, we might have observed some convergence, especially at the beginning of the sequence where the child wearing a

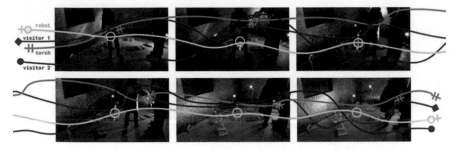

Fig. 8.12 Sequence of 6 images (left to right; image sequence and lines continue in second row) showing 4 actors visualised as threads, as trajectories of movement in a meshwork at Zebrastraat, Gent

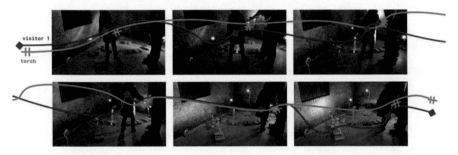

Fig. 8.13 Sequence showing 2 actors visualised as threads, as trajectories of movement in a meshwork at Zebrastraat, Gent

Fig. 8.14 A different sequence showing 2 actors visualised as threads at Zebrastraat, Gent

red coat (and carrying the torch) is directly engaging with the robot, pointing the torch light at it, and making quick rounded gestures around it. This performative aspect just described, is only visible in the video footage, while in the printed sequence this level of detail (gestures and correspondence) is difficult to identify. In the sequence, we can see the paths (trajectories of *becoming)* of visitor 2 and the robot as to be sharing the space, and hence their environments overlap. The trajectories do not converge since visitor 2 and the robot do not engage in direct correspondence yet their shared environment is constituted amongst other things of visible light beams from the other robots and the movements of visitor 1 and the torch.

8.6.2 Latour's Actors in the Network

When looking at threads within *meshes,* we struggle to see how things are deeply interrelated, because as we follow movement, the relationships and associations between elements change and dissolve, and we cannot pay close attention to nuanced inter-relations, and mediators between mediators; things simply move too fast. Hence we are interested in what happens when we put a *meshwork* analysis approach along-side the in-depth analysis of instances that ANT offers. When discussing ANT earlier, we selected two instances and ways of visualising key actors and their associations. Those images give us additional information about what is going on in the *meshwork* at particular points in time. It is impossible to present all actors or their connections, in the form of a graphical representation, therefore we only present a few. Let us discuss them a bit more in depth here.

The *Lichtsuchende* are robotic creatures fixed to the ground, much like Random International's *Audience* [35] installation, where little mirror creatures turn towards visitors while their bases are static, grounded. In both installations, the individual robotic creatures have no feet or wheels to move about in space, but can rotate their heads to face and engage with visitors. Each robotic creature is an actor in the *network*, and is interrelated to all other creatures, through a long list of actors: PCB design and assemblage of components, program and algorithmic structure, acrylic sheets from which they were laser cut, power cables that feed current to them, light present in the room, et cetera. These and other associations can be described in relation to the robots, but many other actors and associations were at play, mediating between actors. When looking at these two images side-by-side (see Fig. 8.15) we can see how some of the strongest actors (i.e. visitors and torch) have moved, and as Latour says drawing on Whitehead, there is a need for "new associations in order to persist in its existence" [27].

Fig. 8.15 Different ways of visualising connections between actors in a network

In that quote, Latour is referring specifically to society, but we choose also to apply it to groups of actors that are societal in the sense of *being* together and influencing each other. In the left, visitor 2 extends their hand to grab the torch and take control over the gestures and actions that visitor 1 (dressed in a red coat) is performing. That action is counteracted by visitor 1, moving their hand away, continuing their direct connection with the torch and in return with the robotic creatures that they together (visitor 1 and torch) are engaging and connecting with. The mediating actor here is clearly and strongly the light that is emitted in all possible directions, from different sources and actors within the space. In the right image, we see an instance where new associations have been established between visitor 1, the torch, and the robot that faces towards the right, away from visitor 1. A new association is being established, yet other actors are also at play, either actively engaged (e.g. robot 2, hidden robot) or ongoingly engaged (e.g. power cables), while visitor 2 is merely standing, refraining from taking direct action, yet with their presence leaving a trace and interjecting the trajectory of different light beams, casting shadows around the space.

Other actors at play which may be less obvious in either of these images are the selection process that led to the curated exhibition, the wheelbarrow used to bring the bricks that acted as ballast, or the smell and humidity of the cellar where the *Lichtsuchende* were installed. These weaker actors have left traces, but are neither present in the diagrams nor visually perceivable at first from the images. As Latour discusses: "If no trace is produced, they [objects] offer no information to the observer and will have no visible effect on other agents" [27]. The issue is that the information provided in the images is insufficient for the observer to note the existence of subtle, less active actors. Contextual information and the trajectories of actions—following paths in *meshworks*—are needed. How else could we account for all the actors at work (no matter how weak or strong), when for their traces to be noticed we have to take into account their paths within the *mesh*, their trajectories of becoming? Things are perhaps more than their relations or qualities [28], because in every attempt to translate them into actors we can never completely unpack them. Thus talking about and taking into account potentialities, and performativity aspects as proposed by agential realism [1] may be a better approach to understand how things emerge from and make environments together, how actors in *meshes* work, get entangled and correspond.

8.7 Bringing Things Together

In this paper, we have brought together four of the theoretical worlds that underpin an understanding of the ways that people are creating digitally. Heidegger's *thingness* gets into the what it is for something to be a thing in the world, whether physical or digital. Bennet's sense of *vibrant materials* along with DeLanda's *fluid assemblages* and Redström and Wiltse's *multi-intentionality* take us into the blurry edges of *thingness*, the ways that they form and reform, coalesce and disperse in their continual *mattering*. Latour's ANT gives a way to read complex situations in terms of their component parts and the relations between them, again extending beyond the immediate to bring in a wide swathe of more than human actors and actants. Finally, Ingold's *meshworks* draw attention to not just the relations between things, but the trajectories and lifeworlds that they follow, the situated and embedded unfoldings that they perform in their constant process of becoming.

These are all joyfully resonant theories in their own rights, pointing at ever more vibrant understandings of the world, that allow us to go beyond any sense of the world as inert clumps of matter occasionally stirred up by human actions. This follows the exhortation that "we need to devise new procedures, technologies and regimes of perception that enable us to consult nonhumans more closely, or to listen and respond more carefully to their outbreaks, objections, testimonies, and propositions" [4]. Digital creation is a strange bedfellow with some of these very material, biological, cultural understandings of the universe, as so many of the technologies that support digital works start by setting themselves up along lines of difference from the surrounding world: creating clean spaces for logic to be enacted, built on extractive practices, that offer little space for traditional life to engage. However, as we have discussed, there are many qualities to digital—and especially physical-digital—creations that speak to these questions. The *thinging* of electronic or computational artworks can be vividly apparent, their changes writ larger than the reorganisation of material. *Vitality* is easy to grasp from the first moment of making a light blink with a microcontroller and the sense of working with something at least somewhat autonomous. Every electronic device is a *fluid assemblage*, bringing together code, computation and physicality for a time, changing its identity through software, networking and interaction. Networks carry through to the interdependencies of hardware and software on people, code, structure and connections, the leaky abstractions of concepts, information and hardware that build the foundations for any kind of digital enactment. When we come to *meshworks*, however, the traditional story does not follow the sense of possibility as closely. A network view of relations is very much in keeping with computer science and the development of digital ideals, as nodes with relations, however complex. To work with a sense that these things have wiggly trajectories, that they entangle and correspond with people's life courses is part of the new frontier of computational thinking, where the fuzziness and politics of computation need to be acknowledged. Every computation runs on a piece of hardware, every piece of data came from a thing in the world, and to see them purely in terms of the relations and abstractions they encode leaves a practice that is unable to engage with humanity, let alone more-than-humanity. Instead, we can look for the mess, for potentialities, the co-dependencies, the *becomings-with* that are the hallmarks of a materially engaged digital practice.

Acknowledgements This work has been supported by The University of Edinburgh's Innovation Initiative Grant, New Media Scotland, SOCIAM: The Theory and Practice of Social Machines (EPSRC grant number EP/J017728/1), Edinburgh College of Art, Delft University of Technology, and the University of Kent. Special thanks also to Sean Williams and Owen Green for the intellectually challenging conversations that helped crystallise some of the ideas presented in this chapter, and to all the venues where the project has been presented (e.g. NTAA, ZKM, Lumen Prize; to name a few) and to the people that visited the installation and interacted with the *Lichtsuchende*.

References

1. Barad, K.: Meeting the Universe Halfway: Quantum Physics and the Entanglement of Matter and Meaning. Duke University Press Books, Durham (2007)
2. Beesley, P.: Orgone reef. In: Castle, H. (ed.) Architectural Design, vol. 75, pp 46–53 (2005)
3. Beesley, P.: Hylozoic Ground (2010). http://www.hylozoicground.com/Venice/index.html. Accessed 27 July 2022
4. Bennett, J.: Vibrant Matter: A Political Ecology of Things. Duke University Press, Durham (2010)
5. Bryant, L., Srnicek, N., Harman, G.: The Speculative Turn: Continental Materialism and Realism. Repress (2011)
6. DeLanda, M.: A New Philosophy of Society: Assemblage Theory and Social Complexity. Continuum, London (2006)
7. Frauenberger, C.: Entanglement HCI the next wave? ACM Trans. Comput.-Hum. Interact. 27(1), 2 (2019). https://doi.org/10.1145/3364998
8. Gibson, J.J.: The Ecological Approach to Visual Perception. Houghton Mifflin, Boston (1979)
9. GLOBALE: Exo-Evolution: Zentrum für Kunst und Medien, Karlsruhe, Germany (2015–2016). https://zkm.de/de/node/26700/m-r#lichtsuchende. Accessed: 27 July 2022
10. Goldberg, K., Santarromana, J.: Telegarden. http://goldberg.berkeley.edu/garden/Ars/ (1995–2004). Accessed 27 July 2022
11. Gondry, M.: The Eternal Sunshine of the Spotless Mind. https://www.imdb.com/title/tt0338013/ (2004). Accessed 27 July 2022
12. Harman, G.: Tool-Being: Heidegger and the Metaphysics of Objects. Open Court, Chicago (2002)
13. Harman, G.: Heidegger Explained: From Phenomenon to Thing. Open Court, Chicago (2007)
14. Heidegger, M.: Being and Time. Blackwell, Oxford (1967)
15. Heidegger, M.: Sein und Zeit (1927), vol. 2. Vittorio Klostermann, Frankfurt am Main (1977)
16. Heidegger, M.: The Thing (Das Zeug first published 1950). In Poetry Language Thought. Penguin Perennial Classics, New York (2001)
17. Ihde, D.: Technology and the Lifeworld: From Garden to Earth. Indiana University Press, Bloomington (1990)
18. Ingold, T.: The Perception of the Environment: Essays on livelihood, making and dwelling. Routledge, London (2000)
19. Ingold, T.: Being Alive: Essays on movement, knowledge and description. Routledge, London (2001)
20. Ingold, T.: Lines: A brief history. Routledge, London (2007)
21. Ingold, T.: Making: Anthropology, Archaeology. Art and Architecture. Routledge, London (2013)
22. Joler, V., Crawford, K.: Anatomy of an AI System (2018). http://www.anatomyof.ai. Accessed 27 July 2022
23. von Jungenfeld, R.: Portable Projections: Analyzing Cocreated Site-Specific Video Walks. Leonardo 53(5), 492–497 (2020). https://doi.org/10.1162/leon_a_01794
24. Jungenfeld, R. von: Rebalancing Media in Environments: analysing flows of action. In the 27th International Symposium on Electronic Art (ISEA). Barcelona (2022). https://doi.org/10.7238/ISEA2022.Proceedings
25. Klee, P.: Pädagogisches Skizzenbuch. In: Gropius, W., Moholy-Nagy, L. (eds.) 2/14 Bauhaus Books. Langen, München (1925)
26. Latour, B.: The Pasteurisation of France. Harvard University Press, Cambridge Mass (1988)
27. Latour, B.: Reassembling the Social: An Introduction to Actor-Network-Theory. Oxford University Press, Oxford (2005)
28. Latour, B., Harman, G., Erdelyi, P.: The Prince and the Wolf: Latour and Harman at the LSE. Zero Books, Winchester (2011)
29. Lozano-Hemmer, R.: Vectorial Elevation, Relational Architecture #4 (2000). http://archive.aec.at/prix/#35289. Accessed 27 July 2022

30. Malafouris, L.: How Things Shape the Mind: A Theory of Material Engagement. MIT Press, Cambridge, Mass (2013)
31. Maslow, A.: A Theory of Human Motivation. Psychol. Rev. **50**(4), 370–396 (1943). https://doi.org/10.1037/h0054346
32. Middleton, J.: Unity Panda. https://www.keir.xyz/work/unity-panda/ (2010–2012). Accessed 27 July 2022
33. Murray-Rust, D., von Jungenfeld, R.: Lichtsuchende: exploring the emergence of a cybernetic society. In: Johnson et al. (eds) Evolutionary and Biologically Inspired Music, Sound, Art and Design. EvoMUSART 2015. Lecture Notes in Computer Science, vol. 9027. Springer, Cham (2015). https://doi.org/10.1007/978-3-319-16498-4_15
34. Murray-Rust, D., von Jungenfeld, R.: Thinking through robotic imaginaries. RTD Conference Proceedings (2019). https://doi.org/10.6084/m9.figshare.4746973.v1
35. Random International: Audience (2008). https://www.random-international.com/audience-2008. Accessed 27 July 2022
36. Redström, J., Wiltse, H.: Changing Things: The Future of Objects in a Digital World. Bloomsbury Academic, London (2018)
37. Schön, D.: The Reflective Practitioner: How Professionals Think in Action. Temple Smith, London (1983)
38. Schrödinger, E.: What is Life? (1944). Cambridge University Press (Canto Classics), Cambridge (1967)
39. von Uexküll, J.: A Stroll Through the World of Animals and Men: A Picture Book of Invisible Worlds (1934). https://monoskop.org/images/1/1d/Uexkuell_Jakob_von_A_Stroll_Through_the_Worlds_of_Animals_and_Men_A_Picture_Book_of_Invisible_Worlds.pdf. Accessed 27 July 2022
40. Verbeek, P-P.: Cyborg intentionality: rethinking the phenomenology of human-technology relations. Phenomenol. Cognitive Sci. **7**, 387–395 (2008). https://doi.org/10.1007/s11097-008-9099-x
41. Watterson, B.: Calvin and Hobbs 25–01–1993 [vignette] (1993). http://www.gocomics.com/calvinandhobbes/1993/01/25. Accessed 27 July 2022

Chapter 9
Generative Video Art

Pedro Alves da Veiga

Abstract Generative art is historically and widely used for the production of abstract images and animations, each frame corresponding to a generation or iteration of the generative system, which runs within the aesthetic boundaries defined by its author. But rather than being limited to image or sound synthesis, generative systems can also manipulate video samples and still images from external sources, and include vectors that can be mapped to the concepts of shot, sequence, rhythm and montage. Furthermore, generative systems need not be limited to the visual plane and can also render audio, either through sound synthesis or by manipulating sound samples. And in this case, since the output is a constant and uninterrupted audio-visual stream, is it not possible to speak of generative video art, as it becomes indistinguishable from its modern-day video art digital counterparts? Within this perspective, this article traces back the historical roots of generative video art, and proposes a theoretical model for generative video art systems, as a creative intersection of two artistic genres, often seen as disjoint.

Keywords Cinematic art · Video art · Generative art · Autonomous systems

9.1 Introduction

Before the advent of video art, avant-garde film—or abstract film—was mostly regarded as a spin-off of visual art movements like Cubism, Suprematism, Constructivism, Dadaism or Surrealism ([60]: 110). Nevertheless, early experimental works dating from as far back as the 1910s, as those by Arnaldo Ginna and Bruno Corra, deserve credit in video art history. What is even more striking is that under the influence of Italian and Russian Futurism, experimental pieces by Hans Richter and

P. A. da Veiga (✉)
CIAC – Research Centre for Arts and Communication, Portuguese Open University, Lisbon, Portugal
e-mail: pedro.veiga@uab.pt

© The Author(s), under exclusive license to Springer Nature Switzerland AG 2023 241
A. L. Brooks (ed.), *Creating Digitally*, Intelligent Systems Reference Library 241,
https://doi.org/10.1007/978-3-031-31360-8_9

László Moholy-Nagy already evidence the use of a structuring device, later recognised as a key characteristic of generative art systems [55] and also used for image production and montage techniques ([47]: 27).

To further reinforce this early ancestry, the Futurist Cinema manifesto posited that cinema should detach itself from reality and from photography, from the graceful and solemn to become anti-graceful, deforming, impressionistic, synthetic, dynamic, free ([32, 47]: 28), all of which came to be trademarks of video art.

Still, video art, as a new branch of art, only assumed its status in the 1960s, at a time when new image recording instruments and technologies migrated from the corporate world to the public. If it can be defined from the technological point of view as an art form that uses video technology—both analogue and digital— video art also introduced counterculture aesthetics, created new concepts and forms that provoked displacements and deconstructions in the ways of relating to moving images. Although contemporary with early experiments in computer art and generative art, video art was less focused on the technological exploration, and more intent on the subversion of the conventional uses of television and film for the masses. Video art "was Paik's attack on the physical interface of a commercial moving image—his first show consisted of televisions with magnets attached to them, and TV monitors ripped out of their enclosures" [29].

If Paik's work was mostly produced by hacking the inner workings of video technologies, through hardware manipulation, its present-day echo can be found in generative video artworks, which now hack video streams from a software perspective.

Metz and Guzzetti [37] posit one formal—and very relevant—distinction between video art and classic cinematic precepts: they characterize classic cinema as being based upon a narrative form and a reality effect, as opposed to an oneiric construct, which is the territory for video art to thrive, where time is disconnected from physical reality. Their stance on experimental film is rather critical, but stresses the absence of a linear narrative—or, as they call it, intelligibility:

> those avant-garde or experimental films which, as the enlightened audience knows, it is appropriate at once to understand and not to understand (...) These films, whose objective social function is to answer the naively puzzled wish of certain intellectuals for nonnaivete, have integrated within their institutional regimen of intelligibility a certain dose of elegant and coded unintelligibility, in such a way that their very unintelligibility is in return intelligible ([37]: 88).

Whereas traditional film delivers a closed universe to the audience, video art triggers sensations and stimulates the imagination. Today video art is a discipline that reconciles different artistic fields, such as the visual arts, music, literature, cinema, dance, and theatre, extracting concepts and methods from different artistic movements, theories, and technological advances [36], and appears to directly inherit the Futurists' characterization. Video art embodies a social and cultural awareness and critique, both in terms of content and of representation, but also an aspect of plurality and the status of art-differently ([21]: 186). In current video art practices, technologies, mediums, media, and genres have converged or became blurred. Moving images can now be recorded on film, on memory cards or smartphones, they can be edited in computers and tablets, projected online or at home, and all at a fraction of the cost when compared with earlier technological equipment used for the same tasks.

The idea of using video as part of a closed system through a feedback loop was already common to several artists during the 1970s, and this approach was actually in line with the rise of systems theory and computer-based art, at a time when the concept of participatory art paved the way to that of interactive art [19].

> We have gone beyond the image, to a nameless mixture, a discourse-image, if you like, or a sound-image ("Son-Image," Godard calls it), whose first side is occupied by television and second side by the computer, in our all-purpose machine society ([3]: 199).

Tracing generative video art's lineage through video art's distinct studies, contexts and theoretical propositions is an intricate task: electronic TV [43], expanded cinema, intermedia and videotronics [61], abstract film [25], calculated cinema [6]; experimental cinema [24], artists' video, experimental video, new television and guerilla TV [36], generative cinema [27], video installations, TV art and projected art [21], algorithmic editing [13] are all designations—among many more, undoubtedly— that reflect not just a diverse range of art movements and artivism, but also their contemporary cultural contexts and technological developments. Generative art, on the other hand, has ensured a somewhat *well-behaved lineage*, even though a proliferation of designations can also be found: computer art, process art, evolutionary art, genetic algorithms, among others, which are thoroughly dissected and historically discussed by Galanter [17].

9.2   Roots

Among the many influential artists and theoreticians that could be summoned into tracing generative video art's lineage, I chose to only highlight a selected few in the following paragraphs, as they represent periods, trends, art movements, or a specific use of technological tools and media.

As Bonet suggests [6] the first point of confluence between filmmakers and computers constitute, in some ways, the two greatest poles of reference in twentieth century cultural history. For him both poles, at the beginning and end of that same century, encouraged the idea of a convergence between art, science, and technology. But video art was born from the need to critique a commoditized culture through its mass-produced technological pinnacle, therefore I posit yet another vector of convergence: society. Video art and generative video art both share this DNA of socio-cultural critique and intervention.

Bonet places the origin of calculated cinema as early as 1916, with the experiments of futurist artists Arnaldo Ginna and Bruno Corra in conceiving the film *Vita Futurista* (Fig. 9.1) in autonomous episodes or sketches.

In the late 1940s John Witney and his brother James won the prize for best sound at the Brussels Film Festival with *Five Abstract Film Exercises*.[1] They called their

---

[1] The film can be viewed here: https://www.youtube.com/watch?v=JdCjwS1OxBU.

Fig. 9.1 Film poster for the
première of *Vita Futurista*
(Wikimedia Commons)

works *audio-visual-music* and their goal was to create films from pure light, by acting directly on the film by means of cutout shapes, instead of capturing images of objects in the world ([45]: 39).

Nam June Paik and Wolf Vostell, two artists linked to the Fluxus movement, first exhibited their innovative artworks in 1963, thus establishing the commonly accepted birth date of this art form. Video art has assumed different formats, such as installations exhibited in galleries or museums, screened videos or performances that incorporate television sets, video monitors and projections, displaying moving images and live or recorded sounds.

Around the same time as Paik and Vostell's works were exhibited, the term *generative art* made its first appearance with Georg Nees' *Generative Computergraphik* exhibitions in 1965, both solo and with Frieder Nake.

> As Constructivism and Futurism attempted to invoke the possibility of a worldview based on industrial processes, so generative art presents us with a *Weltanschauung* of computation. Forms produced by generative systems often take on a complex nature, exploiting principles of emergence to produce structures that could not be made by human hands. Inspiration taken from processes found in nature is common, with the tension between organic and mechanical forms ever-present. A common challenge in computational aesthetics is the simulation of organic behaviour and spontaneous irregularities, phenomena that appear in nature without prompting but which can only be replicated by computers with the explicit encoding of such behaviour ([59]: 1).

In 1967, Michael Noll [41] brought up the connection between computers and the visual arts and the benefits of the artistic exploration of new technologies, such as the reduction of labour intensive and repetitive tasks that required precision. Noll and his peers published hypotheses and models for automating the creation of optical, geometric, kinetic, dynamic, psychedelic, and stereoscopic works of art. Limited by the capabilities of early computers for processing images and sound, as well as contemporary trends in visual arts—abstractionism, cubism, and constructivism—these new approaches were mostly oriented towards concept and procedure, using a rather minimalist approach.

Nowadays using digital computers to tackle generative systems and their iterative nature ensures that several complex operations can take place almost instantaneously, not just in numerical terms, but more specifically in terms of media manipulation, including still and moving image files, sound files, as well as image and sound synthesis, which would otherwise be extremely hard—if not impossible—to accomplish in an analogue environment. And current computers also provide practical solutions for stochastic generative systems, through the introduction of chance (even though I prefer the expression "controlled randomness").

Moles posits the concept of constraint as a revolt against (pure) chance:

> There is no art without constraint. To say that music is an art is to say that it obeys rules. Pure chance represents total liberty, and the word construct means precisely to revolt against chance. An art is exactly defined by the set of rules it follows ([38]: 102).

But what better revolt than that of appropriation, embracing chance through constraints? Among the many definitions of generative art [55] a common trait is the existence of an algorithmic system with some degree of autonomy, iteratively combining structure with controlled randomness. And this set of rules is indeed what defines generative art. The system's output is a complex flow of graphics, text, audio, all of which can be synthesised or manipulated and recombined, either independently or creating cross-media patterns. Thus, a seemingly infinite and iterative sequence of states or combinations is achieved, within a specific aesthetic boundary and intent defined by the artist/programmer [10].

James Whitney's seminal work *Lapis*,[2] from 1966, could be posited as the first generative video artwork, as it was produced with the help of a (analogue) computer, built from surplus World War II anti-aircraft guidance hardware:

> He [John Whitney] transformed this military-spec surplus into a machine for creating experimental animation – literally and metaphorically retooling a device that had itself served to remake human vision for modern war. A twin of this machine would enable John's brother James to create the 1966 film Lapis, a work P. Adams Sitney would describe as "the most elaborate example of a mandala in cinema." ([45]: 37).

Lapis, like most early generative art, was an abstract work, consisting entirely of hundreds of moving points of light, which are given a radiating structure, similar to that of a mandala. *Lapis*, which is Latin for *stone*, winks at the alchemical transformation of the philosopher's stone, turning light and chaos into shape and contrast.

This approach embodied the attitude of the *structural* or *materialist* filmmaker, who took pride in exhibiting not just the film, but the whole apparatus of production

---

[2] For an interview conducted by David Em with John Whitney about his creation processes and work, which also includes excerpts of *Lapis*, see https://www.youtube.com/watch?v=cP5Mj6ZvZJc.

and representation. This went against the conventions of cinema, which sought to conceal the processes to focus mainly on the end result. The structural/materialist approach was intent on engaging the audience at all levels of the creation and production processes, much like a scientific researcher and experimenter—a maker—which confronted, to a certain extent, the ideal of the artist as a transcendent and illuminated individual.

Since those early days several other examples have evolved, dealing with the applicability of generative systems to non-abstract forms, such as L-systems use in 3D software for generating plants [34, 44], generative music [9, 14, 49] and generative literature [2, 40], all of which seek not just to exhibit the artwork as an outcome of a process, but also to detail and discuss the process itself.

From a visual stance, if the generative system's vocabulary consists of lines, geometric figures and patterns, the rapid sequencing of its states—or generations— will deliver an animation. However, if instead of being limited to lines and geometric primitives the system also manipulates images and image sequences, then the output becomes almost indistinguishable from (digital) video art, except for its potentially infinite duration and constant variation.

The recent rise in popularity (or at least widespread visibility) of generative video artworks, can be attributed in part to the COVID-19 pandemic, since their online exhibition—as a video file, resulting from a screen capture or automatic frame calculation and sequencing—was facilitated in the categories of *Video Art, Experimental Cinema and Expanded Cinema*, in a range of events, from festivals to galleries, from contemporary urban architecture with giant public screens to trade shows and conference displays, from entertainment and consumerism centres to multimedia music shows and VJ performances [29]. All these events embraced the virtual form on the Internet in an unprecedented way, in complement to previous physical formats and venues.

Fig. 9.2 Still image from Alchimia (author)

The immense poetic and expressive potentials of film have been barely realized within the cinematic cultural legacy, mainly due to industrialization, commercialization, politicization, and consequent adherence to the pop-cultural paradigms. Unrestrained by commercial imperatives, motivated by unconventional views of film, animation, and art in general, generative artists have started to engage these potentials playfully and efficiently, with explicit or implicit critique of cinema in a broader cultural, economic, and political context ([20]: 384)

And some of those artists have also been embracing non-abstract (even if surreal) aesthetics, often incorporating sound into their artworks, such as Brian Eno's *77 Million Paintings* [33], Jim Bizzocchi's *Re:Cycle* [5], Doug Goodwin's *Mersenne Devil Twister* [18], Rachele Riley's *The Evolution of Silence (V2)* [48], or Pedro Veiga's *Alchimia* [54] (Fig. 9.2). In these works, the concepts of cinematic shot, sequence, sound sample, and timed montage replace—or coexist with—the concepts of point, line, and colour in the generative systems' grammars, as atoms of their alphabet.

Generative video art may thus be perceived as both the offspring of abstract generative art with analogue video art, as well as a specialization of generative cinema. Much like video art, generative cinema in digital art became more accessible—and therefore popular—with the democratization of the technologies for digital video recording and editing. The ensuing development of the area, however, was fostered not just by technical aspects, such as the development of software and hardware for manipulating large numbers of equally large multimedia files and complex databases, but also by methodological and conceptual aspects, going beyond the limitations of image and sound synthesis, embracing different poetics, much like video art did in respect to film and cinema.

Through generative video art, artists gain new insights and are able to explore conceptual, formal, technical, expressive and communicational elements of generative systems, film and animation, enhanced through experimentation and stochastic factors—easily achievable though the use of computational systems. Furthermore, the algorithmic nature of generative artworks along with the non-destructive characteristics of digital art—which allow for the adaptation and evolution of artworks, without the destruction of the previous versions—ensures that they can be repurposed and redeveloped into new projects with potentially distinct intentions, identities, impacts, and outcomes [56], effectively turning them into generative frameworks.

This article thus posits that generative video art has specialised characteristics, inherited from film, generative art and cinema, whose combination determines a specific and distinct identity. Over the next sections these characteristics will be analysed and discussed.

9.3 Process

When designing a generative video art project, the artist will be faced with a set of decisions, ranging from intention to aesthetics, from programming choices to audience impact. Since generative processes are closely linked to representations of

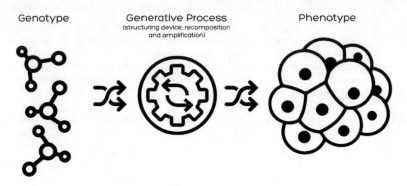

Fig. 9.3 The genotype and phenotype in generative art (author)

natural phenomena, including evolutionary organisms and artificial life models, the use of biological metaphors is also popular [11].

The terms genotype and phenotype are used to represent two distinct stages of the generative process. In biology the genotype is a collection of DNA, whereas the phenotype is the collection of the resulting features and characteristics the DNA determines on the organism. In digital computer systems, the genotype is the data that is used as input into an algorithm, which then produces the phenotype as a calculated result.

The genetic/generative system evolves through the introduction of changes, either deliberate or stochastic to the genotype selection, discarding those combinations that do not fit within the artist's vision, and further breeding genotypes that do. Thus, the enactment of the generative system delivers a phenotype as its output, as one instance of the artwork. This also implies that a finite genotype can originate an exponentially larger number of phenotypes, as is usually the case with generative artworks, not only due to the different combinations of genotypes, but also due to the introduction of stochastic and random elements. Figure 9.3 illustrates these stages.

Whereas in previous work it was proposed that the creative process behind generative art systems could be decomposed into three steps [55], the present article posits an extra step, which tackles in detail the figurative dimension, and can also be a generative process in itself. It is thus the first step to be considered.

9.3.1 Vocabulary Identification

At the genotype identification stage the artist will determine which symbols will be involved in the artistic creation. Usually L-systems, which can be found at the core of most generative art systems, are built from grammars, comprising a vocabulary (made of units or symbols), axioms (combinations of the vocabulary units) and rules (which determine the evolution of each vocabulary unit or their combinations). Each abstract symbol can then be assigned a concrete conversion or interpretation: into spatial positioning, choice of geometric shape, translation, rotation, scale, or even musical note pitch or duration. But this list is as long as the full range of objects that can be manipulated by the programming language, including all primitives that can be used from within the code (to obtain image and sound synthesis, for example) to the manipulation of external files containing video snippets, images (both static and animated), sound samples, text files or any other relevant file types or databases.

Let us then use the word *vocabulary* to designate the set of all possible symbol replacements we can consider in the creative process of a specific generative artwork. This choice of vocabulary is also the first step into defining a subject or theme (if applicable) and also a first step toward the definition of the artwork's aesthetic boundaries.

Once the vocabulary is chosen the artist will engage in a process of deconstruction, decomposing each collected element into their simplest—one would dare suggest atomic—form, so that these atoms can then be manipulated by the structuring device.

Among the many experimental results to which the informational theory of perception makes it possible to attain, the conceptions developed here suggest an experimental aesthetic methodology. Thus, one of the most general methods of aesthetic discovery consists, on the basis of the materiality of the work of art, in progressively destroying it by known quantities, and in following the variations of aesthetic sensations, value, of the knowledge of the work as a function of this destruction. This is a method of concomitant variations. The used method of destruction will depend on I) the nature of the artistic message (audio, visual, etc.), 2) the a priori knowledge that one may already have on it (form, subject, etc.), 3) factors that we seek to highlight (regularity, originality, semantics, etc.) ([39]: 239).

However, vocabulary units may be derived from other simple units. When Paik [42] translated the metrics of space (the frame) into the metrics of time, which he manipulated in order to produce his video installations, he laid the foundations for the current use of temporal (rather than spatial/frame) units of calculation, most particularly in digital media, since clock synchronism is achievable in most programming languages. "These experiments with temporal and iterative strategies set out to expand the material space of the cinematic experience" ([26]: 123). Temporal units and timed iterations may thus be used to describe different types of audio-visual sequences, which will in turn be treated as vocabulary units: frames, flicker forms, loops, and cycles.

The term frame has come to designate the shortest temporal unit present in both film and also generative systems, and the concept of *frame rate* (the number of frames that are shown in one second) is implemented in most programming languages currently used for generative art. Therefore, the frame can be defined in terms of a fraction of a second, the most popular values being directly inherited from video and television: 60, 30, 26 or 25 frames per second.

Flicker forms consist in the rapid flashing of either structured or unstructured frames and are "the vehicle for the attainment of subtle distinctions of cinematic stasis in the midst of extreme speed which can be presented so as to generate both psychological and apperceptive reactions in its spectators" ([53]: 288). The most common and familiar flicker form is the strobe light, closely followed by short animated GIFs, endlessly repeating the same (structured) content. But if the artists so desire, they can introduce considerable randomness into the content, thus making it vary from total structure to (almost) total chaos, and yet retaining the flicker form by obeying the time constraint.

Loops essentially differ from flicker forms in terms of their duration. Whereas the flicker form is characterised by rapid flashing content, the loop may last for several seconds and even minutes. Loops became one the most prevalent modes of presentation in video art ([1]: 14) but in more recent years animated NFTs and GIFs have also contributed to the loop's stable (growing?) popularity.

From an artistic and cognitive perspective, the loop allows for a temporal processing that can lengthen the experience by requiring the observer to attend to and process time-related information, to observe more carefully what was irrelevant beforehand, to deepen their understanding and immersion. In other words, the potentially endless structural repetition of a short scene may lead to an extension of time, at least at the level of judgment, perception, and experience ([50] :19).

Cycles may contain all other temporal units, organised into timed, evolving sequences, potentially including other cycles, like an encompassing meta-generative structure. Cycles are structurally reminiscent of the concept of loop. However, in this context, the author posits the term as an encompassing structure, within which several loops can occur, as well as all other types of temporal units. Cycles can thus provide a timed structure to what may otherwise be perceived as unstructured content, providing both the familiar and the unexpected, seriality and network.

All these vocabulary units are illustrated in Fig. 9.4, with their different temporal and structural characteristics.

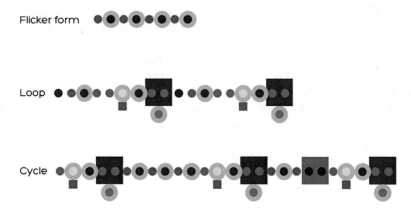

Fig. 9.4 Flicker forms, loops and cycles (author)

9.3.2 Structuring Device

The essence of a structuring device is a set of rules and procedures—an algorithm. It will define strategies for combining the previously selected vocabulary atoms into more complex structures: the actual language of the artwork. This section will present early identifications of structuring devices used in experimental film and will then complement them by a theoretical framework, converging into an appropriate characterization of generative video art's structuring devices.

In the context of generative video art, structuring devices inherit significant technical influences from cinema, most particularly from montage theory. Among the most relevant influences are *calculated cinema* (metric and rhythmic montage), *structural cinema* (fixed camera position, strobing, loop, and rephotography), and *soft cinema* (algorithms running over media databases). All these montage techniques act upon the conversion of input-time to output-time:

> (...) the relationship of input-time and output-time is much more complex – e.g., in some extreme situations or in dreams our whole life can be experienced as a flashback compressed into a split second (the survivors from air crashes or ski accidents tell of it often) ... or, as in the example of Proust, one can brood over a brief childhood experience practically all of one's life in the isolation of a cork-lined room. That means, certain input-time can be extended or compressed in output-time at will... and this metamorphosis (not only in quantity, but also in quality) is the very function of our brain which is, in computer terms, the central processing unit itself. The painstaking process of editing is nothing but the simulation of this brain function ([42]: 98).

Going back to the 1920s, at a time when avant-garde cinema was heavily influenced by soviet filmmakers, such as Eisenstein and Vertov, Hans Richter produced several interesting film works, which can be used to exemplify types of montage that are now associated with both Russian montage film theory and the video art form. His works illustrate not just the rhythmic synchronization achieved between abstract animations and sound, but also the comparison between Eisenstein's concepts on montage to Japanese Haiku poetry, where the gaps in meaning create the space for imagination to fulfill ([12]: 90–103). Richter's animations can still be appreciated in digital transcriptions, and *Rhythmus 23*[3] (from 1923) or *Filmstudie*[4] (from 1926) are remarkable—and very early—precursors of video art, combing direct frame painting with cinematic shots. In these examples the structuring device is the set of rules followed by the artist in determining the origin, order, and duration of the different sequences. Hans Richter characterised his artworks in terms of the organisation of abstract forms, claiming it was founded on "a universal language, which is what abstract art should be about" ([15]: 172).

A few years later, László Moholy-Nagy's *Lichtspiel: schwarz, weiss, grau*, created in 1930, also provides an excellent example of a machinic (non-computerized) structuring device in action. Produced with the help of a mobile sculpture he designed and

---

[3] https://youtu.be/CMd2J9teidY.

[4] https://youtu.be/ZXrjrr6ifME.

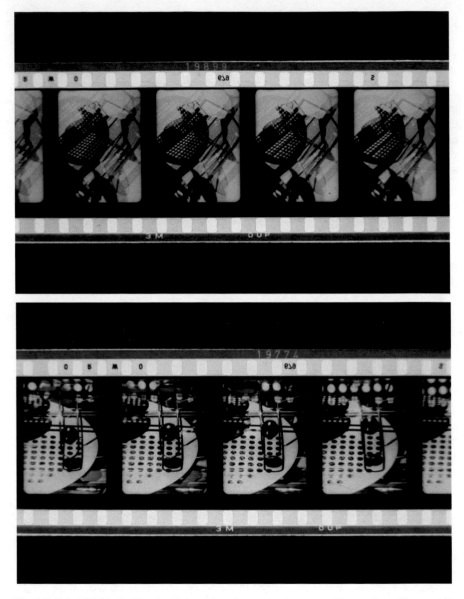

Fig. 9.5 László Moholy-Nagy's Lichtspiel: schwarz, weiss, grau (Centre Pompidou, public domain)

called *light space modulator*,[5] the film essentially contains fixed camera sequences of the device in action, depicting its abilities to modulate light (in black, white, and grey renderings) (Fig. 9.5).

---

[5] https://youtu.be/nVnF9A3azSA.

In 1949 the Whitney brothers produced an experiment in synthetic sound, by inscribing it directly on film, through the use of a structuring device: pendulums of differing lengths were fit with weights and then attached by wire to the film strip. Their movements were used to directly etch lines on the strip, creating patterns that could generate sound through the projector, thus effectively transforming motion into sound ([45]: 39–40).

Directors such as Sergei Eisenstein and Dziga Vertov laid the foundations of five different types of montage: metric, rhythmic, tonal, overtonal and intellectual [51]. However, the algorithmic and calculated nature of generative video art determines that the focus will fall mainly on the first two: metric and rhythmic montage— despite the fact that exercises in addressing tonal and overtonal montage types have also been addressed by Manovich's *soft cinema* [30]. Vertov describes editing as "the combining (addition, subtraction, multiplication, division, and factoring out) of related pieces" and then describes one stage of editing as a "numerical calculation of the montage groupings" ([58]: 90). He then proceeds explaining how an editing table should have "definite calculations, similar to systems of musical notation, as well as studies in rhythm, 'intervals,' etc." ([58]: 100) (Fig. 9.6).

As mentioned, in the current context, metric and rhythmic are the most relevant montage types inherited from calculated cinema, since they imply the consistent application of an algorithm (or automation) to a human activity: that of editing film according to specific and objective rules.

Metric montage is based on the absolute length of shots, regardless of their content, aiming for a constant cadence, often synchronized with music—which is something a digital computational system does with utmost precision and efficiency. Rhythmic montage, on the other hand, seeks an editorial and compositional relationship in which the sense of motion within each sequence is as important as its length, thus reinforcing one another [12]. Determining rhythm and intensity, translated into acoustic pattern detection, or the amount of variation in image and sound sequences is something that computers can also very easily perform, therefore allowing for the automation of those processes. This allows not just for rhythm (duration) and intensity (amount of variation) to be rules-driven, through automated analysis, but also—pending on the artist's decision—to be predefined from the beginning, thus using one to act upon the other.

Another important approach to the field is brought by *structural film,* an expression coined by Sitney ([53]: 348) to describe a formal category of film/video, where form is crucial and content peripheral.

> In structural film, form became content. The viewer's identification with the 'dream screen' was disrupted. The structural film rejected the cinema of pure vision. (...) experimental film shifted into new philosophical territory. Underground sensation gave way to structural investigation ([47]: 72).

Structural film proposes that the shaping of light, time, and process results in a new form of aesthetic pleasure, free of symbolism or narrative. It combines predetermination (for example, camera position, number of frames or exposures, repetition) with chance (the unpredictable events that occur at the moment of shooting) ([47]:

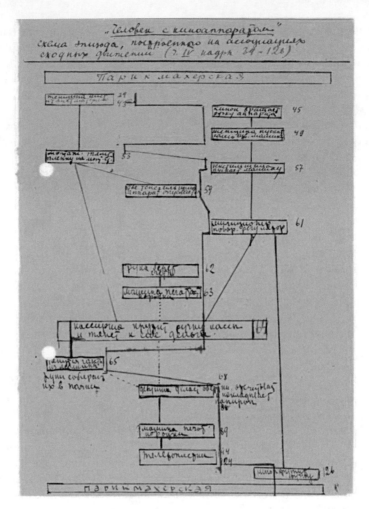

Fig. 9.6 Dziga Vertov's editing diagram for *Man with a Movie Camera* (1929), the *hair salon* sequence. The numerical calculations in the editing process are clearly visible, with the blue boxes describing shot contents, and the red line between them displaying the rhythmic quality of editing ('intervals', as Vertov called them) (cinemetrics.lv)

74). This combination of chaos and order is yet another characteristic that draws structural film closer to generative art, adding to the four properties that Sitney uses to characterize structural film:

- the use of a fixed camera position—or its digital conceptual equivalent;
- the use of flicker forms—also known as *strobing;*
- the use of loops;

- and rephotography from the screen, which could be described as the recursive and generative act of capturing a previously generated frame (or sequence) and using it as part of the vocabulary for generating subsequent frames (or sequences).

To finalize the triad of relevant foundations to the field, now is the time to introduce *soft cinema*. Manovich [30] posits the use of an algorithm at the heart of what they describe as *soft cinema*: whereas *hard* cinema is characterised by story-telling elements, scripting, shooting, and editing guidelines, *soft* cinema is the result of algorithmically sequencing audio-visual units from a database, often incorporating controlled randomness in the selection mechanisms. The algorithm can be used to create the vocabulary—or the database—since "once digitized, the data has to be cleaned up, organized, and indexed. The computer age brought with it a new cultural algorithm: reality → media → data → database" ([31]: 224). And the algorithm is then reversed, as the generative video artist uses it to create the artwork: database → data → media → reality.

> The adjustable combinations and weightings allow generator users to create films edited according to formal criteria similar to those of classical montage, but without regard for the development of a narrative storyline. In this sense, such works are closer to a relational database than to an edited film ([30]: 13).

In soft cinema the screen can be simultaneously occupied by a mosaic of images, not just one, and their succession is not necessarily fixed. Each new image may be drastically different from previous ones, depending on the use of controlled randomness in the selection of the database units, and the database itself.

Generative video art inherits, thus, characteristics of calculated cinema, structural film, and soft cinema. All these characteristics imply the existence of rules, which are at the heart of generative art. These rules allow for systemic autonomy, relative to the conscious decisions of the human artist—a significant phenomenological property of generative artworks. The system, once started, will continue to evolve within its defined boundaries, allowing for unexpected combinations or results, through the introduction of controlled randomness.

The centre of the spectrum that ranges between complete chaos and simplistic order is much richer than either of the extremes. A possible explanation of this property, posited by Davis [8], is that the human mind finds appealing those visual and auditory event combinations where properties of hierarchical complexity and subtle disorder are present, and that combinations of these loosely defined properties tend be maximized at the centre of this spectrum. This is what Galanter [17] describes as a complex system, one that is at the apogee of the complexity curve, interweaving order and chaos through rules and constraints, in an aesthetically pleasing way.

9.3.3 Recomposition and Amplification

Once the structuring device is defined, cognitive extensions are developed through correlations between different media types, involving recursive structures and patterns. This is the stage where the recomposition of the (previously decomposed) atomic elements takes place.

> The system of generative aesthetics aims at a numerical and operational description of characteristics of aesthetic structures (which can be realized in a number of material elements) which will, as abstract schemes, fall into the three categories of the formation principle, distribution principle and set principle. These can be manipulated and applied to an unordered set of elements, so as to produce what we perceive macro aesthetically as complex and orderly arrangements, and micro aesthetically as redundancies and information ([4]: 207).

Through the use of controlled randomness, each time the system runs it delivers a (potentially) heterogeneous realm of different outcomes, instances of the phenotype, with different factors of unpredictability. As these chance procedures induce unexpected forms despite coming from a precise mathematical procedure, the artworks maintain unexpected elements.

The combination of vocabulary atoms extracted from the database can also be dynamically rephotographed, which means that the result of their combination is inserted into the database, thus becoming a new vocabulary atom. As the vocabulary evolves, so does the artwork, and so the next iterations/generations will be created over an increasingly larger vocabulary. The term *amplification* is also used to describe this stage, as the amount of generated information will largely exceed the initial vocabulary. But amplification may also be a result of errors in the code, as they often produce unexpected results, such as novel visuals or combinations. Through the added layer of mystification (as the artist will not expect them) these glitches contribute to the expansion of the otherwise planned artwork.

The core of this stage relies in the iterative identification of the most aesthetically pleasing or meaningful combinations of atomic elements, the creation of new complex vocabulary elements through repeated execution and observation. Adjustments are then—also iteratively—introduced in the structuring device to reflect these choices and observe the new outcomes they imply. This is how the aesthetic boundaries of the artwork are defined.

> (...) another common feature of generative systems: the emergence of new properties that result from local interactions between individual components. These new properties are not specified in the genotype — they emerge via the generative process ([34]: 6).

Each newer generation—which can be a frame or a sequence—depends upon, or inherits, structural and compositional characteristics of the previous generation(s), but through the constrained unpredictability built into the structuring device, each time a complex generative artwork is executed, its outcome will be different.

Due to the fact that interference mechanisms can be built into the system, either derived from information gathered from the system execution itself, from human interaction or external sources, mapping these interferences to semantic options is another important step, as they can be used to influence the direction and evolution of the generative artwork. In this way, sound, image, composition, or movement can be interpreted and manipulated during run-time.

[Concerning generative art] process (or structuring) and change (or transformation) are among its most definitive features, and that these features and the very term 'generative' imply dynamic development and motion ([7]: 117).

Once the artist is pleased with the recomposition and amplification mechanisms that were introduced into the structuring device, it is time to move on to the final stage.

9.3.4 Event Detection

In the fourth stage, the artist has already introduced adjustments to the system, both in terms of structuring device and amplification mechanisms, and is now concerned in identifying the more (aesthetically, meaningfully, or otherwise) relevant occurrences as the system runs. Drawing on the previously presented genotype/phenotype metaphor, the artist will now identify which element (or family of elements) of the phenotype will be selected for public exhibition.

Each execution is called *an event*, and the artist will probably disregard many events, through trial and error. But one of them will eventually be identified as a full-bodied artistic expression of the initial concept and aesthetics, like an epiphany:

I was also imposing a new and more restrictive requirement on my work, i.e. that the structuring process must generate a visual epiphany, or epiphanies, in the painting. As is well known, epiphany is used in the arts and literature to refer to sudden revelations or manifestations of meaning. Speculatively, in science, epiphany might be considered roughly analogous to the sudden emergence of a new and higher level of order through the self-organizing principles of dissipative structures, such as living matter (...). Also, conceptually, it is suggestive of a mutation ([7]: 120).

This epiphany (or epiphanies) will be registered through screen-recording mechanisms or by calculating each individual frame and then converted onto a video format, according to the exhibition requirements. As a video file, its essence will become potentially indistinguishable from other digital video art counterparts, produced by other (non-generative or computational) methods.

Rather than delivering one single, immutable video stream at each showing, a generative video art system may deliver a coherent family of instances or events, sharing common aesthetic traits, but distinct from one another.

Ultimately these instances also share another trait in common with early video art: impermanence. The potentially impermanent and ephemeral nature of video was at one moment considered a virtue by some artists who wished to avoid the influences and commercialism of the art market. For them, working live (or, as we might put it in terms of generative video art, in run time) could be regarded as an artistic statement, and such may still be the case with many VJs or live-cinema practitioners. This implies, however, that many relevant artworks are—and will be—lost, marginalized, or ignored, and other less relevant artworks deemed unworthy of preservation [36].

These four development stages are illustrated in Fig. 9.7.

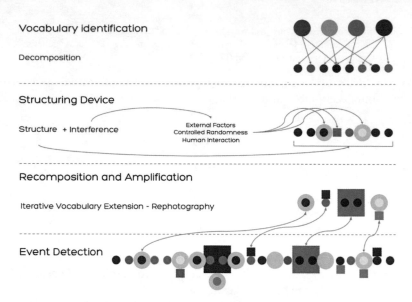

Fig. 9.7 The proposed four development stages of a generative video artwork (author)

9.4 Poetics

As was previously mentioned, generative video art inherits a rich historical background, not just in technical terms, but also in its aesthetics of displacement, deconstruction, and oneirism. Early video art pieces were mostly poetic works in which formal devices were symbolically used to establish complex interrelations between images. Generative video art works have embraced the use of formal devices to predefine the composition and structure of the artwork, defining its boundaries. All the instances (obtained during runtime) evolve within those boundaries, even before deciding what images and sounds populate which sequences. The poetics of generative video art are not just those of content: they are also those of form. They are organic, as one aspect influences the other, closely bound and evolving together.

> The idea of "wholeness" is obviously not new, but recently it has taken on a meaning different than the accepted "organic unity" principle which Eisenstein stated so lucidly: " … in an organic work of art, elements that nourish the work as a whole pervade all the features composing this work." A unified canon pierces not only the whole and each of its parts, but also each element that is called to participate in the work of composition. One and the same principle will feed any element, appearing in each in a qualitatively different form. Only in this case are we justified in considering a work of art organic ([52]: 27–28).

Fluxus is also a significant part of generative video art's inheritance, and the following excerpts of Friedman's Twelve Fluxus Ideas [16] illustrate how (almost naturally) relevant they are in the present context:

Globalism: This is achieved through the encouragement of dialogue among like minds, regardless of nation, and that of unlike minds when social purposes are in tune. Fluxus posits globalism, democracy, and anti-elitism as intelligent premises for art, for culture and for long-term human survival. Generative video art seeks a universal language, one that speaks of human emotions and aesthetics, of ideas and sensation, regardless of nation or spoken language, one that bares the artists and their technique, rather than idolise them.

Unity of art and life: Fluxus' conscious goal was to erase the boundaries between art and life, regarding both as parts of a unified field of reference, a single context. Generative video art embodies the very notion of an *organism*, one that evolves, generation after generation, thriving on the artist's vision. As it handles captured images of life itself, this unity exists in more than just the theoretical plane: it is a part of the artefact itself.

Intermedia: Even though we can still identify separate art forms and their media, intermedia seeks to point out that our time often calls for art concepts that draw on the roots of several media, growing into new hybrids. And generative video art is yet another hybrid, the result of the convergence of video and computers through digital technologies, tackling sampled and synthesized, still and animated images, sound, music, and text.

Experimentalism: Fluxus advocated for experimental research orientation, by trying out new—sometimes iconoclastic—approaches and assessing the results. Experiments that yield useful results cease being experiments and become results, exactly like the event detection stage posits, in the generative video art creative process.

Chance: Randomness is a tradition with a legacy going back to Duchamp, to Dada and to Cage. A new, uncontrolled variable is taken into consideration and changes the previously held worldview. It may stem from chance or be the result of signal interference, or even be the result of a sudden insight. When controlled randomness is incorporated in the artefact, it ceases to be random and becomes evolutionary. And again, controlled randomness is, as previously mentioned, one key characteristic of generative art, and of generative video art as well.

Playfulness: In the Dada tradition, humour is a key topic for Fluxus, but the concept of playfulness extends beyond humour: it refers to the play of ideas, of free experimentation, and free association, and the play of paradigm shifting, found in scientific experiments, as well as in pranks. Generative video art fully embraces these aspects, and adds one more: playfulness through interaction and participation, during runtime.

Simplicity: Another term for this concept is elegance, in the relationship between truth and beauty. Generative video art is usually complex, from a technological and aesthetic point of view, but it seeks elegance, just as much as Fluxus, and not just in terms of the impact achieved on the art audience, but also in the construction, in the elegance of the coding of the artefact itself.

Implicativeness: An ideal Fluxus work implies many more works, as could be expected from the advocated relationship with experimentalism and the scientific method. And the exact proposition holds true for generative video art: an artwork's algorithms may be adapted to newer versions of the same artwork or to a completely new project, in a non-destructive way.

Exemplativism: This is defined as the quality of a work that exemplifies the theory and meaning of its construction. Like so many generative video art projects, which are accompanied by scientific texts detailing their concept and implementation.

Specificity: The tendency of a work to be self-contained and to embody all its own parts. This idea is actually ambiguous, and Friedman acknowledges this fact by stating "this may seem to contradict the philosophical ambiguity and radical transformation of Fluxus. Nevertheless, but it is a key element in Fluxus". There might be a divergence here, since generative video art thrives on the oneiric, on the ambiguous. However, from the artist's point of view there will always be a specific reason for creating the artwork and a specific intention for its exhibition.

Presence in time: The ephemeral quality is key to Fluxus performance works, but also because they embody a different sense of duration: musical compositions with inordinately extended durations or artworks that grow and evolve over equally long spans. Time is a central issue in Fluxus and so it is in generative video art, as was demonstrated by its roots in calculated cinema and its relationship to time-dependent vocabulary (flicker forms, loops, and cycles).

Musicality: This addresses the fact that many Fluxus works are designed as scores, as works that can be realized by artists other than the creator. Musicality is linked to experimentalism and the scientific method, as experiments are expected to be reproducible and deliver a similar outcome. A scientist must be able to reproduce the work of any other scientist for an experiment to remain valid. A generative video art system will deliver an instance of the artwork, even if it is executed in a different environment and by a different operator—as long as the experiment conditions are ensured. The code is the score.

The fact that these ideas so adequately fit generative video art, should cause no surprise, especially when dealing with a time-based art form that not only needs time to be fully appreciated, but also bends and manipulates time and its flow, that maps input-time to output-time, that delivers a *fluxus*, where experimentation and exploration play a crucial role.

> All poetic language is the language of exploration. Since the beginning of bad writing, writers have used images as ornaments. The point of Imagisme is that it does not use images *as ornaments*. The image is itself the speech. The image is the word beyond formulated language [46].

Drawing on the correspondence exchanged between Marshall McLuhan and Ezra Pound about the written form and the image, one can now posit the applicability of key concepts, such as the *mosaic*, the *perceptions of simultaneities* and the *vortex* to generative video art, from a poetics stance.

For [46], an image is that which presents an intellectual and emotional complex in an instant of time. The image does not stand for an idea: it acts as a radiant node or cluster. In generative video art, the image may be deconstructed in the early stages (while defining the vocabulary atoms), only to be reconstructed, recomposed, reorganised, recombined. Thus, the image becomes a *vortex*, "from which, and through which, and into which, ideas are constantly rushing" [46]. Vorticism, as applied to the visual arts, is illustrated in Fig. 9.8, which seeks to instantaneously deliver such a complex construct.

Pound stressed the representation of energy, of mechanical efficiency, of concentrated power, and also justified vorticism as a legitimate expression of life. For him, vorticism brought a new arrangement of forms, and allowed for different and new ways of perceiving, representing, and understanding the changing reality ([28]: 155).

Generative video art, as a time-based art form, not only embodies these concepts, but also adds a time-based vortex and a time-based mosaic to them, as each iteration produces a new vortex, a new mosaic, in rapid sequence.

The expression "perceptions of simultaneities" is used by McLuhan ([35]: 193) in a letter to Ezra Pound, to characterize his work, *The Cantos*. McLuhan's subsequent development of the *mosaic* theory is grounded in Pound's *imagisme* and *vorticism*, and tackles non-linear thinking, *presentness*, and juxtaposition. For Lamberti ([22]: 32) "McLuhan uses his mosaic to question traditional ideas of knowledge and to

Fig. 9.8 Edward Wadsworth, Vorticist Study, 1914 (Wikimedia, public domain)

Fig. 9.9 A frame of *Hello*, 2020, a cinematic generative art piece, illustrative of the poetics of the mosaic and vortex (author)

move the reader from a linear (logical, ordered, exclusive) to an acoustic (non-logical, simultaneous, inclusive) perspective".

This perspective is embraced by video art, and most particularly by generative video art, determined by the absence of a formal logical and linear narrative (replaced by sensations and evocations), by recomposing and recombining its vocabulary into simultaneous renderings of different sources, and by welcoming chance, interaction, and participation.

> (…) the display joins the individual strands of film produced using the generator to form a kind of montage of simultaneity. This program divides the screen into individual segments and within each either a film or abstract animation is shown. The division is done by an algorithm, ensuring that the screen is harmoniously divided up. Thus, differences in size and positioning on the screen take into account the correspondence and relationships between the individual films. The same stage of processing that divides up the screen allocates soundtracks, a looped text and a voice-over channel ([30]: 13).

The coherency and the arrangement of all elements on the screen produce a continuum that alternates between rhythmic, visual structures and descriptive symbols. Narrative elements appear and disappear again, or make room for other elements that are in competition with them. At times, the impartiality and non-interpreted quality of the database entries give rise to aesthetic and meaningful structures that repeatedly fall apart and are transformed into other constellations.

> The Cultural Producer who samples from the raging flows of media detritus – endless satellite feeds, cable and broadcast transmissions, and the sedimentary layers of these through the

past 25-50 years – becomes the heroic Luther, wresting deconstructive (re)form(ations)s out of the desultory, formless industrial wasteland. Deconstructive film- and video-making demonstrate the inherent formlessness of mass media by making it into the "New Nature" ([23]: 84).

Figure 9.9 shows a frame of *Hello*, a generative video art piece where images of Thai trains and suburban landscapes are combined with occasional real-time stills of the audience, and displayed as a backdrop of a text-based conversation being held in Portuguese and Thai. There is also a real-time generated soundtrack mixing sound samples captured in Portuguese trains and suburban public transports. These deconstructive (re)form(ations)s—images, the sounds, and text fragments—all stimulate the audience's senses, allowing for a passive sensorial enjoyment, or an active deciphering of the concept and content [57].

These traits also determine that generative video art should be included in McLuhan's *cool media* category, as it requires audience completion through active viewing and interpretation, with potentially different outcomes for different audience members. A generative video artwork thus becomes a vortex of vortexes, a mosaic of mosaics, a theoretical universe of all its potential outcomes, where each mosaic and each vortex are only instanced at runtime.

One final consideration—very much in line with the aesthetics of Structuralism and Futurism—derives from the fact that in some instances the technical generative system may actually be the object of (poetic and aesthetic) interest, rather than its outcome: "In a fully realized scenario, LeWitt's dictum is extended to the point where a machine does not only make the art, a machine is the art" ([17]: 168).

9.5 Conclusion

Code-based artworks have reached a level of maturity, going beyond simple visual experimentation to expressing more complex visions. (...) While generative art is inextricably linked to the computer as a means of production, the work is not about the computer itself. While screen-based work and the investigation of realtime self-contained systems remain an important aspect of generative art, it would be a mistake to think generative work is primarily expressed in pixels. I for one look forward to an extended rethinking of computational aesthetics that encompasses a much wider range of possible outputs ([59]: 3).

Generative video art inherits a rich tradition of several art currents and movements, fostered by the current phenomenon of media convergence. We saw at the beginning of this article how the Futurist Cinema manifesto [32] still holds its relevance, by advocating the use of analogies, cinematic poems, simultaneity, interpenetration, musical experimentalism, dramatized feelings, and sensations, using objects as metaphors, windows, unreal reconstructions of the human body, among many other traits.

It was also significant to point out how much of the current work falling within the scope of generative video art also embodies Friedman's Twelve Fluxus Ideas [16]. However, the nature of generative video art suggests that the idea of *simplicity* ought to be replaced by *complexity*, as the art of combining order and chaos, symmetry and asymmetry.

> (...) the human mind is itself constrained to find appealing those visual and auditory event combinations that share properties of both symmetry and asymmetry, hierarchical complexity and subtle disorder, and that combinations of these loosely-defined properties tend to place interesting pieces in the centre of this spectrum ([8]: 1).

The use of controlled randomness can be achieved though computational means during runtime (most programming languages provide functions for pseudo-random numbers generation), but also through direct human interaction or other indirect or external sources, such as data from sensors that may read audience behaviours (movement, sound, temperature, among other factors), information feeds (tweets or Instagram posts with specific hashtags, for example) or other provenances, whose control resides outside the system. And this unpredictability will not only be found in the artwork's conceptualization, but also in its production, operation, and presentation ([20]: 8). Generative video art can thus cover the whole spectrum of audience participation and interaction.

As a conclusion to this article, the author now posits the following definition, inspired by Galanter's ([17]: 155) own definition of generative art:

> Generative video art is an art form in which the artist cedes control to a system with functional autonomy. This system is designed with an initial vocabulary containing still and moving images, whose provenance is mostly external (to the system). Using a structuring device and a set of recomposition and amplification rules, the system runs within selected aesthetic boundaries, producing events that contribute to or result in a completed video stream.

References

1. Abramovic, M.: Interview with marina Abramovic. In: Biesenbach, et al. (eds.) Video Acts: Single Channel Works from the Collections of Pamela and Richard Kramlich and the New Art Trust. P.S. I Contemporary Art Center, New York (2002)
2. Balpe, J.P.: Principles and processes of generative literature: questions to literature. Dichtung digital. J. für Kunst und Kultur digitaler Medien **7**(1), 1–8 (2005)
3. Bellour R (1996) The Double Helix. In: Druckrey T (ed) Electronic Culture: Technology and Visual Representation. Aperture, New York
4. Bense, M.: The projects of generative aesthetics. In: Reichardt, J. (ed.) Cybernetics, Art, and Ideas. New York Graphic Society, Greenwich, CT (1971)
5. Bizzocchi, J.: Re: cycle-a generative ambient video engine. In: International Conference on Entertainment Computing, pp. 354–357. Springer, Berlin, Heidelberg (2011)
6. Bonet, E.: Calculated cinema. In: Verbindingen/Jonctions 5. http://archive.constantvzw.org/events/vj5/calculatedLF.html. Accessed 8 Mar 2022 (2017)
7. Clauser, H.R.: Towards a dynamic, generative computer art. Leonardo **21**(2), 115–122 (1988)
8. Davis, M.W.: Complexity formalisms, order and disorder in the structure of art. In: International Conference on Evolutionary Programming, pp. 1–12. Springer, Berlin, Heidelberg (1997)
9. Dean, R.T.: Generative live music-making using autoregressive time series models: Melodies and beats. J. Create. Music Syst. **1**(2) (2017)

10. Dorin, A.: Chance and complexity: stochastic and generative processes in art and creativity. In: Proceedings of the Virtual Reality International Conference: Laval Virtual, pp. 1–8 (2013)
11. Dorin, A., McCormack, J.: First iteration/generative systems (Guest Editor's Introduction). Leonardo **34**(3), 335 (2001)
12. Eisenstein, S.: Film form: essays in film theory. Harcourt, Brace, New York (1949)
13. Enns, C.: The Poetry of Logical Ideas: Towards a Mathematical Genealogy of Media Art. Doctoral dissertation, York University (2019)
14. Eno, B.: Discreet Music Liner Notes. http://music.hyperreal.org/artists/brian_eno/discreet-txt. html. Accessed 8 Mar 2022 (1975)
15. Foster, S. (ed.): Hans Richter: Activism. MIT Press, Cambridge, MA, Modernism and the Avant-Garde (1998)
16. Friedman, K.: Twelve fluxus ideas. Radic. Des. Des. Cult. J. **1**(1), 1–27 (2007)
17. Galanter, P.: Generative art theory. In: Paul, C. (ed.) A Companion to Digital Art, pp. 146–180. John Wiley & Sons Inc. (2016)
18. Goodwin, D.: Mersenne Devil Twister. https://vimeo.com/20657096. Accessed 8 Mar 2022 (2011)
19. Graham, D.: Video-Architecture-Television; Writings on Video and Video Works 1970–1978. The Press of Nova Scotia College of Art and Design, New York University Press (1979)
20. Grba, D.: Avoid setup: insights and implications of generative cinema. Technoetic Arts **15**(3), 247–260 (2017)
21. Hayden, M.H.: Video Art Historicized: Traditions and Negotiations. Routledge (2015). https:/ /doi.org/10.4324/9781315548210
22. Lamberti, E.: Marshall McLuhan's Mosaic: Probing the Literary Origins of Media Studies. University of Toronto Press, Toronto (2012)
23. Lattanzi, B.: We are all projectionists. Millenn. Film J. **39**(40), 84 (2003)
24. Le Grice, M.: Experimental Cinema in the Digital Age. BFI Publishing (2002)
25. Le Grice, M.: Abstract Film and Beyond. MIT Press (1977)
26. Legget, M.: Generative Systems and the Cinematic Spaces of Film and Installation Art. Leonardo **40**(2), 123–128 (2007)
27. Lioret, A.: Galatema: a framework for generative cinema. In: Generative Art 2010 Proceedings, Milan (2010)
28. Macedo, A.G.: Wyndham Lewis's literary work 1908–1928: vorticism, futurism and the poetics of the avant-garde. Universidade do Minho, Centro de Estudos Humanísticos (CEHUM) (2014)
29. Manovich, L.: The poetics of augmented space. Vis. Commun. **5**(2), 219–240 (2006)
30. Manovich, L.: Soft Cinema. TKM, Karlsruhe (2002)
31. Manovich, L.: The Language of New Media. MIT Press, Cambridge (2001)
32. Marinetti, F.T., Corra, B., Settimelli, E., Ginna, A., Balla, G.H., Chiti, R.: The Futurist Cinema. L'Italia futurista, Milan. https://tinyurl.com/322f8rb9. Accessed 8 Mar 2022 (1916)
33. Marshall, K., Loydell, R.: Thinking inside the box: Brian Eno, music, movement and light. J. Vis. Art Pract. **16**(2), 104–118 (2017)
34. McCormack, J.: Art and the mirror of nature. Digit. Create. **14**(1), 3–22 (2003)
35. McLuhan, M.: Letters of Marshall McLuhan, Molinaro M, McLuhan C, Toye W (eds). Oxford University Press, Oxford (1988)
36. Meigh-Andrews, C.: A History of Video Art: The Development of Form and Function. Berg, Oxford (2013)
37. Metz, C., Guzzetti, A.: The fiction film and its spectator: a metapsychological study. New Lit. Hist. **8**(1), 75–105 (1976). https://doi.org/10.2307/468615
38. Moles, A.: Information Theory and Esthetic Perception. University of Illinois Press (1966)
39. Moles, A.: Théorie de l'information et perception esthétique. Revue Philosophique de la France et de l'Étranger, Presses Universitaires de France **147**, 233–242 (1957)
40. Montfort, N.: World clock. Bad Quarto (2013)
41. Noll, A.M.: Computers and the visual arts. Des. Q. **66**(67), 65–71 (1967)
42. Paik, N.J.: Input-time and output-time. In: Video Art: An Anthology, p. 98 (1976)
43. Paik NJ, Ippolito J, Hanhardt JG (2004) Nam June Paik: Global Groove. Guggenheim Museum

44. Parish, Y.I., Müller, P.: Procedural modelling of cities. In: Proceedings of the 28th Annual Conference on Computer Graphics and Interactive Techniques, pp. 301–308 (2001)
45. Patterson, Z.: From the gun controller to the mandala: the cybernetic cinema of John and James Whitney. Grey Room (36), 36–57 (2009). https://doi.org/10.1162/grey.2009.1.36.36
46. Pound, E.: Vorticism. Fortnight. Rev. **96**, 461–471. https://fortnightlyreview.co.uk/vorticism/. Accessed 8 Mar 2022 (1914)
47. Rees, A.L.: A History of Experimental Film and Video. British Film Institute (1999)
48. Riley, R.: The evolution of silence. ACM SIGGRAPH 2014 Art Gallery, pp. 400–401 (2014)
49. Rodrigues, A., Costa, E., Cardoso, A., Machado, P., Cruz, T.: Evolving l-systems with musical notes. In: International Conference on Computational Intelligence in Music, Sound, Art and Design. Springer, Cham, pp. 186–201 (2016)
50. Ross, C.: The temporalities of video: extendedness revisited. Art. J. **65**(3), 82–99 (2006). https://doi.org/10.2307/20068483
51. Salvaggio, J.L.: Between formalism and semiotics: Eisenstein's film language. Dispositio **4**(11/12), 289–297 (1979)
52. Sharits, P.: Words per page. Afterimage **4**, 26–42 (1972)
53. Sitney, P.: Visionary Film: The American Avant-Garde 1943–1978. Oxford University Press (1979)
54. Veiga, P.A.: Alchimia: an inexplicable or mysterious transmutation, a seemingly magical process of transformation, creation, or combination. In: Proceedings of Artech 2017, the 8th International Conference on Digital Arts, Macau, China, ACM, pp. 175–178 (2017a)
55. Veiga, P.A.: Generative theatre of totality. J. Sci. Technol. Arts **9**(3), 33–43 (2017). https://doi.org/10.7559/citarj.v9i3.422
56. Veiga, P.A.: A/r/cography: art, research and communication. In: Proceedings of Artech 2019, the 9th International Conference on Digital and Interactive Arts, Association for Computing Machinery, pp. 1–9 (2019)
57. Veiga P.A.: Hello, an interactive cinematic generative artwork. In: Proceedings of the Eleventh International Conference on Computational Creativity ICCC'20, pp. 469–475 (2020)
58. Vertov, D.: Kino-Eye: The Writings of Dziga Vertov. University of California Press, Berkeley (1984)
59. Watz, M.: Closed systems: generative art and software abstraction. In: Watz, M., Doms, A., de Lavandeyra, S.E. (eds.) MetaDeSIGN—LAb[au], pp. 1–3. Les Presses du Réel, Dijon (2010)
60. Weibel, P.: Expanded cinema, video and virtual environments. In: Weibel, P. (ed.) Shaw J, pp. 110–125. The Cinematic Imaginary After Film. MIT Press, Future cinema (2003)
61. Youngblood, G.: Expanded Cinema. EP Dutton & Co., New York (1970)

Chapter 10
A Guide to Evaluating the Experience of Media and Arts Technology

Nick Bryan-Kinns and Courtney N. Reed

Abstract Evaluation is essential to understanding the value that digital creativity brings to people's experience, for example in terms of their enjoyment, creativity, and engagement. There is a substantial body of research on how to design and evaluate interactive arts and digital creativity applications. There is also extensive Human-Computer Interaction (HCI) literature on how to evaluate user interfaces and user experiences. However, it can be difficult for artists, practitioners, and researchers to navigate such a broad and disparate collection of materials when considering how to evaluate technology they create that is at the intersection of art and interaction. This chapter provides a guide to designing robust user studies of creative applications at the intersection of art, technology and interaction, which we refer to as *Media and Arts Technology* (MAT). We break MAT studies down into two main kinds: *proof-of-concept* and *comparative studies*. As MAT studies are exploratory in nature, their evaluation requires the collection and analysis of both qualitative data such as free text questionnaire responses, interviews, and observations, and also quantitative data such as questionnaires, number of interactions, and length of time spent interacting. This chapter draws on over 20 years of experience of designing and evaluating novel interactive systems to provide a concrete template on how to structure a study to evaluate MATs that is both rigorous and repeatable, and how to report study results that are publishable and accessible to a wide readership in art and science communities alike.

N. Bryan-Kinns (✉)
University of the Arts London, Mile End, London, UK
e-mail: n.bryankinns@arts.ac.uk

C. N. Reed
Queen Mary University of London, Mile End, London, UK
e-mail: creed@mpi-inf.mpg.de

N. Bryan-Kinns
Creative Computing Institute, University of the Arts London, London, UK

C. N. Reed
Max Planck Institute for Informatics, Saarland Informatics Campus, Saarbrücken, Germany

10.1   Introduction

Media and Arts Technology (MAT) research exists in an interdisciplinary space at
the intersection of artistic and creative practice, technology innovation, and research
on human cognition and interaction. With MATs, the design of the creative application itself and the study of its use through user studies are intertwined and may
take the form of a scientific intervention or an examination. This poses a challenge
to you, as a researcher: on one hand, it offers opportunities for novel and engaging
exploratory research and yet, at the same time, needs to be undertaken and framed
in a way that works within current academic research discourse and vernacular.
There is a natural tension between the focus of science and the arts when examined
independently: science tends to focus on building and testing generalisable models
and adding knowledge to our understanding of the world [12, 17]. Artistic practice
approaches the understanding of the world slightly differently, often focusing toward
creativity and the individuality and subjectivity of the human condition, whether or
not that produces any particularly novel understanding of the world [12, 13]. However, these fields are intertwined and inseparable, having encouraged each other's
advancement since the earliest ventures in philosophy and understanding of the world
[1, 16]. In order to conduct good quality research in MAT, we must acknowledge
both components: you as the researcher are responsible for honoring the inventive-
ness, innovation, and adaptability of the arts in a way which adheres to the structure
and procedure of scientific research.

10.1.1   Principles of Quality Research

In order to make your work understandable to, and valued across research communities, it should meet two critical requirements: it must be both **rigorous** and
repeatable:

Rigorous means that your research is conducted using tried and tested scientific
methods and practices to collect and analyse data. MAT research is interdisciplinary,
exploring new forms of digital Media and Arts, and yet is rooted in the science and
engineering of the Technology. With these fields being extremely broad in their own
right, MAT research must be diligent in conducting research that is appropriate for,
and consistent with, existing bodies of knowledge. It is therefore vital to collect and
analyse data in a structured way using established methodologies from scientific
fields, to ensure that data collected and results analysed are considered to be valid
and reliable by other researchers.

Repeatable means that your study plan and methodology are written in enough detail
that someone else could run the study without you being present to explain it. This
is important to maintain consistency of your study (each time you run the study it is

done the same way), and to allow others to be able to reproduce your results if they want to.

Keeping these two core requirements in mind, you can design and evaluate compelling research which use MATs to contribute to the different fields in interdisciplinary research. The goal of this chapter is to outline practices for conducting and presenting research in a rigorous and repeatable way. In addition to providing guidelines for structuring your research, we illustrate these through examples of sound research practices used in existing MAT research.

This chapter starts by introducing the two types of MAT study—proof-of-concept and comparative—and provides example studies of both varieties, ranging from interactive music technology to playful tangible interaction. Referencing these existing projects, we introduce a series of guidelines for designing MATs as part of an user study and explain how to define relevant research questions for your research. Then, we outline methods to determine appropriate data to collect from quantitative and qualitative sources and how these are combined as 'mixed methods'. A guide to questionnaire and interview design is provided, which focuses on how to elicit further insights on experience from participants. Tools and techniques for qualitative and quantitative data analysis are introduced, including thematic analysis [7] for interview data and statistical analysis for questionnaire and interaction data. The chapter concludes with guidance on how to report the results of the user studies so that others may be able to refer to your work and findings for their own design and technology.

In implementing the robust study design practices discussed in this paper, we hope that MAT researchers will find more common ground and understanding of each other's multidisciplinary work. As such, this chapter serves as a guide for researchers from many disciplines studying MATs, leading to richer collaboration and dissemination of knowledge across research communities.

10.1.2   HCI, UX, and Interactive Arts

Before moving on, we would like to briefly discuss the historical link between the more computer science fields of HCI and user experience (UX) and digital arts, media, and creativity. The development of these fields has always been tightly intertwined [17, 28], with the arts providing a source of inspiration for technology and creativity in its evaluation [16]. The earliest uses of computers to create art were in the 1950s, for example, in 1951 one of the earliest example of computer arts was the use of the Ferranti Mark 1 computer at the University of Manchester (there would only have been a handful of computers in the UK at the time) to play simple tunes. At a similar time Human Factors and Ergonomics, which are the origins of contemporary HCI and UX, emerged in the 1940s s and 1950s s to help designers design machines which were easier to use, for example, to improve the design of airplane controls to make flying safer and less error prone. The focus for Human Factors was really on the functionality of the machine. A lot of it was to do with how to layout, or configure, the controls in order to reduce chance of human error. This relationship is reciprocal and has worked in a collaborative way over the decades—typically, technology from other fields is adopted into arts, where it is used creatively. From these applications, new practices are developed and learned, stimulating the study and further expanding the technology itself. Interactive Art tends to place importance on what the user or listener cares about [16, 20], as well as challenging the status quo to provide room for novel ideas [1]. This provides a space for HCI to move beyond more traditional HCI topics such as task completion and efficiency to understanding and communication between humans and computational agents [20].

10.2   Types of MAT Study

There are two broad types of MAT study which we will discuss in this chapter:

Proof-of-concept studies examine people's responses to a single MAT. This kind of study asks *"What if..."* questions such as "What if I make this MAT, how do people respond to it, and what is their experience of it?" In this way, these studies often involve creating a MAT which is a digital intervention and then evaluating people's response to it. These studies allow early-stage investigation of how people respond to a new form of interaction, for instance when there is little existing research in the area and it is therefore unclear how people might respond to the interaction. Results of such a study can inform further studies by identifying broad kinds of response to the MAT and identifying possible interesting and challenging avenues for interaction design.

Comparative studies do as their name indicates—they compare two or more variations of a particular MAT. These studies compare the effects of specific design features and ask *"What effect do particular design feature(s) have on people's experience?"* Rather than trying to determine whether one design is better than another, these studies are interested in *how* the experience is different between MATs. In this way, comparative studies involve in-depth examination of MATs and differences are typically restricted to one design feature. Results inform the development of theories about how design features might contribute to enjoyment, creativity, and engagement, and generate interaction design questions for future studies.

Often, proof-of-concept studies are used to research a general interaction question, which is then further refined and explored through specific questions in a comparative study. However, this is not necessarily the case and it should be made clear that one kind of study is not better or more rigorous or repeatable than the other—they simply ask different kinds of questions. As Sect. 10.3 will elaborate, it is important to keep your research goals in mind in order to decide and justify the methods you use.

In this chapter, papers published about existing MAT research projects are used to illustrate and bring to life these different study types, how they can be conducted, and how they are presented. These projects are briefly introduced in the following subsections to give a flavour for different kinds of MATs and user study approaches.

10.2.1   Example Proof-of-Concept Studies

The proof-of-concept studies we refer to throughout this chapter are very different from one another but share the same type of research focus and question: a MAT is designed in order to conduct an open-ended exploration of users' behaviour and interaction. As mentioned above, although the research questions are very specific, they ask more of *"What if?"* or *"How do...?"* questions and are aimed at exploring the general effects of a MAT's use as illustrated below.

10.2.1.1   Mazi

Nonnis and Bryan-Kinns [32] designed and studied Mazi, a Tangible User Interface (TUI) which uses haptic and auditory feedback to encourage spontaneous and collaborative play between children with high support needs. Mazi was developed through an iterative prototyping process and used in a proof-of-concept style study to explore how principles of TUI design along with theories of social interaction could be used to encourage social play [31]. Mazi's final design features a dome-like shape to facilitate the circular configurations found naturally in communicative behavior and uses soft yet durable materials to allow the children to play in their own way with sensors embedded in Mazi which generate music. The core proof-of-concept research question of Mazi was:

> **What if** I make this MAT a large, soft circular shape and make it create music when multiple children play with it at the same time, **how do** children with autism respond to this, and what playful and social interaction does it prompt?.

Because the authors wanted to determine how the tangible and auditory feedback of the design influenced the communication between the children, it was most suitable to conduct a proof-of-concept study; indeed, the wide variety of abilities and interests of children involved makes it unsuitable for comparative study. The proof-of-concept study took place over five weeks with five children aged between 6 and 9 years old at a Special Education Needs (SEN) in London.

The study process needed to be flexible to the needs of the children and teachers. In the proof-of-concept study, the researchers collected observational data about how the children interacted with each other and with Mazi, and analysed these using existing behavioural science models. The results of the study demonstrated that working with Mazi helped the children to master basic social skills and engage with different sensory interactions [31]. This open-ended strategy allowed the researchers to focus on the most salient elements of the interaction in-context, and yet at the same time produced a study method and results which could be replicated by other researchers (Fig. 10.1).

(a) Mazi, made of wool and featuring inflatable
bubbles for triggering sounds.

(b) Children playing together with Mazi.

Fig. 10.1 Mazi, a tangible user interface for stimulating interaction and participation for autistic children. Images used with permission from the Authors: http://isam.eecs.qmul.ac.uk/projects/Mazi/mazi.html

10.2.1.2   VoxEMG

Reed and McPherson [33] design and explore the use of novel vocal interaction through surface electromyography (EMG) using an autobiographical approach [29]. The authors developed a system, the VoxEMG, for gathering the electrical neural impulses which cause muscular contractions (Fig. 10.2). These EMG signals are used in real-time sound design to allow a singer to interact with very low-level movements in their practice through auditory feedback. The auditory feedback allows the vocalists to interact with their existing, embodied understanding of their action and "hear" movements which would not normally produce definable sound. The work was formed through an autobiographical approach in Reed's extended experience with the setup lasting over a year. The fundamental research question was:

What if I sonify movements and actions which singers are not normally consciously aware of, and **how will** they react, change, and/or perceive their movements when they receive this new feedback?.

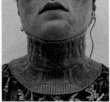

(a) The VoxEMG board developed for
sensing activation of the laryngeal muscles.

(b) Reed wearing the Singing Knit wearable
collar for vocal EMG interaction.

Fig. 10.2 Sensing and interacting with laryngeal muscular activations through VoxEMG integrated into wearable designs (referenced from [35])

Autobiographical methods and first-person accounts are extremely useful as they demonstrate how lived-experience and understanding of a system's use can improve and inform its design [19]. In this case, Reed applied her experience working as a semi-professional vocalist to the long-term interaction with the EMG system. Through detailed interaction notes, journaling, debugging, and an iterative design and testing process, Reed was able to uncover subtle understanding of the interactions with it [29, 34]. This study again is more suitable as a proof-of-concept because it requires an exploratory approach; although there are some hypotheses as to how the musicians would respond, the researchers wanted to keep the interaction open-ended and see how the singer's behaviour changed while using this MAT.

10.2.1.3   Polymetros

Polymetros is a collaborative music system designed as an in-person audience experience for multiple participants (Fig. 10.3) [2]. The design is purposefully simple, using minimal music, and allowing participants to create short loops (8 notes long with 8 possible pitches and only one instrument sound). There are 3 physical instruments in Polymetros (Fig. 10.3a) which are synchronised together so that the loops of each instrument are synchronised with each other. One person can play Polymetros on their own, but the sound becomes richer and more interesting as 2 or 3 instruments are played at the same time (Fig. 10.3b). Essentially the research question of this MAT was:

What if I make a MAT musical instrument which requires three people to play it, **how do** people respond to it, and what is their experience of it?

Bengler and Bryan-Kinns [2] wanted to explore how people would respond to this novel collaborative music system, how they would engage with it, and how they might engage with each other. With these open-ended questions, a proof-of-concept study is the most appropriate. A study was conducted over two days at the Victoria

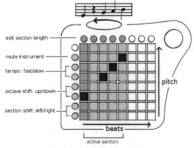
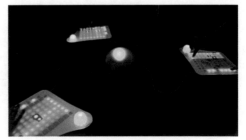

(a) The layout of each Polymetros instrument.

(b) Creating collaboratively between three musicians with the Polymetros system.

Fig. 10.3 The polymetros collaborative music system (referenced from [2])

& Albert Museum (London, UK) with random members of the public (150+), to see how they would respond to the novel form of music making. Data was collected with many different tools, including questionnaires, observations, video recordings, and data logs from the Polymetros system. Later studies then took on a more comparative study structure, comparing how Polymetros was perceived by people in different cultural contexts—in the UK, Spain, and China [3]. Results indicate the significance of ownership and supporting individual participation in collaborative creativity; the physicality of the system's design assisted in non-verbal communication and understanding of the other players' actions and structure roles in the compositional process [2].

10.2.2 Comparative Studies

In contrast, the comparative studies we will discuss in this chapter are focused on particular aspects of interaction or effects of the MAT's use. These studies are different in the questions they ask, which are more focused on "How does X effect Y" or "Can we do...". They focus on the design of an interactive system and specific design features can change or impact users' interaction as introduced in the following examples.

10.2.2.1   Keppi

Keppi [4] is a Digital Musical Instrument (DMI) designed to explore the effect that disfluency in a musical instrument's design might have on performers' and audiences' perception of skill and risk in performance (Fig. 10.4). The research question here started as a more open question of "What if I make this MAT musical instrument which is risky to play in performance as its musical properties degrade in real time" and then moved to a more focused comparative study question:

Fig. 10.4 The Keppi, a cylindrical instrument designed with intentional disfluency characteristics (referenced from [4])

What effect does increasing the disfluency in the design of a MAT musical instrument have on audience and performer perception of skill and risk in live performance?

In this comparative study Keppi was designed and produced incorporating a disfluent design characteristic: It would turn itself off if not constantly moved. Six percussionists then performed live on stage with different versions of Keppi which each had different levels of disfluency. Audience feedback was collected in real time during the performances through an app on their mobile phones, and also through post-event questionnaires. Performer feedback was collected through survey questions, and the style of music created and performed using Keppi was analysed. The results of the study suggested that whilst different levels of disfluency in the design of Keppi did not have an effect on audience enjoyment of the performance, it did have an effect on their recognition of the skill of the performers. Moreover, performers noted that the disfluent behaviour of Keppi was viewed as a positive design feature, which contradicts conventional Human-Computer Interaction design guidelines which stress the importance of intuitive and reliable user interfaces.

10.2.2.2   Daisyphone

Daisyphone is an online collaborative music editor which allows groups of people to edit a shared loop of music 48 notes long (Fig. 10.5). The design is purposefully very simple, and the interaction is restricted to adding notes (12 pitches are possible, and 4 instrument sounds) and removing notes, and drawing to communicate [10]. The circular area shows the shared musical loop and drawn annotations can be seen around the outside. The circular representation of the loop was chosen purposefully to be different to most music sequencers, thereby reducing familiarity with the inter-

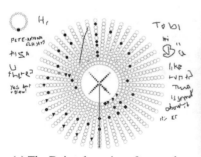

(a) The Daisyphone interface on the desktop, showing player input and annotations.

(b) Compositional collaboration with Daisyphone on mobile.

Fig. 10.5 The Daisyphone interface for collaborative music-making (**a** referenced from [10], **b** © EPSRC, available to the public: https://www.flickr.com/photos/epsrc/3340524049/)

face. Support for shared drawing and shared editing was drawn from HCI research which stated that this would improve collaboration (in-text document editing). [10] explored the effects of having a sense of personal identity and a shared way to communicate would have on people's mutual engagement [9]. Mutual engagement being "it involves engagement with both the products of an activity and with the others who are contributing to those products" (ibid.). The research then focused around the question of:

> **What role** does personal identity play in collaborative musical composition, and **how do** different representations of personal identity **compare** when users complete a specific task together?

The authors undertook a comparative study to see what effect providing cues to personal identity in the interface and providing a shared area to communicate in would have on people's mutual engagement. The study involved 39 participants collaborating online (participants could not see or hear each other in person) to create short loops of music. Each participant spent about 1 h in the study. Data was collected from questionnaires, and logs of interaction with both the music and the shared drawing area. Questionnaires were used to gather feedback from participants and to make comparisons between different versions of the Daisyphone interface.

10.2.2.3 Smart Trousers

Skach et al. [37] designed a pair of smart trousers which were able to sense posture changes (Fig. 10.6). This MAT involved an interdisciplinary approach combining electronics and e-textile design with behavioural and social science. The use of a wearable and the authors' relevant fashion design background allowed for the creation of a wearable that could be used to study wearers' behaviour without being disruptive to a social environment. Fabric pressure sensors were integrated into the garment and measured contact between points on the wearer's body and the legs as well as the surface of a chair. Through this garment, the authors aimed to explore different behaviour related to emotional and social communication. The study focused on exploring non-verbal communication through posture and gestures and inquired specifically as to whether gestures on the legs could be classified through pressure sensing. The research was based around a question of:

> **Can we** use pressure sensing to gather data about the movement of the lower body when seated, and **can this data be used** to classify behaviours of the wearer as they engage in conversation?

This work used a comparative study to gather data about participants wearing the smart trousers while they performed a number of actions. In this sense, the comparison is not between one MAT and another, but rather between different users and the same wearable. The interaction with the trousers showed that, even though individual participants followed the movement directions differently, in their individual interpretation, it was possible to classify different gestures using the data gathered.

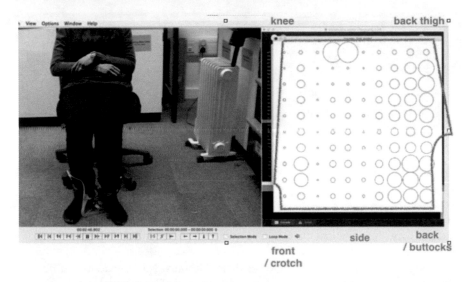

Fig. 10.6 Visualising the pressure measurements (right) in a participant's seated position (left) while wearing the smart trousers (referenced from [37])

10.2.3   From Proof-of-Concept to Comparative Study: The Chaos Bells

Research into the interaction with large musical instruments by Mice and McPherson [23] provides an excellent example of how exploratory proof-of-concept studies can inform further comparative research.

This work began with a proof-of-concept interview study which explored musicians' relationship with the physically large instruments they had already been trained on [23]. Interview questions focused on gestures and the precision of movements, fatigue during performance, and improvisation. In addition, participants reviewed *Cello Suite no. 1 in G Major* (J. S. Bach) and discussed difficulties they would have playing the piece transposed for their own instrument. This exploration revealed a number of insights into timbral control and the embodied relationships between the musicians and their large instruments.

From this study, the authors conducted a series of comparative studies using the Chaos Bells (Fig. 10.7), a large-scale digital musical instrument (DMI) developed by Mice [24]. The design features a set of pendulums which use accelerometers which drive a Karplus-Strong algorithm as the performer strikes, raises, and swings them (Fig. 10.7a). In further study of the instrument, the authors compare different pitch mappings on the pendulums to compare how different tonal layouts [24, 26], instrument size, and the performer's body influence the idomatic gestures and patterns during improvisation, as well as the performers' perspectives of their own bodies [25]

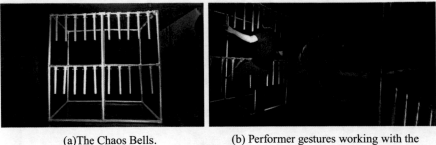

(a)The Chaos Bells.   (b) Performer gestures working with the
                                             large instrument.

Fig. 10.7 The Chaos Bells, with height and width of 2 metres, designed to study the impact of interface size on interaction and gestures. Photos © Lia Mice, used with permission

(Fig. 10.7b), which show that the size of the instrument determines the gestures which can be used.

Through this work, we can see how the different study types can address different types of research questions using different approaches. The authors move between different focuses:

Q1: **How do** musicians perceive their interaction with their large instruments and **how might they act/ feel** about performing different music when playing them? (proof-of-concept)
Q2: **Do different layouts** of the tones on a large instrument impact the gestures and movements used by musicians while playing them, and **does this change** their perception in practice? (comparative)

The proof-of-concept study used more open-ended approaches to gather a set of feedback and ideas surrounding the design and performance with large instruments. This was done to get a better sense of how performers work with instruments they have been trained on and know well. By using an exploratory approach, the authors determined key factors in the interaction with such instruments. This information informed the development of a new DMI based around these principles, where more specific questions of interaction such as tonal layouts could be examined in an appropriate context.

10.3 Designing a MAT Study

The first step in conducting a robust MAT study is defining your research questions— Having specific questions in mind will help to structure the rest of the research into either a proof-of-concept or comparative study. As discussed in Sect. 10.2, both kinds of study are equally valid but will focus on different kinds of questions and scenarios. Having clear questions will drive consistency in the study and its analysis and focus the reporting of your results in a way which is understandable to others. This will also help to justify your design choices for both the study procedure, analysis, and the MAT— attention must be given to details of the Media and Arts themselves (e.g., musical theory, visual arts practices, traditional craft) as well as the Technology driving the MAT interaction. The media and art should be researched in detail before beginning. For instance, when working with wearable MATs, it would be important to familiarise yourself with artists' practices in fashion design, wearable technology and integration, and so on. In the presentation of both Skach et al.'s smart trousers [37, p. 117] and Reed et al.'s VoxEMG wearable implementation [35, pp. 171–172], significant consideration and direction was first drawn from previous work in textile sensing technology. The design and use of the MAT will very likely rely on background and existing work from other science backgrounds as well, in these cases fields such as cognitive and behavioural sciences, bio-mechanics, and sensorimotor interaction.

10.3.1 Choosing a Study Type

If your research interests are more exploratory, for instance wanting to know how people might interact with a new MAT in their typical performance practices (as done with VoxEMG and the Chaos Bells), proof-of-concept studies can provide a suitable format for your research questions. Similarly, proof-of-concept studies are good for questions such as *How do people collaborate with this MAT? How do participants' perceptions of their movement change if they use this MAT over a long time period? How do participants choose which notes to play/ which sounds to include/ which roles to take on in a duet performance?* These questions are specific, but they do not seek to understand a specific difference—rather, they are open-ended and aim to gather information about a specific context. This can be beneficial as well if you do not have any existing knowledge or want to prompt any particular behaviours. Proof-of-concept studies can be useful to observe the impact of and attitude towards a new design in-context—for instance Mazi worked with the children and their routines in their day-to-day environment [32].

On the other hand, research questions that directly involve comparisons naturally warrant a comparative study. For instance, the Daisyphone study's questions revolved around a comparison of features included in an user interface. Comparative studies are more appropriate if your questions are something like *Do participants prefer one collaboration method over another? Which of these interface features is more important to experienced users, compared to novice users?* or *How does participants' accuracy score change after using this MAT for a month?* These research questions compare specific elements of experience and design and maybe compare responses from two or more groups of people (e.g., experts and novices). Additionally, you may be looking to gather information about different, specific moments in an interaction; for instance, with the smart trousers, examining the data gathered by the sensors at different moments in the interaction.

10.3.2 Designing the MAT Itself

With these research questions in mind, you can decide on what MAT you will create and then how you will study it. When designing a MAT, it is also important to consider again that the methods are rigorous and repeatable—someone else should be able to understand and recreate your design. With the growing movement of open-science, many researchers working in fields such as HCI are making their designs open-source so that they can be easily shared and accessed by other researchers and communities that will benefit from their use. Regardless of whether or not the MAT design is to be open-source, it is important to carefully document the design process to support the research in publication. Some key elements which should be included in this design documentation include:

Design Inspiration: MATs are results of inspiration-led design, often combining ideas and theories from different disciplines, and often the design focus is driven by from personal interest. For example, the collaborative music making system Daisyphone [10] was created as a result of a personal passion for performing music in combination with an interest in exploring new forms of networked music performance which were emerging at the time. If you are studying the design process or working in an autobiographical use case, you should note important details about your own experience and background which are relevant to the design (e.g., with VoxEMG [34]).

Identify Design Features: Key design features of your MAT should be informed by previous research—your study is likely to explore a new and novel context, but it should be based in established knowledge from the research community. For example, the design of Daisyphone was informed by Computer Supported Collaborative Work (CSCW) research, which stated that sense of identity was important in collaboration; therefore, identity was represented by colour in the user interface. This one design feature was then varied across different versions of the MAT to understand how sense of identity affected people's collaborative creativity and engagement. The research questions therefore focused on this particular aspect.

Reduce Complexity: The complexity of the interaction design should be reduced as much as possible whilst still allowing for fun, creative, and engaging interaction. This makes the MAT easier to design and build and also makes the study more focused on the novel interaction itself rather than, say, learning effects. For example, Mazi was designed to support only a very small range of musical notes making it more of a 'sound toy' than a collaborative composition system. This allowed the studies to focus on how children interacted with each other rather than focusing on, say, the composition process. Additionally, keeping the design focused on novel interaction, rather than other elements of the design, will also help to reduce confounding factors—those parameters that influence both the study conditions and the results but are not accounted for and likely not the intended parameters you want to observe.

For instance, if you give participants two interfaces with different layouts to use but they involve dozens of unexplained buttons and knobs, the difficulty of the systems will confound the relationship between the instruments' layout and the participants' interaction. The results will be unclear and you will likely learn more about the participants' frustration than their preference for layout.

Decide on the Study Setting: Evaluations of MATs take place in settings ranging from controlled laboratory settings to ad-hoc settings in public. These settings give different levels of control of the study, realism of conducting the activity in-context, and appropriate data collection methods and study structures. Whilst a laboratory study gives the most control, it is the least realistic setting for creative, fun, and engaging interaction. There are trade-offs between different settings, and you should be prepared to justify the decisions you make in planning your study when discussing your research.

Focus the Activity: Even for creative activities there needs to be some focus to the interaction. Whilst evaluating MATs is not concerned with, say the efficiency or productivity of carrying out specified tasks with an interface, it is nonetheless important to focus peoples' activity in order to be able to understand their responses to the design's features. For example, with musical MATs, people need to be provided with a motivation for creating music. This might be in the form of a compositional brief; e.g., in Daisyphone studies, participants were asked to create a jingle for the Olympic Games, and in the Keppi study performers were asked to create a musical performance using the MAT.

10.4   Conducting and Reporting Your MAT Study

Table 10.1 outlines the necessary components for conducing proof-of-concept and comparative studies. We have structured this section to serve as a template to both guide the design of the research as well as its presentation (e.g., in a research paper). Following this structure and ensuring you have all of the key components in your study planning and presentation of your work will help to make the research rigorous and repeatable, and understandable by other researchers. The template sections you see here are used in the majority of research papers across scientific disciplines. This will help you to structure your presentation in a variety of venues for the topics related to your interdisciplinary work.

10.4.1   Section: Background

The first step should be to conduct a literature review of existing research surrounding your topic area. The related work should create a coherent idea driving the current research. This should be done to help define research questions (e.g., whether there

Table 10.1 Components necessary for both types of MAT study

		Proof-of-concept	Comparative
	Background	✓	✓
	Research questions	✓	✓
Study methodology	Aims	✓	✓
	Hypotheses		✓
	Independent variables		✓
	Dependent variables		✓
	Conditions		✓
	Participants	✓	✓
	Tools	✓	✓
	Procedure	✓	✓
	Data collection	✓	✓
	Analysis	✓	✓
	Results	✓	✓
	Discussion	✓	✓

a gap in the existing knowledge or previous publications suggest further exploration you can address in your work) and to provide a state-of-the art for methods being used in similar studies. A strong background will help you to justify your choices for study design, methods, and data analysis, thus increasing the robustness of your work.

You should provide a brief overview of the relevant literature when presenting your research to contextualise what you are doing within an existing body of knowledge and provide rationale for your decisions. With MAT research, the background may consist of relevant work from different disciplines and so you will want to spent time to make sure that you connect literature coming from different fields. This might include defining terminology you will use, with respect to different definitions in other research, or uniting concepts from multi-disciplinary work so you can discuss them together. For instance, in Reed and McPherson [34], the authors use the Background section of the paper to unite concepts surrounding mental imagery and embodiment from cognitive science and design practice within the context of the paper.

10.4.2 Section: Research Questions

As mentioned in Sect. 10.3, you should define specific research questions you wish to address through your work. With existing literature in mind, these questions will help to focus the research and the design of your MAT and study. A research question, while targeting a specific area for exploration, is usually something broad such as

"How do people engage with each other when playing a three-person musical instrument?" (as in Polymetros). It is common, when writing a research article, to include the research questions and the main contributions your work makes to the existing literature after presenting your Background. This helps to connect the Background to your current work and focus a reader on the key points you will address in the current study.

10.4.3   Section: Study Methodology

Referring to the above table, you should include the following components in your study design and when presenting your research. Methodology is important in order to ensure the study is robust and reproducible. The methodology should include detailed information about the procedure so that another researcher may reproduce it exactly as you did. In your presentation, you may wish to use subsections for each component of the methodology.

10.4.3.1   Aims

You should define and list the aims of the study. There should be a small number of aims - one or two, and definitely less than five.

For a proof-of-concept study, the aim is usually to find out people's conceptualisation or understanding of a novel piece of interaction, e.g., "The aim of the study is to explore how people engage with each other when they play a three-person musical instrument." (Polymetros).

A comparative study may examine a specific interaction paradigm, perhaps further exploring an observation from a proof-of-concept study. A comparative study example: "The aim of the study is to test whether a shared music making system with support for shared communication channels supports greater levels of mutual engagement than one without." (Daisyphone).

10.4.3.2   Hypotheses

If you are doing a comparative study, you will have at least one hypothesis which you test. There should be a small number of hypotheses, usually between one and five hypotheses is fine. For example, in the Daisyphone comparative study there were two hypotheses. The study examined the effect that providing a graphical annotation function as an additional communication channel has on mutual engagement:

- **H1**: Mutual engagement would be greater where an additional channel of communication was provided - graphical annotation.

You might have a more exploratory kind of comparative study and just predict that there would be some difference (but not know what kind of difference it is):

- **H2**: Mutual engagement would be different where an additional channel of communication was provided compared to when there is no additional channel of communication.

Make sure that, if not already done in the Background, that you define the terminology you will use here; in these hypotheses, you will want to clearly state what is meant by "mutual engagement," and how it might be measured. This should be informed by and connected to your Background section. For example, in the Daisyphone study "mutual engagement" was described as a collaboration in which there is both "Evidence of engagement with the product of the joint activity, i.e. music in our domain. For example, participants' reports of feeling engaged with the product, a high quality product, focused contributions, or demonstrations of skills and expertise in creating contributions." and "Evidence of engagement with others in the activity. For example, more reports of feeling engaged with the group, coherent final joint products, colocation of contributions, mutual modification of work, discussions of quality of the joint product, repetition and reinterpretation of others' contributions" (as described in [10]).

10.4.3.3 Variables

Defining variables is an important part of a comparative study plan—variables are things that are changed in your study, or things that are measured in your study (i.e., they are *variable* in the study). There are two main kinds of variables to define: *independent variables* and *dependent variables*.

You should also be aware of other elements which might act as *confounding variables*. These are things that might change between each participant in your study and which might have an effect on their performance; for instance, the skills and expertise of participants might be a confounding variable and therefore need to be controlled. If some participants are skilled musicians, their experience might have an effect on how they play a 3-person musical instrument. The time of day might be a confounding variable if people are tired in the evening versus the morning, etc. In MAT studies, it is often difficult to control these confounding variables and so they need to be included in the Discussion section of your report and you should elaborate on how they may have impacted the findings and outcomes of your study. As mentioned previously, limiting the complexity of the study will help to keep confounding variables under control.

Dependent Variables Dependent variables are things that you measure in a comparative study. They *depend* on the interaction and what happens during the study. You should describe these as concretely as possible; for example, in conventional HCI studies, a dependent variable could be the time in seconds it takes to complete a task. For more exploratory MAT studies, your dependent variable might be more subjective; for example, in the Daisyphone study there were 9 dependent variables including: (1) "Quality measure: participants' reports of their assessment of the quality of the final product and the collaboration itself," (2) "Contribution to joint production measure: number of notes contributed," and (3) "Proximal interaction measure: closeness of participants' contributions to each other's contributions" (as described in [10]).

For each of these, you will need to define how the dependent variable is measured or how a quality is assessed. The best way to do this is to find good definitions in existing research literature and either use those definitions or modify them to suit your studies.

Independent Variables Independent variables are things that you change in the study; for example, changing features of the interface or the app which is provided to see their effect on user experience (as measured by your dependent variables). You define the independent variable along with the "levels" of variable, or what things you change. For example, to test the effect of providing shared annotation in the Daisyphone study, an independent variable would be "Annotation" and the 2 levels would be "Annotation" or "No Annotation" (Daisyphone): "In the Annotation condition, participants could 'draw' on the Daisyphone, and these graphical annotations were shared with other participants. In the No Annotation condition, no graphical annotation was supported, and so communication could only occur through the music."

10.4.3.4   Conditions

In your comparative study you will test different variants of the interaction—each of these variants is called a *condition* or treatment in the study.

In the Daisyphone study, two hypotheses were tested: **H1** "Mutual engagement would be greater where participants had explicit cues to identity" and **H2** "Mutual engagement would be greater where an additional channel of communication was provided - graphical annotation".

Two independent variables were used to test these hypotheses: (1) cues to identity (for H1), and (2) communication channel (for H2). There were therefore 4 conditions in this study (see Table 10.2).

10.4.3.5   Participants

Then, decide who your participants will be—are they general public, or do they need certain skills and experience? If so, what skills and experience? You will also need to decide how you will recruit the participants and if they will be paid any incentives? This information should be described in detail when you document your research. For example, the participants for the Daisyphone study are described in [10]:

Table 10.2 The four conditions examined in the Daisyphone study, showing the combination of the two independent variables. An X indicates where the variable was included in the condition

		Independent variables	
		Communication channel	Identity cues
Conditions	1		
	2	X	
	3		X
	4	X	X

Table 10.3 Example table for presentation of participant demographics and background (adapted from [10])

Participant	Age	Gender	Musical ability	Musical preference
P1	25	M	Novice	Hip hop
P2	24	M	Intermediate	Hip hop
P3	20	F	Intermediate	Hip hop
P4	22	M	Intermediate	Rock
P5	21	F	Expert	Latin
P6	22	M	Novice	Classical
P7	22	M	Intermediate	Rock
P8	21	F	Intermediate	Ambient

Final year Computer Science students at the first author's institution were recruited through advertisements to take part in the experiment as part of their course, but not offered any incentives to take part. Thirty-nine of a possible 80 participants took part (28 males, 11 females; aged from 20 to 29 years old, mean age: 22; average computer literacy: expert; average musical ability: intermediate; none were professional or trained musicians; none had used Daisyphone before). Participants' musical preferences ranged from Hip Hop (most popular) to Latin (least popular).

If you report individual responses in your results (e.g., interview answers), then you should give a table to provide brief demographic of each participant, as this may have an impact on their answers. Give each person a participant ID (e.g. P1, P2, etc.) which you can then use in the results. For example, see Table 10.3.

Sample Size As a rule of thumb, you will need at least 10 participants in a proof-of-concept study. Ideally you would have much more than 10 participants for a proof-of-concept study, but it is worth noting that the most common sample size in papers in the leading HCI conference (ACM CHI) is 12 participants [11]. For a comparative study, you need at least 10 participants for each condition, e.g. in the Daisyphone example you would need at least 40 participants as there are 4 conditions (assuming

Table 10.4 An example of a balanced Latin square for the four conditions of the Daisyphone study; here, the conditions are in a different order for each participant and each condition precedes another only once, ensuring that potential effects from order are removed from the study

		Trial order			
Participant	1	1	2	4	3
	2	2	3	1	4
	3	3	4	2	1
	4	4	1	3	2

that each person only does one condition—this is called an unrepeated-measure, also referred to as 'between groups').

Data suggests that, in order to discover all of the usability problems in testing, 15 participants are needed; through iterative testing, this could be done in as little as 5 participants [30]. You can reduce the number of participants needed in a comparative study by designing the study so that each participant does multiple conditions (*repeated-measures*, also called within-groups), but this makes the study longer (therefore increasing participant boredom and fatigue), increases the chance of learning effects, where participants learn something about the interaction in one condition which then either makes the other conditions easier to use or harder to use due to confusion, and potentially reduces the power of the statistical tests you can do on the data after the study.

Repeated-measures study design is beneficial in that you can ask participants comparative questions between the conditions which can help to understand how participants react differently to different conditions e.g., to compare the experience of condition 1 to their experience of condition 2.

For example, you could run the Daisyphone study with 10 participants and have each person use all four conditions. As another option, you could run the study with 20 participants and 10 of the people use conditions 1 & 2, and the other 10 participants use conditions 3 & 4. In this case, you would need to decide why it is not some other combination (e.g., 1 & 3 + 2 & 4) and again be prepared to make this justification when presenting the research. When using repeated measures, you also need to make sure to counterbalance the ordering of the conditions so that the ordering does not cause an effect in your results. You can use a balanced Latin square to figure out different orderings for your conditions to ensure that your study is balanced[1]; for instance, if you had a study with four conditions of the Daisyphone study, you could divide your measures as in Table 10.4.

However, it is important to note that you may need more participants depending on what kind of data is being collected and what analyses need to be done. In order to do some statistical testing with accuracy, larger sample sizes might be needed. This should also be considered when recruiting participants and when considering how many variables are examined at one time.

---

[1] Balanced Latin Square Generator: https://cs.uwaterloo.ca/ dmasson/tools/latin_square/.

10.4.3.6 Tools: The Media and Arts Technology Itself

If you need to build a MAT to be used in your study—for instance, an app, a VR experience, tangible interface, etc., you will need to document the design and be able to explain how it works. This would include detailing your expectation of how people would interact with it, and how it works on a technical level. You will need to connect your design to the Background to be able to define and discuss how it is similar or different to existing tools. If you are using existing software, describe what the software is and how it will be used in this specific research context (making sure to credit the original authors/ developers and cite any relevant publications).

10.4.3.7 Procedure

With your materials and tools ready, you should then define the study procedure. Provide a step by step description of what participants in the study will be asked to do, and how long is spent on each step. Include *everything*—include the step at the beginning when you explain the structure of the study to participants (e.g. 2 min), include the part where you ask participants to fill in questionnaire (e.g. 5 min), etc. In the pursuit of repeatable studies and reproducible findings, it is important that you are detailed enough for another researcher to conduct the research exactly as you have done.

   All studies should start with an introduction where you explain the purpose of the study (but don't mention what results you are looking for as this may bias their views) and that participants are free to stop at any time they like. You should also collect demographic data such as age, gender, and other relevant information such as musical experience. Specialist skills such as musical experience can be assessed using questionnaires such as the Goldsmiths Musical Sophistication Index (Gold-MSI) [27]—using existing questionnaires helps to strengthen the rigor of your study. Do not collect anything that could directly identify the participant; for instance, do not collect their name, address, or email address in the demographic information. If you are studying or working in an institution such as a University you will likely need to ask participants to complete a consent form and ensure that you study has ethical clearance from your institution to proceed.

10.4.3.8 Data Collection

One critical component of study design is the decision on what kind of data will be needed and collected, either quantitative, qualitative, or some mixture of both. *Quantitative data* is data which is in a numerical form can be assigned a value (e.g., it has a quantity—a numerical value—and is able to be measured as such). It may be participants' ratings of their enjoyment on a Likert scale from 1 to 5, or it may be more conventional HCI measures such as speed of completing an activity, or number of errors, etc. *Qualitative data* takes the form of words and is more descriptive (e.g., it is more about a specific quality or descriptor of whatever is being examined). Many studies incorporate mixed methods, where both quantitative and qualitative approaches are taken. This is especially the case in your MAT research, where you might need to both collect concrete data, for instance on the MAT's computational performance or on the interaction, and also understand the emotional and aesthetic aspects of the design in a qualitative sense. It is important to consider what form your data will take when planning a study to know what kind of analyses you might use to interpret the data, and to ensure that you are able to collect enough data to get an appropriate and well-rounded analysis. As with types of MAT study, one kind of data is not a better option, but likely one will be better at addressing which perspectives you wish to explore and you should be prepared to justify your choices in your research presentation.

For proof-of-concept studies you will most likely collect data by writing down your observations of people's interaction with the MAT (and each other), video recording their interaction, and then having questionnaires and interviews at the end of the study to get some idea of how people responded to your MAT. For a comparative study you should describe what data you will collect for each dependent variable. Decide whether the data will be collected whilst the participant is interacting with the MAT, or after they have used it. Some examples of data collection are outlined below.

It is important to note any possible ethical considerations with your data collection and how they will be addressed. Check with your affiliated institution's General Data Protection Regulation (GDPR) or other privacy and ethics procedures. For example, if you video record people interacting with your MAT, then does your institution allow you to publish images of people's faces? If so, will you request people's consent to use their photos in papers? If they don't give consent, can you still get the data you need for your study?

Interviews When conducting interviews with participants, consider what kinds of responses you are seeking. You might use an existing interview structure or a semi-structured interview where you decide prompts based on the topics you wish to explore. It is best to audio record interviews and then transcribe them later—people will be much more descriptive and reflective in spoken answers than in written answers.

Interviews are particularly important for proof-of-concept studies in order to get participants' feedback on novel experiences. They are also important in comparative studies to help understand why people behaved the way they did and responded to questions the way they did. You should start your interview with open questions, where you try to get participants to explain their understanding of the experience (e.g., ask: "Please could you describe what you just experienced to me?", or "Please tell me how you would describe what you just experienced to a friend?", followed by more specific question such as "What did you find most engaging about the system?", or "What did you find most challenging about the experience?").

Then, you can start to probe for more specific feedback; for example; "Please could you tell me about your experience of playing music on the three-person music instrument?", followed by "Did you find that other people responded to your musical contributions?". Make sure to follow any questions that could be answered with a yes/ no answer with probing questions, (e.g., "Can you tell me why that was?" and/ or "Can you give me a specific example?"). More specific and targeted questions and answers can help to explain people's responses to particular features of your MAT.

Questionnaires It is best to use existing questionnaires, such as the Gold-MSI mentioned earlier [27], the Creativity Support Index (CSI) [15] or the NASA Task Load Index (TLX) [18]. These questionnaires are validated through existing research—this means that they have been shown to be reliable in testing for certain kinds of feedback. When using such questionnaires, it is best not to select a subset of questions—use the whole questionnaire. You need to be careful to not have too many questions otherwise participants will become bored or frustrated, especially in online studies.

Make sure to choose your questions and questionnaires carefully to address your research questions. You will need to balance your time: not too long for participants to complete without losing focus, and not too short that you don't get any useful data. As a rule of thumb, you are likely to need at least 10 questionnaire questions to get useful data. Try to restrict the length of questions to one A4 side if possible. In the Polymetros example a 7-question questionnaire was printed on one side of A5 paper, but this was due to very short time for people to complete the questionnaire in a high traffic public venue and resulted in very little useful data.

Video-Cued Recall You might also ask participants to watch a video recording of their interaction experience and provide a commentary on it. This provides some reflective data on what they did and how they responded to the interaction. To provide results which are more comparable between participants, you might focus on getting participants to identify particular pieces of behaviour you are interested in, (e.g., points at which they learnt a new aspect of the interaction, felt frustrated, introduced new ideas, or they felt most immersed in the experience).

Observations and Retrospective Video Analysis Observations are usually carried during the interaction to get an overview of the forms of interaction with the MAT. You would then go over video recordings of the interaction to annotate the video with descriptions of the interaction and then code the interaction. For example, you could analyse video in terms of the following coding schemes depending on your study, or develop your own coding scheme:

- Participants mirroring, transforming, or complementing each other's contributions or actions. This is referred to as evidence of Mutual Engagement [10].
- Changes in participants performative interaction—whether they are simply observing, or participating, or performing [36].
- Number of new ideas generated e.g., ideas in a brainstorming session.
- Topics of conversation between participants—what are the conversations predominantly about? The system, the creative act, each other, the organisation of the activity, the weather? If there is a lot of talking between participants, you could use Thematic Analysis (see Sect. 10.4.4) to identify the common themes.

10.4.4   Section: Data Analysis

Data analysis will generally be broken into either quantitative or qualitative analysis. Generally, decide and describe what kinds of data analysis you plan to do with the data collected. Often, this will not be perfectly clear before you get the results, but you should provide an indication of the kinds of data analysis you plan to do so that you can collect the right kind of data for the kinds of analysis you want to be able to do. This will be largely dependent on what you want to know from your study.

Data Measure Types When collecting data from participants through questionnaires or other feedback mechanisms, you need to specify what kind of measure it is. You will come across the following terms:

Nominal data, or categorical data, uses labels for variables without giving a quantitative value; for instance, having participants indicate gender or using arbitrary categories (e.g., Interface A or Interface B). This data is separated in distinct categories which cannot be ranked.

Ordinal data is similar, but the categories follow a natural order; for instance, asking participants the highest education level they have achieved. There is an order from compulsory schooling to undergraduate to postgraduate studies. This could also include ranking where things labeled as first, second, third preference.

Interval data, or integer data, is measured quantitatively but there is no zero point and it can be negative. The difference between the values on the scale should be measurable and comparable. Age is a common interval data value.

Ratio data is similar but there is a zero-point restricting the range. For instance, asking a participant how many times they use a software during the course of a day. Time duration is also a common ratio data, for example, measuring how long a participant took to complete an activity.

Statistical Testing For quantitative data, you should decide the statistical tests to perform on the data to see if your observed results match your hypothesis. We use statistical tests to determine whether there are significant relationships between variables or difference between groups. By *significant* here we mean that the differences are not likely to be just due to chance. Ensure that you have enough participants with the variables you are examining, as mentioned in Sect. 10.4.3.5.

You can use *Regression Testing* to check potential cause-and-effect relationships, comparison tests to check for differences or similarities in groups and their behaviour, or correlation analyses to see whether variables are related. When performing statistical analysis, make sure that your data fits into the assumptions of the test being used (if you use a parametric test). Transforming your data can help to achieve normal distribution and variance. A common comparative test used in MAT work is an *Analysis of Variance* (ANOVA); this kind of test examines the difference in two or more groups based on the mean of their data.

Thematic Analysis With qualitative data, you will need to decide on an evaluation method for the interviews and open-ended questionnaires. Interviews should be transcribed so that you can anonymise and analyse the text data. In MAT research, we commonly use Thematic Analysis on interviews and open-ended questionnaires to identify key themes in people's responses to MATs [6, 7]. This method is particularly useful for qualitative research as it produces results which are useful for further understanding, rather than qualitative results such as frequency of responses [8]. For instance, a single participant might make an important comment about their interaction which is different from the others' perspectives and this would be included as an important part of the analysis, rather than as an outlier. You will also need to consider how you will analyse people's behaviour, such as gestures and movement; for instance, coding movement in Laban notation [21, 22].

Analyses should be presented referencing the specific method used. This Methods section should only include how the analyses were conducted, saving the results until the subsequent Results section. Here, it is important to describe different measures taken from the data; for instance, how participant involvement was measured, how musical complexity was determined, and so on.

Triangulation Additionally, the Methods section is an ideal place to discuss why different analysis methods were were chosen; for instance, justifying the choice of a statistical test. If you want to know about more users' qualitative perceptions of their interaction, it is probably not necessary to conduct statistics testing (and you may not have data which you could test in this way). Therefore, analysing participant responses is more appropriate. If you want to know about more quantitative differences between one interface or another, you could use statistical tests; for instance, seeing if there are more annotations made on one kind of interface than the other (as with Daisyphone).

In most studies, you will want to have a bit of both kinds of data—statistical results are used to support the qualitative observations made, and qualitative results are used to contextualise the numerical data gathered. This practice is referred to as *triangulation*—triangulation is an analysis practice which uses multiple analysis techniques and data streams to get a clear, robust, and well-rounded picture of the study[2]. In the Daisyphone example, quantitative information was collected about the participant's annotation behaviour (number of annotations, locations marked, timing of annotation, etc.) through the program itself. Interviews were also used to gather open-ended response data about the participants' experiences with Daisyphone. Together, the triangulation of these results tell us not only what was going on, but also why. For instance, qualitative analyses of survey responses after working with Polymetros, which suggest controllability and appropriate interest by broader participant groups, was highly associated with participants' responses of feelings of being in control and enjoying the collaborative creative process [2, pp. 238–239].

---

[2] Note that, although it is called *triangulation*, this does not mean that specifically three methods must be used.

10.4.5 Section: Results

Results should be presented based on the types of analyses done. If you performed multiple analyses, for instance a statistical analysis of study outcomes and then a Thematic Analysis of participant feedback, the results section should be divided into separate parts. This can also help to explain the triangulation of the analysis, making sure to explain each approach and analysis separately before connecting them in the subsequent Discussion section.

10.4.5.1 Reporting Statistical Tests

Depending on which format your publication is in, the presentation of statistical results will be slightly different, but they generally follow the same formats (make sure you check the formatting for the style of your presentation). For instance, with an ANOVA we would report something like:

- F(between group DoF,[3] within group DoF) = the F statistic, p = p-value
- Example: $F(2, 26) = 8.76, p = 0.012$.

The p-value The p-value is the probability that the result of the statistics test was due to chance. In HCI, we usually consider something to be statistically significant if the p is less than 0.05 ($p < 0.05$).[4] You will also see other p-values of <0.01 and <0.001—these are not typically used in HCI research (see below for more information on the confidence interval).

It is worthwhile to mention that there is some debate about p-values being viewed as the end-all-be-all in results reporting. When you examine the results of your statistics tests, it is important to use critical thinking about your analyses and interpret the p-value in an appropriate way. As mentioned before, your research will be much stronger if you can connect the results of your statistics testing to other data you collected—in a way, the statistics results are meant to support the validity of the observations you made in your study (e.g., in the Polymetros example of triangulation [2]).

10.4.5.2 Questionnaire Results

Generally, when reporting questionnaire results, you want to report either a data count or a range-mean-standard deviation set. What you report depends on the kind of data. For nominal and ordinal data, you will generally want to report a count; for

---

[3] Degrees of freedom, the maximum number of independent values than can logically occur in your dataset.

[4] NB: In APA formatting, you should not include a leading 0 when reporting p-values because they exist only between 0 and 1.

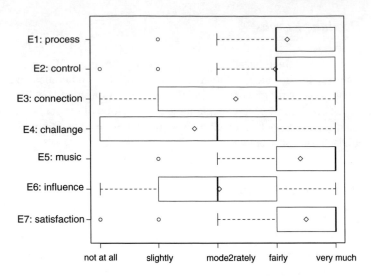

Fig. 10.8 Presentating results using a box-and-whisker plot (referenced from [2])

example, "All participants indicated the interface was Enjoyable (18 participants) or Very Enjoyable (30 participants) to use." For interval and ratio data, you will want to include the range of the response, the mean, and the standard deviation; for instance, "Participants reported that they used the new interface for a longer duration of time, ranging between 45–65 min (M = 50, SD = 2.5)."

You can visually report the results of Likert scale questions using a box-and-whisker plot. These are good at showing many aspects of your data—they show the minimum, the maximum, the median, the upper quartile (top 75%) and lower quartile (lower 25%) of your results. For example, the box-and-whisker plot in Fig. 10.8, referenced from the Polymetros study, shows responses given for elements of playing experience [2].

Participant Quotes It is often helpful to provide quotes from participants to illustrate your results, especially when reporting the results of questionnaires and interviews. This provides evidence that you have objectively collected opinions about your MAT. In addition, referencing specific data points and codes when describing a theme from a thematic analysis can help to better contextualise the material and the meaning of the theme. When referencing specific points, make sure to include the participant's ID number.

Quotes should be short and concise to support specific points, only choosing the relevant portion of the feedback, e.g., if a participant said or wrote in their response: "Well, I enjoyed the experience a lot and found it quite engaging. I was thinking about it the other day whilst walking to work and thought that it was quite nice. That is just my opinion of course, but in general that is my feeling about it. Yes." (P2), you should not quote the whole paragraph, but instead quote the key points, e.g., "I enjoyed the experience a lot and found it quite engaging" (P2). The other points are just repeating this point. If you are really short of space, you can reduce it further, e.g., "enjoyed the experience a lot... quite engaging" (P2).

Make sure to include a range of participants' responses—don't just report one or two participants. If you have 5 participants, make sure to provide quotes from each of them in your results. If you have more than 5 participants then try to report a balance of responses—positive as well as negative. Remember that negative responses are still useful in terms of research, and give you something interesting to discuss in your discussion section e.g., to discuss why you think that they gave negative responses etc.

Codes from Thematic Analyses When reporting a qualitative method such a Thematic Analysis (TA), it is useful to include the level of detail which was transcribed and analysed (e.g., did you include non-verbal utterances such as laughter, facial expressions, etc.). Report how many codes you developed in the analysis and provide a general overview of the themes (you do not need to list the codes, but this may be helpful in getting a clearer view of the data). List the themes that you identified in the TA. You can also list the number of codes within each theme to give an idea of how many times the theme was found in participant responses:

> E.g., The 949 coded segments were clustered into the codes: effort; entanglement; characteristics of the compositions; reflections on the instrument; gestures and techniques; performing perception; performer's body; movement; learning the instrument over time; and 'edge-like interactions' [25, p. 5].

Then provide a description of each of the themes. The description is usually at least a couple of sentences, and if you have space it could be several paragraphs. In the description make sure to illustrate your description with quotes from participants and refer to the participant ID. Remember that TA is about providing a cohesive picture of the data from the codes, so the themes should be linked together and discussed thoroughly. For instance, the *performer's body* theme from Mice & McPherson is further elaborated as:

> E.g., During the course of the study, we noticed examples of participants feeling differently about their bodies while performing the instrument. Some comments were overwhelmingly positive, for example P5 said the instrument makes her body feel powerful, while other comments implied that participants would like to change their bodies to be more suitable for performing such an oversized instrument. P5 said performing the instrument "makes me want more arms", and P8 commented "I need bigger arms". P9 said "I wish I had 3 hands". [25, p. 12].

Remember that the number of codes within a theme does not dictate relevance; the themes are meant to create a clear picture of the data and capture important points and similarities. If the theme contains only one or two codes but provides a critical observation from the research, it is still a valid theme [8]. With subjective research, participant responses when working with MATs will often demonstrate a wide diversity of interaction perspectives and techniques, and it is important to give

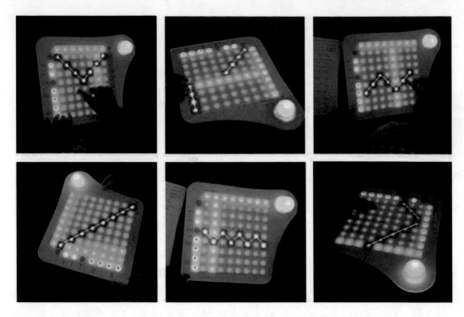

Fig. 10.9 A vignette of typical contributions with Polymetros (referenced from [2])

attention to this variability; for instance, participant responses when working with the Keppi [4, pp. 49–50] or Chaos Bells [25, pp. 9–12] are discussed in detail to represent the varying viewpoints when working with the instruments.

10.4.5.3   Other Observations

It may be beneficial to provide other observations made during the study which do not fall into a specific category of analysis. For example, providing vignettes of observations that support or illustrate findings from your other data. Vignettes are short descriptions of some action and interaction and would often be accompanied by an image from your video recording if you have it.

E.g., from [2]: A prevalent input strategy was the creation of musical patterns characterised by simple geometric properties. The most common phrases consisted of horizontal and upward or downward diagonal lines whereas in most cases all available notes were used (Figure 10.9). Resulting in 'closed musical figures,' this approach was applied by many players providing a clear audio-visual correlation between the representation on the interface and the musical result.

10.4.6   Section: Discussion

In the Discussion, you should reflect on your results—what results were unexpected, why do you think that might be? What explanations can you think of for the results you found? Connect your results back to the Background and the aims of your study—did your results match with the results of other researchers you mentioned in the literature review? If not, why not? What do your results tell you about the things you wanted to find out in the aims of your study?

It is quite usual to find things that you did not expect, or which are surprising and counter intuitive. This is part of why we conduct these exploratory MAT studies. The Discussion can explore why the results were unexpected and suggest future work to further explore the findings in future work (hopefully using your well-reported and rigorous methods). The most important thing is to clearly and objectively report the results, and then reflect on your results, connecting them to your literature review, your study aims, and your intuition about what happened.

10.4.6.1   Limitations

The Discussion is also the place to discuss limitations to your research. In terms of *internal* and *external validity*. Ideally, your research will have both; in the case it does not, you will need to highlight where the reported results are limited.

Internal validity is the extent to which the study accurately and confidently depicts the relationship between the variables being examined. Some factors which might influence the internal validity are uncontrolled or confounding variables, small sample size, repeated testing and learning effects and potential experimenter bias.

External validity refers to the the extent that the findings of the study can be applied in other settings—how generalisable the findings are to the world outside of the study. Some factors which might reduce the external validity are selection bias (e.g., the participants who take part in a MAT study might be interested in or already using technology in artistic settings, so might react differently than others who are not as likely to participate), situational factors such as time of day and location, and limited examined factors, such as only looking at music within the Western canon.

In either case, you must identify these factors in validity, not only to lead future work but also to demonstrate your awareness and acknowledgement of potential limitations in the work. For instance, in [34, pp. 8–9], the autoethnographic approach means that, while the results are useful for design and interaction research, the specific interactions discussed in the paper may not be generalisable to other singers with their individual perspectives:

> It is of course critical to again state that, while the interaction observed provides a detailed account of a prolonged interaction with biofeedback through sEMG, this interaction is highly specific to the user... As mentioned previously, further studies to conduct similar trials and autobiographical use cases with the system will be necessary to validate the universality or differences in the experiences.

10.4.6.2 Future Work

After providing the results and having discussed some limitations, it is worthwhile to suggest future research studies which will further explore the results you have achieved or address the limitations. You might also include some suggestions for research based on interesting or exciting findings, to expand or apply what was learned through the study to additional research questions.

10.4.7 Other Presentation Components

When presenting your research, you will likely need to include an Abstract, Introduction, and Conclusion to bookend the components outlined here. These portions of a paper are sometimes written last, after all of the other information is in place, to better summarise everything in the presentation together.

After the main body of the paper, you may also wish to present your materials alongside your research and results in an appendix. This will help with reproducibility and allow others to follow your methods in their own work. In your appendix you could therefore include all questionnaires, exactly as given to participants and lists of interview questions.

10.5 Conclusion

Research in Media and Arts Technology (MAT) must strike a careful balance between arts and science practices and norms. This can be a difficult and yet rewarding balance to achieve. However, the two are not mutually exclusive and the interaction between HCI, User Experience, and Interactive Arts and Media have led to the advancement of both fields. While scientific approaches make the research grounded and extendable to other fields, arts approaches offer new inspiration, contextualisation, and opportunities to explore many facets of the human condition. Through this guide, we have introduced you to scientific practices to make the design of MATs, their study and evaluation, and their presentation and dissemination to others both rigorous and repeatable. Through this guided approach, we hope to increase the accessibility and validity of MAT research in wider artistic and scientific communities and to further explore the human condition of being in the world.

Acknowledgements Work on this chapter was supported by the EPSRC and AHRC Centre for Doctoral Training in Media and Arts Technology (EP/L01632X/1). We would like to thank all the students and staff at the Media and Arts Technology Centre and the Centre for Digital Music at Queen Mary University of London who have contributed to ongoing discourse about the nature of Media and Arts Technology which inspired many of the thoughts in this chapter.

References

1. Benford, S., Greenhalgh, C., Giannachi, G., Walker, B., Marshall, J., Rodden, T.: Uncomfortable user experience. Commun. ACM **56**(9), 66–73 (2013)
2. Bengler, B., Bryan-Kinns, N.: Designing collaborative musical experiences for broad audiences. In Proceedings of the 9th ACM Conference on Creativity & Cognition (C&C'13), pp. 234–242. ACM (2013)
3. Bengler, B., Bryan-Kinns, N.,: In the wild: evaluating collaborative interactive musical experiences in public settings. In *Interactive Experience in the Digital Age*, pp. 169–186. Springer International Publishing (2014)
4. Bin, S. M. A., Bryan-Kinns, N., McPherson, A. P.: Risky business: Disfluency as a design strategy. In *Proceedings of the International Conference on New Interfaces for Musical Expression*, pp. 45–50. Blacksburg, Virginia, USA (2018)
5. Blythe, M.A., Overbeeke, K., Monk, A.F., Wright, P.C. (eds.): Funology: From Usability to Enjoyment. Springer (2004)
6. Braun, V., Clarke, V.: Using thematic analysis in psychology. Qual. Res. Psychol. **3**(2), 77–101 (2006)
7. Braun, V., Clarke, V.: Thematic analysis. In: Cooper, H., Camic, P.M., Long, D.L., Panter, A.T., Rindskopf, D., Sher, K.J. (eds.) PA Handbook of Research Methods in Psychology, volume 2: Research Designs: Quantitative, Qualitative, Neuropsychological, and Biological. American Psychological Association, Washington (2012)
8. Braun, V., Clarke, V.: One size fits all? What counts as quality practice in (reflexive) thematic analysis? Qual. Res. Psychol. **18**(3), 328–352 (2020)
9. Bryan-Kinns, N.: Daisyphone: The design and impact of a novel environment for remote group music improvisation. In: Proceedings of the 5th Conference on Designing Interactive Systems: Processes, Practices, Methods, and Techniques, DIS '04, pp. 135–144. Association for Computing Machinery, New York, NY, USA (2004)
10. Bryan-Kinns, N., Hamilton, F.: Identifying mutual engagement. Behav. Inform. Technol. **31**(2), 101–125 (2009)
11. Caine, K.: Local standards for sample size at CHI. In: Proceedings of the ACM CHI Conference on Human Factors in Computing Systems, pp. 981–992 (2016)
12. Candy, L.: Research and creative practice. In: Edmonds, E.A., Candy, L. (eds.) Interacting: Art, Research and the Creative Practitioner, pp. 33–59 (2011)
13. Candy, L., Edmonds, E.: Practice-based research in the creative arts: foundations and futures from the front line. Leonardo **51**(1), 63–69 (2018)
14. Candy, L., Ferguson, S. (eds.): Interactive Experience in the Digital Age. Springer (2014)
15. Cherry, E., Latulipe, C.: Quantifying the creativity support of digital tools through the creativity support index. ACM Trans. Comput.-Hum. Interact. **21**(4) (2014)
16. Duarte, E.F., Merkle, L.E., Baranauskas, M.C.C.: The interface between interactive art and human-computer interaction: exploring dialogue genres and evaluative Practices. J. Interactive Syst. **10**, 20 (2019)
17. England, D.: Art.CHI: curating the digital. In: Curating the Digital, pp. 1–7. Springer International Publishing (2016)
18. Hart, S.G., Staveland, L.E.: Development of NASA-TLX (Task Load Index): results of empirical and theoretical research. In: Advances in psychology, pp. 139–183. Elsevier (1988)
19. ...Höök, K., Caramiaux, B., Erkut, C., Forlizzi, J., Hajinejad, N., Haller, M., Hummels, C.C.M., Isbister, K., Jonsson, M., Khut, G., Loke, L., Lottridge, D., Marti, P., Melcer, E., Müller, F.F., Petersen, M.G., Schiphorst, T., Segura, E.M., Ståhl, A., Svanæs, D., Tholander, J., Tobiasson, H.: Embracing first-person perspectives in soma-based design. Informatics **5**(8), 1–26 (2015)
20. Jeon, M., Fiebrink, R., Edmonds, E.A., Herath, D.: From rituals to magic: interactive art and HCI of the past, present, and future. Int. J. Hum.-Comput. Stud. **131**, 108–119 (2019)
21. Laban, R., Ullmann, L.: The Mastery of Movement (1971)

22. Loke, L., Larssen, A.T., Robertson, T.: Labanotation for design of movement-based interaction. In: Proceedings of the Second Australasian Conference on Interactive Entertainment, IE '05, pp. 113–120. Creativity & Cognition Studios Press, Sydney, AUS (2005)

23. Mice, L., McPherson, A.: Embodied cognition in performers of large acoustic instruments as a method of designing large digital musical instruments. In: Proceedings of Computer Music Multidisciplinary Research Conference, Marseille, France, Marseille, France (2019)

24. Mice, L., McPherson, A.: From miming to NIMEing: the development of idiomatic gestural language on large scale DMIs. In: Michon, R., Schroeder, F. (eds.) Proceedings of the International Conference on New Interfaces for Musical Expression, pp. 570–575. Birmingham City University, Birmingham, UK (2020)

25. Mice, L., McPherson, A.: Super size me: interface size, identity and embodiment in digital musical instrument design. In: Proceedings of the 2022 CHI Conference on Human Factors in Computing Systems, CHI '22, pp. 1–15. Association for Computing Machinery, New York, NY, USA (2022)

26. Mice, L., McPherson, A.: The M in NIME: motivic analysis and the case for a musicology of NIME performances. In: Proceedings of International Conference on New Interfaces for Musical Expression, Waipapa Taumata Rau, Auckland, Aotearoa New Zealand (2022)

27. Müllensiefen, D., Gingras, B., Stewart, L.: The musicality of non-musicians: an index for assessing musical sophistication in the general population. PLoS ONE **9**(2), e89642 (2014)

28. Nam, H.Y., Nitsche, M.: Interactive installations as performance. In: Proceedings of the 8th International Conference on Tangible, Embedded and Embodied Interaction-TEI '14. ACM Press (2013)

29. Neustaedter, C., Sengers, P.: Autobiographical design in HCI research: designing and learning through use-it-yourself. In: Proceedings of the DIS 2012, June 11–15, 2012, Newcastle, UK, pp. 514–523 (2006)

30. Nielsen, J., Landauer, T.K.: A mathematical model of the finding of usability problems. In: Proceedings of ACM INTERCHI'93 Conference, Amsterdam, The Netherlands, 24–29 April, pp. 206–213 (1993)

31. Nonnis, A., Bryan-Kinns, N.: Mazi: a tangible toy for collaborative play between children with autism. In: Proceedings of the 18th ACM International Conference on Interaction Design and Children, IDC '19, pp. 672–675. Association for Computing Machinery, New York, NY, USA (2019)

32. Nonnis, A., Bryan-Kinns, N.: Mazi: Tangible technologies as a channel for collaborative Play. In: Proceedings of the 2019 CHI Conference on Human Factors in Computing Systems, CHI '19, pp. 1–13. Association for Computing Machinery, New York, NY, USA (2019)

33. Reed, C., McPherson, A.: Surface electromyography for direct vocal control. In: Michon, R., Schroeder, F. (eds.) Proceedings of the International Conference on New Interfaces for Musical Expression, pp. 458–463. Birmingham City University, Birmingham, UK (2020)

34. Reed, C.N., McPherson, A.P.: Surface electromyography for sensing performance intention and musical imagery in vocalists. In: Proceedings of the Fifteenth International Conference on Tangible, Embedded, and Embodied Interaction, TEI '21. Association for Computing Machinery, New York, NY, USA (2021)

35. Reed, C.N., Skach, S., Strohmeier, P., McPherson, A.P.: Singing knit: soft knit biosensing for augmenting vocal performances. In: Augmented Humans 2022, AHs 2022, pp. 170–183. Association for Computing Machinery, New York, NY, USA (2022)

36. Sheridan, J.G., Bryan-Kinns, N.: Designing for performative tangible interaction. Int. J. Arts Technol. 1(3–4) (2009)

37. Skach, S., Stewart, R., Healey, P.G.T.: Smart Arse: posture classification with textile sensors in trousers. In: Proceedings of the 20th ACM International Conference on Multimodal Interaction, ICMI '18, pp. 116–124. Association for Computing Machinery, New York, NY, USA (2018)

Chapter 11
haptic HONGI: Reflections on Collaboration in the Transdisciplinary Creation of an AR Artwork

Mairi Gunn, Angus Campbell, Mark Billinghurst, Wendy Lawn, Prasanth Sasikumar, and Sachith Muthukumarana

Abstract This chapter explores the complexities of collaborative digital creation through *haptic* HONGI, an Augmented Reality (AR) project that is both transdisciplinary and intercultural in its conception. *haptic* HONGI aims to use contemporary digital technologies to bridge troubling intercultural relationships in Aotearoa New Zealand by creating a face-to-face tabletop encounter. The project both addresses the cultural divide between Māori and non- Māori in Aotearoa New Zealand, but also explores how the tension between artists and technologists can be overcome to

M. Gunn (✉)
Te Waka Tūhura Elam School of Fine Arts and Design, and Empathic Computing Laboratory, Waipapa Taumata Rau, University of Auckland, Auckland, New Zealand
e-mail: mairi.gunn@auckland.ac.nz

A. Campbell
Te Waka Tūhura Elam School of Fine Arts and Design, Waipapa Taumata Rau, University of Auckland, Auckland, New Zealand
e-mail: angus.campbell@auckland.ac.nz

M. Billinghurst
Auckland Bioengineering Institute, Waipapa Taumata Rau, University of Auckland, Auckland, New Zealand
e-mail: mark.billinghurst@auckland.ac.nz

W. Lawn
Product, Disguise, London, England
e-mail: wendy.lawn@disguise.one

P. Sasikumar
Empathic Computing Laboratory, Auckland Bioengineering Institute, Waipapa Taumata Rau, University of Auckland, Auckland, New Zealand
e-mail: prasanth.sasikumar.psk@gmail.com

S. Muthukumarana
Augmented Human Lab, Auckland Bioengineering Institute, Waipapa Taumata Rau, University of Auckland, Auckland, New Zealand
e-mail: sachith@ahlab.org

create a transdisciplinary piece. In this chapter we describe the technology developed and reflect on lessons learned that could be useful for other teams creating transdisciplinary artwork using AR technologies.

11.1 Introduction

This chapter explores the complexities of collaborative digital creation through *haptic* HONGI, an Augmented Reality (AR) project that is both transdisciplinary and intercultural in its conception. *haptic* HONGI aims to use contemporary digital technologies to bridge troubling intercultural relationships in Aotearoa New Zealand by creating a face-to-face tabletop encounter. The project both addresses the cultural divide between Māori and non-Māori in Aotearoa New Zealand, but also explores how the tension between artists and technologists can be overcome to create a transdisciplinary piece. In this chapter we describe the technology developed and reflect on lessons learned that could be useful for other teams creating transdisciplinary artwork using AR technologies.

We begin by describing the background and motivation for the work (Sect. 11.2), then the participant experience with the piece and how it was created (Sect. 11.3), and the feedback collected (Sect. 11.4). In Sect. 11.5 we examine *haptic* HONGI from a design perspective, and the lessons learned from developing the project (Sect. 11.6). Finally, in Sect. 11.7, we end with conclusions and directions for future work.

11.2 Background and Motivation

Like many countries of the British Commonwealth, Aotearoa New Zealand is grappling with messy colonial leftovers. Contemporary NZ society bears witness to the results of colonisation: dispossession, disenfranchisement, and entrenchment in binary oppositions, especially amongst its indigenous Māori people. The aim of *haptic* HONGI was to address this and provide opportunities to discuss what society is currently like, whilst enabling participants to collaboratively contemplate and practise what it could be like in the future.

In Māori society, the hongi is part of a traditional greeting. It involves the pressing together of noses, and often foreheads, so that two people in greeting share the breath of life as a gesture of unity (see Fig. 11.1). Māori Psychologist Cleave Barlow [1] described this as:

> ...peace and oneness of thought, purpose, desire, and hope; and such is the desire of the hosts and visitors when they greet (hongi) one another [...] the second meaning of the hongi is as a sign of life and immortality, and it symbolises the action of the gods in breathing into humans the breath of life. By this action, the life-force is permanently established in the spiritual and physical bodies become a single living entity.

Fig. 11.1 Still frame of a hongi from short film Glory Box by the main author. https://www.you
tube.com/watch?v=GpZuD3PAupE

One goal of *haptic* HONGI was to recreate the traditional hongi greeting experience and use it as a catalyst to trigger reflection on the social division and awkwardness, or discomfort, between Māori and non-Māori. Another was to give visitors a safe experience of close proximity to and interaction with people from a different culture. Its particular focus is the moment of first contact between different people.

11.3 The *haptic* HONGI Experience

In *haptic* HONGI, the visitor sits at a table adorned with a checked tablecloth and a vase of fresh foliage to connote a convivial domestic scene and to connect them to Papatūānuku (Mother Earth) (see Fig. 11.2). They wear an AR head mounted display (HMD) and through the AR HMD, they can see a volumetric virtual avatar of Tania Remana (see Fig. 11.3), a Māori multimedia artist and performer from the Ngāpuhi tribe. The pre-recorded 3D virtual video of Tania appears across the table, is visible to the visitor, and locks them in a mutual gaze.

When the visitor sits at the table, Tania greets them in a re-imagined, contemporary first encounter between Indigenous Māori and newcomers from shores distant to Aotearoa New Zealand. She says:

Kia ora! Hi! Are you all right? Can I help you? You just seem lost. Where are you from?

No hea koe? Where are you from? Oh! Wow! What's your name? Ko wai koe? Ko Tania ahau. I'm Tania. Welcome to Aotearoa. This is my home.

And when we see someone we've never met before, we often hongi and we greet each other with a hongi, we close our eyes, we press our noses together and we feel the wairua.

Fig. 11.2 *haptic* HONGI setup with table, tablecloth, vase of foliage, black wall and HoloLens 2 with actuator

Fig. 11.3 Tania's virtual avatar greets the visitor "Ko Tania ahau. I am Tania", then leans forward to hongi

I'll show you...

At which point her virtual avatar appears to lean forward and press her nose against the visitor, who at the same time feels real pressure against their nose and forehead. Although the encounter is only brief, the combination of a virtual person superimposed over the real world, and real touch sensation provide an engaging experience.

11.3.1 *The Technology Behind* haptic *HONGI*

From a technical standpoint, *haptic* HONGI is a volumetric capture and playback system that can accommodate peripheral devices for haptics. The key components are; (1) a volumetric capture system, (2) an AR display, (3) a haptic feedback system, and (4) a playback system. Each of these are described in more detail in the rest of this section.

Volumetric Capture

Tania's performance was recorded with an Intel RealSense D435 depth sensing camera. This camera captures both colour (at 1920 × 1080 pixel resolution and 30 frames per second) and depth information (at 1280 × 720 pixel resolution and 90 frames per second). We used DepthKit (depthkit.tv) software to record the volumetric data from the RealSense camera and stored it as point cloud data. This was chosen due to its light-weight encoding that is relatively easy on computing resources. It enabled us to have a completely stand-alone and portable solution. Prior to using DepthKit, we used a variety of off-the-shelf and in-house built tools used for recording and play-back with the Unity game engine. The output from DepthKit was a three-dimensional point cloud of Tania speaking, ready to be included in the AR viewing application.

AR Display

The current version of *haptic* HONGI is a standalone application viewed on the Microsoft Hololens 2 AR HMD.[1] The Hololens 2 is an optical see-through AR display that allows the wearer to see virtual content directly superimposed over the real world. It provides a high resolution AR display with 54 degree diagonal field of view display and 2048 x 1080 pixel resolution per eye. It also has integrated cameras and inertial sensors that can precisely track the HMD relative to the real world and ensure that the virtual content appears fixed in space.

Haptic Feedback

A key part of the experience is the haptic feedback which is provided by a mechanical actuator that sits in the HoloLens 2 and applies pressure to the wearers' forehead and nose. This was achieved by using Shape-memory Alloys (SMAs) and flexible 3D printed structures, and inspired by earlier work on Cloth Tiles [12]. We used a Shape Memory Alloy strip from Toki Corporation (Japan), which is a bimetallic material that can change its length through applied electrical current or heat.

We designed a 3D-printed housing that can hold an SMA loop inside, which can pull and flip both ends of the housing structure. The SMA wire strip can warm up to

---

[1] See https://www.microsoft.com/en-us/hololens/.

Fig. 11.4 The SMA actuator set into the HoloLens 2 visor

60°C when activated. The SMA wire inside the actuator generates the internal force to drive the actuator. The 3D printed design holds the SMA wire and determines the motion of the actuator based on its properties such as thickness, shape, and orientation. We attached small sculpted pieces of makeup sponge to both moving ends to realize gentle haptic feedback and to prevent direct contact of the SMA to the visitors' face.

Several versions of the actuator were implemented and tested to determine the ideal shape and size. The finalised version of the actuator was designed to be placed above the forehead and nose area of the user with two moving elements to mimic the touch sensation in two distinct locations (see Fig. 11.4). Since the SMA wire requires a comparatively high current (1.4 A) to activate, a custom-made circuit board was designed to interface the actuator with the AR application. A serial command from the application could control the actuator, and trigger time and activation duration were the controlling parameters.

Playback System

An AR viewing application was developed in the Unity game engine to view the virtual volumetric video of Tania and trigger the haptic feedback at the end of her speech.

Creating the viewing application in Unity required positioning the recorded point cloud of Tania, adding lights to the scene, and writing some shader code to correctly render the volumetric video. One of the most challenging aspects was making sure that the audio of Tania's speech was correctly synchronized with the visual recording.

The haptic feedback unit is completely independent of the main application. This ensures that a demonstration can proceed without haptics even if the haptic component failed. To enable a smooth running of the installation for the exhibition, we connected a Bluetooth keyboard to the Hololens 2, enabling the person showing *haptic* HONGI to use key presses to test out individual components like playback, termination, and haptic feedback and so on.

Once all devices are in operation, a command to activate the actuator, issued by the *haptic* HONGI application running in the HoloLens 2, is picked up by the microcontroller (connected to the same network via Wi-Fi hotspot). The microcontroller then activates the SMA actuator.

11.4   User Feedback from the Experience

The latest version of *haptic* HONGI was shown during the Ars Electronica Garden Aotearoa exhibition (Wellington 2022) and feedback was collected from a diverse range of people.

Some smiled and laughed, whereas others sat stock still and a significant number of people wept. After trying the experience, some were offered the short written survey that allowed space for further comments, or a longer voice-recorded interview depending on the time pressures and the inclination of the visitors. These interviews were each about 15 minutes duration. The discussion was rich, honest and sometimes emotional. It was exciting to have such lively and positive interactions with members of the public and the professional community.

Likert scales benefit from larger sample sizes and are highly dependent on the relevance of the questions. Discursive design processes are more aligned to long form discussions that may include the lived experiences of participants and their subjective, even contradictory impressions and observed reactions, including body language and tone of voice. What is valued is the stimulating discussions that are not necessarily quantifiable.

11.4.1   Quantitative Assessment

Visitors to *haptic* HONGI at the Ars Electronica exhibition were asked the following three questions:

1. *How confident would you be to recognise Tania if you saw her on the street?*
2. *How comfortable would you be to say "hi" to Tania if you saw her?*
3. *How similar was haptic HONGI to a meeting with a real person?*

These were answered using a Likert scale from 1 to 5, where 1 = Not Very, and 5 = Very. In order to explore the impact of the haptic feedback on the user experience, we collected 8 completed questionnaires from people who tried *haptic* HONGI with the haptics turned off (No Haptic condition), and 18 questionnaires from people with the haptics turned on (Haptic condition). Figure 11.5 below shows the average scores for the responses for each of the three questions, for each condition.

Although the overall number of questionnaires was not large, the results from questions one and two seem to show that providing haptic feedback in the visual hongi helped visitors connect more strongly with Tania.

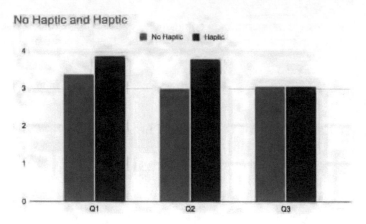

Fig. 11.5 Average scores for each of the user survey question, for each condition

11.4.2 Qualitative Assessment—A Discursive Approach

To collect qualitative feedback, as part of the questionnaire, there was space for visitors to write additional comments. These are telling, and some of the key themes from the comments are summarised below.

Firstly, there was a high level of presence, and Tania appeared very real to some visitors. For example, one person wrote: *"It felt like it could have been a live feed and she was really there"*, and another wrote *"I felt I stood in front of a real person named Tania. I wasn't sure how to respond but ended up talking to her and tried answering her questions"*. Several other people mentioned how it *"felt like an in-person experience"*, and that they felt that they should respond to Tania.

Visitors appreciated the human interplay, the 'conversation' between cultures. In particular one person wrote: "I found this really moving. I feel usually when Pākehā meet with Māori it is a matter of interfacing on Pākehā terms. We see them through our lens. I think this changes that dynamic. Suddenly Pākehā are in a foreign cultural context." For this person, *haptic* HONGI clearly achieved its goals. Another person left the message: "Just so awesome that AR can facilitate kōrero (discussion) like this!".

The actuator, the haptic component of the hongi, added to the feeling of connection for some. For example, one person wrote *"The hongi was amazing—certainly helped to break the ice and establish a connection with her"*, while another wrote *"Feeling the hongi was quite dramatic"*, and a third wrote that they *"Enjoyed the light touch of the hongi."* For many the hongi was the highlight of the experience.

Many of the visitors were able to respond by using a few words from the Māori language. Perhaps the wall poster of the exhibit, including a Māori word in large font and a portrait of an obviously Māori woman performing part of a Māori ritual, acted as a filter by inviting those drawn to te ao Māori (the Māori worldview) and repelling those who are not so.

However, there were a few remarks from people whose experience was hampered by ill-fitting HoloLens or problematic lack of aligned registration. For example, one person wrote *"It would be better if we could see a bit more of Tania"*, which was due to incorrect alignment of the AR HMD. The haptic feedback was also weak for some, with one person leaving the message *"… my hongi was a thin pressure across the bridge of my nose, which gave me a fright!"*. It is clear that improvements could be made in these areas.

Those who had experienced a real life hongi could, naturally, point to the deficiencies of this virtual experience. One comment was: *"Sharing a hongi is about sharing the breath, the "ha", so that was missing, and the warmth of another person"*, and another wrote about breathing the same air being more important than the haptic element. It is important to note that *haptic* HONGI was never intended to replace real life hongi, but to practice and prepare for engagement with Māori.

11.4.3 Māori Responses

For Māori the hongi has different significance than for non-Māori, and so an important question was how Māori visitors would react to the experience What follows are excerpts from a recorded discussion with a female Māori visitor.

Designer: You moved forward with your eyes closed… You were in it.

Visitor 1: I did. I found it incredible. It was really… enriching. I felt the wairua (spiritual aspect) when I did the hongi. It was really weird… how much I felt that. Thank you.

I didn't know where it was going, but actually to feel the hongi… the way in which you did that, invited me. I felt like I was being invited in to a really special place and I felt that.

Designer: So, this isn't the idea that technology will replace real life, but can you see a use for this technology?

Visitor 1: There's huge amounts of potential. So, what I do know is…with a lot of older people… the biggest health issue is loneliness. And so… a few months ago I was just discussing technology to have in people's homes, so that they can actually have, you know like, connection with people.

This visitor was thankful for what the team was doing to help people be more open to engaging with Māori. Another Māori visitor said *"So, I suppose this is the way it's going to be now"*. Not necessarily… Real in-person, kānohi ki te kānohi (eye to eye), will always be preferable, if there is that option. However, Tania's recorded presence simply IS open and welcoming. She consciously chose that stance, rather than vehemently telling everyone to go away.

[A video of Tania and her visitors responding to *haptic* HONGI at Ars Electronica Garden Aotearoa 2022 https://youtu.be/wsE8JfZHZ7s]

11.4.4 Immigrant Responses

It is interesting that people from different cultures may have vastly different responses to the experience than local people. Foreign visitors seemed especially moved and affected by the experience. The following is a voice-recorded discussion with a German visitor, resident in Aotearoa New Zealand.

> Visitor 2: That was an absolutely fantastic experience. That was so engaging, I was so immersed and the interesting thing is it's AR but it's real. So it's the, you know, the thing is that I am looking at the wall but I am not just looking at the wall.
>
> And the interaction and how the body language is and how it draws me in to react to it and not feel stupid about it. This subtle… introducing te ao Māori (the Māori world).
>
> And then, at the end, the hongi, where you have the physical touch on my nose… It blew me away! This is fantastic! It's one of the best experiences I have, I tell you. It's… It's deeply moving.
>
> Designer: Brilliant. And so, you had the actuator actually on your nose when you started. But you could still feel it pushing?
>
> Visitor 2: Yeah, absolutely.
>
> And the actuator is also built in a way that it has the right softness and the right scale so that you do not feel that there is like a, just a pinch, or like a fingertip…. a, a touch. It's actually… it's a nose touch, so there is a certain length to it, there is a certain softness, and warmness to it…
>
> And the interesting thing is, I mean perhaps minute or so long. But it didn't take long to build this human connection.

Our German visitor was certainly enthusiastic about the experience, including the softness and warmth of the actuator. Interestingly, many visitors responded to Tania's avatar as though it was an actual person from the start of the experience. They waved, nodded, and spoke aloud, telling 'her' their name and saying where they were from. The hongi and actuator action came at the end, so could not be entirely credited for the elevated presence that visitors felt from the start.

Another English visitor was still unsure whether 'Tania' was present even after the experience came to an end. He seemed flabbergasted. How was he feeling?

> Visitor 3: Very on edge of wondering what you're talking to… That's what I would say, yeah. Like, whether they are listening to what you're saying. That's the biggest takeaway. It's like I don't know if she was actually able to… so I was like, I froze on the spot. I didn't know what… I couldn't say what I wanted to say back… very strange. But cool.
>
> Designer: She felt three dimensional to you?
>
> Visitor 3: Yes, yeah kind of eerily so, because I wanted to converse. That's the thing. That's how it felt. Like I wanted to make conversation.
>
> Designer: So how similar was the haptic HONGI experience to meeting with a real person?
>
> Visitor 3: Like, you're still aware that it was digital, but um pretty close.

I would say the sound was most impressive. Like the voice, it felt like it was, with the visual...

It felt like a conversation, like a real live person.

Another woman, a Canadian, commented on the warmth of Tania's performance.

Visitor 4: Well, I think she just was so warm... "And this is what we do... ", and the way that her facial was just so gentle and calm, so that made me comfortable.

Designer. What did you enjoy most about the experience?

Visitor 4: That I can be so moved by... (weeping)

Designer: By the avatar?

Visitor 4: Yeah.

Clearly, in line with filmmaking tradition, the quality of the script, the performance, the lighting, audio recording and the resolution of the video capture and output are defining factors relating to the audience experience. One visitor appreciated the technology and its application to human connection, while acknowledging that it also threw him out of the experience

Designer: How much did the hongi help you to connect with Tania?

Visitor 5: It was yeah it was nice. But I will say that when the 'dial' (actuator) makes its noise and then it's like a soft pad just hitting, it's like you're having a connection and then you're just repelled after being... and then you're... "OK, I know this is a fiction".

Designer: Yeah, yeah. So, you know, the sound and the pad hitting here... it's an artificial...

Visitor 5: Yeah but it's good, though, like, because it's also awkward.

There's an emphasis on technology to make things so hyper-realistic and I guess so realistic that it feels real, but I think it's also interesting to have imperfections. I wouldn't say clunky... but it's also intellectually interesting to experience and I think it's good, it's not a bad thing.

Designer: Good. That's interesting. I think is a bit clunky.

Visitor 5: No... it was good. It was authentic and nice and you're trying to deal with thematically authentic topics. It was lovely.

11.5   Reflections on Design

Haptic HONGI was a transdisciplinary work with a production team comprised of a community-focused Māori improvisational performer and multi-media artist from outside the academy, a Pākehā (descended from European Settler/colonisers) designer, a UK-based lecturer, character designer, animator and creative technologist, and AR developers and software engineers from Aotearoa New Zealand, India, Japan, China and Sri Lanka, who specialise in human-computer interaction, including haptics. Crossing epistemological boundaries is encouraged in the academy yet

Fig. 11.6 The relationship
between XR, MR, VR and
AR © Mark Billinghurst

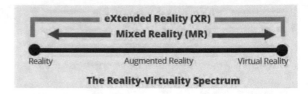

remains challenging for those who are steeped in our own disciplines' conventions. We hope this project will highlight where some common ground lies so we can use those overlapping territories as a foundation on which to build truly transdisciplinary practices.

Our starting point was the desire to support relationship building between cultures, including between designers and computer engineers. While working together on this project, we employed a discursive design approach [18], to explore whether Extended Reality (XR) technologies (see Fig. 11.6), more particularly Augmented Reality in the middle of the Reality-Virtuality Spectrum, could help to create cognitive and behavioural shifts in our local reality.

Design as a discipline emerged out of a need to continue economic growth post-World War 2, therefore it focused on increasing the sales of products by enhancing consumer appeal [9]. Many designers and design schools challenged this superficial role of design, noting that the discipline offered far more for the positive improvement of society [19]. In contemporary society, design is pervasive, but, due to this omnipresence, it has also been explored as a discipline to much greater depth [20]. Beyond a commercial paradigm, one branch of theoretical exploration emerged from critical, speculative and fictional approaches to design; this has been described as "discursive design" by Tharp and Tharp [18]. They explained that:

> Design has the opportunity for intellectual service. Discursive design's primary agenda is to convey ideas... Discursive design asks its audience to take an anthropological gaze and seek understanding of its artefacts beyond basic form and utility.

Development of *haptic* HONGI is well-aligned to this design framework, and we made use of Tharp and Tharp's "nine facets" to narrate and explore the research project through their series of sequential themes of Intention, Understanding, Message, Scenario, Artifact, Audience, Context, Interaction, and Impact [18]. In the rest of this section we describe how *haptic* HONGI addressed these nine themes in more detail.

11.5.1 Intention: What's a Discursive Designer to Do?

Tharp and Tharp write that a discursive designer should have five specific aims: to remind, to inform, to provoke, to inspire, and to persuade [18]. A provocation can come in many forms. In this case, we sought to create a short XR experience designed to remind us all of the potential for reconnection and collaboration from a time before

Covid-19 and beyond. In Aotearoa New Zealand, such provocation could inform the visitor about Māori rituals of encounter as the piece provokes interaction as it inspires and uplifts the visitor with a warm welcome and an inkling of potential. Imparting information via Tania's script was instructive, thereby creating more certainty in the minds of the visitors. Certainty is one way to allay fears. While the persuasive aspect may be present, it must be subtle because in today's Aotearoa, some are resistant to the use of te reo (the Māori language) and tikanga (Māori philosophies of best practice, or rightness).

"As a designer, the discursive practitioner is also prescriptive—planning and creating artifacts—but rather than driving toward usefulness, usability, and desirability, their goal is communicative." [18]. The designer initiated and facilitated meetings (conversations) with and between all members of the team. *haptic* HONGI itself took a conversational form and provoked further discussion that is ongoing. This was the main focus. The filmmaking practice that backgrounded this research prioritised people and context, and sought to make technical artefacts, such as cameras, lights, microphones, wires, cables and so on, vanish. The developers, however, were building the primary tool without which the project and the brief could never have been fulfilled.

11.5.2 Understanding: What's a Discursive Designer to Know?

The genesis of this research arose out of moving image-making practices and historical investigation into societal currents that resulted in separation and alienation. Left unaddressed, such a state of affairs could be destructive in the extreme. We know this. While coverage of local hate-crimes is outside the scope of this chapter, we have seen their aftermath and have taken part in organised events aimed at countering their destructive forces. We have been asked to share the heavy-lifting, to work with our own communities to shine a light on and work to bridge gaps between people from different cultural backgrounds. Our work was backgrounded by some deeply challenging historical events.

While many details were left unspoken, our knowledge and understanding led to a commitment on all of our parts to create the best possible artwork to stimulate discussion and initiate cognitive shifts.

haptic HONGI is a consciously intercultural experience. We use the term 'intercultural' to highlight the enmeshed and interactional nature of our alliances. Expanding on this set of relationships, we describe our research and research community as 'transdisciplinary' to highlight our expansion beyond the university, to include members of non-academic communities. The Organisation for Economic Co-operation and Development (OECD) definition of transdisciplinary is:

...a mode of research that integrates both academic researchers from unrelated disciplines

- including natural sciences and SSH—and non-academic participants to achieve a common goal, involving the creation of new knowledge and theory. In drawing on the breadth of science and non-scientific knowledge domains such as local and traditional knowledge, and cultural norms and values, it aims to supplement and transform scientific insights for the good of society [13].

Transdisciplinary research

For this work to bear fruit, its completion and success is totally reliant of the knowledge and experience of engineers and developers from the Empathic Computing Laboratory[2] and others, not least the artist Tania Remana. Our collaboration began as a transdisciplinary[3] project and ended with implementation of bleeding edge[4] technology that our creative endeavours rested upon. Contrary to assertions in papers such as Driver et al.'s [4] that empirically explores scientists' perceptions of design and designers, the lab always seeks to engage designers and artists as valued creative partners. Although we are situated in different disciplines, honest communication supports our collaborative impetus and helps us rise above or drive through our differences.

In the case of *haptic* HONGI, some of the technical aspects of programming took place in an organic process within sub-teams which were then integrated into more and more granular prototypes that the full team could then provide input on. While the coding and electronic development was in process, the role of a designer was to be a project manager, a provider of tangential information, a procurement worker to find and provide resources, and a grant writer. The team offered up ideas and discussed opportunities, while the designer tried to elicit expressions of concern, then backed off to let people get on with their work.

There were some non-negotiable fundamentals, such as—all work is based around a table, all work supports human interaction, all work includes a direct, mutual gaze, all work respects yet encompasses cultural difference. Other than that, there was a lightly held wish- list. Overly specific or unrealistic expectations could have created real problems. In this way, we avoid putting undue pressure on our co-creators with an emphasis on gratitude for whatever they manage to produce.

However, one issue that arose, was the invisibility of our tacit knowledge. We did not know and could not see what our team members knew, or did not know. Only time, shared experience and trusting relationships can help this surface. Pushing co-creators to perform can result in myriad stressors so we must all learn to be honest with ourselves and others. This is not easy for everyone and it helps to acknowledge that we are entering unknown territories together. Discoveries are made during and because of the practice. This is the magic of research, and particularly practice-led research. It is the doing, rather than the knowing that reveals hidden treasure. The knowing, or understanding is secondary. It comes later.

---

[2] http://www.empathiccomputing.org/.

[3] https://hci.auckland.ac.nz/2020/09/10/abi-and-cai-work-together-to-show-their-arty-science-side/.

[4] Bleeding edge refers to a product or service that is new, experimental, generally untested, and carries a high degree of uncertainty. From https://www.investopedia.com/terms/b/bleeding-edge.asp Accessed 11th August 2022.

Such revelations add to the lived experience of all participants who perceive and experience personally degrees of intercultural awkwardness or connection. When do we feel alienated by difference? When do we feel connected? What supports or hinders connection? What we know can be a yearning for an experience, rather than a memory of one. In a broader context, what we know includes existing knowledge of other XR experiences and the relevant corpus of literature. In practical terms, it serves this work to start from the ground up, from pragmatic domestic wisdom. Is there a better place for human engagement than the dining table?

Commensality

At its most elemental, commensality means sharing a table,[5] whereas others use it to mean a shared meal [16]. Anthropologist Tan Chee-Beng [17] provided a deeply considered interpretation:

> …[commensality] is the expression of the value of hospitality, of expressing care and love or valuing a relationship. This institution of hospitality has helped in human social evolution and organising and maintaining social relations beyond the domestic unit or a small human group. It continues to be useful for organising social relations in this even more globalised and cosmopolitan world. Commensality [...] is a way of inclusion in the human world that is differentiated by ethnicity, and nationality, religion and class.

As can be seen, the *haptic* HONGI installation literally has people sharing a table with Tania, and so it is a stage set for commensal encounter. *haptic* HONGI is the latest in a suite of XR experiences collectively entitled *common/room*. Each of these XR experiences is situated at a dining table, signifying the place where ideas are fomented, connections are deepened and where lively philosophical conversations might crystallise into strategic, collaborative action. The table is the ground that becomes a crucible, not in the sense of a melting pot, but an alchemical tool and platform, a representation of common ground, where we may initiate potential relationships between strangers as a first step towards commoning.[6]

11.5.3   Message: What's a Discursive Designer to Say?

The process leading to *haptic* HONGI has been experimental and iterative. However, the intention was always for technology to play a bridging role across cultures—for the technology to encourage broader social discourse. The desire to reach across

---

[5] From https://www.merriam-webster.com/dictionary/commensal Accessed 29th September 2021 The etymology of *commensal*... derives from the Latin prefix *com-*, meaning "with, together," jointly and the Latin adjective *mensalis,* meaning "of the table".

[6] From https://www.onthecommons.org/work/what-commoning-anyway accessed 2nd August 2022.
  The act of commoning draws on a network of relationships made under the expectation that we will each take care of one another and with a shared understanding that some things belong to all of us—which is the essence of the commons itself. The practice of commoning demonstrates a shift in thinking from the prevailing ethic of "you're on your own" to "we're in this together."

from the virtual/digital and into the real meant that a large part of the scripting of the message, was resting on the shoulders of the women who were captured, to later appear in AR at the table across from the visitors. They are the voices of the experience.

Tania Remana was raised by a Pākehā (a white person of European descent). She is an artist who is comfortable with reconstruction and imagining the new. As a performer, she played the role of someone charged with welcoming visitors. Although she had never seen an AR experience, she gleaned what was being asked of her and acted accordingly. Her performance and timing lifted the degree of presence in the work. However, Tania was also challenged by having to construct her own ritual because she is still learning aspects of Māoritanga (Māori cultural practices), including the language, te reo.

As this excerpt from an interview with Tania shows, the playfulness of the project and the company of members of the team from across the globe excited her:

> Tania Remana: I am a middle-aged woman who now identifies as Māori. It's taken me a while to take that culture on as my own, knowing that I had other blood strains in me, which I knew little of, as well as my Māori side.
>
> I was quite intrigued when you approached me about being part of it, be a part of the conversation around a table, 360 degrees. I was intrigued because it was different. So it was an opportunity to be out there and put my fingers in a few little puddles, pies, whirlpools, who knows, it's great. I loved it.
>
> Designer: You're a performer. But in this role you're kind of being put in the role of...
>
> Tania: Tangata whenua (people of the land—in a jokey way).
>
> You know, in the day... I hated pushing noses up against snotty old men and I just couldn't do it. So go around the back in the kitchen, there I was, in the kitchen. That's where it all happened... in the kitchen with all those kuikuis (old ladies).
>
> Designer: But in a way, you're safer than being at a real pōwhiri where you do have to greet strangers and hongi...
>
> Tania: I'm just actually reflecting now and that's the kaupapa (the principle idea) of it. I wouldn't hongi anyone back in the day, I wouldn't. I wouldn't be part of that tikanga of doing the hongi. And this is actually a kaupapa where it's haptic hongi. I've just clicked. So, it's perfect. No wonder I felt so at ease with it.

It was only during this discussion that Tania realised why she had been so relaxed—the technology had afforded her a large measure of safety. We had often discussed this aspect of AR in the Empathic Computing Laboratory (ECL), but, activist and academic, Tina Ngata, has advised to never make assumptions about the safety of Māori based on one's own non-Māori experience.

11.5.4 Scenario: How Does a Discursive Designer Set the Stage for Discourse?

"The designer can vary what we refer to as the clarity, reality, familiarity, veracity, and desirability of the scenario." [18]. In developing new iterations of the work, some aspects were non-negotiable. The dining table, the vase of flowers or foliage, and the cheery tablecloth. On every count of the above set of requirements, the dining table is effective. Visitors commented, unprompted, that they see mealtime discussion as fundamental to a good life. Māori have previously commented that important discussions happen at the tables in the wharekai (the dining room). Unlike formal discussion in the wharenui, the big ancestral house, women can join in and take charge of discussion around the dining table. This is exactly what Tania was referring to in the above conversation.

This domestic setup is a conscious counter to the sometimes unsettlingly abstract and confusing confections some designers and artists come up with in efforts to generate a futuristic or technologically savvy display. Giving priority to the kitchen table, the humble vase and the company of older women, is a conscious decision to insert a domestic construct, populated by mature women from different ethnicities, into the Metaverse. The Metaverse is defined by Dripke et al. [3] as:

> ... a perpetual and persistent multiuser environment merging physical reality with digital virtuality. It is based on the convergence of technologies that enable multisensory interactions with virtual environments, digital objects and people such as virtual reality (VR) and augmented reality (AR). Hence, the Metaverse is an interconnected web of social, networked immersive environments in persistent multiuser platforms. It enables seamless embodied user communication in real-time and dynamic interactions with digital artifacts.

In plain language, the Metaverse is a manmade virtual world mediated by technology. As such, it only contains entities and ideas created and invited in by people with sufficient resources. It therefore risks being subject to ideological, geographical, and wealth-based exclusivity.

11.5.5 Artifact: What's a Discursive Designer to Make?

To provide the conditions for a virtual meet and greet, we use digital information as an interface between real and virtual people. We are generating a table-top encounter between a real-life visitor to a gallery/library/exhibition and a volumetric avatar of women who, by virtue of not being 'really there' can be kept safe from 'audience' demands and reactions in real time/place. The basic configuration of the technical preparation for these table-based experiences includes depth sensing cameras and AR headsets to display the captured data.

The current version of the *haptic* HONGI is the latest in several versions of the experience. The first iteration of *haptic* HONGI included a Computer Generated (CGI) avatar, created by UK-based character creator and academic Wendy Lawn, and the Māori co-creator, Tania. Wendy followed her own process for virtual character creation. The creation of Tania's virtual avatar relied on software from Reallusion— Character Creator 3, and iClone 7. Photos were taken of Tania in NZ as reference, and the Headshot application used AI to transfer the photo of Tania's head to a 3D human template in Character Creator. From there, the appearance of the 3D human model was edited to become a virtual replica of Tania.

Discovering suitable clothing and hair assets took trial and error. Motions and lip sync animations were then added to the character in iClone. Once animated, virtual Tania was exported from iClone and imported into the Unity game engine.

Assembling the character in Unity required positioning the character, setting up shaders and then applying the animation. The Unity project was then packaged and sent from the UK to NZ for team member Prasanth to introduce virtual Tania to the volumetric capture project. The challenges were largely around understanding how Tania would look and how virtual Tania would transition from the volumetric capture as visualised through the HoloLens 2 AR HMD. Wendy's challenges were around the inferred design brief and realising it without being present in the key capturing and developing sessions, making an avatar, then dressing and bringing to life someone she had never met in person. Distances can bring a disconnect that only regular and deep communication can remedy.

Dreaming up the idea of temporarily replacing the virtual avatar created by volumetric capture with a computer generated avatar and thinking it would be 'interesting' to compare audience reactions to the two different styles of virtual avatar, was conceptually alluring. But there was no knowledge of the degree of difficulty the teammates would face. In the mock-up/documentation for the online exhibition while the visitors, the manuhiri, consider a response to Tania's spoken invitation, the volumetric video transforms into a computer-generated avatar (see Fig. 11.7) of the wāhine Māori (the Māori woman), who sings a waiata (a song), about human connection. In this version, the virtual avatar vanishes, and the volumetric Tania returns.

The defining feature of *haptic* HONGI was created in response to a concept identified and brought to the table by Tania herself. When asked what she thought should be included in a heartfelt, authentic, meaningful greeting, she identified the hongi.

Fig. 11.7 Tania's computer-generated avatar sings a song about social cohesion

During development, the *haptic* HONGI system was built and run on a variety of devices as shown in the table below (Table 11.1).

The first versions used the NED+Glass X2 Pro as an AR display and added an intel T265 camera for tracking. The requirement was for a lightweight device that had SLAM tracking and has a good field of view, and this setup met the criteria.

Due to the Covid-19 pandemic, the early experiences had to be tailored for a digital experience for a global audience. Hence, we streamed the experience as a video to Mozilla Hubs. A high-level system overview of this version is shown below (see Fig. 11.8).

Table 11.1 Overview of technical requirements for the different iterations of the system

Venue/Iteration	Year	Playback	Capture	Portability
Come to the Table! Demonstration Siggraph Asia	2019	Meta II	Intel Realsense	PC
Siggraph Asia	2019	HTC VIVE	Intel Realsense	PC
First Contact -take 2 Ars Electronica (online Mozilla Hubs)	2020	AceSight H2 + T265(tracking)	Depthkit (Intel Realsense)	PC
haptic HONGI Ars Electronica (Online – mockup)	2021	Hololens 2	Depthkit (Azure Kinect) + iClone7(for CGI)	PC, Standalone
Ars Electronica Physical Exhibition	2022	Hololens 2	Depthkit (Azure Kinect)	PC, Standalone

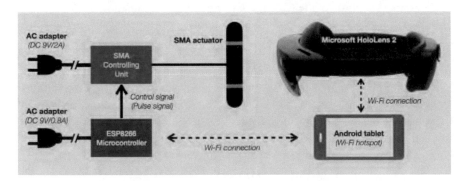

Fig. 11.8 High-level system over-view © Prasanth Sasikumar. [The video outlining our AR research to date can be accessed here: https://youtu.be/4Tu2l4vcLJI]

11.5.6 Audience: To Whom Does a Discursive Designer Speak?

In the very early stages of the development process, the designer engaged with the technical team and introduced them to Tania who would later appear, virtually, to address the visitors. The quality of this relationship between the team and the host determines the quality of the whole experience because, in the gallery/exhibition, the visitor, by putting on the headset, steps into the team's shoes to receive what the host/performer presented to them.

Tania has no control whatsoever over who visits her table. Anyone who enters the exhibition space is welcomed. This, in fact, mirrors the real world because the pōwhiri, the ritual of encounter led by Māori, is totally inclusive [14]. Any doubts or conflict with certain visitors can be addressed after the welcome in further ritualised discussions designed to hear people out and respond in a safe process.

How does one create a visceral, empathic bridge between subject and viewer? Work in the field of XR makes space for and necessitates different kinds of collaborators. Theatrical performances require actors to project and emote to reach every member of an audience. In contrast to this, because of audience proximity to cinema performers enabled by ubiquitous microphones along with telephoto lenses and close-up shots, acting for film must be more internal and subtle. Our kind of art installation that is more akin to documentary filmmaking sits somewhere in-between, by inviting real people to put themselves forward while asking a great deal of them. "Be yourself" is an inadequate kind of direction. Although, in the spirit of decolonisation and true collaboration, we need to afford our co-creators agency. It is a difficult balance between putting words into performers mouths, thereby leaving no room for self-determined actions, and leaving them to flounder in the unknown.

Tania, as an experienced improvisational theatre performer, was no stranger to self-direction (see Fig. 11.9). Once she realised that no strong direction would be forthcoming, she pushed offers of interaction with a 'director' aside, opting instead for the empty space in front of her that she could populate with an imagined audience. She seized the opportunity, wrote her own script, based on foreshortened traditional Māori rituals of encounter, and took control. Her performance was expressive. It reached across the table and through the technological veil to her audience.

11.5.7 Context: How Does a Discursive Designer Disseminate?

The team demonstrated the first iteration of the AR experience (titled *Come to the Table! Haere Mai ki te Tēpu!),* at SIGGRAPH Asia (Brisbane, 2019) and at Women in HealthTech at the University of Auckland (2020). In that same year, we were invited to exhibit the second AR iteration *First Contact—take 2* in Garden Aotearoa, a local spinoff of Ars Electronica, the global festival for art, technology and society.

Fig. 11.9 Tania is a performer

Covid put paid to the physical exhibition in 2020[7] and again in 2021.[8] It was only in mid 2022 that we were finally able to see and share with others our completed version of *haptic* HONGI in the real world.[9]

Because these experiences are designed to engage and bring to our tables people from every community, the Gallery, Library and Museum sector (GLAM) is their natural habitat. However, to reach everyday members of the public, it is important to include more community-focused centres and local venues that are frequented by those who might feel daunted by or out of place in our grander institutions.

11.5.8 Interaction: How Does a Discursive Designer Connect?

Although it was hoped that the technology would act as a mere veil between the virtual and the real people, it needs to be said that what AR affords us as designers, developers, artists and visitors is substantial. Without the technology, the designer/demonstrator would be sitting at a table alone, hoping that Tania could spare some time to join her and the visitors. As it was, at Ars Electronica, *haptic* HONGI could run for six hours a day for a whole week; day in and day out, interacting with visitors who could meet with Tania virtually, while Tania pleased herself by enjoying the exhibition (see Fig. 11.10), sightseeing or visiting friends.

She was not bound to the table, although she was able to watch people reacting to her "performance". She was amazed how happy she was making them.

---

[7] https://www.ars.nz/first-contact/.

[8] https://www.ars.nz/haptic-hongi/.

[9] https://www.ars.nz/haptic-hongi-2022/.

Fig. 11.10 Tania, wearing her korowai, her feather cloak, at the Ars Electronica exhibition opening. Song Bai Photographer for Te Herenga Waka—Victoria University of Wellington

The project itself drew technologists to it (see Fig. 11.11). The process of creating the *haptic* HONGI experience gave us all a real-world opportunity to practise intercultural engagement. No one needed convincing about the potential value of a work that sought to grapple with intercultural discomfort. We were all in agreement about that. Each time Mairi booted the Hololens 2, she smiled to herself while selecting the application created by the team that had been entitled "Counter-racism". The technology was therefore a catalyst for collaboration and creative production around a shared concern.

The HoloLens 2 headset sitting ready for action on the table communicated to exhibition visitors that this was some kind of virtual experience. It aroused curiosity and excitement. The headset was an attraction for some and a hinderance for others who felt rather embarrassed about it being their first time. This is where the tablecloth and the vase of foliage feature. The wires and technology are offset by the homely familiarity of a setup that speaks of domesticity, comfort, and conviviality. Even so, some visitors had to be invited and enticed to the table. The designer is there to help put the HoloLens on and to engage the visitors in conversation after they have seen and heard Tania welcome them to Aotearoa. Tania communicates virtually; the designer connects directly with the visitors.

Fig. 11.11 Team members Prasanth, Mairi and Ryo in the lab after preparation for Ars Electronica was complete

11.5.9 Impact: What Effect Can a Discursive Designer Have?

The value of discursive design interventions can be measured by "the impact designers can have with their work upon society, other institutions, the design profession, and the designers themselves" [18]. What designers call 'impact' also encompasses "User feedback from the experience" in Sect. 11.4, above.

The effects of our AR experience exceeded our expectations due to the reliability of the technology, the high resolution of both the visual and audio components, the curation of the setting, the appropriateness of Tania's performance and the readiness of the 'audience' for such an encounter.

Ours is not a vehicle for mass audiences. It is a one-on-one experience; a slow burner—unspectacular, subtle and yet emotional for some. There are no loud sounds or flashing lights. Some visitors were concerned that we were situated in an exhibition backwater. In fact, at Ars Electronica, it was in a perfect position—a quiet spot with black walls and down lights. The original intention of the wider practice-led research was to determine whether XR affords greater support for deep interhuman connection, between subject and audience in a filmmaking/moving image framework. As a moving image work, *haptic* HONGI can only ever be seen/felt/experienced by a small number of people. This aspect of AR production in addition to the lack of an explicit, wider narrative was something new to contemplate.

Considering that we had no certainty that *haptic* HONGI would even function, the overall visitor response was very positive. Whether the experience was simply fun or deeply moving and engaging, visitors appeared to enjoy the mix of technology and real people. Above all, because of the form that the experience takes, as a conversation between two people at a dining table, many visitors were primed for and

Fig. 11.12 A smiling visitor

open to a lively chat about technology and about connection between people. These conversations were able to be accommodated at the table due to the slow trickle of exhibition attendees.

A major impact of *haptic* HONGI is within the research community itself. Aside from the technical and creative discoveries, we have also established a level of trust and pleasure in each other's company that is characteristic of lifelong friendships. These friendships span the globe. Certainly, Tania's experience was powerful. Advice from a Māori elder, to always bring Māori co-creators with you, has been hugely impactful. By virtue of Tania attending the exhibition, she was able to watch people enjoy her welcome (see Fig. 11.12) from a distance. This was uplifting and interesting for all concerned.

11.6   Lessons Learned

Several design decisions were made in creating *haptic* HONGI. We had been advised that including text/titles/credits can work against the magic of XR experiences. A choice was made not to use a computer graphic virtual avatar. The addition of a CGI avatar would have introduced an additional virtual character that would likely have raised further complex ideas and relationships that could have disrupted the integrity of the experience—the "low threshold", as a visitor called it, "diminishing the feeling of presence". The decision to focus on bringing the hongi to life, by inserting a haptic actuator into the headset and using a volumetric video avatar, opened new avenues for experimentation and discussion. The developers should be recognised for their contribution to local and global knowledge, especially as it pertains to supporting cultural understanding.

This was a complex coming together of creatives, researchers, technicians, academics and participants, in the making of a virtual-physical bridge to traverse social isolation, time zones, knowledge gaps, life experiences, expectations and multiple realities. What some might call "the cutting edge", or "the bleeding edge", a more appropriate phrase could be the "precipice of ignorance", in that we were always only a breath away from the unknown. The coding and thinking of the computer scientists and engineers was almost totally opaque to others. Working with new and developing technology meant that all co-creators were working with the unknown. This is exciting but risky. The individual and combined contributions could all fail at any moment. Such risk-taking is what it takes to generate new knowledge.

Forming creative relationships with people from different cultural backgrounds can also feel risky for some. Our technical team was already drawn from different cultural backgrounds, but, although 'diverse', we were all non-Māori. Linda Tuhiwai-Smith [15] pointed out that Māori are among the most researched people on earth. We are urged to question what Māori communities might gain from taking part in our research. In this instance, our work is more about forming bonds and working together, than putting Māori under the microscope. We also sustain connections to wider communities so we had no need to treat one person as the fount of all knowledge, to be interrogated at will, or to be held up as a representative of all Māori. We have learned that knowledge is not there for the taking.

Our advice would be to listen, then reflect and relate. Resist the temptation to seize upon assumptions based on stereotypes or fear. Keeping an open mind can be a conscious exercise. This is something we practiced and enabled through *haptic* HONGI, because visitors could never dominate, silence or abuse Tania. Their role was to look and listen.

Relationship and reciprocity are paramount. In our colonised society, it has become starkly evident that these two practices have not been prioritised. Rather than being depleting, both, with time, can bring learning and creative, authentic friendship. This is a core concept in our work.

Advice from Māori elders formed the basis of our approach. Involve Māori from the very start to avoid inviting participants in at the last minute. This prevents tokenism, a kind of objectification through which the selection of collaborators rests on a shallow appreciation of the attributes of the person in question. The best way to avoid such an unfortunate situation is to work with and through existing relationships. In this way, we are introduced to each other by people who already know both parties. This supports ongoing relationships and creates broader networks by bringing groups of people together. The success of a project can be shared by all, instead of some standing in the sunshine while others are jettisoned or marginalised.

Invitation from Māori to work with them as individuals or as a community is ideal, but even then, ongoing clarification and negotiation between disparate world-views may still be necessary. The inherent challenges of navigating such deep, honest conversations are off-putting for some. But if, during these discussions, that go beyond mere politeness, shared values are unearthed, the rewards of mutual understanding and friendship are immeasurable. This is a current state of affairs of which our team can be proud. Opportunities for in-depth dialogue and evaluation are advantageous on many levels, including personal, intrinsic discoveries in addition to learning about others and the outside world from your point of view, and theirs. Pākehā academic, Alison Jones, [8] writes about Pākehā-Māori relationships and collaborations.

> To those Pākehā researchers who would collapse the Māori-Pākehā hyphen into 'us' there is one harshly pragmatic response: it does not work. A research approach to Māori, whether as research collaborators or as subjects, that assumes a mutual interest, minimal difference, and the set of shared assumptions, is doomed to practical failure. [...] ... No mere exercise of a 'duty of care' or 'cultural sensitivity' or attempts at 'sharing' by Pākehā will in themselves create a Māori-Pākehā collaboration where differences are largely erased and shared work becomes easy; this, in my experience, can only be seen as a fantasy.

This approach to difference is to live and let live. It is important to be embedded in reality, to keep in mind the inordinate imbalance regarding access to equipment, funding and status when entering transdisciplinary collaborations with women, older people, people of colour, Indigenous other marginalised communities. An exploration of these complexities are beyond the scope of this chapter, but the certitude of this disparity should never be far from our minds. Covering costs and supplying food need to be routine, while taking participants to events and celebrations is a way to show respect.

In contemplating the reaction of people to the work, it might be helpful to refer to the thoughts of author and farmer Wendell Berry that appeared in *Feminism, the Body and the Machine* from his slim volume *Why I Am Not Going to Buy a Computer* [2]:

> The body characterizes everything it touches. What it makes it traces over with the marks of its pulses and breathings, its excitements, hesitations, flaws, and mistakes. On its good work, it leaves the marks of skill, care and love persisting through hesitations, flaws, and mistakes. And to those of us who love and honour the life of the body in this world, these marks are precious things, necessities of life.
>
> I know that there are some people, perhaps many to whom you cannot appeal on behalf of the body. To them disembodiment is a goal and they long for the realm of pure mind or pure machine; the difference is negligible.

However, at the experimental stage of this work, almost all of the above, the flaws, mistakes, hesitations, excitements and so on are fully present. We are quite distant from the pure machine of which Berry spoke. It makes sense, out of interest's sake, to navigate between extreme or pure positions. At the very least, we might consider the assertion of computer scientist and philosopher Jaron Lanier who does not share the zeolotry of those who anticipate the singularity, when machines will to be able to reproduce themselves and take over the world [11].

> When my friends and I built the first virtual reality machines, the whole point was to make this world more creative, expressive, empathic, and interesting. It was not to escape it.
>
> Will trendy cloud-based economics, science, or cultural processes outpace old-fashioned approaches that demand human understanding? No, because it is only encounters with human understanding that allow the contents of the cloud to exist.

Lanier is a polymath, a musician and a scientist, whose own life exemplifies transdisciplinarity. He sees improvements in VR as a way to hone our perception in the real world. The interesting part of a VR experience is when we remove the headset [10].

> There will always be circumstances in which an illusion rendered by a layer of media technology, no matter how refined, will be revealed to be a little clumsy in comparison to unmediated reality. The forgery will be a little courser and slower; a trace less graceful. [...] When confronted with high quality VR, we become more discriminating. VR trains us to perceive better, until that latest fancy VR setup doesn't seem so high quality anymore. [...] Through VR, we learn to sense what makes physical reality real. We learn to perform new probing experiments with our bodies and our thoughts, moment to moment, mostly unconsciously. Encountering top quality VR refines our abilities to discern and enjoy physicality.

Therefore, the question is not whether an AR experience is close to and therefore might replace reality, but how AR and reality together might enhance our social lives.

11.7 Conclusion and Directions for Future Work

In this chapter we describe *haptic* HONGI, an interactive piece designed to explore intercultural relationships in Aotearoa New Zealand. Within a transdisciplinary discursive design framework we have generated discourse *about* our design, outlining discourse *for* our design and, most importantly perhaps, discourse *through* our design that uses a conversational form to provoke discussion about overcoming intercultural discomfort. A more particular focus with this work is the potential for a playful interaction with Indigenous Māori in Aotearoa New Zealand. Since successful commoning practices rest on productive conversations, the outcome was gratifying.

Although the piece was successful, there are a number of improvements that could be made in the future. The development of the ability to automatically lock the gaze of the viewer to that of the avatar would be a major improvement. This will enable visitors to have the feeling that they are easily able to make direct eye-contract with the virtual avatar, and so reliably increase connection.

There have been suggestions about using machine learning to facilitate life-like conversations between the visitors and the volumetric video avatar of the host. This might be useful if ever such an experience is introduced into airports, museums or other public venues for short cultural exchanges. However, further layers of artificiality might run counter to a desire for an authentic experience of connection.

There is a an issue with the current haptic element, in its prototypical form, in that it does not offer a uniformly successful follow-through. This means that visitor experiences are somewhat haphazard. So there is an opportunity to explore future designs for the haptic elements. This is challenging though, and other researchers in the field drew our attention to failed attempts by other XR researchers to develop a haptic hongi experience.

The placement of the HoloLens 2 headset could also be improved. Currently it is quite variable as it can be tilted forward or back, or rotated left or right. The angle of the transparent visor and the position of the actuator can vary on different shaped and sized heads (see Fig. 11.13). Since the latter is beyond the control of the developers, this will always create a range of audience experiences. This could be addressed by exploring different AR display options. The anticipated release of the Magic Leap 2 display, for example, could be a step in the right direction because it would make exhibition of *haptic* HONGI more flexible by masking backgrounds, therefore avoiding Tania appearing as a semi-transparent ghost. However, the high cost (USD $3,299) and the restricted potential to embed an actuator might be limiting factors.

Considering future research, we wonder when AR will become more user-friendly. What will newly developed headsets offer the experience? When will we be able to edit the volumetric data? When will it be more straight forward to synchronize the picture with the audio? What about the number of people excluded from the

Fig. 11.13 Variable headset placement

experience and the expensive and rare equipment and expertise required to develop such experiences as *haptic* HONGI. How can access to this technology become more democratic and widely available?

Taking into account responses from the visitors to *haptic* HONGI at Ars Electronica Garden Aotearoa, there seems to be a future for this technology in intercultural settings. This could include training diplomats seeking information about people they might encounter when they are posted to foreign countries, introducing visitors to a new people in place, and providing a safe interface between cultural opposites who need to find unity. Therefore, it has universal appeal because disparity between ages, genders, and worldviews has never been so great.

Maybe in this sense, the COVID-19 pandemic, while depriving us of human contact, helped us to define what sociability, conviviality and commensalism affords us, and how intercultural relationships can immeasurably enrich our lives.

References

1. Barlow, C.: Tikanga whakaaro—key concepts in Māori culture. Oxford University Press, Auckland, New Zealand (1991)
2. Berry, W.: Why I am not going to buy a computer. Penguin, UK (2018)
3. Dripke, A., Ruberg, M., Schmuck, D.: Metaverse **2**, 486–497 (2022). https://doi.org/10.3390/encyclopedia2010031
4. Driver, A., Peralta, C., Moultrie, J.: An exploratory study of scientists' perceptions of design and designers (2010). https://dl.designresearchsociety.org/cgi/viewcontent.cgi?article=1813&context=drs-conference-papers
5. Gunn, M., Bai, H., Sasikumar, P.: Come to the Table! Haere mai ki te tēpu. Presented at the November 17 (2019)
6. Gunn, M., Billinghurst, M., Bai, H., Sasikumar, P.: First Contact—Take 2: Using XR technology as a bridge between Māori, Pākehā and people from other cultures in Aotearoa, New Zealand **11**, 67–90 (2021). https://doi.org/10.1386/vcr_00043_1
7. Gunn, M.J., Sasikumar, P., Bai, H.: First Contact—Take 2 Using XR to Overcome Intercultural Discomfort (racism) (2020). https://search.datacite.org/works/https:/doi.org/10.2312/egve.20201281
8. Hoskins, T.K., Jones, A.: Critical conversations in kaupapa Māori research. Huia Publishers, New Zealand (2017)
9. Krippendorff, K.: The semantic turn. Taylor & Francis, Boca Raton, Fla (2005)
10. Lanier, J.: Dawn of the new everything: a journey through virtual reality. Vintage Digital (2017)
11. Lanier, J.: You are not a gadget. Vintage Books, New York (2011)
12. Muthukumarana, S.: Clothtiles: A prototyping platform to fabricate customized actuators on clothing using 3d printing and shape-memory alloys. In Proceedings of the 2021 CHI Conference on Human Factors in Computing Systems. Presented at the CHI Conference on Human Factors in Computing Systems May (2021)
13. OECD: Addressing societal challenges using transdisciplinary research. OECD Publishing, Paris (2020)
14. Salmond, A.: Hui: A study of Maori ceremonial gatherings. Reed Methuen, Auckland (1985)
15. Smith, L.T.: Decolonizing methodologies. Zed Books, London (1999)
16. Spence, C., Mancini, M., Huisman, G.: Digital Comm. Eating Drink. Comp. Technol. **10**, 2252 (2019). https://doi.org/10.3389/fpsyg.2019.02252

17. Tan, C.-B.: Commensality and the organization of social relations. In: Kerner, S., Chou C., Warmind, M. (eds.) Commensality: from everyday food to feast, 1st edn, Bloomsbury Publishing Plc, London (2015)
18. Tharp, B.M., Tharp, S.M.: Discursive design. The MIT Press, Cambridge, MA (2018)
19. Whiteley, N.: Design for society. Reaktion, London (1997)
20. Willis, A.-M.: The design philosophy reader. Bloomsbury Visual Arts, London, New York , oxford, New Delhi, Sydney (2019)

Chapter 12
Connecting Past and Present Through a Multisensory Toolkit—A Non-pharmacological Intervention for People Living with Dementia

Esther Olorunda and Rachel McCrindle

Abstract Agitation and passivity are two of the behavioural and psychological symptoms exhibited by many people with dementia especially in the later stages of the disease. These behaviours are often attributed to a lack of stimulation from their physical and social environments and can lead to a reduced quality of life. The use of non-pharmacological (NP) interventions such as art therapy, music therapy, aromatherapy and reminiscence therapy have proved successful in the mitigation of these symptoms. This chapter describes the development of AMuSED (**A**ctive **Mu**lti-**S**ensory **E**nvironment for people living with **D**ementia) as an innovative non-pharmacological intervention that builds upon NP best practice to provide engagement, stimulation, and social interaction to people living with dementia. AMuSED combines, in a tabletop toolkit, the use of cognitive, visual, aural, tactile and gustatory stimulation with a set of themed elements, activities, and reminiscence tools and in doing so crosses the boundaries between art and technology. It can be used in home, community, care home and hospital environments by individuals or groups to help people connect with their pasts and discuss their life lived experiences with their friends and family or those caring for them. AMuSED is affordable, portable, and adaptable to fit differing stages of dementia and can be personalised to an individual's interests or themed to encourage discussion and engagement in social groups.

Keywords Non-pharmacological intervention · Activity co-ordinator · Dementia therapy · Dementia activities · Multi-sensory engagement · Reminiscence therapy

E. Olorunda (✉) · R. McCrindle
School of Biological Sciences, University of Reading, Readings, UK
e-mail: olorundaesther@yahoo.com

R. McCrindle
e-mail: r.j.mccrindle@reading.ac.uk

© The Author(s), under exclusive license to Springer Nature Switzerland AG 2023
A. L. Brooks (ed.), *Creating Digitally*, Intelligent Systems Reference Library 241,
https://doi.org/10.1007/978-3-031-31360-8_12

12.1   Introduction

Dementia is an overall term used to represent a range of progressive disorders that cause damage to the brain. It is a health condition that can affect the memory, thinking capability, mood, behaviour, orientation, and lifestyle of a person living with it, with symptoms manifesting differently in each person affected. Dementia predominantly affects older people, therefore as the world's population ages, the number of people living with the condition continues to increase, especially in low- and middle-income countries [67]. For this reason, dementia care has become a key societal challenge across the globe with the number of people living with dementia projected to increase from 58 million people in 2020 to 82 million by 2030 and 152 million people by 2050 [2].

The progressive nature of dementia means that symptoms tend to worsen over time and although there is no cure for the illness, there are pharmacological and non-pharmacological (NP) interventions and methods available that aim to maintain quality of life, increase life expectancy, and prolong independence of those living with dementia for as long as possible post diagnosis. The use of non-pharmacological interventions for dementia care, presents opportunities for creative methods and interventions that combine art and technology in different forms to engage and stimulate people with dementia and encourage them to interact with other people. This chapter examines one such creative outcome that brings together reminiscence therapy and multisensory stimulation to form the AMuSED (Active Multi-Sensory Environment for people living with Dementia) toolkit as an engaging and exciting fusion of art, sound, sensory elements, interactive activities, digital technology and reminiscence prompts for use by activity coordinators and caregivers to provide person centred care to people living with dementia.

12.2   Impact of Dementia

Dementia not only causes cognitive decline in people living with the disease, but also reduces their capacity to perform daily tasks. Finkel et al. [34] report that many older people living with dementia also develop behavioural issues such as agitation, aggression, delusion and wandering, and according to Colling [37], a further manifestation of dementia especially in its advanced stages is the exhibition of passive behaviours such as a decline in motor skills, apathy, and a lack of interaction with the environment. Passivity has been shown to be even more of a problem than aggression [80], causing disruption to activities of daily living (Vernooij-Dassen 2007) as well as decline in physical and social functions (Harwood et al. 2000). Although agitation and passivity seem like opposites, their causes may both be attributed to a lack of stimulation from the physical and social environment (Kolanowski et al. 2005).

Non-pharmacological interventions are safe, non-invasive ways of treating some of the behavioural and psychological symptoms of dementia [72], and are aimed at maintaining the function and participation of people living with dementia for as long as possible as the disease progresses, thereby reducing disability, improving both patient and caregiver quality of life [30, 95], and maintaining independence [92]. Non-pharmacological interventions can be delivered in both informal and formal settings [93]. Informal interventions in the form of support groups or dementia cafés [7] can also incorporate more formal approaches such as psychological and reminiscence therapies [31, 50]. Non-pharmacological avenues for engagement include interventions and pleasurable activities such as art [17], plants and green spaces [5], and therapies such as music, food, exercise programmes, aromatherapy, light therapy, cognitive stimulation, multisensory stimulation, and reminiscence therapy [13, 31]. These therapies and activities are usually administered by professional activity coordinators and/or care workers in community or care home settings, but also by friends and family who take on the care role informally [77] as many people continue to live in their homes post diagnosis [6]. A focus on non-pharmacological activities not only benefits those living with dementia but can also help strengthen the bond between the person with dementia and their caregivers and reduce stress for both parties [18]).

Reminiscence in particular is not only useful in the mitigation of the symptoms of dementia but is also a highly effective way to connect and blur the boundaries between the past and present especially when using multimedia cues such as photos and videos. In doing so it can also promote inter-generational connections by allowing people living with dementia to share their stories and memories with families, loved ones and caregivers. This connection of the past and the present can be further enhanced by harnessing the crossover between art and technology and the creative fusion of bringing traditional and digital elements together in a single intervention.

12.3 Reminiscence Therapy

Developed by Butler [16] as a form of treatment for dementia using the process of "Life Review" reminiscence therapy is one of the most widely used non-pharmacological interventions for people living with dementia [90]. O'Philbin et al. [68] and Woods et al. [89] describe it as the discussion of past events, activities, and experiences, usually with the aid of memory triggers or prompts from the past such as music, photographs, household items, or other familiar objects.

Reminiscence therapy serves as a means of bringing memories, especially long-term ones, into the present and may be organised individually or in groups. Ching-Teng et al. [21] suggests that topics which tend to provide positive influence are considered for use in individual reminiscence therapy, and one-to-one therapy sessions are conducted to help people recall their long-term memories. It is important for the social worker or therapy conductor to understand the psychological conditions, social conditions, cultural backgrounds, and life experiences of older adults before providing reminiscence therapy, because this form of therapy requires the use of skills such as listening, communication, empathy, timely reaffirmation, and positive feedback [20].

Group based reminiscence therapy is also widely used as it allows participants to stimulate one another through communication, and increases their attention span [69]. This form of reminiscence therapy normally involves topics that are seen as non-threatening and has been found to have positive effects on older people with depressed moods [70] including those living in long term care environments. This style of reminiscence therapy can take many forms with different components, the use of themes, and family members sometimes being incorporated into sessions [68].

Conductors of group-based reminiscence therapy sessions [54] focus on maintaining an atmosphere of harmony within the group throughout the sessions to prevent unnecessary loss of energy by the attendees [71]. Li et al. [55] describe how conductors observe the interactions among participants to ensure that the goal of group reminiscence is achieved and the way they guide the reminiscence process through using open questions, supplementary items, questioning and timely in-depth clarification [82] to assist the group in expanding and deepening their recollection. Jones [48] also explains how sometimes, focus is placed on each phase of life experience to steer the participants into a comprehensive recollection.

Reminiscence therapy can be very beneficial for people living with dementia as it places an emphasis on long-term memories, which are usually remembered more than short-term or recent memories [61]. According to Hilgeman et al. [43], the psychological improvement induced by this form of therapy can be beneficial to both people living with dementia and their caregivers. It is also a strengths-based approach that can provide person-centred care because it draws on people's long-term memories which are often preserved in dementia [91]. However, Lazar et al. [52] note that reminiscence therapy might not always be cost-effective because the person leading the sessions must be trained and the items used as memory triggers must be sourced and prepared.

12.4 Multisensory Stimulation

Multisensory stimulation involves the stimulation of different senses following a non-directive and facilitative approach [57]. It is the combination of lights, sounds, smells, tactile objects and sometimes taste to stimulate the different senses and provide stimulation, relaxation and reduced stress [78]. This form of stimulation can be carried out using different elements such as aromas and scents, music and sounds, fibreoptic sprays and lights, bubble tubes, photos, tactile elements, and occasionally foods [66] or by incorporating sensory stimulation into daily care routines [83]. The advantage of multisensory stimulation is that different elements can be variously combined to give enhanced experiences, for example by combining traditional elements such as photos with digitally recorded descriptions or messages in recordable photo albums, or by combining musical recordings and sounds with different aromas and fragrances.

According to Zaree [94], a distinguishing element of multisensory stimulation is the adoption of a non-directive approach that encourages the participants to explore and engage with the sensory stimuli of their choice. This makes this form of stimulation compatible with person-centred care and places emphasis on the acknowledgement of the personhood of the person living with dementia [63], the personalisation of their care and the shared decision-making process which prioritises the relationship between the person living with dementia and their caregiver [32]. Finnema et al. [35] explain that multisensory stimulation is suitable for people with severe dementia, especially those with limited verbal communication because it does not rely on cognitive abilities and provides stimulation without the need for intellectual activity while also providing an atmosphere of trust and relaxation [15].

The advantage of multisensory stimulation over other forms of therapies and stimulation is that people can engage with elements that stimulate their different senses: visual, audio, olfactory, tactile, gustatory, as well as cognitive functions. The stimulation of each of these senses offers unique benefits to people living with dementia and the combination of these different forms of stimulation can produce positive effects on the mood and wellbeing of a person with dementia. Visual stimulation can be delivered through media such as photos and videos. Incorporating photos into reminiscence therapy provides a strong link between past and present events, as looking at external pictures can create internal pictures in the mind of a person and bring memories to life through the reminiscence process [49]. The use of videos in multisensory stimulation also provides audio stimulation as well as visual, the combination of which has been reported to promote expression and communication in people living with dementia [33]. Other avenues for delivering audio stimulation include the use of sounds and music. Olfactory stimulation can be achieved using different smells, essential oils and fragrances which can be delivered through traditional means such as aroma capsules and boxes or through digital technology like diffusers. Tactile stimulation can be achieved using twiddle muffs which are knitted or crocheted bands or hand warmers with tactile items attached to them, or through items made from differently textured materials such as plush, rough, smooth, crinkly etc., while gustatory stimulation can be provided using foods such as popcorn, biscuits, candy floss etc.

A unique approach to multisensory stimulation is the development of thematic multisensory stimulation [84]. This form of multisensory stimulation involves simultaneously stimulating multiple senses using stimuli related to positive themes that could awaken positive feelings and memories in people living with dementia. This approach is based on the premise that positive memories are preserved better than neutral memories and are therefore recalled more easily. Thematic multisensory stimulation is observed in a study by Goto et al. [36] which compared the effects of a Japanese garden room to the effects of stimuli from a generic multisensory room. The results of this study showed better reduced stress levels and positive behavioural changes in the participants who visited the Japanese garden room than in the participants who visited the traditional multisensory room. This study concluded that thematic multisensory stimulation might lead to improvements in psychological and behavioural outcomes [37]. It was also further suggested in Goto et al. [38] that exposure to nature should be included in dementia care, after it was shown to not only relieve psychological stress, but also improve verbalisation and memory retrieval. However, Lancioni et al. [51] highlight a drawback of multisensory rooms stating that although the equipment required in their creation may be easily acquired, they are generally expensive and therefore might not be the most cost-effective form of stimulation for people living with dementia. They also lack portability and usually require the dedication of a separate room. Therefore, it is imperative that multisensory stimulation be delivered to people living with dementia through more cost-effective and portable avenues.

Therapies and activities such as art, music and aromatherapy have been shown in literature to have numerous benefits for people living with dementia and are therefore widely used in practice to deliver these benefits. The incorporation of these different therapies and activities into multisensory interventions will not only multiply these benefits but also provide avenues for different forms of stimulation to be delivered within an activity session. When combined with reminiscence therapy, multisensory stimulation can deliver much stronger memory cues due to the combination of sensory elements used. For example, the combination of both visual and olfactory stimulation during a reminiscence session will result in much stronger reminiscence cues than if only one form of stimulation is used. Therefore, in connecting the past to the present through reminiscence, it is important that non-pharmacological interventions incorporate multisensory stimulation to deliver better reminiscence sessions and improve the benefits of reminiscence to people living with dementia.

12.5   The AMuSED Toolkit

The AMuSED toolkit is an **A**ctive **Mu**lti-**S**ensory **E**nvironment for people living with **D**ementia. It is an exciting and innovative non-pharmacological intervention that provides engagement, stimulation and social interaction to people living with dementia using traditional and digital media elements that aim to mitigate the behavioural and psychological symptoms of dementia. It is an affordable, adaptable, and portable toolkit that can be used by caregivers and activity coordinators to provide engaging and stimulating multi-sensory and multi-activity sessions to people living with dementia in a range of care provisions.

Developed in conjunction with experts from the Reading Dementia Friendly Steering Group using a design thinking approach [12] that incorporates user feedback and prototyping [28], AMuSED incorporates the use of known successful therapies and activities for people living with dementia such as art therapy, music therapy, aromatherapy, and reminiscence therapy into a themed intervention. It especially focuses on reminiscence therapy as a way of connecting the past to the present and can be used in both single user environments such as personal homes, and multi-user environments such as hospitals, care facilities and community settings like chatty cafés where people gather informally to engage with one another and discuss different topics while enjoying tea or coffee (see Sect. 12.5.1.2). People living with dementia benefit from reminiscence therapy as the positive memories evoked help to improve their mood and build relationships with their caregivers and family members. The AMuSED toolkit helps to bring these positive memories to life using a themed mix of photos, physical and digital elements, and engaging activities.

Visually, the toolkit comprises a cube formed from four side panels with reminiscence images printed on both sides slotted into a bottom panel bordered by colourful Lego bricks that not only provide visual stimulation but act as a border and a means of keeping the four side panels upright and intact (Fig. 12.1). The resulting 36 cm by 36 cm cube is completed by a brightly coloured top panel also bordered by Lego bricks with "AMuSED Toolkit", the corresponding theme and the dementia forget-me-not flower branded on it. The AMuSED toolkit includes different forms of technology to promote reminiscence and communication using various digital elements such as sound buttons which are combined with traditional elements such as reminiscence photos and activities following from feedback provided by the dementia experts that people living with dementia perceive traditional elements as familiar and relatable. This fusion of themed digital and traditional elements provides an engaging and stimulating toolkit that promotes reminiscence and social interaction among people living with dementia, as well as connections between the past and the present by using reminiscence images and elements aimed at provoking positive memories and encouraging trips down memory lane.

Aside from the themed reminiscence images on the box panels each AMuSED toolkit contains in excess of twenty elements and activities that can be used by activity coordinators and caregivers in one or more sessions to deliver strong memory cues aimed at reminding people with dementia of their long-term memories, positive experiences and adventures such as trips to the seaside, countryside, fairgrounds, and festive days such as Christmas. The AMuSED toolkit promotes the discussion of these different memories and adventures by creatively incorporating the use of digital technology such as sound buttons, recordable photo albums, QR codes, videos,

Fig. 12.1 Example of an AMuSED toolkit box

etc., with more traditional elements such as colouring, word searches, games and music, and other sensory objects such as aromas, food, lights, and tactile objects. All elements are brought together in a well thought out themed manner using carefully planned activities displayed in the activity booklet (see Fig. 12.11), which when used alongside meaningful question prompts (see Fig. 12.12), helps to deliver engaging and interesting reminiscence sessions to people living with dementia. This combination of reminiscence images, engaging activities, and visual, aural, olfactory, tactile, and gustatory stimulation is discussed in the remainder of the chapter with examples taken from the development and evaluation of the Seaside, Countryside, Christmas, and Entertainment themed AMuSED toolkits.

12.5.1   Themes

The AMuSED toolkits are realised as different themed editions as the inclusion of themes in interventions has been proven to stimulate the mind and make activities more enjoyable [79]. Themes can be added to fit a group of intended users to make activity sessions more interesting and engaging and can be further customised if created for an individual. They are especially useful in reminiscence sessions because they help to focus the sessions or discussions around a particular topic or subject matter. The AMuSED toolkit currently has four main themes namely, Seaside, Countryside (MERL), Christmas, and Entertainment. The elements incorporated into each theme are combined to form a multisensory environment as each theme includes elements that stimulate the visual, audio, olfactory, gustatory, and tactile senses.

12.5.1.1   Seaside Edition

The AMuSED Seaside toolkit (Fig. 12.2) is aimed at bringing real life seaside experiences to people living with dementia. This toolkit displays bright and colourful reminiscence photos of the seaside such as people building sandcastles, people playing with beach balls, rock pools, arcades at the pier, Butlin's family holiday, animals commonly found at the beach (seagulls, starfish, jellyfish etc.), seashells, message in a bottle, cruise ships and foods consumed at the seaside such as fish and chips. These photos serve as memory triggers and help to induce positive memories of days at the seaside.

The toolkit contains a combination of digital and traditional elements such as jellyfish lamps, sound buttons, kinetic sand, postcards, message in a bottle, windmills, seashells, and aromas such as sea breeze, suntan lotion, out at sea, seashore, smoked fish, and rock pools. These are carefully combined in the activity booklet to enable caregivers and activity coordinators to deliver multisensory and multiactivity sessions with the aid of the prompt questions. Session attendees are encouraged, through the prompt questions to reminisce about their past experiences and exciting adventures at the seaside and for one conversation to lead to another.

Fig. 12.2 The AMuSED seaside edition toolkit showing lid of box; two outside panels; inside panels and box contents

Due to the diverse nature of people and their different lived experiences, the images used in the AMuSED toolkit cover a range of topics. These images not only cover the beach and animals at sea, but also different foods, activities, and locations. The aim of providing this form of diversity is to try and ensure that each person using the toolkit can relate to at least one of the images even if from different perspectives. In doing so it encourages people to reminisce about and share their stories relating to the different images on the toolkit. Although appearing random, the images on the toolkit work together to provide visual aids to reminiscence therapy. Especially when combined with the prompt sheet, they help to portray a whole day at the seaside for people living with dementia, starting from what seaside location they remember visiting, how they got to the seaside, who they went with, what activities they partook in, what animals they saw and what they ate whilst there. This makes for smoother conversations and better reminiscence sessions.

The reminiscence images on the AMuSED Seaside toolkit can be used to start discussions on, for example, animals at the seaside, which people can list individually or in groups, the sea animals they are familiar with and narrate their experiences at the seaside with these animals. Additional stimulation can be included into this discussion by playing the sounds of the sea, seagulls and other appropriate sounds using the recordable sound buttons and including the different aroma capsules such as sea breeze, seashore and rockpools. This reminiscence session can be further strengthened by including tactile sea animals and not only discussing these but also letting people feel their different materials. The fusion of the various parts of the toolkit allows for different cohesive and themed activities to be developed from its multiple sensory elements and activities including making sandcastles, aqua painting (brushing water on special paper that reveals images when wet), and for the more energetic playing beach quoits or beach ball games.

12.5.1.2 Countryside (MERL) Edition

The Countryside AMuSED toolkit (Fig. 12.3) was developed in conjunction with The Museum of English Rural Life (MERL). The MERL is part of the University of Reading and boasts a diverse collection that explores how the skills of farmers and craftspeople, past and present, can shape our future lives. The museum also caters to people living with dementia and partners with dementia organisations on projects that support people living with dementia. One such avenue of support is the chatty café they organise on the second Tuesday of every month with the aim of getting people living with dementia engaged and communicating with each other, with their caregivers, and with The MERL staff while also enjoying the tea and cakes provided by The MERL.

The AMuSED Countryside edition incorporates reminiscence images from the MERL collections dating from 1900 to 1960 which cover topics such as: farm harvests, farm animals (sheep, birds, cows and pigs), transport vehicles, childhood, cooking utensils, ration book, MERL artifacts such as Morris dancing bells, squash racket, ping pong paddle, picnic hamper, cricket balls, old fashioned quoit game (see

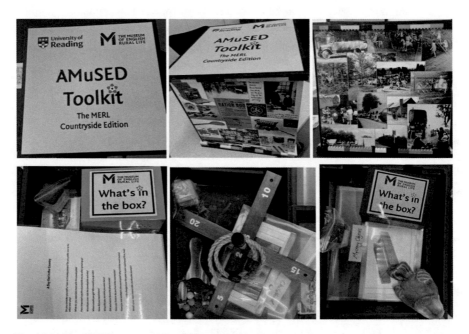

Fig. 12.3 The MERL countryside toolkit and its elements

Fig. 12.3 bottom middle image for a modernised version of quoits with rope circles for people to throw around the pegs), road milestone, youth hostel badge, cycling badge, as well as maps of the river Thames through Berkshire and Oxfordshire and onto the Cotswolds. These images were carefully curated to serve as visual cues for memories related to the countryside, farming, and rural life especially for people who grew up in the Berkshire and Oxfordshire area due to the chatty café attendee's demographic, but which could be readily tailored to other parts of the UK or wider afield.

The toolkit contains a combination of digital and traditional elements such as a digital photo album that narrates a story about each photo, reminiscence items, quoits game (Fig. 12.3 bottom middle image), countryside related aqua paints (when images are painted with water their colours appear and then disappear again when dry enabling easy painting as well as reuse), plush birds with authentic bird sounds, jigsaw puzzles, aroma capsules related to the farmyard such as stables and horses, potatoes, cut grass and summertime strawberries and mystery boxes presented as a "What's in the Box" quiz that contains tactile elements such as sheep's wool, hops, and dancing bells together with aural, visual and sometimes olfactory cues.

Once again, the reminiscence photos on the box panels are linked to the elements and activities in the toolkit such that they can be combined for use in an activity session. For example, the photos of farm animals on the panels can be used to start a conversation about farms, being on a farm and farm animals such as cows, pigs, birds, horses, sheep etc. which can be progressed by including the aromas of farmyard, horses, and stables into the conversation through the introduction of the aroma capsules. The birds can also be included to provide tactile stimulation as well as audio recordings to remind people of the bird sounds common in the countryside. To include an activity and extra forms of stimulation, the mystery box containing tactile farmyard elements such as sheep's wool can be passed around to all the session attendees with hints provided to help them guess "What's in the box", and the audio on the box describing the process of obtaining wool from sheep is automatically played as soon as the box is opened. Finally, a number of more physical and traditional activities are included such as the garden quoits and catapult (with soft pellets!).

12.5.1.3   Christmas Edition

The Christmas AMuSED toolkit (Fig. 12.4) was developed for use over the Christmas season to help people living with dementia reminisce about their past Christmas holidays, how they spent them and who they spent them with. It was aimed at connecting past Christmas traditions to modern practices using reminiscence cues. This toolkit incorporated the use of Christmas related images such as Christmas trees, carolling, presents, snowmen and sledges, Santa Claus, reindeer, Christmas foods and toys, as well as activities and elements from the Christmas season including Santa hoopla games, festive bingo, musical snow globes, wreath decorations, snowmen decorating, Christmas trees, tinsel, and lights. The inner sides of this toolkit display single large images such as Christmas pomanders (traditional decoration made from oranges

Fig. 12.4 The AMuSED Christmas edition toolkit

studded with cloves to give off a distinctive aroma, see Fig. 12.8), the twelve days of Christmas to encourage singalongs, well known shops, and Christmas decorations.

Collectively the images on the toolkit panels were aimed at providing memory cues to help people living with dementia reminisce about their childhood Christmas holidays and activities. For example, one of the panels has a large visually stimulating photo of a Christmas pomander which can be used within a reminiscence discussion of childhood Christmas activities, and then made in an activity session using the oranges, cloves and ribbons included in the toolkit to further strengthen the reminiscence experience, as attendees can touch and feel these items for tactile stimulation, smell them, and then hang them as a decorative piece just like they did during their childhood. Remaining oranges can be consumed for gustatory stimulation, and the festive spirit enhanced with Christmas carols played by the snow globe. This reminiscence example uses just a small number of the elements in the AMuSED toolkit but can provoke strong childhood memories and result in people sharing their family Christmas traditions, activities, and foods.

12.5.1.4 Entertainment Edition

This edition of the AMuSED toolkit (Fig. 12.5) was developed as a combination of entertainment and life themes. It includes reminiscence images related to entertainment such as cinemas, movie tickets and posters, festivals and events, dancing, fairgrounds, arcades, circus, and various music and movie stars. It also includes reminiscence images related to life in Reading and places where people would have experienced different forms of entertainment locally such as shops from the 70s, notable locations in Reading e.g., Union Street, cinemas from the 70s and 80s, Reading Football Club teams from different eras, and other sporting events like Wimbledon, greyhound racing, Henley Regatta, Royal Ascot etc. The elements featured in this toolkit correspond with the theme and were included to promote reminiscence about different forms of entertainment within and outside Reading. They include for example, recordable sound buttons, QR code access to video clips,

Fig. 12.5 The AMuSED entertainment edition toolkit

CDs of wartime tunes, masks of famous movie stars, sound effects machine, light box, clapperboard, bingo machine, hooks and ducks and a popcorn machine.

One of the ways in which this toolkit can be used is by creating a reminiscence discussion using the panels and the prompt sheet. Attendees of this reminiscence session can interact with the panel displaying reminiscence images related to entertainment in Reading and discuss images such as which cinemas in Reading they used to frequent, when they visited, and what movies they enjoyed watching. This conversation can lead to further discussion about their favourite movies and movie stars, and the QR code cubes can be introduced into the session for the attendees to watch short videos from some of their favourite movies while eating freshly made popcorn made with the popcorn machine (Fig. 12.10) also included in the toolkit. This discussion and reminiscence session can lead to an activity where the attendees roleplay as movie producers and describe the type of movies they would make if given the chance. The names of the movies can be set up by the activity coordinator using the light box (Fig. 12.6) to mimic the film displays commonly used in cinemas in the past, photos of which are also displayed on the panels. There is even a red-carpet backdrop and film photo props to recreate winning moments like movie premiers and award nights such as the Oscars.

Each of the different AMuSED toolkits, although differently themed, were developed with the same aim of promoting reminiscence in people living with dementia by providing strong memory cues related to the corresponding theme. The elements

Fig. 12.6 Left to right: Jellyfish lamp, snow globe, recordable photo album, and light box

used in these different toolkit editions were also carefully selected for the various forms of stimulation they provide, resulting in a toolkit that provides multisensory stimulation by delivering activities that offer visual, audio, olfactory, tactile, and gustatory stimulation to people living with dementia.

12.5.2   Stimulation

To decide what elements would be relevant to include in each toolkit, the AMuSED toolkit was developed following an extensive review of literature of the technologies and methods used to provide engagement to people living with dementia in practice, and alongside consultation/co-design with dementia experts. It was found that different elements can provide different forms of stimulation such as visual, audio, olfactory, gustatory, tactile, and cognitive stimulation, the effective delivery of which can help improve the quality of life of people living with dementia. Each AMuSED toolkit thus aims to include elements that can deliver multisensory stimulation, cater to visual, aural, olfactory, gustatory, and tactile senses and enhance reminiscence sessions as discussed below with examples taken from across the different toolkits.

12.5.2.1   Visual Stimulation

Visual stimulation is a key factor in reminiscence therapy as images are heavily used in this form of therapy due to their ability to evoke long-term memories as well as emotions in people living with dementia [93]. Therefore, the use of images spanning the different decades and life forming events of people's lives were paramount in the development of the AMuSED toolkit.

Visual stimulation in the AMuSED toolkit is provided using reminiscence images on both sides of each box panel. Images used are large, clear, bright, vibrant, and easy to look at to encourage ease of use. Whilst outside panels provide collections of items, inside panels focus on single, larger sized images to cater to people who might have some difficulty viewing smaller images and who might prefer a smaller number of larger images. Each image used is related to the corresponding theme. The images are selected to cover a range of related topics and are also carefully chosen so that the different topics related to the theme come together to portray a full experience especially when used in combination with prompt questions to promote smoother conversations and better reminiscence sessions. The larger images also often relate to a practical activity or physical object in the box.

In addition to the reminiscence images, digital elements such as visually stimulating lamps or fibre optic displays (Fig. 12.6) are included in the toolkit as a substitute for the more expensive bubble tubes and fibreoptic cables used for visual stimulation in dedicated multisensory rooms and which have been directly linked to the provision of calm feelings and the reduction of anxiety levels in people living with dementia [58]. By using smaller and cheaper versions of these light sources, this is in line with the portability and affordability attributes of the AMuSED toolkit which aims to incorporate multisensory stimulation into a portable tabletop toolkit and eliminate the need for a large space and separate room for multisensory stimulation.

12.5.2.2 Audio Stimulation

Sensory stimulation has been used for many years to provide engagement to people living with dementia, especially those in their later stages [23] and the provision of audio stimulation has been shown to increase their focus and comprehension as well as help them concentrate on a task [29]. Audio stimulation is sometimes combined with other forms of stimulation such as visual stimulation. This combination strengthens the overall experience of the participants in a session and is especially useful for people who might have other forms of impairments.

The AMuSED toolkit provides audio stimulation through recordable sound buttons, CDs for singalongs, sound effect machines and other digital elements related to the corresponding theme (Fig. 12.7). These can be used with the images or incorporated into activity sessions to deliver audio cues and improve the reminiscence experience. The audio stimulation in the AMuSED toolkit is further expanded in some of the toolkit editions through the inclusion of QR code cubes which link short videos corresponding to the theme; a recordable photo album which showcases as well as narrates information about photos and other elements; and the mystery boxes that combine audio with tactile and at times olfactory stimulation.

Fig. 12.7 Top (left to right): sound machine, QR code cube, sound buttons, recordable photo album. Bottom (left to right): mystery box and mystery box containing sheep's wool

Fig. 12.8 Left to right: aroma capsules, Christmas pomanders and scent kits

12.5.2.3 Olfactory Stimulation

Olfactory stimulation such as aromatherapy and the use of smells and fragrances has gained popularity in the engagement of people living with dementia. This form of stimulation has been known to cause an improvement in cognitive impairment due to the stimulation of the limbic system and hypothalamus [81]. The use of smells and aromatherapy is one of the main complementary forms of therapy used by healthcare professionals in hospitals and care homes [14]. It is usually administered to people living with dementia with aroma capsules, diffusers, vaporisers, or massages and frequently results in a positive response stemming from the pleasant aromas [86].

The use of olfactory stimulation in reminiscence therapy has shown positive results as the autobiographical memories recalled using olfactory cues are reported as stronger, clearer, and more emotional than those recalled by verbal cues [22]. Olfactory stimuli tailored to themes have also been found to be successful in the alleviation of depression in older adults and people living with dementia [40]. The use of this form of stimuli has also been reported to produce the strongest emotional sensations with the feeling of relieving past experiences when compared to other cues such as language label cues [42].

The primary aim of reminiscence therapy for people living with dementia is the remembrance of positive long-term memories, therefore olfactory stimulation is deemed appropriate as it has been shown to evoke even older memories—including those before the age of 10 years—when compared with visual stimulation [88]. A study which used magnetic resonance imaging to examine the brain activity of the participants at the point where they recalled autobiographical memories using olfactory stimulation cues, showed that the emotional regions of the brain such as the amygdala are more activated by olfactory stimulation cues [4].

Olfactory stimulation is provided by the AMuSED toolkits through the inclusion of different elements such as fruits, essential oils, and aroma capsules (Fig. 12.8), each of which were carefully selected with the aim of evoking memories related to the corresponding theme such as farmyard and cut grass for the countryside theme, pomanders and pine needles for the Christmas theme, and gardening scent kits for

Fig. 12.9 (Left to right): Plush bird, fuzzy balls, kinetic sand and plastic sea animals. *Source* Amazon [3]

the entertainment and life theme. These aromas can be used alongside other forms of stimulation such as visual, audio, and tactile stimulation to promote engagement and create a cohesive reminiscence experience.

12.5.2.4 Tactile Stimulation

The use of items such as twiddle muffs (knitted or crocheted bands or hand warmers with tactile items attached to them) to provide tactile stimulation is very popular in dementia care especially in the implementation of sensory rooms [46]. Tactile stimulation has been proven to promote relaxation and calmness and to contribute to a sense of trust in people living with dementia. It is also known to lead to a reduction in anxiety and an increase in cognition [19]. Studies have shown that when people living with dementia are offered opportunities to experience different touch sensations, it results in a positive impact on their physical, emotional, and social lives [11].

The tactile elements in the toolkit are made of different materials such as soft material, plastic, sand, shells, and foam (Fig. 12.9). These are all related to the theme and can be combined with the reminiscence images and other elements of the toolkit to provide increased engagement especially in group sessions. For example, the sand included in the Seaside toolkit links to the reminiscence images of sand on the toolkit panels and can be used when discussing trips to the seaside, building sandcastles, picking up seashells etc. Discussions can be enhanced by replicating the sandcastle building experience physically. The birds included in the MERL toolkit combine tactile and audio stimulation by making the sound attributed to real birds. This ties in with the reminiscence images of the Countryside theme and reminds people of experiences related to living or being in the Countryside.

12.5.2.5 Gustatory Stimulation

The implementation and administration of multisensory environments to people living with dementia is often focused on visual and tactile stimulation with little to no consideration given to gustatory stimulation or elements that stimulate the sense of taste [24]. Dementia, especially Alzheimer's Dementia is categorised by

Fig. 12.10 Popcorn and popcorn machine

Example Activity
Trips to the Seaside

Set the scene for your session using the images on the insides and outsides of the cube, the sound buttons of the seagulls and the waves and the scents of the sea. Use the question prompts if you want some starter questions.

Let people tell their stories and reminisce about the times they spent at the beach. The photos on the inside and outsides of the box provide great discussion points. Have the whole box as a central point or encourage conversation by distributing the different sides of the box around the table so that people can discuss the pictures. Add the brochures of Prestatyn and entertainment posters for more directed discussion. Have a knobbly knees competition!

Press the buttons to play the sounds of the seagulls and the waves, read the message in the bottle (from the Police song 'Message in a Bottle) and ask people to say what messages they would put in the bottle. Send postcards!

Ask people to sniff the scent cubes and try to identify the smells of the sea, extending this into reminiscence about the smells of the sea they remember.

Fig. 12.11 A page from the seaside activity booklet

Fig. 12.12 The prompt questions of the seaside AMuSED toolkit

Example Prompt Questions

- Have you ever been to the seaside? Where did you go?
- How did you travel there?
- Where did you stay?
- What activities did you do while you were there?
- Did you ever ride on a donkey?
- Did you collect seashells?
- Did you have a favourite beach to visit?
- What is your most memorable moment at the seaside?
- If you were writing a message in a bottle to put into the sea, what would you write?
- Have you ever been on a cruise? Where did you go?
- What's your favourite seafood or fish?
- Who would you send a postcard to?
- Did you enjoy swimming or sailing?

extensive neuronal loss in the brain, including the taste cortex and therefore people living with the condition can experience a reduced sense of taste [67]. However, Prins et al. [74] explains that the gustatory sensory systems can still be activated even in the advanced stages of the condition. Non-pharmacological interventions such as food therapy are widely used in the management of the behavioural and psychological symptoms of dementia and may involve the preparation, tasting and eating of food by people living with dementia, usually with the aid or supervision of their family members or caregivers. When combined with other non-pharmacological interventions, gustatory stimulation through food therapy has been shown to promote positive mood and engagement in people living with dementia [76].

To incorporate gustatory stimulation, food elements were included in the AMuSED toolkit as a way of bringing the reminiscence sessions to life. For example, eating popcorn while at a reminiscence session about movies and cinema can help to create a movie scenario and make people feel like they are at the cinema (Fig. 12.10). This helps to strengthen the theme by acting as an additional memory cue and can improve the reminiscence sessions. Similarly, a candyfloss making machine was intended for inclusion in the Seaside box, however this was removed for health and safety reasons during the pre-evaluation phase due to occasional 'spitting' of hot sugar whilst it was in operation. Sticks of rock are now seen as a safer alternative!

12.5.2.6   Cognitive Stimulation

Cognitively stimulating elements such as quizzes and jigsaw puzzles were also included in the AMuSED toolkit to fit each of the themes. A range of jigsaw puzzles with twelve to thirty-five pieces were sourced from dementia-friendly vendors to ensure that they would provide the right amount of cognitive stimulation from this AMuSED toolkit activity. Quizzes relating to 'old' movies, television programmes and Christmas also proved to be very popular. Dementia friendly word search sheets themed to each box and mostly including four to six letter words in the search were also included in the toolkits to provide an element of challenge into the sessions which also included activities such as writing messages in bottles or making decorations. Maps were included for their ability to provide stimulation and improve orientation [9], and colouring and painting sheets of varying levels of difficulty were also included for variety.

The design of the toolkit allows the activity coordinator or caregiver to either use it as a single tool for communication within a group or to deconstruct it into separate panels that can be used by individuals for more focused discussions. The toolkit incorporates different sensory elements; therefore, a selected panel can be used alongside elements and activities that correspond with it. For instance, the panel on the Christmas toolkit with reminiscence images of snowmen can be used in a session and combined with a "build your own snowman" activity, along with conversations about building snowmen in the past and remembering the years when it snowed at Christmas. The elements, activities and reminiscence images in the AMuSED toolkit are all connected in a way that helps the activity coordinator to bring these different aspects of the toolkit together to create an experience for people living with dementia.

12.5.3   Activity Booklet and Prompt Questions

An activity booklet is included in the AMuSED toolkit to give information about the toolkit and provide the caregiver or activity coordinator with suggested activities. This booklet gives an initial introduction to the AMuSED toolkit and includes different activities and prompt questions which have been shown to promote recall, stimulate response and enrich the overall reminiscence experience [50] as well as help people living with dementia to engage in storytelling [75]. Prompt questions have also been proven useful in triggering reminiscence in people living with dementia as well as helping this group of people focus on positive questions which usually lead to positive memories [26].

The main element of the activity booklet is the "Example Activity" section which contains suggestions of activities that could be carried out by activity coordinators using the toolkit elements (Fig. 12.11). Each of the activities included in this section incorporate the use of different multisensory elements and are aimed at recreating past experiences and adventures. The activity booklet gets people living with dementia involved in activities that they would have participated in when younger such as building sandcastles, finding, and playing with seashells, listening to the sounds of the ocean and seagulls, and identifying animals at the seaside. Engaging in these activities in the present will spur memories of being at the seaside and engaging in the same activities either during childhood or at a different stage of life. This will also encourage discussion and storytelling among people especially when prompt questions such as "Have you ever been to the seaside?" "What activities did you do while you were there?" "Did you have a favourite beach to visit?" are used by the activity coordinator during the session.

The prompt questions used in the AMuSED toolkit were carefully selected to promote only positive memories to achieve positive reminiscence (Fig. 12.12). When conducting reminiscence sessions it is important to focus on positive memories and avoid putting pressure on the participants to provide correct answers or recollections as this could awake feelings of distress or lead to the remembrance of unpleasant memories [53]. The AMuSED prompt questions are leading questions that the activity coordinators can ask the participants during the sessions to get them to narrate their experiences and stories. All the questions are also linked to the different reminiscence images and elements in the toolkit and help to link the activities being performed in the present to the memories of those performed in the past.

12.6 Evaluation and Feedback

This section describes the design, evaluation and feedback associated with development of the AMuSED toolkit. It provides details of the design thinking methodology used in the research, and the collaboration with dementia experts used to create a user-centred Active Multi-Sensory Environment for people living with Dementia. It explains the different types of feedback received throughout the process including getting feedback about the AMuSED concept from dementia experts, developing different prototyping iterations, and the final evaluation conducted in different dementia care facilities. It gives examples of some of the comments received from activity coordinators when using the toolkits, as well as the improvements made based on the expert feedback during the development and evaluation process.

12.6.1 Development Process and Expert Input

AMuSED was developed using the Design Thinking methodology [12], a series of processes used in the identification of problems and creation of useful solutions. In doing so, this approach combines an iterative yet structured development process with the consideration of user needs. At its core is the principle that development should always start by building an understanding of the group of people a solution is being created for and what their needs are [1, 8].

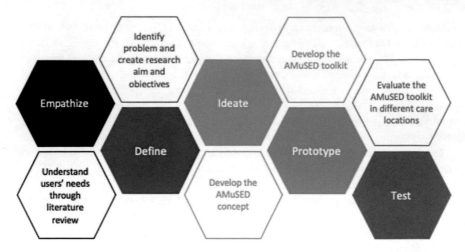

Fig. 12.13 Design thinking diagram reflecting the research and evaluation of AMuSED

Design thinking is an approach to problem solving that produces relevant solutions through ideation [10], placing emphasis on the development of possibilities rather than the satisfaction of constraints [62]. By facilitating brainstorming, design thinking presents itself as an iterative approach and serves as a catalyst for collaboration, innovation, and enablement that helps to bring products, experiences and services to the market [87]. Comprised of various stages including empathy, ideation, prototyping and testing, it provides a framework for the identification of problems and how they can be approached as well as potential solutions [47, 56].

The use of the design thinking methodology ensured that the creation of AMuSED was built on proven research literature and best practice in care homes and that it was evaluated in an iterative manner at different stages of the development process as follows (Fig. 12.13):

Empathize: a literature review was conducted to gain an understanding of people living with dementia, their care givers and the challenges associated with providing engaging and stimulating activities to people with dementia in various care provisions.

Define: the problems to be solved and what had previously been done by others to solve them were identified. Identifying gaps in literature and practice allowed the problems and challenges associated with the use of non-pharmacological interventions to provide engagement and stimulation to people living with dementia to be defined in more detail. This helped the aims and objectives of this research and the initial concept and requirements for the AMuSED toolkit to be defined.

Ideate: prior to Covid we observed care crew at a hospital in Reading to better understand the relationship between caregivers and their patients with dementia and see some of the activities they used to provide engagement and stimulation. Subsequently throughout covid and its lockdowns we worked virtually with the Reading Dementia

Friendly Steering Group, to refine and realise the AMuSED concept as a multisensory, multiactivity themed toolkit that could be delivered in a single portable box and used by care givers, care teams and dementia activity coordinators to engage and stimulate people living with dementia physically, mentally, and socially through a combination of sensory elements and themed activities.

Prototype: to gain feedback and design out any potential issues, a series of iterative prototypes were developed and evaluated at different stages of the research (Fig. 12.14). For every prototype iteration, feedback was provided by the dementia experts on both the design of the AMuSED toolkit, the reminiscence images on the panels, and the elements to be included in the toolkit.

Test: this stage involved transferring the final AMuSED toolkits into different dementia care provisions such as care homes, sheltered housing facilities and dementia chatty cafes to be evaluated by caregivers and activity coordinators. Feedback was also received on how the AMuSED toolkit could be further improved to benefit people living with dementia and their caregivers.

12.6.2   Evaluation of Use in Care Facilities

The differently themed editions of the AMuSED toolkit (Seaside, Countryside, Christmas, Entertainment) were evaluated at various care provisions for people living with dementia including care homes, sheltered housing facilities, personal homes, and chatty cafés.

It was found that the AMuSED toolkit was used in different ways across these locations. In care homes for example the toolkits were mainly used by activity coordinators and care staff to conduct reminiscence sessions for their residents. In the sheltered housing facility, where residents regularly get together for activity sessions, the AMuSED toolkit contents were combined with the coordinators own dress-up costumes, props and activities to further enhance the experience, for example, by dressing up as a cinema usher and handing out popcorn, ice cream, and movie tickets before residents watched a film.

A different mode of use was adopted by a home care facility where care staff go into the personal homes of people living with dementia to provide care and stimulation. Here, the Seaside and Entertainment toolkits were separated into different panels and one or two panels were taken into individuals' homes along with a small number of themed elements to allow the care staff to focus on one or two activities and not overwhelm or overstimulate their care recipient and to ensure suitability of the activities selected for an individual. This care facility also used their Seaside AMuSED box as a complete reminiscence experience in Friday sessions with larger groups of people in community settings.

The Countryside AMuSED toolkit was also evaluated in a community setting where it was incorporated into the chatty café sessions of The MERL. In this location, the Countryside edition toolkit was deconstructed and each of the panels and elements

Fig. 12.14 Different prototype iterations of the AMuSED toolkit

were separated into different sections and combined with the exhibits from The MERL to fit the themes normally used in the chatty cafés such as food, mystery objects, childhood, dressing up and going out, journeys and transport, waking lives, landmarks of life, and makers at work. All the AMuSED panels and elements used in these sessions were divided into separate trays to display the different chatty café themes and were combined with the prompt sheets developed by The MERL which contained questions related to life in the countryside such as "have you lived in the countryside?", "what do you associate most with the countryside?" etc. The toolkit was also used in a Seasons themed session where the different attendees were tasked with separating the elements of the toolkit into their corresponding seasons while also using the reminiscence images, aroma capsules and digital elements such as mystery boxes as clues. This led to an animated discussion among the attendees about their favourite seasons and the different seasonal activities they participated in during their childhood such as collecting and roasting chestnuts in the winter, picking fruits from the farm in the summer, picking bluebells in the spring, building haystacks in autumn etc.

12.6.3 Positive Feedback

The AMuSED toolkit was successfully able to bring positive memories of past experiences and adventures into the present during all the activity sessions conducted at the different evaluation sites, as well as deliver multisensory stimulation using a combination of digital and traditional elements. The feedback received during the evaluation showed the relevance of the use of images in connecting the past and the present [27, 45] by highlighting that the images were able to promote engagement [44], communication [64] and reminiscence in people living with dementia by creating discussion points and even allowing conversations among group attendees to evolve on their own through various discussions.

The different ways in which the AMuSED toolkits were used during the evaluation also showed the versatility and adaptability of the toolkit to the different locations and audiences. The toolkits received highly positive feedback from all evaluation sites, with different managers and activity coordinators providing additional feedback on its usefulness and novelty with examples being:

> For The MERL our reminiscence practice has been heavily staff led as it has involved accessing collections. More recently we have been piloting different models of working e.g. The Chatty Café, which is an informal drop in opportunity. The AMuSED toolkit is fundamental to this offering allowing an engagement with the contents whilst having refreshments, and in a more self-directed, unsupervised manner (something that could not happen if real collections were involved). It has enabled a flexibility of our service

> What is really positive about the AMuSED toolkit is that it contains affordable digital elements (such as the 'What's in the Box, the audio photo frame and audio buttons, the tweeting birds) which are well judged, enhancing but not detracting from the collection elements. These have been very well received by audiences and can be used in a much more undirected way.

> Everyone can be involved, as if someone isn't able to physically participate, they can still be involved through reminiscence and scents/sands.

> I plan to take my time and keep referring back to the toolkit, I see many different opportunities. I would like to reminisce about cinema and film, dress up and follow with a film, using the masks and red-carpet scene. I would turn the whole activity into a cinema experience dressing up as an Usher, (selling/giving) popcorn or ice-creams with cinema tickets.

> A great tool for the facilitator. It takes the strain of having to think of activities for each session. It provides a variety of activities to appeal to all the trainees

12.6.4 Challenges and Improvements

The positive experiences at the user sites are due largely to the design thinking approach described in Sect. 12.6.2 which involved dementia experts and the evolutionary series of prototypes created throughout the development process of the AMuSED concept and toolkit. This enabled many of the issues and deficiencies of AMuSED to be designed out during the sequence of prototypes. Improvements made included:

- The evolution of the AMuSED toolkit from an individual sensory experience to a social group activity and multisensory experience, something requested by many of the care home managers and activity coordinators.
- The inclusion of themes, themed activities, and images with local interests to make the AMuSED toolkit more relatable to care home residents.
- Incorporating more traditional elements into the toolkit due to feedback from dementia experts stating that people living with dementia respond better to traditional and familiar elements such as printed photos, whilst at the same time including more music and video elements in the toolkits by incorporating QR code cubes in the later versions.
- Incorporating low-cost elements to provide the music and media rather than expensive technology due to the risk of loss or damage.
- Developing a more robust and faster way of slotting the side panels into the base of AMuSED box so that activity organisers could put it together in under 2 min at the start of a session and without the need to have any ties around the entire structure to keep it rigid. This problem was solved through using Lego brings as a border to the top and bottom panels with built up tracks at the corners.
- Using laminated panels to ensure that they could be wiped clean and sanitised with alcohol.
- Improving portability and preserving the quality and stability of the AMuSED box by providing a neat and efficient method of storing the box and its contents rather than storing them all in the AMuSED box. This involved opting for a large plastic tub with a foamboard layer created to separate the AMuSED panels and its toolkit contents.

Feedback from the activity coordinators during the evaluation period on how the AMuSED toolkit could be further improved included introducing wheels to enhance portability. One of the evaluation sites also included a teddy bear in their AMuSED sessions as part of their activity experience. This addition was so popular with attendees who all wanted to cuddle it, that any future AMuSED toolkits will include an appropriately themed or AMuSED branded mascot.

12.7 Conclusion

Reminiscence therapy has been proven as a highly effective form of stimulation for people living with dementia. A very useful tool in connecting the past and the present, this form of therapy utilises memory cues to engage people living with dementia and allows them to not only remember positive memories but also share interesting stories with others.

The AMuSED toolkit makes use of digital and traditional elements to deliver reminiscence therapy as well as multisensory and multiactivity stimulation to people living with dementia. This novel approach to reminiscence therapy brings the reminiscence experience to life and provides a multisensory environment in which people living with dementia can be engaged and stimulated. The toolkit provides a novel approach to reminiscence therapy by bridging the gap between the past and the present through multisensory stimulation. Its creative fusion of both digital and traditional elements results in an intervention that not only incorporates technology for reminiscence but also still maintains familiarity for people living with dementia through the incorporation of photos and other traditional elements. The development of the AMuSED toolkit is underpinned by current research and current best practice in activity provision for dementia care. The toolkits are designed to be affordable and personalisable to an individual's or group's interests. Elements of the toolkit such as music, fragrances and essential oils for aromatherapy can also be personalised based on the mood or emotional requirements of a particular user or themed to encourage group social interaction and activity.

The AMuSED toolkit takes from best practices in dementia care such as reminiscence therapy and its ability to help people living with dementia recall memories in the present [21], prompt questions which have been shown by [50, 75] to promote recall, encourage storytelling and stimulate response during reminiscence therapy, and multisensory stimulation which not only emphasises the personhood of the person with dementia [63], but also provides multiple benefits including encouraging communication and interaction especially in people living with moderate and severe dementia [69].

The AMuSED toolkit combines these practices in a novel format which has been used by activity coordinators and caregivers with people living with dementia in care homes, sheltered housing, dementia community and personal homes with positive results especially in providing engagement and promoting discussion among people living with dementia. The elements and activities contained in the toolkit have been shown to effectively connect the past and the present during reminiscence sessions by promoting positive memories for people living with dementia and encouraging communication and storytelling during activity sessions.

Work is continuing on the AMuSED toolkits to take on board evaluation feedback; expand the AMuSED themes to include, for example, sports, fashion, weddings, and cultural events; translate the AMuSED concept to different user groups such as adults with specific learning; and to scale up the concept to make AMuSED more widely available (Fig. 12.15).

Fig. 12.15 The seaside edition AMuSED toolkit

References

1. Ahmed, B., Dannhauser, T., Philip, N.: A lean design thinking methodology (LDTM) for machine learning and modern data projects. In: 2018 10th Computer Science and Electronic Engineering (CEEC), pp. 11–14 (2018). https://doi.org/10.1109/CEEC.2018.8674234
2. Alzheimer's Disease International: Numbers of People with Dementia Worldwide (2020). https://www.alzint.org/resource/numbers-of-people-with-dementia-worldwide/. Accessed 21 Feb 2022
3. Amazon. Amazon.com. Spend Less. Smile More (n.d.). https://www.amazon.com/. Accessed 24 Feb 2022
4. Arshamian, A., Iannilli, E., Gerber, J.C., Willander, J., Persson, J., Seo, H.S., Hummel, T., Larsson, M.: The functional neuroanatomy of odor evoked autobiographical memories cued by odors and words. Neuropsychologia **51**(1), 123–131 (2013). https://doi.org/10.1016/j.neuropsychologia.2012.10.023
5. Ashton, J.: Plants and green spaces provide more than just aesthetic benefits. Perspect. Public Health **135**(4), 178 (2015)
6. Banerjee, S., Murray, J., Foley, B., Atkins, L., Schneider, J. and Mann, A.: Predictors of institutionalisation in people with dementia. J. Neurol. Neurosurg. Psychiatry, **74**(9), pp.1315–1316 (2003). https://doi.org/10.1136/jnnp.74.9.1315
7. Bannan, N., Montgomery-Smith, C.: 'Singing for the brain': reflections on the human capacity for music arising from a pilot study of group singing with Alzheimer's patients. J. R. Soc. Promot. Health **128**(2), 73–78. https://doi.org/10.1177/1466424007087807
8. Bell, S.: Design Thinking. Temple University Libraries, pp. 44–50 (2008)
9. Bertrand, E., Naylor, R., Laks, J., Marinho, V., Spector, A. and Mograbi, D.C.: Cognitive stimulation therapy for brazilian people with dementia: examination of implementation'issues and cultural adaptation. Aging & mental health, **23**(10), pp.1400–1404 (2019). https://doi.org/10.1080/13607863.2018.1488944

10. Black, S., Gardner, D.G., Pierce, J.L., Steers, R.: Design thinking. Organ. Behav. (2019)
11. Bray, J., Atkinson, T., Latham, I., Brooker, D.: Practice of Namaste Care for people living with dementia in the UK. Nurs. Older People **33**(3) (2021). https://doi.org/10.7748/nop.2018.e1109
12. Brown, T.: Design thinking. Harv. Bus. Rev. **86**(6), 84 (2008)
13. Buettner, L., Ferrario, J.: Therapeutic recreation-nursing team: a therapeutic intervention for nursing home residents with dementia. Annu. Ther. Recreat. **7**, 21–28 (1998)
14. Buckle, J.: Clinical Aromatherapy-e-Book: Essential Oils in Practice. Elsevier Health Sciences (2014)
15. Burns, I., Cox, H., Plant, H.: Leisure or therapeutics? Snoezelen and the care of older persons with dementia. Int. J. Nurs. Pract. **6**(3), 118–126 (2000). https://doi.org/10.1046/j.1440-172x.2000.00196.x
16. Butler, R.N.: The life review: an interpretation of reminiscence in the aged. Psychiatry **26**(1), 65–76 (1963). https://doi.org/10.1080/00332747.1963.11023339
17. Camic, P.M., Chatterjee, H.J.: Museums and art galleries as partners for public health interventions. Perspect. Public Health **133**(1), 66–71 (2013). https://doi.org/10.1177/1757913912468523
18. Carbonneau, H., Caron, C.D. and Desrosiers, J.: Effects of an adapted leisure education program as a means of support for caregivers of people with dementia. Arch. Gerontol. Geriatr. **53**(1), 31–39 (2011). https://doi.org/10.1016/j.archger.2010.06.009
19. Chang, C.: Nonpharmacological management of BPSD: agitation and behavioral problems in dementia. In: Geriatric Practice, pp. 253–265. Springer, Cham (2020). https://doi.org/10.1007/978-3-030-19625-7_21
20. Chao, S.Y., Chen, C.R., Liu, H.Y., Clark, M.J.: Meet the real elders: reminiscence links past and present. J. Clin. Nurs. **17**(19), 2647–2653 (2008). https://doi.org/10.1111/j.1365-2702.2008.02341.x
21. Ching-Teng, Y., Ya-Ping, Y., Chia-Ju, L., Hsiu-Yueh, L.: Effect of group reminiscence therapy on depression and perceived meaning of life of veterans diagnosed with dementia at veteran homes. Soc. Work Health Care **59**(2), 75–90 (2020). https://doi.org/10.1080/00981389.2019.1710320
22. Chu, S., Downes, J.J.: Proust nose best: odors are better cues of autobiographical memory. Mem. Cognit. **30**(4), 511–518 (2002)
23. Cohen-Mansfield, J., Marx, M.S., Dakheel-Ali, M., Regier, N.G., Thein, K.: Can persons with dementia be engaged with stimuli? Am. J. Geriatr. Psychiatry **18**(4), 351–362 (2010). https://doi.org/10.1097/jgp.0b013e3181c531fd
24. Collier, L., Jakob, A.: The multisensory environment (MSE) in dementia care: examining its role and quality from a user perspective. HERD: Health Environ. Res. Des. J. **10**(5), 39–51 (2017). https://doi.org/10.1177/1937586716683508
25. Colling, K.B.: A taxonomy of passive behaviors in people with Alzheimer's disease. J. Nurs. Scholarsh. **32**(3), 239–244 (2000). https://doi.org/10.1111/j.1547-5069.2000.00239.x
26. Dai, J., Moffatt, K.: Making space for social sharing: insights from a community-based social group for people with dementia. In: Proceedings of the 2020 CHI Conference on Human Factors in Computing Systems, pp. 1–13 (2020). https://doi.org/10.1145/3313831.3376133
27. De Groot, W., Kenning, G., van den Hoven, E., Eggen, B.: Exploring how a multisensory media album can support dementia care staff. In: Dementia Lab Conference, pp. 51–61. Springer (2021). https://doi.org/10.1007/978-3-030-70293-9_4
28. Deininger, M., Daly, S.R., Lee, J.C., Seifert, C.M., Sienko, K.H.: Prototyping for context: exploring stakeholder feedback based on prototype type, stakeholder group and question type. Res. Eng. Design **30**(4), 453–471 (2019). https://doi.org/10.1007/s00163-019-00317-5
29. Dixon, E., Lazar, A.: The role of sensory changes in everyday technology use by people with mild to moderate dementia. In: The 22nd International ACM SIGACCESS Conference on Computers and Accessibility, pp. 1–12 (2020). https://doi.org/10.1145/3373625.3417000
30. Dyer, S.M., Laver, K., Pond, C.D., Cumming, R.G., Whitehead, C., Crotty, M.: Clinical practice guidelines and principles of care for people with dementia in Australia. Aust. Fam. Phys. **45**(12), 884–889 (2016)

31. Dyer, S.M., Harrison, S.L., Laver, K., Whitehead, C., Crotty, M.: An overview of systematic reviews of pharmacological and non-pharmacological interventions for the treatment of behavioral and psychological symptoms of dementia. Int. Psychogeriatr. **30**(3), 295–309 (2018). https://doi.org/10.1017/S1041610217002344
32. Edvardsson, D., Winblad, B., Sandman, P.O.: Person-centred care of people with severe Alzheimer's disease: current status and ways forward. Lancet Neurol. **7**(4), 362–367 (2008). https://doi.org/10.1016/S1474-4422(08)70063-2
33. Ellis, P., Van Leeuwen, L., Brown, K.: Visual-music vibrations. Digit. Creat. **19**(3), 194–202 (2008)
34. Finkel S.I., Burns, A., and Cohen, G.: Behavioural and psychological symptoms of dementia (BPSD): a clinical and research update. Int. Psychogeriatr. **12**(s1), 13–18 (2000)
35. Finnema, E., Dröes, R.M., Ribbe, M., Van Tilburg, W.: The effects of emotion-oriented approaches in the care for persons suffering from dementia: a review of the literature. Int. J. Geriatr. Psychiatry **15**(2), 141–161 (2000). https://doi.org/10.1002/(SICI)1099-1166(200002)15:2%3c141::AID-GPS92%3e3.0.CO;2-5
36. Goto, S., Kamal, N., Puzio, H., Kobylarz, F., Herrup, K.: Differential responses of individuals with late-stage dementia to two novel environments: a multimedia room and an interior garden. J. Alzheimer's Dis. **42**(3), 985–998 (2014). https://doi.org/10.3233/JAD-131379
37. Goto, S., Gianfagia, T.J., Munafo, J.P., Fujii, E., Shen, X., Sun, M., Shi, B.E., Liu, C., Hamano, H., Herrup, K.: The power of traditional design techniques: the effects of viewing a Japanese garden on individuals with cognitive impairment. HERD: Health Environ. Res. Des. J. **10**(4), 74–86 (2017). https://doi.org/10.1177/1937586716680064
38. Goto, S., Shen, X., Sun, M., Hamano, Y., Herrup, K.: The positive effects of viewing gardens for persons with dementia. J. Alzheimer's Dis. **66**(4), 1705–1720 (2018). https://doi.org/10.3233/JAD-170510
39. Guss, R., Middleton, J., Beanland, T., Slade, L., Moniz-Cook, E., Watts, S., Bone, A.: A guide to psychosocial interventions in early stages of dementia. The British Psychological Society, Leicester (2014)
40. Hanaoka, H., Muraki, T., Ede, J., Yasuhara, K., Okamura, H.: Effects of olfactory stimulation on reminiscence practice in community-dwelling elderly individuals. Psychogeriatrics **18**(4), 283–291 (2018). https://doi.org/10.1111/psyg.12322
41. Harwood, D.G., Barker, W.W., Ownby, R.L. and Duara, R.: Relationship of behavioral and psychological symptoms to cognitive impairment and functional status in Alzheimer's disease. Int. J. Geriatr. Psychiatry, **15**(5), 393–400 (2000). https://doi.org/10.1002/(SICI)1099-1166(200005)15:53.0.CO;2-O
42. Herz, R.S.: A naturalistic analysis of autobiographical memories triggered by olfactory visual and auditory stimuli. Chem. Senses **29**(3), 217–224 (2004). https://doi.org/10.1093/chemse/bjh025
43. Hilgeman, M.M., Allen, R.S., Snow, A.L., Durkin, D.W., DeCoster, J., Burgio, L.: Preserving identity and planning for advance care (PIPAC): preliminary outcomes from a patient-centered intervention for individuals with mild dementia. Aging Ment. Health **18**(4), 411–424 (2014). https://doi.org/10.1080/13607863.2013.868403
44. Holloway, J., Rick Voight, M.B.A., Hayley Studer, C.P.A., Evans, A., BC-DMT, B.C: Cognitive benefits of photo reminiscence therapy for dementia patients. Natl. Inst. Dementia Educ. **18**(22), 1–6 (2021)
45. Houben, M., Engen, V.V., Kenning, G., Brankaert, R.: Smile: capturing and sharing personal photos to stimulate social relations and support self-identity in dementia. In: Dementia Lab Conference, pp. 83–93. Springer, Cham (2021). https://doi.org/10.1007/978-3-030-70293-9_7
46. Jakob, A., Collier, L.: Sensory enrichment through textiles for people living with dementia. In: Proceedings of Intersections: Collaborations in Textile Design Research Conference, Loughborough University London, U.K. (2017)
47. Johansson-Sköldberg, U., Woodilla, J., Çetinkaya, M.: Design thinking: past, present and possible futures. Create. Innov. Manag. **22**(2), 121–146 (2013). https://doi.org/10.1111/caim.12023

48. Jones, E.D.: Reminiscence therapy for older women with depression: effects of nursing intervention classification in assisted-living long-term care. J. Gerontol. Nurs. **29**(7), 26–33 (2003)
49. Jönsson, B., Philipson, L., Svensk, A.: What Isaac Taught Us. Certec, LTH, Lund University, Sweden (1998). http://www.english.certec.lth.se/doc/whatIsaac/. Accessed 8 July 2022
50. Kelly, L., Ahessy, B.: Reminiscence-focused music therapy to promote positive mood and engagement and shared interaction for people living with dementia. In: Voices: A World Forum for Music Therapy, vol. 21, no. 2 (2021)
51. Lancioni, G.E., Cuvo, A.J., O'reilly, M.F.: Snoezelen: an overview of research with people with developmental disabilities and dementia. Disabil. Rehabil. **24**(4), 175–184 (2002). https://doi.org/10.1080/0963828011007491 1
52. Lazar, A., Thompson, H., Demiris, G.: A systematic review of the use of technology for reminiscence therapy. Health Educ. Behav. **41**(1_suppl), 51S–61S (2014). https://doi.org/10.1177/1090198114537067
53. Lazar, A., Thompson, H.J., Demiris, G.: Design recommendations for recreational systems involving older adults living with dementia. J. Appl. Gerontol. **37**(5), 595–619 (2018). https://doi.org/10.1177/0733464816643880
54. Lee, K.: Inclusion of people with dementia in research can help nurses understand how to deliver successful reminiscence interventions. Evid. Based Nurs. (2020). https://doi.org/10.1136/ebnurs-2020-103271
55. Li, M., Lyu, J.H., Zhang, Y., Gao, M.L., Li, R., Mao, P.X., Li, W.J., Ma, X.: Efficacy of group reminiscence therapy on cognition, depression, neuropsychiatric symptoms, and activities of daily living for patients with Alzheimer disease. J. Geriatr. Psychiatry Neurol. **33**(5), 272–281 (2020). https://doi.org/10.1177/0891988719882099
56. Lockwood, T.: Design Thinking: Integrating Innovation, Customer Experience, and Brand Value. Simon and Schuster (2010)
57. López-Almela, A., Gómez-Conesa, A.: Intervención en demencias mediante estimulación multisensorial (snoezelen). Fisioterapia **33**(2), 79–88 (2011). https://doi.org/10.1016/j.ft.2011.02.004
58. Lorusso, L., Park, N.K., Bosch, S., Freytes, I.M., Shorr, R., Conroy, M., Ahrentzen, S.: Sensory environments for behavioral health in dementia: diffusion of an environmental innovation at the veterans health administration. HERD: Health Environ. Res. Des. J. **13**(4) 44–56 (2020). https://doi.org/10.1177/1937586720922852
59. Macleod, F., Storey, L., Rushe, T., McLaughlin, K.: Towards an increased understanding of reminiscence therapy for people with dementia: a narrative analysis. Dementia **20**(4), 1375–1407 (2021). https://doi.org/10.1177/1471301220941275
60. Maseda, A., Cibeira, N., Lorenzo-López, L., González-Abraldes, I., Buján, A., de Labra, C., Millán-Calenti, J.C.: Multisensory stimulation and individualized music sessions on older adults with severe dementia: effects on mood, behavior, and biomedical parameters. J. Alzheimers Dis. **63**(4), 1415–1425 (2018)
61. Morris, R.G.: Recent developments in the neuropsychology of dementia, Int. Rev. Psychiatry. **6**(1), 85–107 (1994). https://doi.org/10.3109/09540269409025245
62. Neumeier, M.: The Designful Company: How to Build a Culture of Nonstop Innovation. Peachpit Press (2009)
63. Norberg, A.: Sense of self among persons with advanced dementia. In: Wisniewski, T. (ed.) Alzheimer's Disease, Chapter 13. Codon Publications, Brisbane (AU) (2019). https://doi.org/10.15586/alzheimersdisease.2019.ch13. https://www.ncbi.nlm.nih.gov/books/NBK552152/
64. Nordgren, L., Asp, M.: Photo-elicited conversations about therapy dogs as a tool for engagement and communication in dementia care: a case study. Animals **9**(10), 820 (2019). https://doi.org/10.3390/ani9100820
65. Nyman, S.R., Szymczynska, P.: Meaningful activities for improving the wellbeing of people with dementia: beyond mere pleasure to meeting fundamental psychological needs. Perspect. Public Health **136**(2), 99–107 (2016). https://doi.org/10.1177/1757913915626193

66. O'Connor, D.W., Ames, D., Gardner, B., King, M.: Psychosocial treatments of behavior symptoms in dementia: a systematic review of reports meeting quality standards. Int. Psychogeriatr. **21**(2), 225–240 (2009). https://doi.org/10.1017/S1041610208007588

67. Ogawa, T., Irikawa, N., Yanagisawa, D., Shiino, A., Tooyama, I., Shimizu, T.: Taste detection and recognition thresholds in Japanese patients with Alzheimer-type dementia. Auris Nasus Larynx **44**(2), 168–173 (2017). https://doi.org/10.1016/j.anl.2016.06.010

68. O'Philbin, L., Woods, B., Farrell, E.M., Spector, A.E., Orrell, M.: Reminiscence therapy for dementia: an abridged Cochrane systematic review of the evidence from randomized controlled trials. Expert Rev. Neurother. **18**(9), 715–727 (2018). https://doi.org/10.1080/14737175.2018.1509709

69. Park, K., Lee, S., Yang, J., Song, T., Hong, G.R.S.: A systematic review and meta-analysis on the effect of reminiscence therapy for people with dementia. Int. Psychogeriatr. **31**(11), 1581–1597 (2019). https://doi.org/10.1017/S1041610218002168

70. Pinquart, M., Duberstein, P.R., Lyness, J.M.: Effects of psychotherapy and other behavioral interventions on clinically depressed older adults: a meta-analysis. Aging Ment. Health **11**(6), 645–657 (2007). https://doi.org/10.1080/13607860701529635

71. Pittiglio, L.: Use of reminiscence therapy in patients with Alzheimer's disease. Prof. Case Manag. **5**(6), 216–220 (2000)

72. Poulos, C.J., Bayer, A., Beaupre, L., Clare, L., Poulos, R.G., Wang, R.H., Zuidema, S., McGilton, K.S.: A comprehensive approach to reablement in dementia. Alzheimer's Dement. Transl. Res. Clin. Interv. **3**(3), 450–458 (2017). https://doi.org/10.1016/j.trci.2017.06.005

73. Prince, M., Guerchet, M., Prina, M.: The Global Impact of Dementia 2013–2050. Alzheimer's Disease International (2013). http://www.alz.co.uk/research/G8-policy-brief. Accessed 24 Feb 2022

74. Prins, A.J., Scherder, E.J.A., Van Straten, A., Zwaagstra, Y., Milders, M.V.: Sensory stimulation for nursing-home residents: systematic review and meta-analysis of its effects on sleep quality and rest-activity rhythm in dementia. Dement. Geriatr. Cogn. Disord. **49**(3), 219–234 (2020). https://doi.org/10.1159/000509433

75. Purves, B.A., Phinney, A., Hulko, W., Puurveen, G., Astell, A.J.: Developing CIRCA-BC and exploring the role of the computer as a third participant in conversation. Am. J. Alzheimer's Dis. Other Dement.® **30**(1), 101–107 (2015). https://doi.org/10.1177/1533317514539031

76. Quail, Z., Carter, M.M., Wei, A., Li, X.: Management of cognitive decline in Alzheimer's disease using a non-pharmacological intervention program: a case report. Medicine **99**(21) (2020). https://doi.org/10.1097/MD.0000000000020128

77. Roland, K.P., and Chappell, N.L.: Relationship and stage of dementia differences in caregiver perspectives on the meaning of activity. Dementia, **16**(2), 178–191 (2017). https://doi.org/10.1177/1471301215586287

78. Sánchez, A., Millán-Calenti, J.C., Lorenzo-López, L., Maseda, A.: Multisensory stimulation for people with dementia: a review of the literature. Am. J. Alzheimer's Dis. Other Dement.® **28**(1), 7–14 (2013). https://doi.org/10.1177/1533317512466693

79. Shoesmith, E., Charura, D. and Surr, C.: Acceptability and feasibility study of a six-week person-centred, therapeutic visual art intervention for people with dementia. Arts & Health, **13**(3), 296–314 (2021)

80. Stewart, E.G.: Art therapy and neuroscience blend: working with patients who have dementia. Art therapy, **21**(3), 148–155 (2004)

81. Takahashi, Y., Shindo, S., Kanbayashi, T., Takeshima, M., Imanishi, A., Mishima, K.: Examination of the influence of cedar fragrance on cognitive function and behavioral and psychological symptoms of dementia in Alzheimer type dementia. Neuropsychopharmacol. Rep. **40**(1), 10–15 (2020). https://doi.org/10.1002/npr2.12096

82. Thomas, J.M., Sezgin, D.: Effectiveness of reminiscence therapy in reducing agitation and depression and improving quality of life and cognition in long-term care residents with dementia: a systematic review and meta-analysis. Geriatr. Nurs. **42**(6), 1497–1506 (2021). https://doi.org/10.1016/j.gerinurse.2021.10.014

83. Van Weert, J.C., Van Dulmen, A.M., Spreeuwenberg, P.M., Ribbe, M.W., Bensing, J.M.: Behavioral and mood effects of snoezelen integrated into 24-hour dementia care. J. Am. Geriatr. Soc. **53**(1), 24–33 (2005). https://doi.org/10.1111/j.1532-5415.2005.53006.x
84. Verkaik, R., van der Heide, I., van Eerden, E., Spreeuwenberg, P., Scherder, E., Francke, A.: Effects of enriched thematic multi-sensory stimulation on BPSD in a beach room: a pilot study among nursing-home residents with dementia. OBM Geriatr. **3**(4), 1–1 (2019). https://doi.org/10.21926/obm.geriatr.1904092
85. Vernooij-Dassen, M.: Meaningful activities for people with dementia. Aging & Mental Health, **11**(4), 359–360 (2007). https://doi.org/10.1080/13607860701498443
86. Watson, K., Hatcher, D., Good, A.: A randomised controlled trial of lavender (Lavandula angustifolia) and lemon balm (Melissa officinalis) essential oils for the treatment of agitated behaviour in older people with and without dementia. Complement. Ther. Med. **42**, 366–373 (2019). https://doi.org/10.1016/j.ctim.2018.12.016
87. Welsh, M.A., Dehler, G.E.: Combining critical reflection and design thinking to develop integrative learners. J. Manag. Educ. **37**(6), 771–802 (2013). https://doi.org/10.1177/1052562912470107
88. Willander, J., Larsson, M.: Smell your way back to childhood: autobiographical odor memory. Psychon. Bull. Rev. **13**(2), 240–244 (2006)
89. Woods, R.T., Bruce, E., Edwards, R.T., Elvish, R., Hoare, Z., Hounsome, B., Keady, J., Moniz-Cook, E.D., Orgeta, V., Orrell, M., Rees, J.: REMCARE: reminiscence groups for people with dementia and their family caregivers-effectiveness and cost-effectiveness pragmatic multicentre randomised trial. Health Technol. Assess. **16**(48) (2012). https://doi.org/10.3310/hta16480
90. Woods, R.T., Orrell, M., Bruce, E., Edwards, R.T., Hoare, Z., Hounsome, B., Keady, J., Moniz-Cook, E., Orgeta, V., Rees, J., Russell, I.: REMCARE: pragmatic multi-centre randomised trial of reminiscence groups for people with dementia and their family carers: effectiveness and economic analysis. PLoS ONE **11**(4), e0152843 (2016). https://doi.org/10.1371/journal.pone.0152843
91. Woods, B., O'Philbin, L., Farrell, E.M., Spector, A.E., Orrell, M.: Reminiscence therapy for dementia. Cochrane Database Syst. Rev. (3) (2018). https://doi.org/10.1002/14651858.CD001120.pub3
92. World Health Organization: International Classification of Functioning, Disability and Health. Geneva (2001). Accessed Feb 2022
93. Wu, D., Chen, T., Huang, X., Chen, L., Yue, Y., Yang, H., Hu, X., Gong, Q.: The role of old photos in reminiscence therapy in elderly women with depressive symptoms: a functional magnetic resonance imaging study. Biol. Res. Nurs. **22**(2), 234–246 (2020). https://doi.org/10.1177/1099800420908002
94. Zaree, M.: Multisensory stimulation in dementia. Funct. Disabil. J. **3**(1), 123–130 (2020). https://doi.org/10.32598/fdj.3.19
95. Zucchella, C., Sinforiani, E., Tamburin, S., Federico, A., Mantovani, E., Bernini, S., Casale, R., Bartolo, M.: The multidisciplinary approach to Alzheimer's disease and dementia. A narrative review of non-pharmacological treatment. Front. Neurol. **9**, 1058 (2018). https://doi.org/10.3389/fneur.2018.01058

Chapter 13
Extended Digital Musical Instruments to Empower Well-Being Through Creativity

Elena Partesotti

Abstract At this time of considerable technological development, interdisciplinarity assumes a key role in the advancement of research and the design of virtual reality and immersive environments. In this chapter, I will start by reporting some significant examples of technologies designed to generate new creative and therapeutic spaces that use the user's movements, such as *CARE HERE*, *e-mocomu* and *Sentire*, and I will then describe the design prototyping of *BehCreative*. At the basis of these technologies are a few common features, including a sense of control, a multimodal design, and sensor-free performative movement. I will also address music therapy and rehabilitation, empowerment and Creative Empowerment based on the embodied cognition paradigm—a concept already present in the research of other authors yet never studied in depth or connected with Extended Digital Musical Instruments (EDMIs).

Keywords Creative empowerment · BehCreative · Interaction · DMI · EDMI · Music therapy · Rehabilitation

13.1 Extended Digital Musical Instruments

From 1980 up to now, the use of interactive digital technologies has grown extensively in several fields; in addition, interdisciplinarity between scientific and humanistic subjects has increasingly proved beneficial. A dialogue has emerged from the need to develop "methodologies for designing systems of easier and more immediate use" in [1, 2]).

Over the years, this has led to the design of experimental technologies that are grounded in the interconnection between action and the subject's perception of it

---

"[...] the essence of technology is by no means anything technological" Heidegger, 1977.

E. Partesotti (✉)
NICS, Unicamp University, Campinas, Brazil
e-mail: eparteso@unicamp.br

and are easy to use, even in the health field, such as in the area of rehabilitation and music therapy. When combined with the music sector, these technologies enable users without any pre-existing musical abilities and with physical challenges to play an instrument properly, producing a highly proficient, expressive and creative outcome. As a consequence, the use of technology itself continues to transform, enhancing and reprogramming traditional learning schemes by introducing new, subjective stimuli. This is possible through the use of motion capture (MOCAP) technology, which enables the sensorless body tracking of the user. This is the case with Extended Digital Musical Instruments (EDMI), which means that we use and control these DMI as artefacts "in order to extend our abilities" [3]. In other words EDMI are instruments that allows a distribute interaction within the space on a practical level, and—as property—promote a sense of agency and control on a cognitive one. In this way the user can create new way to cognise with such tools [4]. EDMI are thus extended in a cognitive, sensorimotor and perceptual level.

From this perspective, it appears that the creative and cognitive dimensions are interdependent, since MOCAP technologies promote immersive, performative environments in which the user creates and extends new connections, or new ways to pursue a goal through their movements. And it is precisely through these movements, presented in the context of an exploratory experience, that the user can gain full control over the technology, thus achieving *Creative Empowerment*.

With this term, I refer to the subject's mastery of technology, which—once achieved—facilitates the subject's creative self-expression [5].

Movements are regulated by the person's exploration of the environment through their ability to master sensorimotor contingencies—a form of control that depends on practice [6]. This proficiency can lead to the *flow* that is associated with the creative opportunities that such an experience can offer the user.

Moreover, everyone has equal opportunities for artistic creation and self-expression, as referenced in the psychology of empowerment, and this feature enables us to achieve goals through a collaborative relationship with a therapist, in which the therapist works together with the client rather than acting as an expert. I will describe these concepts in the following paragraphs.

In order to offer the user an EDMI that helps them achieve full creative expression and self-awareness, we need to empower these very users in EDMI design. The design of assistive technology is, in fact, paramount given that the effectiveness of treatment will depend on it. Participatory and co-design methods include different steps in the construction of a user-friendly technology, including users in the design process. These tools should provide users with a structure and scaffold for creativity and, more importantly, these processes should involve users with a diverse grade of disabilities in order to offer assistive technology that is effective yet inclusive [7].

13.1.1 Music Therapy and the Mirroring Phase

Music is a medium for describing how to move, think and socialise [8], acting as a resource for the body that can be thought of as a technological prosthesis, or a material that extends the capabilities and actions of the body. In other words, "[…] music is an accomplice of body configuration. It is a technology of body building, a device that affords capacity, motivation, coordination, energy and endurance" [8]. As with musical instruments and technological systems, music itself must be understood for its non-verbal abilities, activation and imitation of the user (2000). As DeNora points out

> [m]usic's role as a resource for configuring emotional and embodied agency is not one that can be predetermined […] Music not only affects how people feel emotionally; it also affects the physical body by providing a ground for self-perception of the body, and by providing entrainment devices and prosthetic technologies for the body (2000).

Already considered a means for medical treatment and psychophysical well-being in primitive societies (2,000 BC), nowadays music is used as a therapeutic tool that brings psychophysiological and physical benefits to the individual and continues to be studied for its neurophysiological effects. Today, for example, we know that music influences the immune and neurohormonal systems [9], modulating, in particular, the function of the autonomic nervous system and metabolism [10] and increasing self-perceived well-being [11].

Music therapy has been officially recognised at different times in different cultures. In the United States, it only became official in 1998, when the American music therapy Association was formed. In Europe, however, the European music therapy Confederation was established in 1990. The discipline consists of the systematic application of music, directed by a professional in a therapeutic environment, with the goal of activating biological, psychological, intellectual, physiological, social and spiritual mechanisms through the proposed musical activities [12]. The music therapy approach includes different methodologies that professionals can apply, depending on their training, in sessions with clients.

Within the clinical improvisation of the Nordoff-Robbins methodology, for example, music therapy stimulates interaction on various levels. Indeed, one of the most important aspects in music therapy is the concept of innate musicality, taken from the Nordoff-Robbins methodology, which is taught mainly in the United States and applied through clinical improvisation. Also referred to as the *music child*, this concept conveys the idea that any person—regardless of their pathology—keeps their musical intelligence intact and is able to respond naturally and spontaneously to sound stimuli. In fact, the Nordoff-Robbins methodology emphasises the peculiarity of innate music, thus opening up the possibility of direct communication with the client [13]. Improvisation is a fundamental aspect that occupies a central part, as for the therapist, "the music improvised in sessions is an embodiment or expression of the client in any one moment" [14]. Indeed, during improvisation, emotional creativity emerges in the communicative musical act (2000) and in "rebuilding a new schema after the shattering of one's belief system […] and of rediscovery of the mind–body connection" [15]. Therefore, the term "clinical or creative improvisation" is applied as a technique, which represents "specific therapeutic meaning and purpose in an environment facilitating response and interaction" (APMT 1985).

This type of improvisation is free: it is a constant observation of the type of material that emerges between therapist and client, establishing a listening-response dialogue [14]. In Nordoff-Robbins, the therapeutic process involves a period of time consisting of distinct phases, which, as Lorenzo [16] reports, are usually divided into three steps:

1. *Reflection.* The therapist acts as a *mirror* to the client, reflecting their reactions and attitudes, ensuring the client feels listened to and accepted and helping them to achieve self-awareness.
2. *Identification.* In this phase, the client has already reached a certain level of awareness of their own attitudes and reactions.
3. *Contact.* This stage involves the consolidation of the client-therapist bond of trust, in which the subject accepts the therapist's interventions and aid, and positively advances in the therapeutic process.

Regarding the importance of the first phase, in which the music therapist acts as a *mirror* to reflect the client's musical behaviour, Trevarthen and Malloch [14] underline that this process enables clients to see a reflection of themselves to aid self-understanding.

This crucial aspect also connects music therapy with current body philosophy, in particular the *embodied simulation* model, a concept proposed by [17] and the specific phenomenal state called *intentional consonance* [18]. Intentional consonance or *attunement* generates a sort of familiarity with individuals and happens when we recognise others as being similar to ourselves, making non-linguistic interpersonal communication possible through our mirror mechanisms[1] [17]. The mirroring concept—i.e. the exercise that the music therapist practises during the sessions, which consists specifically in acting as a mirror for the client—connects to the mirror mechanism. While this mechanism supports *corporeality*—an intersubjective[2] dimension of our subjectivity, "providing a new cognitive dimension that assists us in defining our nature" (2020)—on the other hand, being connected to the motor system enables the music therapist and the client to organise how the action is executed, and how to perceive and imitate it. The mirroring phase works because, as outlined by the authors:

> When we are present while others are doing something, we immediately comprehend most of their [sensorimotor] and emotional intentions without the need to explicitly represent them linguistically (2020)

In this way, the therapeutic bond is strengthened, making the therapeutic sessions more profitable. Therefore, through music, the innate musical potential present in

---

[1] FOR further reading, see Lorenzo [19] and Rizzolatti and Sinigaglia [20].

[2] The conjunction of the dimension of *otherness* with that of identity. For further reading, see Gallese and Guerra [17].

every human being is used and developed, and through the connection between musical technology and musical therapy, these characteristics can be reinforced. With an EDMI, it is possible to access this innate potential by acting on each phase from which the session is composed, and by structuring interventions that aim to reinforce social skills and expressiveness, regardless of physical and/or psychological challenges, technical skills or the musical training of each person [21].

Although the importance of using virtual and immersive technologies in sessions is starting to be understood, we are still far from being able to systematically apply immersive technologies in this field. This is due to an outdated misunderstanding that considers the implementation of the technological instrument too difficult and its scarce potential if compared to the traditional musical approach. Fortunately, this 'dated' consideration is being overcome, even through the period of isolation we have experienced in the last 3 years. As different authors have outlined [5, 22, 23, 24], technology should be implemented in music therapy considering its efficacy and the remarkable demand for applied technology in clinical practice sessions.

As already mentioned, this limitation is motivated by the inadequate technical and theoretical training of professionals. In fact, both research and the systematic application of EDMIs and DMIs in music therapy are scarce and largely consist of pilot experiments or technological prototypes from research projects, which— once such projects are completed—are hardly available or unavailable for sale, are prohibitively expensive, or cannot be accessed in the places where therapy is carried out.[3]

13.1.2 EDMIs with Therapeutic Purposes

Several scholars have pointed out the importance of technology within music therapy practice in the past few decades. One of the conclusions that emerged from a review by Hahna et al. [25] was that digital technology had been successfully applied for assessment and evaluation purposes in music therapy, as well as providing clients with a new, more creative approach. Although dated, this review is valuable in that it highlights the importance of the evaluation component in music therapy and the benefits of technology in enhancing the client's expressivity. This last observation anticipates significant changes in the coming years in music therapy sessions. Some of these have been accelerated by the new need for virtual interventions provoked by COVID-19, while other more gradual changes have witnessed the inclusion of music technology in school curricula or in the latest editions of music therapy manuals.

In the music therapy field, there are still major problems in terms of the flexibility to accommodate diverse learning abilities in the technological area, as these devices require extensive training in computer science [5, 26]. From the client's perspective,

---

[3] For an example of a pilot experiment with an EDMI for Music therapy, see Motion Composer™ in [21],for an example of a research project on multisensorial technologies, see the RHYME project in Cappelen and Anderson 2016.

however, one of the greatest obstacles is musical expression itself: not everyone can play a musical instrument or follow a rhythm by coordinating their movements. In fact, the majority do not have the adequate notions or skills to facilitate communication during therapeutic sessions for either party. This presupposes an indefinite number of sessions in which to develop a relationship based on free expression insofar as it will be limited by the knowledge and possibilities of the client. In other words, this lack in musical knowledges poses a threat to the mutual collaboration, making it more likely that a hierarchical one will be established to the detriment of the customer's expressiveness.

In the digital era, technology is a medium in the relationship between a person's inner world and the external world—an extension of ourselves. From this point of view, there are DMIs that can mediate the needs of the patient and the therapist (whether it is through music therapy or physical rehabilitation). This mediation takes place through proper engagement. DMIs can take the shape of traditional instruments, or something quite different, as they contain a sound generation unit and a control surface that can be separated [27]. DMIs can be used in multiple contexts, providing benefits for both experts and non-experts in artistic and musical fields (2006). When referring to DMIs, I distinguish between DMI and Extended DMIs (EDMIs), where the body itself becomes the digital musical instrument, as the interaction occurs through a correlation between the environment and the agent, making the distributive nature of the space offered to the user essential as its property of manipulation and incorporation. In this sense, motor skills are strictly linked to abilities of sound expression and, in other cases, visual expression. Therefore, given the nature of this performative space, therapeutic and pedagogical learning takes place through movement. This highlights the importance of considering proprioception as a bodily basis for consciousness in music [28]. Indeed, Peñalba outlines that the internal simulation of movement (mimesis) and exploration enable us to understand musical aspects that we would not otherwise understand. This concept can shift within an EDMI, as it implements MOCAP or sensorless systems alongside a performative space where the subject may use their body to explore and play around. One subcategory of motion analysis is the above-mentioned MOCAP, which involves camera-based sensors and systems—Kinect by Microsoft, for example [29]—and is usually preferred in the design of new instruments for musical expression due to its low cost and relative ease of use. Moreover, as Mulder points out [30], alternate controllers include three further subcategories: touch controllers, in which the surface has to be touched physically; extended-range controllers, in which it is not necessary to touch the surface although there are limitations in the range of effective gestures; and immersive controllers, which place only a few restrictions on the user's performance.

Thanks to MOCAP systems, an EDMI can become an appropriate instrument for self-awareness, making movements easier to perform and being equally available to everybody. Moreover, the possibilities that EDMIs encompass reside not only in the therapeutic field but also in the pedagogical and artistic ones, which is why I prefer to refer to an extended DMI rather than just an assistive DMI. In this sense, technology becomes a social instrument for equality, regardless of age, race or above all, disability.

Below, I will propose three examples of EDMIs that rely on sensory integration. Two of them implement audiovisual feedback, while the third audio feedback.

CARE HERE

CARE HERE stands for *Creating Aesthetically Resonant Environments for the Handicapped, Elderly and Rehabilitation* and stems from a European project. The rationale behind the project was to create an environment that enables the user to express the experience they perceive in it [31]. This concept is applied in technology with the aim of encouraging a greater awareness of one's body and movements in children with neuromotor deficits.

As outlined by Brooks and Hasselblad [32], CARE HERE is based on the concept of *Aesthetic Resonance* [33], which occurs when technology responds to a user's effort immediately and in an aesthetically pleasing way. As a consequence, the subject forgets the intention that prompted them to enact a physical movement, and therefore an effort, to receive feedback [32]. The framework underpinning this project is based on "[…] open architectural algorithms for motion detection, creative interaction and analysis, including the proactive libraries of interactive therapeutic exercise batteries based on multimedia manipulation in real time."

Behind CARE HERE there is another project by Brooks himself called *Soundscapes* (Fig. 13.1). The project consists of an "interactive virtual space" (2004) in a motion capture library and a collection of programs capable of providing indirect feedback with respect to the user's audiovisual gestures to trigger a *feedforward* (*homeostatic control system*, 2004)—in other words, a continuous retrofeeding mechanism that anticipates the result associated with the motor action on the feedback. CARE HERE applies the concept of back-feeding by offering the opportunity to

Fig. 13.1 A user during the authors' experiment and the type of visual feedback related to her movement. Reproduced from "Creative aesthetically resonant environments for handicapped, elderly and rehabilitation: Sweden" by Brooks and Hasselblad [32, p. 195] with permission from Brooks

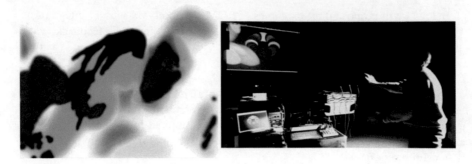

Fig. 13.2 Head/hand within an Eyesweb Bodypaint Aesthetic Resonant Environment on the left and the prototype enabling gesture control on the right. Reproduced from "Soundscapes: the evolution of a concept, apparatus and method where ludic engagement in virtual interactive space is a supplemental tool for therapeutic motivation" by Brooks [31, pp. 17, 12] with permission from Brooks

select the type of feedback according to the user's preferences. For Brooks and Hasselblad [32], this opportunity makes it possible to bind the user to the same psychological dynamic that underlies games and video games as it evokes an: "understanding of the causality involved and is analogous to the *flow state* involved in play and game psychology—exhibited when a child is engrossed in a computer game."

The CARE HERE technology is based on a sensor system. An algorithm was applied in collaboration with the InfoMus Lab in Genoa, Italy, using the EyesWeb software, which activates movement to visual feedback, otherwise known as *Silhouette Motion Images* (Fig. 13.2). In addition, the movement was segmented through a special program in order to distinguish the phases of motion from the user's pauses [32].

Sentire

Sentire is a body-machine interface [34] that sonifies proximity and touch to enhance body perception and social interaction. It uses a digital system that mediates body movements and musical sounds in the form of two bracelets, which enables two (or more) people to interact with each other. Audio feedback is heard when proximity and touch are detected. The artwork is thus thought of as a participatory performance in which a guiding performer invites the spectators (one at a time) to interact with him/her while wearing the bracelets for 6–10 min. The interaction develops either standing (with eyes open) or sitting (with eyes closed), yet always in a non-verbal context (Fig. 13.3).

Driving this technology is the belief that auditory feedback has the potential to increase embodied experiences and coordination for therapeutic purposes [34]. Distance and touch between the users can be measured and mapped in real time to an algorithmic sound environment [35], making it possible to digitally design real-world proxemics [36]. Through this multimodal experience, an awareness of the self and the other is enhanced on bodily—especially kinaesthetic—levels, i.e.

Fig. 13.3 Two users of Sentire, reproduced from https://www.interaktive-technologien.de/pro jekte/sentire by Lussana 2019, with permission from Lussana

movement perception [37, 38]. During the interaction, a process of co-determination occurs, creating a so-called perception–action loop [39]. Since 2019, Sentire has been embedded in a research project at the Humboldt University of Berlin. It is now developed for therapeutic purposes, taking into consideration the increasing incidence of social isolation, chronic stress, and diminished body awareness caused by psychosomatic illnesses and mental disorders.

E-mocomu

The e-mocomu prototype denotes e-motion, colour and music, and is an EDMI developed under an interdisciplinary perspective with the primary aim of enabling users to control sounds and colours through their movements in space. E-mocomu stemmed from an interest in synaesthesia—in particular, chromaesthesia (sound-colour)—and the search for a theoretical correspondence between sounds and colours mediated by emotions and to be applied as feedback for the user [40]. In chromaesthesia, areas of the brain associated with the auditory system of the fusiform gyrus are activated together with those involved in colour processing [41]. Diverse applicable theories have been studied—Newton, Munsell, Wells, Scriabin, Marion, Kandinsky and Veronesi. Eventually, in the pilot experiment, the correspondence theory of the Italian artist from the Viennese Secession, Luigi Veronesi, was applied (Fig. 13.4). The artist conducted an in-depth study of music and colours from a scientific-mathematical point of view by graphically representing the duration of notes, pauses and silence and attributing an appropriate colour with a specific brightness variation to each tone and semitone [42]. He also highlighted the correlation between colour and pitch, with lower notes represented by darker colours and higher notes by lighter ones.

The first prototype was composed using a Kinect quartz composer and synapse software. The latter version implemented processing software for an audiovisual

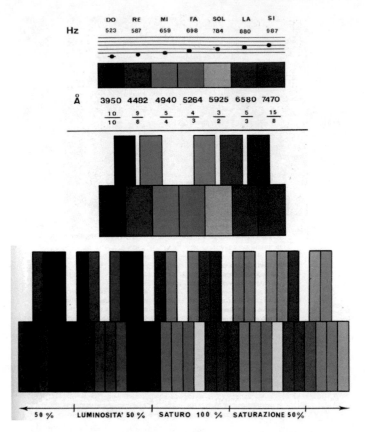

Fig. 13.4 The chromatic and natural musical scale according to the frequency ratios calculated by Veronesi. Below: the piano octaves represented chromatically. Reproduced from "Proposal for a theory on the relationships between sounds and colours" by Luigi [42, pp. 19–21] with the permission of the Luigi Veronesi Committee

output. Once in the immersive space, the user could receive three kinds of audiovisual feedback linked to the body movement within it. No movements corresponded to stopping the EDMI. In an initial study, we analysed the responses of three groups of people to audiovisual stimuli, their learning capacity, and their emotional and improvisational feedback after performing several audiovisual and motor tasks [43]. The results showed a positive self-perceived outcome of the performance as well as a positive boost in self-confidence. A more recent version of e-mocomu was introduced at an *OltreMusica* event in Padua, Italy, in a workshop with people with disabilities from the association (Fig. 13.5). During this workshop, children and people of different ages with and without disabilities—such as autism spectrum disorder (ASD) among others—were able to enjoy a free improvisation session with e-mocomu. The experience was highly rated by the users.

Fig. 13.5 Images taken during the workshop within *OltreMusica* event at Cooperativa Nuova Idea, Padua, 2019

In the examples proposed here, an EDMI was able to boost progression and development during therapeutic sessions, both in music therapy and rehabilitation. This happens when the user is motivated to work from inside out, which refers to the internal motivation of the user to independently take control of the situation [44]—a condition that occurs when there is freedom of interaction in an interactive environment.

CARE HERE and e-mocomu, in particular, rely on the creative interaction of users, who manipulate the audiovisual feedback with their bodies.

In yet another study, which dates back to 2010, Camurri et al. [45] tested the *stanza logo-motoria*—an interactive multimodal environment—demonstrating the efficacy of a multimodal system in treating dyslexia within an educational context, and advocating its implementation with ASD clients.

Peñalba et al. [46] note that some interactive musical technologies reduce the differences between users with and without disabilities, bringing both classes of users to the same level of competency as that shown by skilled musicians. In other words, these technologies enable users without musical skills and with physical difficulties to play an instrument properly, and to produce a highly proficient learning and expressive outcome [5, 21].

13.1.2.1   On Virtual Reality for Rehabilitation

The digital landscape has changed radically in the last 35 years, paving the way for new trajectories for expression and creativity in the many fields in which these media are implemented [47]. In this panorama, virtual reality (VR) made its way first and foremost as an innovative tool for entertainment, before entering the pedagogic field known as *playful learning* [48], thus "improving student engagement, promoting creative thinking towards learning and developing approaches towards multi-disciplinary learning." [48]. Moreover, its ecological validity means VR is also widely used in rehabilitation, particularly in motor rehabilitation. As Gallese and Guerra [17] underline, VR has absorbed the models of intersubjectivity and agency experienced in cinema. Cinema, in fact, is an excellent example of a mirror mechanism—a mechanism that supports our intersubjective dimension and provides us with a new cognitive dimension—to understand the direct experience of emotions in the era of interactive technologies and what the authors define as *embodied simulation*:

> [...] a basic functioning mechanism of the brain-body system [...] By activating sensorimotor and visceromotor maps in the brain of the observer this system can facilitate the construction of a direct and non-linguistic relationship with space, objects, actions, emotion and sensations of others. (p. xix)

Therefore, while the application of VR-based rehabilitation can be effective through the embodied simulation mechanism, we should consider it in conjunction with other therapeutic interventions [49, 50]. VR is a valuable option for neuro-functional recovery rehabilitation used with different methods and purposes, such as post-stroke rehabilitation of the upper limbs, for balance, or to improve spatial orientation/navigation and stimulate movement in the lower limbs [51, 52], and through Rehabilitation Gaming Systems (RGS) [53]. The core aim in using RGS is to convince patients that they can perform therapy-related movements, thereby leading the brain to believe that they can perform the simulated movements, while in reality such movements are limited by their disabilities. In this way, it is possible to accelerate the neurofunctional recovery process. As several authors underline [54, 55], rehabilitation in the post-stroke patient is concentrated mainly in the upper limbs because the patient's quality of life depends on them, but also because VR is more effective in this sector. Indeed, despite the wide fields of application, it appears that VR and gaming VR are effective in both neurological rehabilitation and in the performance improvement of older adults [56] in upper extremity treatment [57]. Serious game applications—game design for application in fields other than pure entertainment, such as in the learning and rehabilitation context—are still growing and are typically based on MOCAP. This category is the most extensive and most widespread in the current panorama, enabling the extraction of features such as acceleration, velocity and positioning in space in order to understand the behaviour of the subject for therapeutic purposes. It is important to note that accelerometers are still the most used in the technological and artistic-performative field and that, for the most part, they are implemented in other systems [29]. The increasingly widespread use of implementing sensors over other motion analysis techniques—despite being insufficient in certain aspects—is preferred due to their low cost and ease of use (2014).

Among the VR technologies implemented in this field, there is little use—if any—of technologies based on and created with the musical trajectory in mind, such as an EDMI, which uses musical elements with a creative and expressive background. In their research, Stephan et al. [58] claim that "[...] exposure to a movement-related tone sequence can crossmodally and specifically affect subsequent performance of a motor sequence that has never been physically practiced" (p. 7). The authors also highlight the need to understand auditory-motor system interactions in order to contribute to the development of new strategies using sound as a neuromodulatory tool for motor rehabilitation. In another study, Särkämö et al. [59] noted that regular, self-directed music listening during the early post-stroke stage can enhance cognitive recovery, increasing dopamine levels and preventing a negative mood. The authors highlight the importance of everyday music listening during early stroke recovery by providing an individually targeted, easy-to-conduct and inexpensive means to facilitate cognitive and emotional recovery (2008). The use of music in physiotherapy can therefore provide multiple benefits, as it stimulates both sensory and cognitive integration—that is, performing multiple tasks at once, such as singing, tapping, moving, clapping, etc.—which is essential in everyday life. Furthermore, music therapy focuses on the emotional process that psychotherapy usually fails to address. In this way, its implementation would provide additional benefits. The link between motor entrainment and emotional processes is underlined by several authors [60, 61]. In this field, EDMIs could appeal to and be used in post-stroke patients, tackling a rehabilitation issue linked to the monotony of treatment sessions [62]. As expressed by Zagalo and Branco [47], a creative technology must spark interest and be easy to interact with, like a toy, or engage, motivate and maintain concentration while pushing for mastery (2021, 14) among other aspects. These prerequisites should also be considered when designing an EDMI for rehabilitation. By adding a musical component, such as rhythm, we could also consider its therapeutic application in people with ASD, Parkinson's disease, or post-stroke patients who currently demonstrate good responsiveness to treatment with VR. The studies currently available on people with these types of special needs help us to understand that, during the design phase of an EDMI, it is possible to tailor software to the needs of the user [5], boosting potential strengths and disregarding any challenges imposed by the disease [63] to avoid potentially deleterious effects [64].

13.1.3 Embodiment, the Environment and EDMIs

EDMIs are technologies applicable in the field of therapy, but which are closely linked to artistic performance and therefore to creative expression [65], the father of *embodiment*, has left us a great philosophical legacy to reflect on—one that helps us to decode interaction and performance from a perceptual-phenomenological perspective. Our body enables us to interact in and with the outside world through movement, and in this digital age in which we encounter more and more interactive experiences [66], this interaction is benefited by technology. Experience is the interpretative key for us to understand abstract domains and is based on the body and on the perception of its exterior. Here, it is the public—who is also the user—who is involved in the creation of their own interactive experience.

When discussing interactive technologies such as EDMIs, we know there is an open dialogue about the meaning of *interaction*, both as a noun and as an adjective [67]. In this chapter, since the EDMI is an interdisciplinary product stemming from both physical and mathematical design alongside a sociological and philosophical concept of the theory of embodiment, the reader will observe that both these aspects are considered.

Working with immersive multimodal environments, it is worthwhile considering the embodied cognition paradigm in order to situate the interaction that occurs between the agent and the environment, such as the sensorimotor contingency theory (SCT). According to the SCT [6], perception and action are interconnected since they influence the configurations of the agent's sensorimotor system (here, the user).

So, when dealing with immersive audiovisual technologies that enhance the user's self-awareness, we need to consider the role of *proprioception*. As indicated by Laskowski et al. [68], proprioception is one of the somatic senses that collects sensory information from the body, although it does not belong to one of the five basic senses. Thanks to proprioceptors, we are aware of the movements of our body in space and are able to control them. In this sense, proprioception can be conscious and voluntary, or it can be unconscious and allow muscle function to stabilise the joints that need to be modulated (1997). It is possible to use the different types of proprioceptors to classify EDMI technologies; for example, in CARE HERE and e-mocomu motor proprioceptors are used, while in Sentire mainly passive tactile proprioceptors and secondarily motor ones play a part.

In the embodied cognition paradigm, on the other hand, the environment is the result of a series of relationships produced by the activity of the subject. According to the enactive approach, the internal world of the subject and the external world in which the interaction takes place are not separate but rather domains that complement each other [69]. The continuous designation of the external and internal domains mentioned above influences perception and action. From the enactive point of view, what the subject is able to experience, know and manipulate is influenced by their own structure of being [70]. Therefore, within a responsive environment in which the perception and action of the subject are linked and influenced by the relationship with the space, this perception can increase and stimulate certain behaviours through precise sensorimotor stimulation. Indeed, in the SCT, the interaction between subject and environment is regulated by sensorimotor laws. The ability to master sensorimotor contingencies in every sensory modality derives from practical knowledge [6]. Thus, perception corresponds to changes in the sensorimotor contingencies in relation to the movement of the subject. Within this performative environment, the user's perception is grounded on their self-awareness of bodily changes as a result of their actions. The same actions modify how sensory stimuli are perceived. In this dynamic, the practice leads to a certain degree of control over the object, or mastery. In fact, as explained by Caruana and Borghi [71], the sensorimotor system automatically extracts the 'offers' from the observed objects and encodes them in terms of potential actions.

Among the features of musical creation outlined by Clarke [72], one is particularly meaningful when referring to EDMIs: its intimate connection with environmental affordance. And since we are dealing with EDMIs in which the movement of the subject corresponds to audio and/or visual feedback, the aforementioned concepts lead us to consider the Gibsonian concept of affordance (1998) and to project it into the virtual domain. Indeed, *affordance* refers to the characteristics of the environment offered to the subject: "[…] a point of contact between the creature and its environment, an environment in which the creature moves around and within which it acts" [73]. By transferring these concepts to an EDMI, the affordance is virtual in the sense that it recalls a series of Sensorimotor Maps that the user activates through their exploration of space and the feedback received from the immersive environment (mapping). Through them, the subject is embedded and even extended within the environment. In an EDMI, the client's extended embodiment is implied during the interaction—the user can feel themselves directly projected into the environment from a holistic perspective. In this way, the empowerment process appears as a social construct that links the user's inner and outer worlds through technology, which is both social and technical since it is implemented in both settings [8].

13.1.4   Sensorimotor Contingencies, Creativity and Sensory Integration

Despite studies on creativity, what has become apparent in recent years is the need for investigations that focus on the experience, since our body is "the ultimate source of our experience of ourselves and our relationship with the world" [17, p. 11]. From this perspective, studies that establish a connection between immersive technologies, creativity and user empowerment are still scarce.

Creativity, in fact, is usually associated with an increase in sensory gating and attention control [74, p. 202], while emotional (and mental) processes that regulate goal-oriented behaviour are tied to cognitive control [75]. It follows that an enhancement of sensory gating and attention control is tied to emotional processes and *Creative Empowerment*, a concept described further on, regarding EDMI. Cognitive control "can be understood as emotional process" (2015, p. 8); that is to say that emotion and conflict lead to cognitive control. Thus, emotion is paramount in eliciting and regulating conflict. In fact, as the authors outline: "There are many reasons why emotion would play an important role in the engagement of cognitive and behavioural resources to resolve conflicts [..]" (2015, p. 6). That is to say, emotion regulates goal-oriented behaviour (and cognition) by capturing attention and mobilising an organism for action, as well as helping promote adaptive behavioural responses to handle conflict.

Various authors [76–78] claim that brain imaging studies reveal an apparent connection between mental disorders and high levels of creativity as a result of shared brain characteristics. In fact, creativity appears to be influenced by a multitude of genetic variations [79]. On the other hand, music and music-making can be considered creative actions. Music is a universal language and represents a multisensory experience—both perceptive and physical. It has already been demonstrated that it affects the nervous system by altering levels of dopamine, norepinephrine and serotonin [80]. Moreover, exposure to music enhances brain plasticity and facilitates a wide variety of emotional and cognitive functions besides having a generally positive effect on post-stroke patients [59]. In the rehabilitative field, Riley et al. suggest [81] that technologies aimed at improving the daily life of people with dementia can foster creativity and help them perform different tasks that empower them. This statement is also valid for people with various kinds of disabilities, as well as in the pedagogical sphere, for clients and professionals alike [82]. Riley et al. [81] also highlight the importance of active music-making as an enjoyable tool and a potentially empowering experience. An effective way to apply these concepts to EDMIs, considering the mechanisms underlying cognitive control and the reward system, is reinforcement learning. Reinforcement learning is based on reward prediction error (the difference between the outcome of a choice and the chosen value) and value functions (estimates for the sum of future rewards) where the actions of the organism are chosen probabilistically and which ultimate goal is in fact to maximize future reward [83].

In light of these considerations, an EDMI could also be effective in rehabilitative recovery and in music therapy sessions, for example, through rhythm. The importance of rhythm in this field has been addressed in the past 15 years for its efficacy in treating people with diverse disabilities [64, 84] both in physiotherapy and in music therapy rehabilitation. Within music therapy itself, there are different methods that apply rhythm in the sessions, such as functional music therapy and neurologic music therapy, whereby the latter focuses on improving the client's speech and timed movements [84]. At the same time, neuroscience has shown that the integration of various sensory modalities occurs in the human brain [85]. Thus, the kind of performative experience that users have with technology should be aligned to provide them with a form of *coherent multisensory integration* close to what they experience in the real world, and one that could be strengthened through music. This situation—coherent multisensory integration—may occur and be strengthened within a performative technology with multimodal feedback, which reproduces a synaesthetic experience. As Custodero [86] points out, clear feedbacks are paramount for achieving a state of *flow* and thus a creative experience. For [87], however, sensory integration helps develop an adaptive response on both the subcortical and cortical levels, which involves cognitive and emotional processes, providing the material upon which to work in therapy. Its inherently multisensory characteristics mean that music can already give us diverse feedback such as auditory, visual (the score) and bodily feedback (2011). The studies conducted by Koelsch [88] and the reviews by Lin et al. [89] indicate that music therapy improves non-verbal communication, in addition to social interaction and self-expression, which are key factors in controlling the client's depression and anxiety. Somatic and cognitive anxiety can indeed be reduced in people with a moderate and high trait through music treatment [90]. From this perspective, interactive technology aimed at clients with SLD should typically offer programs that can be modified in real time, according to the participant's behaviour; ones that are capable, for example, of recognising the direction of the subject's eyes for a continuous (non-aggressive) interaction; and ones that offer a risk-free environment for the client. The ECHOES project [91], for instance, proposed learning activities for clients with ASD with different objectives and which change according to the attention of the participant and their exploration of the environment. Considering these statements and given the importance of sensory integration in children with ASD, existing interactive systems could become a valuable tool in enhancing expression and creative play. Moreover, music technology has great potential by offering a multimodal approach with tangible interaction design, oriented music therapy and empowering thinking [92].

According to Clarke [72], a creative process occurs through sensorimotor engagement, among other aspects. Likewise, for the SCT [6, p. 572] the experience appears to be a temporally extended exploratory activity mediated by the subject's sensorimotor contingencies. Therefore, sensory perception is the "ability to explore one's environment in ways mediated by implicit knowledge of patterns of sensorimotor contingency that govern perceptual modes of exploration" (p. 569). Indeed, it is an association of different perceptual-sensorial modalities of which the subject is aware, even at an implicit level. In other words, perception is formed through changes in sensorimotor contingencies that the subject perceives during their movement to reach the initial stimulus (i.e. moving towards an object). While moving, however, the perception of the subject will change, since the information provided by the stimulus will change in relation to the movement itself. Movement and proprioceptive perception are therefore the basis of a co-determination process that is generated during perception and action—through *Multiple Affordances* and *Sensorimotor Maps*, which I will explain in the next paragraph—and which occurs within an EDMI where the user's capacities are empowered and enhanced [5].

Hence, as also emphasised by the SCT, we experience the integration of different sensory modalities through movement, while different physical explorations encourage different perceptual modalities. In this regard, Ruggieri [93] argues that the interaction in space is always synaesthetic since it is linked to the interaction of different sensory modalities in real time. For the author, space and movement are intertwined and the body image is processed based on perceived sensory information, which can be of a different nature (tactile, proprioceptive, sound, etc.). In "constructing the image of their own body, the different subjects can favour one sensory modality over the others, using, for example, mainly the visual channel over the cenesthetic or acoustic one and vice versa" (1997). In this way, it is the subject who unconsciously chooses what sensory information to prioritise.

In the examples provided in paragraph 1.2 and in the description of the BehCreative pilot experiment in 1.5.1, the designers pursued the goal of integrating different sensory modalities within an EDMI, to offer a complete user experience. This way, the user encounters a feeling of *non-mediation*: in other words, the technology disappears [94] or becomes an extension of the body [8]. Effectiveness, however, will depend on the subject's creative and emotional processes, aided by the design of the EDMI itself as adequate tool for Creative Empowerment to occur. Furthermore, according to Gallese and Guerra [17], contemporary cognitive neuroscience is able to investigate the fundamental role of the body in creative expression and its reception—a current field of research, according to the authors.

For these reasons, a technology-driven shift in performative spaces needs to be encouraged, based on general theories on cognition, emotions, creativity and music-related neural activation. To do so successfully, it should engage the field of neuroscience to better understand the underlying mechanisms.

13.1.4.1 Sensorimotor Maps and Multiple Affordances

The body assumes great importance in the process of physical and phenomenological understanding. Together with the body, technology as an art form becomes a useful tool in understanding the relationship between internal and external dimensions. Therefore, in the performative act and within an immersive technological space, the subject continuously changes their body map by restructuring it. This also happens through their incorporated perception, as underlined by Kozel: "Performance is never one-directional [...] performance involves the awareness of being in a state of reception and initiation between inside and outside, modulation and response" [95].

Alongside the concept I refer to as *sensorimotor map* is that of *cognitive map*, developed by Benford and Giannachi [96], which indicates the continuous adaptation undertaken by the user in a mixed environment—i.e. both real and virtual—which obliges them to continuously monitor and engage with the two realities through the perception and use of these cognitive maps. Here, the concept of a sensorimotor map already incorporates the cognitive dimension. Within an EDMI, proprioception plays a fundamental role in the creation of Sensorimotor Maps since the participant is offered not only a free choice of movements based on their exploration of the environment through sensorimotor contingencies, but also free artistic expression, of which they are the direct protagonist. According to the SCT and the concept of the sensorimotor map proposed here, the sensorimotor contingencies within an EDMI will depend on the proficiency acquired by the subject through practical knowledge. That is to say, the user will perceive changes in the different sensory modalities through their movement, heightened by the technology in which they are immersed, and will learn to master them. Furthermore, when using an EDMI, the individual should know which movements produce specific feedback, and it is precisely through the SCT that the user constantly verifies that their actions are the ones required to produce the expected result, thus satisfying their expectations. In this sense, the user charts their own Sensorimotor Maps, stemming from exploratory movement and which will enable the creation and use of Multiple Affordances. They will also allow for the storage of information and experiences encountered within the EDMI. It is precisely through these movements, provided by an exploratory experience, that the user will achieve full control over the technology, reaching *Creative Empowerment*.

Being aware of sensorimotor contingencies means knowing which movements are necessary to produce certain changes, even in technological design, and which sensory modalities guide us towards the movements aimed at obtaining them. Therefore, depending on the type of technological instrument that is used, there will be a different sensorimotor pattern.

These reflections express the concept of *Multiple Affordances*. With this term, I intend to define the presence of more than one affordance, whether potential or real, which are available within the interactive space to the user. As summarised by Anderson and Sharrock [97], the affordances of objects are "constructed and reconstructed in and through the courses of actions in which we engage" (1993). Within an EDMI, however, there are no physical objects, so affordances depend on the body parameters identified through Sensorimotor Maps and mapping. Hence, these affordances represent the possibilities of the subject to interact with the immersive space through the Sensorimotor Maps constructed internally. In particular, Multiple Affordances continue to develop and change depending on the level of control and experience that the subject obtains. Therefore, the more the user reacts and interacts with the feedback, the more the affordances will transform. In other words, the level of experience can pave the way for much more complex actions than those 'suggested' by the object at the beginning of the exploration. Furthermore, Multiple Affordances can be used and explored voluntarily by the user.

Ultimately, since our body is "a priori the ultimate source of our experience" [17], to study the relationship between the body and the immersive environment—without forgetting the concept put forth by the authors of *Embodied Simulation* (see 1.2.1)—we should remember the importance of our motor system, which is activated even when we are stationary, but are incorporating a simulation mode of what we have seen into an inner understanding (2020). In fact, within an EDMI, the user moves with intent to explore the space and to interact through different phases, thus creating new Sensorimotor Maps or repeating those already internalised organically or learned during the explorative experience. As the authors [17] argue, "a movement is a simple dislocation of body parts, like flexing and stretching fingers. A motor act, however, consists of using these movements to attain a motor goal [...]" while "the premotor neurons are much more sensitive to the goal behind the action than to the individual movements necessary to attain it" (2021). A central point of this discourse concerns the goal of this interaction or the goal of the user specifically, which, in the case of EDMIs like those described here, are based on movement for expressive and creative purposes. Therefore, the act of learning through a creative experience within an EDMI is fuelled by rewarding emotional expectations and triggers Creative Empowerment.

13.1.4.2 Empowerment

The term *empowerment*, coined by American psychologist Rappaport, dates back to the second half of the twentieth century and was later adopted by other disciplines such as politics, psychotherapeutic treatment, pedagogy.

There are several definitions of empowerment. Among these, it is important to report that of Kiefer [98], for whom it represents a process of acquiring skills, abilities and information useful for the social and individual aspects of people. According to Zimmerman [99], "Empowerment suggests a distinct approach for developing interventions and creating social change".

At the basis of empowerment is collaboration, which sets the psychology of empowerment apart since the professional cooperates with the client rather than acting as an expert, thus avoiding passive acceptance by the client.

In the field of music therapy, is the concept of empowerment proposed by the music therapist Rolvsjord, who in 2004 argued that there are different types of interaction—subordination, collaboration, domination—according to the client's needs. The author talks about psychological empowerment and the importance of this within music therapy, recognising the various advantages it brings to sessions, referring in particular to cases of senile dementia she has treated [100].

According to Rolvsjord, different definitions of this process exist precisely because of this multi-level construct. According to the music therapist, one should employ a resource-oriented, collaborative and participatory approach. Furthermore, the concept of empowerment raises questions about general and individual intellectual health as well as therapeutic practices. This concept invites us to observe the therapeutic relationship and its power dynamics, and otherwise implies a political dimension within clinical practice, stimulating a discussion between subjectivity, objectivity, research and politics. Finally, it concerns the recognition of the client's rights in terms of music. Some of these advantages entail a discussion about the nature of music therapy and the health that empowerment affects, as well as the recognition of the client's rights in music therapy. Indeed, focusing therapy on empowerment encourages the client's resource-oriented development and practice, as well as collaboration with the therapist.

13.1.4.3 Creative Empowerment

This concept is the result of an investigation in the field of music technology applied to the music therapy domain, beginning with the design of e-mocomu and following with the BehCreative environment, where the subject experiences complex audio-visual modalities according to the type of movement they undertake, and in which the movements are related to Gibson's notion of affordance [101]. In fact, as Clarke reminds us [72], Gibson [101] himself defined affordance as the property of the object in relation to the capacities of the user. Hence, it is not a static attribute we simply perceive. As he explains: "affordances are discovered through active engagement" (2021). They are reciprocal, which means that the kind of aesthetic perception we experience in an EDMI can be extremely different between users. Creative Empowerment is the result of specific processes—exploration, Sensorimotor Maps, and multiple trajectories—that happen within the EDMI, between the user and the environment, and which will be described here. To understand this concept, we should adapt the SCT to within the context of an EDMI.

As explained earlier, the sensorimotor contingency within an EDMI occurs through the manipulation and exploration of the immersive space through the subject's movements. This implies the accomplishment of a state of control over technology, or a *flow*[4]—a psychological state associated with creativity (Csikszentmihalyi 1996 in Wilkie et al. [102]), which enables the subject to express themselves autonomously and without reservation from a creative point of view. In this case, we will discuss *Creative Empowerment*, which derives from a "feeling of control and self-determination in the client, gained through the exercise of creative expression" [5]. In order for this control to take place, a second person—in this case, the figure of the therapist—may be present to put forth ideas and help the client in the exploratory process, according to the mapping of the technology.

In this way, the subject receives pleasant feedback from the environment and their own body during Creative Empowerment through the self-awareness of their movements.

According to Ruggieri [93], motor activity is the basis of the pleasure-displeasure dynamic, referring in particular to "pleasure as being closely linked to bodily activity". One such example is attention, which is followed by an expectation that corresponds to muscular tension, leading in turn to the muscular resolution triggered by the onset of the stimulus. This "internal feedback" of pleasure, associated with the subject's positive thoughts about their performance, allows for the creation of Sensorimotor Maps following the positive expectations created during the interaction. In this way, the subject can express themselves freely through movement in the performative space.

---

[4] This state is attained in a technological context when: "[i]n order to remain engaging, consuming and flow-like, activities that involve musical instruments must offer continued challenges at appropriate levels of difficulty: not too difficult, and not too easy" (see Csikszentmihalyi, quoted in Wilkie et al. [102]).

Benford and Giannachi [96] also emphasise the creative aspect in achieving control, particularly in "command-and-control" technologies (p. 7), as they facilitate creativity and play. While Swingler [103] and Ellis [104] highlight the possibilities offered by interactive environments for children with SLD through the concept of aesthetic resonation, Brooks and Hasselblad [32] refer to aesthetic resonance as "a situation where the response to intent is so immediate and aesthetically pleasing as to make one forget the physical movement (and often effort) involved in the conveying of the intention".

These definitions share a common theme in terms of achieving a state of control, which may be otherwise driven by the therapist's mediation.

It is useful to note that in Creative Empowerment the music therapist or physiotherapist will intervene to adapt an EDMI to their client, supporting them, albeit without influencing the user's awareness of movements, exploration, expression, composition or retrofeeding, or the creation of Sensorimotor Maps, which can take place within the same.

Therefore, Creative Empowerment is a condition that is fostered by these extended digital technologies through the self-awareness of movements [5] as well as physical instantiation and situatedness in an environment [67]—that is, interaction. In addition, it is a creative process that does not rely solely on the body-awareness of the user but on the subjective belief of self-efficacy [105] since it comes from the empowerment theory. Moreover, the role of emotions in creative processes is also fundamental and has been studied extensively at an experimental level [106], although none of these studies took place in immersive technological environments. Indeed, given that emotion affects and fuels the creative process [76, 106], intrinsic motivation plays a vital role, based on enjoyment and challenge. The concept of Creative Empowerment is of course grounded in intrinsic motivation, considering that it is a matter of transforming "general and specific skills into creative behaviour" [107].

In this manner, Creative Empowerment is grounded in:

1. A sense of control over the technology; a mastery that leads to a state of flow;
2. The self-awareness connected to proprioception, resulting from a cyclical process of co-determination between the different parts involved (Multiple Affordances, Sensorimotor Maps);
3. A sense of self-efficacy that emerges on a psychological level and arises from the subjective belief that the user perceives about their own creative abilities (2019);
4. The intrinsic motivation that appears when the experience is enjoyable yet challenging.

13.1.4.4   The Conceptual Framework of Creative Empowerment

In the previous paragraphs, I have described the theory of Creative Empowerment, its foundations, and the importance it assumes in various fields such as in music therapy and pedagogy [108].

It is worth remembering that perception, which in this context concerns sound, image and proprioception, occurs through sensorimotor contingencies; that is, the relationship between the informational changes produced in the environment and the movements produced by one's body. For this reason, there is an interdependence between perception and action, experienced by the user during the interaction. In fact, in the conceptual framework proposed below, action (as an exploratory activity) and perception are two constant and closely linked elements.

In the diagram in Fig. 13.6 we find:

1. Perception of the environment linked to the exploratory experience between agent and environment in an action-perception cycle described by the SCT;
2. Proprioception as the perception of self-awareness;
3. Exploratory experience or the exploration of the environment through the user's movements;
4. Sensorimotor Maps;
5. Multiple Affordances;
6. Mapping.

Sensorimotor Maps are determined by the user's exploration of the environment during their interaction with EDMI technology, by the changes perceived during this

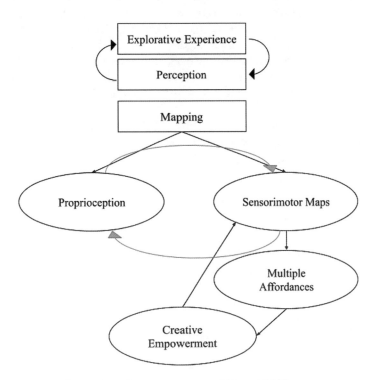

Fig. 13.6 Conceptual map of creative empowerment within an EDMI

exploration of the environment, which also depend on perception, and finally, by the mapping of the technology. The mapping itself changes the relationship between gestures, sounds and/or visual feedback [5] and it is the relation between "the physical interfaces—directly linked to the user—and the logical interface—indirectly linked to the user—forming a unity of mediation, is extremely rich to work on to define the way we would like the user to perceive the experience" [109, p. 532].

In particular, Sensorimotor Maps depend on proprioception as they are created through the movement of the subject and the different perceptual modalities that are activated (visual-proprioceptive, audio-proprioceptive and proprioceptive). In addition, the subject creates their own sensorimotor map through an awareness of their movements in the environment, based on proprioception and their perception of self-efficacy. Therefore, Sensorimotor Maps derive from the subject's perception and their self-awareness of the exploratory movements performed.

Multiple Affordances, however, derive from mapping. They may or may not be pre-established at first glance yet, in any case, they depend on the interaction of the user, who can therefore change the way they interact and use them. In this case, they are virtual and are strictly connected to Sensorimotor Maps, as they are created and recreated continuously (they represent different paths/material that the subject uses and applies for expressive purposes).

The concepts of Sensorimotor Maps and Multiple Affordances converge in Creative Empowerment. The Creative Empowerment of a subject during their interaction with an EDMI allows for the creation of a sensorimotor map once the appropriate degree of control has been reached, as described above. Therefore, while Sensorimotor Maps and Multiple Affordances also depend on the mapping itself, they must be present for Creative Empowerment to occur. As such, co-determination takes place between the action of the subject, with their repertoire of actions, and feedback from the environment.

Mapping is a fundamental parameter because it enables the user to control and play the EDMI, conveying their expressiveness through the feedback set, and allows for data analysis.

13.1.5   BehCreative

The BehCreative EDMI stands for *Behave Creatively* and aims to investigate the physiological, neurological and emotional processes that underlie behavioural changes in the user. In order to investigate functional brain changes, resting-state functional magnetic resonance imaging (rs-fMRI) data was applied.

Behind this decision was the fact that cognitive control is tied to mental and emotional processes that regulate goal-oriented behaviour through conflict [75], mobilising the user's action while promoting adaptive behavioural responses to handle a conflict-driven virtual situation. The fMRI technique is based on nuclear magnetic resonance (NMR) and measures the BOLD (blood oxygenation level-dependent) response of the brain—a combination of variations in blood flow, volume and oxygenation—secondary to neuronal activity.

The BehCreative concept relies on the client's motivation, which should be of the utmost importance when designing an EDMI to treat specific pathologies.

As already mentioned in 1.2, an immersive performative space such as an EDMI is a performative space where the subject becomes an instrument, using their body and senses to explore the space offered for different purposes (artistic, therapeutic or pedagogical). Within it, the user feels directly projected into the environment from a holistic perspective. Hence, the performative space or *environment* is crucial.

Another goal of the BehCreative investigation was to set a common protocol by investigating how audiovisual and motor learning generates and strengthens new Sensorimotor Maps based on crossmodal encoding in healthy subjects, measuring the effects in the brains of these subjects, with a view to its future use in the rehabilitation process of people with physical disabilities, such as stroke patients.

The subjects' behaviour within interactive performative spaces and the brain's reaction/response to audiovisual and motor stimuli can be measured by conducting appropriate studies using functional magnetic resonance imaging (fMRI). The investigation also aims to measure how subjects engage with an interactive environment on the assumption that their actions are mediated by their motivation and perceptions.

Since it is an immersive environment, and due to its flexibility, BehCreative can be tailored to people with a vast spectrum of health issues or special needs.

In designing this performative space, we attributed audiovisual feedbacks to the subject's positioning of their left and right arms. The visual was then projected onto the walls in front of and to both sides of the subject. We also implemented pure data and processing programs connected through the open sound control (OSC) protocol. Another music software was applied for sound synthesis and audio tracks, while Kinect 2 was used for body mapping. We video recorded the performance of each subject to retrieve qualitative data. For quantitative data, we collected the *jerk*[5]—the derivative of body acceleration—to calculate the motion fluidity of their performance and provide the appropriate audio feedback. It is motion fluidity that influences the degree of dissonance or consonance that subjects hear from six of the eight loudspeakers while interacting with the EDMI during each session. For this reason, we collected the jerk of the body to calculate the motion fluidity of the users. Furthermore, we collected and established beforehand six specific arm movements (three per arm) for the two subjects, which we included in the multiple affordance category and named *virtual affordance* (VA). The type of sound corresponding to these movements came from two of the eight speakers in the studio, located behind the performing subject.

---

[5] Jerk is defined as the rate of change of acceleration; that is, the derivative of acceleration with respect to time and, as such, the second derivative of velocity or the third derivative of position.

13.1.5.1   Creative Empowerment in BehCreative

Previous studies with fMRI outlined how patients presented significant changes in their perception of emotions and modulations of the connectivity of the cerebral network [110, 111]. Emotion can be described as a perception-valuation and action process [112] in which an input from an internal or external world is perceived and then triggers an action that alters it. It depends on stimuli that can have innate value or acquire value and has a process of valuation (2015). At a neural level, emotion engages the amygdala, the ventral striatum and the periaqueductal grey (PAG) matter, besides a set of cortical regions such as the anterior insula and the dorsal anterior cingulate cortex (dACC). Hence, multiple anatomical regions are associated with emotions. As explained in paragraph 1.4, emotional process fuels creative process, which is thus the result of an intrinsic motivation such as a reward. For this reason, we can consider Creative Empowerment as depending on the reward feeling pursued by a goal-oriented behaviour, regulated by emotional process and cognitive control.

Here, I describe the data in a former pilot study of two volunteers. The study was structured in two phases: exploration and improvisation. The users' movements were tracked with Kinect 2 and the corresponding visual feedback was displayed through three screens around the user. As for the audio feedback, an octophone system of loudspeakers surrounded the user. Six specific sound sets linked to arm movements were produced by two loudspeakers to offer the users virtual affordance, while the others were linked to the jerk. Audiovisual feedback was bonded, with the colours displayed during certain precise movements (VA) linked to specific audio feedback. More specifically, colours themselves were linked to consonant sounds, while the absence of colour (white) was linked to dissonant sounds. Figure 13.7 shows some of the visual feedback.

Data from the rs-fMRI was analysed in the same way as that described in Feitosa et al. [113]. The degree of the obtained graphs was computed for the two moments data was acquired, as well as the degree of changes between these moments, analysing degree (K), clustering coefficient (C) and betweenness centrality (BC) over the cortex in ten fMRI evaluations and displaying here the first and the fifth ones. Through qualitative analyses, we see a different albeit markable transition from the 1st to the 10th session. The two participants demonstrated a distinct first approach and general interaction with BehCreative technology [114].

As described elsewhere (2020), the graphs in Fig. 13.8 show the varied activation of brain areas other than the visual cortex (peristriate of Broadmann area 19). The first subject shows consistency of clustering coefficient (C) through activation of the visual cortex, underlying alertness and high attention; the prefrontal cortex (PC) for the motor system; and minor activation of the area responsible for the cognitive process of decision-making (orbitofrontal cortex). In the second subject, the most significant increase occurs in the anterior-prefrontal cortex, responsible for the strategic processes in memory recall and cognitive control that facilitate the attainment of a chosen goal. This is particularly important when referring to the rewarding system. Moreover, other significant increases occur in the extrastriate cortex, which has multimodal integrating functions, and in the temporal lobe

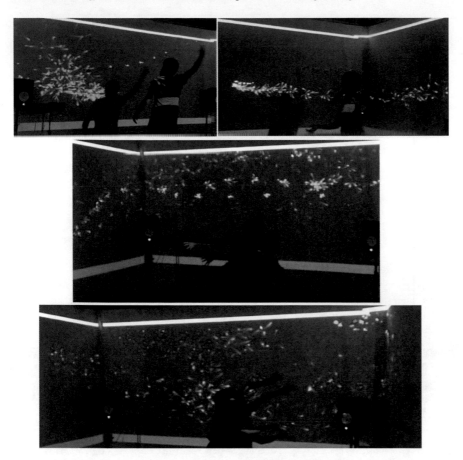

Fig. 13.7 Three users during the experiment: coloured feedback is linked to consonant sounds and VA; white visual feedback (the absence of colour) is linked to dissonant sounds and no VA

(area 37), which processes sensory input into deriving meanings for the appropriate retention of emotional association with others.

Overall, C underlines the largest changes in the second subject—the one with more bodily knowledge. Furthermore, we observe new brain connections over the sessions and a contingency of data between virtual affordance, the questionnaire on self-perceived performance, and fMRI data, revealing an increase in the calibration of motion fluidity (control over the environment) and self-awareness in the daily practice of BehCreative (2020).

These findings can be correlated with the concept of Creative Empowerment linked to the reward circuit (2020). Nevertheless, we are running a more in-depth experiment so further data will be published.

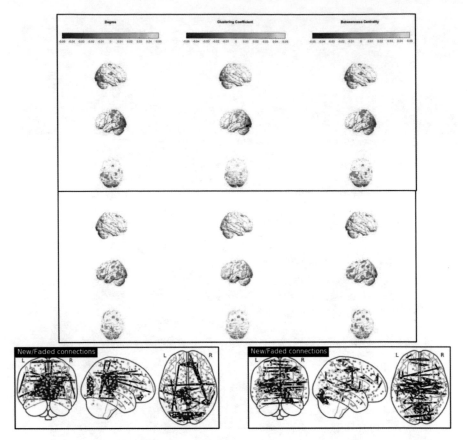

Fig. 13.8 Graph metrics detailing relative changes between fMRI sessions and new versus faded neural connection of 2 subjects from sessions 1–5

13.1.6   Phases of the User's Creative Involvement Within an EDMI

From what has been described so far, it appears that our perception is essential within an EDMI in order to plan a movement, regulate it, perceive it and store it in our memory, and to be able to reproduce the same movements at a later time through what we experienced and the resulting Sensorimotor Maps.

Below, I present the movement of the subject as being key to the improvisation and self-expression that occurs within an EDMI. I summarise this performance in four stages of implication that enable the creative expression of the user:

1. *Environmental exploration.* This phase involves an initial aleatory exploration of the subject to understand the type of interaction that exists with the technology. This experience is connected to the theory of sensorimotor contingencies and enables us to learn the rules that determine how EDMIs function.

2. *Programming (and re-programming) of Sensorimotor Maps.* This phase involves the creation of Sensorimotor Maps through interaction with Multiple Affordances and the subsequent consolidation of these as a result of the exploratory practice.
3. *Memorisation.* This phase is strictly connected to the previous phase given that it involves the subject's memorisation of the experience gained within an EDMI that is useful for a new or future interaction with technology, drawing on [115] concept of a motor prototype of predefined affordances that are activated when observing an object. The Sensorimotor Maps created in the anterior phase are linked to the concept of motor prototype, since they are stored as predefined so as to be automatically re-proposed in future interactions with the same technology.
4. *New programming.* The subject will be able to search and use new affordances, returning to point 2. In other words, the subject will use the material available to them in ways that could differ from the first exploratory experience.

In terms of these points, it appears there is a certain circularity between phases 2, 3 and 4.

Therefore, within a creative space such as an EDMI, co-determination is generated by the interaction between the agent and their environment. Here, embodiment is necessary for interactivity to happen [67] in an action-perception cycle.

13.1.7   Future Perspectives

In the digital era, programming and designing new interfaces for creative musical expression is easier than in the past. The technological period in which we live has changed not only how music is listened to, produced and distributed, but also and above all the ease with which it is used for creative purposes. Let us consider, for example, low-cost technologies such as the Microsoft Kinect camera-based motion tracking system, created in the Microsoft Xbox package for entertainment and later implemented for multiple applications by programmers around the world. In this interdisciplinary panorama, it is clear that if we merge the humanistic field with the therapeutic and pedagogical ones—that is, by offering immersive and extended musical technologies for creative purposes—users may benefit from this potential and develop it in a multitude of ways.

As evidenced in the literature, music technology for therapeutic purposes embraces a vast range of fields and discourses, from an organic perspective in motor rehabilitation and RGS [116] to an emphasis on identity construction in music therapy sessions [117]. In light of the global pandemic we just experienced, it would also be logical to design effective EDMIs that can be applied remotely between the user and the therapist, both in music therapy and in post-stroke rehabilitation, in order to provide a continuum in the patient's rehabilitation process outside the clinical setting [62, 82]. This is not only because of the low cost and accessibility of MOCAP system technologies, but rather because they make it possible to shape the design to the client's needs.

Ultimately, there is a need today for a common protocol for a holistic evaluation of the subject [87, 118] and a research project on creativity, especially within immersive environments such as EDMIs.

As music becomes more accessible through creative technology, the instrument is no longer just an object. In fact, it is now a possibility in the hands of the therapist or caregiver to help assist and develop the user's ideas, enabling their self-expression free from the limitations associated with traditional musical instruments and thus tightening the therapeutic bond for effective intervention. In this way, the client can trust in their abilities since it is through creative expression in a performative process that they immerse themself in this interactive technology. So far, I have mentioned the possibility that the user, as protagonist of the experience, is supported by a therapist. Clearly, in the case of remote rehabilitation, it would be the caregiver who assists in the session. An EDMI is essentially a musical instrument that expands the possibilities of: the user becomes the musical instrument itself. On that account, we could also invite more than one user at a time, as typically happens in music therapy sessions. In this case, and again, depending on the field of application, we could gain further benefits from social interaction. Indeed, collaborative creativity in music making helps people to communicate together [119]. Since music has an intrinsically social nature, working in groups can produce a beneficial effect (2021). Cappelen and Andersson [120] underline the importance of facilitating musical co-creation between users by offering them the opportunity to shift roles dynamically (2011). In this way, the interactive technology designed is open to a myriad of interpretations, interaction forms and roles to try out (2011). This approach highlights the nature of the creative process in music, which can be both social and collaborative as well as individual [121, 122].

Indeed, on the therapeutic side, a clinical approach also involves the implementation of new technologies, and professionals in these fields should be open to its implementation and development in order to stay at the forefront of innovation [5, 123]. This implementation calls for co-design, which should involve users with a varied grade of disabilities in order to create a flexible, equitable environment and stimulating sensory experiences [7, 87, 124, 125] also focusing on the expertise that stakeholders bring into the design process and on a mutual learning approach between adults and children[6] [126, p. 10]. All of these considerations are paramount for reaching Creative Empowerment [5] and a failure-free system [81] by the user.

To conclude, by means of a sensorless EDMI tailored to their strengths, users can interact with their environment and create new forms of expression, finding the motivation they need to recover their well-being through a positive, rewarding experience.

---

[6] For more information, see the FUBI method in Schaper et al. [126].

13.2   Note

In the chapter the author refers to "subject, client, user" as to the agent that performs within the immersive reality environment such as BehCreative and other EDMIs.

Acknowledgements I would like to extend my gratitude to Prof. Tony Brooks for the precious advices and for sharing his thoughts, Dr. Jeroen Gevers for the important suggestions as well as Giuseppina Zarantonello, Michele Partesotti and Pau Cordellat Navarro. To Noa, who patiently waited, and to Dahlia, who was born shortly after.

Funding The author receives support by FAPESP - Fundação de Amparo à Pesquisa do Estado de São Paulo - grant n. 2016/22619-0.

References

1. Faggioli, M.: Tecnologie per la didattica. Apogeo Editor, Milan (2011)
2. Camurri, A., Volpe, G., Mazzarino, B.: Ambienti multimodali interattivi: spazio, espressività e corporeità nella interazione. Forum Editrice Universitaria Udinese. (2006)
3. Thompson, E., Stapleton, M.: Making sense of sense-making: reflections on enactive and extended mind theories. Topoi **28**(1), 23–30 (2009)
4. Clark, A.: Supersizing the Mind: Embodiment, Action and Cognitive Extension. Oxford University Press, Oxford, UK (2008)
5. Partesotti, E., Peñalba, A., Manzolli, J.: Digital instruments and their uses in Music therapy. Nord. J. Music. Ther. **27**(5), 399–418 (2018)
6. O'Regan, J.K., Noë, A.: A sensorimotor account of vision and visual consciousness. Bchav. Brain Sci. **24**(5), 939–973 (2001). https://doi.org/10.1017/S0140525X01000115
7. Cappelen, B., Andersson, A.P.: Trans-Create—co-design with persons with severe disabilities. In: Verma, I. (ed.) Universal Design 2021: From Special to Mainstream Solutions, pp. 87–101. Online: IOS Press (2021)
8. DeNora, T.: Music in Everyday Life. Cambridge University Press, Cambridge (2000)
9. Koelsch, S., Fuermetz, J., Sack, U., Bauer, K., Hohenadel, M., Wiegel, M., Kaisers, U., Heinke, W.: Effects of music listening on cortisol levels and propofol consumption during spinal anesthesia. Front. Psychol. **2** (2011). https://doi.org/10.3389/fpsyg.2011.00058
10. Yamasaki, A., Booker, A., Kapur, V., Tilt, A., Niess, H., Lillemoe, K.D., Warshaw, A.L., Conrad, C.: The impact of music on metabolism. Nutrition **28**(11–12), 1075–1080 (2012)
11. Fernandez, E., Partesotti, E.: Tibetan singing bowls as useful vibroacoustic instruments in Music Therapy: a practical approach. Abstract from the 10th European Music Therapy Conference published in the Nord. J. Music Ther. (sup1):126–127 (2016). https://doi.org/10.1080/08098131.2016.1180157
12. Partesotti, E.: Cos'è la Musicoterapia? In: Cultura da Vivere al Museo Storico del Bottone, pp. 56–57 (2013). ISBN: 78889656235
13. Nordoff, P., Robbins, C.: Creative Music Therapy: A Guide to Fostering Clinical Musicianship. Barcelona Pub, Barcelona (2007)
14. Trevarthen, C., Malloch, S.N.: The dance of wellbeing: defining the musical therapeutic effect. Nordisk Tidsskrift for Musikkterapi **9**(2), 3–17 (2000). https://doi.org/10.1080/08098130009477996
15. Smyth, M.: Culture and society. In: Sutton, J. (ed.) Music, Music Therapy and Trauma: International Perspectives, pp. 57–82. Jessica Kingsley Publishers, London (2002)

16. Lorenzo, A.: Mi aportación a la musicoterapia: metodologia y practica. *Musica, terapia y comunicacion: Revista de musicoterapia* 20, (CIM) Bilbao (2000)

17. Gallese, V., Guerra, M.: The empathetic screen. Translated by F. Anderson. (2020)

18. Gallese, V.: Intentional attunement: a neurophysiological perspective on social cognition and its disruption in autism. Brain Res. **1079**(1), 15–24 (2006)

19. Gallese, V., Keysers, C., Rizzolatti, G.: A unifying view of the basis of social cognition. Trends Cogn. Sci. **8**(9), 396–403 (2004)

20. Rizzolatti, G., Sinigaglia, C.: So quel che fai: il cervello che agisce ei neuroni specchio. Milano: R. Cortina

21. Peñalba, A., Valles, M.J., Partesotti, E., Sevillano, M.Á., Castañón, R.: Accessibility and participation in the use of an inclusive musical instrument: the case of MotionComposer. J. Music Technol. Educ. **12**(1), 79–94 (2019)

22. Hadley, S., Hahna, N., Miller, V., Bonaventura, M.: Setting the scene: an overview of the use of music technology in practice. In: Magee, W.L. (ed.) Music Technology in Therapeutic and Health Settings, pp. 25–43. Jessica Kingsley Publishers, London (2014)

23. Kontogeorgakopoulos, A., Wechsler, R., Keay-Bright, W.: Camera-based motion tracking and performing arts for persons with motor disabilities and autism. In: Kouroupetroglou, G. (ed.) Disability Informatics and Web Accessibility for Motor Limitations, pp. 294–322 (2013). https://doi.org/10.4018/978-1-4666-4442-7.ch009

24. Magee, W.L., Burland, K.: Using electronic music technologies in music therapy: opportunities, limitations and clinical indicators. Br. J. Music Ther. **22**(1), 3–15 (2008)

25. Hahna, N.D., Hadley, S., Miller, V.H., Bonaventura, M.: Music technology usage in music therapy: a survey of practice. Arts Psychother. **39**(5), 456–464 (2012)

26. Keay-Bright, W., Eslambolchilar, P.: Imagining a digital future: how could we design for enchantment within the special education curriculum? Cardiff Met archive. http://hdl.handle.net/10369/10800 (2019). Accessed 5 Feb 2020

27. Miranda, E.R., Wanderley, M.M.: New Digital Musical Instruments: Control and Interaction Beyond the Keyboard. AR Editions Inc., Middleton (2006)

28. Peñalba, A.: Towards a theory of proprioception as a bodily basis for consciousness in music. In: Clarke, D., Clarke, E.F. (eds.) Music and Consciousness: Philosophical, Psychological, and Cultural Perspectives, pp. 215–230. Oxford University Press, Oxford (2011)

29. Medeiros, C.B., Wanderley, M.M.: A comprehensive review of sensors and instrumentation methods in devices for musical expression. Sensors **14**(8), 13556–13591 (2014)

30. Mulder, A.: Towards a choice of gestural constraints for instrumental performers. In: Wanderley, M.M., Battier, M. (eds.) Trends in Gestural Control of Music, pp. 315–335 (2000)

31. Brooks, A.: Soundscapes: the evolution of a concept, apparatus and method where ludic engagement in virtual interactive space is a supplemental tool for therapeutic motivation (Ph.D. Thesis). http://vbn.aau.dk (2011). Accessed 9 April 2022

32. Brooks, A., Hasselblad, S.: Creating aesthetically resonant environments for the handicapped, elderly and rehabilitation: Sweden. In: Proceedings 5th International Conference on Disability, Virtual Reality and Associated Technologies, pp. 191–198. Oxford (2004)

33. Brooks, A.L., Hasselblad, S.: Creating aesthetically resonant environments for the handicapped, elderly, and rehabilitation: Sweden. In Proceedings of 5th International Conference on Disability, Virtual Reality, and Associated Technologies (pp. 191–198). Oxford (2004)

34. Rizzonelli, M., Kim, J.H., Staudt, P., Lussana, M.: Fostering social interaction through sound feedback: Sentire. Organised Sound (2022). https://doi.org/10.1017/S1355771822000024

35. Hunt, A., Wanderley, M.M.: Mapping performer parameters to synthesis engines. Org. Sound **7**(2), 97–108 (2002). https://doi.org/10.1017/S1355771802002030

36. McArthur, J.A.: Digital Proxemics. Peter Lang, Bern (2016). https://www.peterlang.com/view/title/23055

37. Effenberg, A.O., Fehse, U., Schmitz, G., Krueger, B., Mechling, H.: Movement Sonification: Effects on Motor Learning beyond Rhythmic Adjustments. Front. Neurosci. **10** (2016). https://www.frontiersin.org/article/10.3389/fnins.2016.00219

38. Sigrist, R., Rauter, G., Riener, R., Wolf, P.: Augmented visual, auditory, haptic, and multimodal feedback in motor learning: a review. Psychon. Bull. Rev. **20**(1), 21–53 (2013). https://doi.org/10.3758/s13423-012-0333-8
39. Tajadura-Jiménez, A., Väljamäe, A., Bevilacqua, F., Bianchi-Berthouze, N.: Principles for designing body-centered auditory feedback. In: Norman, K.L., Kirakowski, J. (eds.) The Wiley Handbook of Human Computer Interaction, pp. 371–403. Wiley (2018). https://doi.org/10.1002/9781118976005.ch18
40. Partesotti, E., Tavares, T.F.: Color and emotion caused by auditory stimuli. In: Proceedings of ICMC-SMC, pp. 900–904. Athens (2014)
41. Specht, K.: Synaesthesia: cross activations, high interconnectivity, and a parietal hub. Transl. Neurosci. **3**(1), 15–21 (2012)
42. Veronesi, L.: Proposta per una ricerca sui rapporti fra suono e colore. Luigi Veronesi Archive (1977). Accessed 21 September 2013
43. Partesotti, E., Peñalba, A., Manzolli, J.: Interactive musical technology enhances creativity: a case study with e-mocomu technology. In: Proceedings of INTED, Valencia 2017, Spain. ISBN: 978-84-617-8491-2
44. Ellis, P.: The music of sound: a new approach for children with severe and profound and multiple learning difficulties. Br. J. Music Educ. **14**(2), 173–186 (1997)
45. Camurri, A., Volpe, G., Canazza, S., Canepa, C., Rodá, A., Zanolla, S., Foresti, G.L.: The 'Stanza Logo–Motoria': an interactive environment for learning and communication. In: Proceedings of Sound and Music Computing Conference, pp. 353–360 (2010)
46. Peñalba, A., Valles, M.J., Partesotti, E., Castañón, R., Sevillano, M.A.: Types of interaction in the use of MotionComposer, a device that turns movement into sound. In: Proceedings of *ICMEM* Sheffield (2015)
47. Zagalo, N., Branco, P.: The creative revolution that is changing the world. In: Creativity in the Digital Age, pp. 3–15. Springer, London (2015). https://doi.org/10.1007/978-1-4471-6681-8_1
48. Rice, L.: Playful learning. J. Educ. Built Environ. **4**(2), 94–108 (2009)
49. Brandão, A.F., Colombo-Dias, D.R., Reis, S.: Biomechanics sensor node for virtual reality: a wearable device applied to gait recovery for neurofunctional rehabilitation. In: Gervasi, O. et al. (eds.) Computational Science and Its Applications ICCSA 2020, Lecture Notes in Computer Science, vol. 12255. Springer, Cham (2020). https://doi.org/10.1007/978-3-030-58820-5_54
50. Turolla, A., Dam, M., Ventura, L., Tonin, P., Agostini, M., Zucconi, C., Kiper, P., Cagnin, A., Piron, L.: Virtual reality for the rehabilitation of the upper limb motor function after stroke: a prospective controlled trial. J. Neuroeng. Rehabil. **10**, 85 (2013). https://doi.org/10.1186/1743-0003-10-85
51. Brandão, A.F., Colombo-Dias, D.R., de Castro Alvarenga, I., Garcia de Paiva, G.: E-street for prevention of falls of the elderly an urban virtual environment for human–computer interaction from lower limb movements. In: Iano, Y., Arthur, R., Saotome, O., Vieira Estrela, V., Loschi, H.J. (eds.) Proceedings of the 3rd Brazilian Technology Symposium: Emerging Trends and Challenges in Technology, BTSym 2017. Springer, Cham (2019). https://doi.org/10.1007/978-3-319-93112-8_25
52. Moens, B., Muller, C., van Noorden, L., Franěk, M., Celie, B., Boone, J., Bourgois, J., Leman, M.: Encouraging spontaneous synchronisation with D-Jogger, an adaptive music player that aligns movement and music. PLoS One **9**(12), e114-234 (2014). https://doi.org/10.1371/journal.pone.0114234
53. Ballester, B.R., Nirme, J., Camacho, I., Duarte, E., Rodríguez, S., Cuxart, A., Duff, A., Verschure, P.F.M.J.: Domiciliary VR-based therapy for functional recovery and cortical reorganization: randomized controlled trial in participants at the chronic stage post stroke. JMIR Serious Games **5**(3), e15 (2017). https://doi.org/10.2196/games.6773
54. Pallesen, H., Brændstrup Andersen, M., Mo Hansen, G., Biering Lundquist, C., Brunner, I.: Patients' and health professionals' experiences of using virtual reality technology for upper limb training after stroke: a qualitative substudy. Rehabil. Res. Pract. **2018** (2018). https://doi.org/10.1155/2018/4318678

55. Perez-Marcos, D., Chevalley, O., Schmidlin, T., Garipelli, G., Serino, A., Vuadens, P., Tadi, T., Blanke, O., Millán, J.D.R.: Increasing upper limb training intensity in chronic stroke using embodied virtual reality: a pilot study. J. NeuroEng. Rehabil. **14**(119), 1–14 (2017)
56. Feitosa, J.A., Fernandes, C.A., Casseb, R.F., Castellano, G.: Effects of virtual reality-based motor rehabilitation: a systematic review of fMRI studies. J. Neural Eng. **19**(1) (2021). https://doi.org/10.1088/1741-2552/ac456e
57. Rutkowski, S., Kiper, P., Cacciante, L., Cieślik, B., Mazurek, J., Turolla, A., Szczepańska-Gieracha, J.: Use of virtual reality-based training in different fields of rehabilitation: a systematic review and meta-analysis. J. Rehabil. Med. (2020). https://doi.org/10.2340/16501977-2755
58. Stephan, M.A., Brown, R., Lega, C., Penhune, V.: Melodic Priming of Motor Sequence Performance: The Role of the Dorsal Premotor Cortex. Front. Neurosci. **10**, 210 (2016). https://doi.org/10.3389/fnins.2016.00210
59. Särkämö, T., Tervaniemi, M., Laitinen, S., Forsblom, A., Soinila, S., Mikkonen, M., Autti, T., Silvennoinen, H.M., Erkkilä, J., Laine, M., Peretz, I.: Music listening enhances cognitive recovery and mood after middle cerebral artery stroke. Brain **131**(3), 866–876 (2008). https://doi.org/10.1093/brain/awn013
60. Ogden, R., Hawkins, S.: Entrainment as a basis for co-ordinated actions in speech. In The Scottish Consortium for ICPhS 2015 (ed.), Proceedings of the 18th International Congress of Phonetic Sciences. Glasgow: The University of Glasgow (2015)
61. Trost, W., Vuilleumier, P.: Rhythmic entrainment as a mechanism for emotion induction by music: a neurophysiological perspective. J. Music Educ. **60**(4), 436–459 (2013)
62. Scudeletti, L.R., Brandão, A.F., Dias, D., Brega, J.R.F.: KinesiOS: a telerehabilitation and functional analysis system for post-stroke physical rehabilitation therapies. In: Gervasi, O. et al. (eds) Computational Science and Its Applications—ICCSA 2021. Lecture Notes in Computer Science, vol. 12950. Springer, Cham (2021). https://doi.org/10.1007/978-3-030-86960-1_13
63. Porayska-Pomsta, K., Rajendran, G., Smith, T.J., Avramides, K.: Blending human and artificial intelligence to support Autistic children's social communication skills. ACM Transact. Comput. Hum. Interac. (TOCHI), **25**(1), 1–27 (2018)
64. Dalla Bella, S.: Music and movement: towards a translational approach. Neurophysiol. Clin. **48**(6), 377–386 (2018)
65. Merleau-Ponty, M.: Phenomenology of Perception. Routledge, London (1998)
66. Burnett, R.: How images think. MIT Press (2005)
67. Kim, J.H., Seifert, U.: Embodiment and agency: towards an aesthetics of interactive performativity. In: Proceedings of the 4th Sound and Music Computing Conference, pp. 230–237 (2007)
68. Laskowski, E.R., Newcomer-Aney, K., Smith, J.: Refining rehabilitation with proprioception training: expediting return to play. Phys. Sportsmed. **25**(10), 89–102 (1997)
69. Varela, F., Thompson, E., & Rosch, E.: The embodied mind: cognitive science and human experience. MIT Press (1991)
70. Colombetti, G., Thompson, E.: The feeling body: towards an enactive approach to emotion. In Developmental psychopathology (vol. 20, pp. 641–670) (2008)
71. Caruana, F., Borghi, A.M.: No embodied cognition, no party. Giornale Italiano di Psicologia **1**, 119–128 (2013). https://doi.org/10.1421/73987
72. Clarke, E. F. (2020). The Psychology of Creative Processes in Music. In N. Donin (Ed.), The Oxford Handbook of the Creative Process in Music (pp. 1-20). Oxford University Press
73. Dourish, P.: Where the Action Is. The MIT press, London (2004)
74. Benedek, M., Jauk, E.: Creativity and cognitive control. In: Kaufman, J., Sternberg, R. (eds.) Cambridge Handbook of Creativity, pp. 200–223. Cambridge University Press, Cambridge (2019)
75. Inzlicht, M., Bartholow, B.D., Hirsh, J.B.: Emotional foundations of cognitive control. Trends Cogn. Sci. **19**(3), 126–132 (2015)

76. Carson, S.H.: Creativity and mental illness. In: Kaufman, J., Sternberg, R. (eds.) Cambridge Handbook of Creativity, pp. 296–318. Cambridge University Press, Cambridge (2019)
77. Kéri, S.: Genes for psychosis and creativity: a promoter polymorphism of the Neuregulin 1 gene is related to creativity in people with high intellectual achievement. Psychol. Sci. **20**, 1070–1073 (2009)
78. Purcell, S.M., Wray, N.R., Stone, J.L., Visscher, P.M., O'Donovan, M.C., Sullivan, P.F., Sklar, P.: Common polygenic variation contributes to risk of schizophrenia and bipolar disorder. Nature **460**, 748–752 (2009)
79. Kozbelt, A.: Musical creativity over the lifespan. In D. K. Simonton (ed.), The Wiley handbook of genius (pp. 451–472). Wiley Blackwell (2014)
80. Koelsch, S.: A neuroscientific perspective on music therapy. the neurosciences and music iii—disorders and plasticity. Ann. N. Y. Acad. Sci. **1169**, 374–384 (2009)
81. Riley, P., Alm, N., Newell, A.: An interactive tool to promote musical creativity in people with dementia. Comput. Hum. Behav. **25**(3), 599–608 (2009)
82. Brandão, A.F., Colombo-Dias, D.R., Castellano, G., Parizotto, N.A., Trevelin, L.C.: Telemed. e-Heal. 584–589 (2016). https://doi.org/10.1089/tmj.2015.0139
83. Lee, M.R., Chen, T.T.: Digital creativity: research themes and framework. Comput. Hum. Behav. **42**, 12–19 (2015)
84. Thaut, M.: Rhythm, Music, and the Brain: Scientific Foundations and Clinical Applications. Routledge, New York and London (2005)
85. Gallese, V., Sinigaglia, C.: What is so special about embodied simulation? Trends in Cognitive Sciences, **15**(11), 512–519 (2011)
86. Custodero, L.A.: The call to create: flow experience in music learning and teaching. In Hargreaves, D., Miell, D., Macdonald, R. (eds.), Musical Imaginations: Multidisciplinary Perspectives on Creativity, Performance, and Perception. (2011)
87. Berger, D.S.: Eurhythmics for Autism and Other Neurophysiologic Diagnoses: A Sensori-motor Music-Based Treatment Approach. Jessica Kingsley Publisher, London (2016)
88. Koelsch, S.: Investigating emotion with music: neuroscientific approaches. Ann. N. Y. Acad. Sci. **1060**, 412–418 (2005). https://doi.org/10.1196/annals.1360.034
89. Lin, S.T., Yang, P., Lai, C.Y., Su, Y.Y., Yeh, Y.C., Huang, M.F., Chen, C.C.: Mental health implications of music: insight from neuroscientific and clinical studies. Harv. Rev. Psychiatry **19**(1), 34–46 (2011)
90. Mallik, A., Russo, F.A.: The effects of music & auditory beat stimulation on anxiety: a randomized clinical trial. PLoS One **17**(3), e0259312 (2022). https://doi.org/10.1371/journal.pone.0259312
91. Porayska-Pomsta, K., Frauenberger, C., Pain, H., Rajendran, G., Smith, T., Menzies, R., Foster, M.E., Alcorn, A., Wass, S., Bernadini, S., Avramides, W., Keay-Bright, W., Chen, J., Waller, A., Guldberg., K, Good., J., Lemon, O.: Developing technology for autism: an interdisciplinary approach. Pers. Ubiquit. Comput. **16**(2), 117–127 (2012)
92. Cappelen, B., Andersson, A.P.: Designing four generations of 'Musicking Tangibles .In Stensæth, K. (ed.) Music, Health, Technology and Design, pp. 1–20. (2014)
93. Ruggieri, V.: L'esperienza estetica. Fondamenti psicofisiologici per un'educazione estetica. Armando editori, Italia (1997)
94. Leman, M.: Embodied Music Cognition and Mediation Technology. The MIT Press, London (2008)
95. Kozel, S.: Closer: Performance, Technologies, Phenomenology. The MIT Press, London (2007)
96. Benford, S., Giannachi, G.: Performing Mixed Reality. The MIT Press, London (2011)
97. Anderson, R., Sharrock, W.: Can organisations afford knowledge? Comput. Support. Cooper. Work (CSCW) **1**(3), 143–161 (1992)
98. Kiefer, C.: Citizen empowerment: a developmental perspective. Prev. Hum. Serv. **3**, 9–36 (1984)
99. Zimmerman, M.A.: Empowerment theory. In: Rappaport, J., Seidman, E. (eds.) Handbook of Community Psychology, pp. 43–63. Springer, Boston (2000)

100. Rolvsjord, R.: Therapy as empowerment: clinical and political implications of empowerment in mental health practises of music therapy. Nord. J. Music Ther. **13**(2), 99–111 (2004)
101. Gibson, J.J.: The Ecological Approach to Visual Perception. Houghton Mifflin, Boston (1979)
102. Wilkie, K., Holland, S., Mulholland, P.: What can the language of musicians tell us about music interaction design? Comput. Music J. **34**(4), 34–48 (2010)
103. Swingler, T.: The invisible keyboard in the air: An overview of the educational, therapeutic and creative applications of the EMS Soundbeam™. In: Proceedings of the 2nd European Conference for Disability, Virtual Reality & Associated Technology (1998)
104. Ellis, P.: Vibroacoustic sound therapy: case studies with children with profound and multiple learning difficulties and the elderly in long-term residential care. Stud. Heal. Technol. Inform. **103**, 36–42 (2004)
105. Forgeard, M.: Creativity and healing. In: Kaufman, J., Sternberg, R.J. (eds.) The Cambridge Handbook of Creativity, pp. 319–331. Cambridge University Press, Cambridge (2019)
106. Ivcevic, Z., Hoffmann, J.: Emotions and creativity: from process to person and product. In: Kaufman, J., Sternberg, R.J. (eds.) The Cambridge Handbook of Creativity, pp. 273–295. Cambridge University Press, Cambridge (2019)
107. Amabile, T.M.: Creativity in Context: Update to the Social Psychology of Creativity. Westview Press, Boulder (1996)
108. Partesotti, E., Hebling, E.D., Micael, Pereira, A.C.A., Dezotti, C.G., Moroni, A.S., Manzolli, J.: Analysis of affective behavior in the artistic installation MovieScape. In: Brooks, A.L. (eds) ArtsIT, Interactivity and Game Creation. ArtsIT 2022. Lecture Notes of the Institute for Computer Sciences, Social Informatics and Telecommunications Engineering, vol. 479. Springer, Cham (2023). https://doi.org/10.1007/978-3-031-28993-4_23
109. Parés, N., Parés, R.: Towards a model for a virtual reality experience: the virtual subjectiveness. Presence **15**(5), 524–538 (2006)
110. Ruiz, S., Lee, S., Soekadar, S.R., Caria, A., Veit, R., Kircher, T., Birbaumer, N., Sitaram, R.: Acquired self-control of insula cortex modulates emotion recognition and brain network connectivity in schizophrenia. Hum. Brain Mapp. **34**(1), 200–212 (2013)
111. Brühl, A. B., Scherpiet, S., Sulzer, J., Stämpfli, P., Seifritz, E., Herwig, U.: Real-time neurofeedback using functional MRI could improve down-regulation of amygdala activity during emotional stimulation: a proof-of-concept study. Brain Topogr. **27**, 138–148 (2014)
112. Etkin, A., Büchel, C., Gross, J.J.: The neural bases of emotion regulation. Nat. Rev. Neurosci. **16**(11), 693–700 (2015)
113. Feitosa, J.A., Stefano Filho, C.A., Casseb, R.F., Camargo, A., Martins, B.S.G., Ballester, B.R., Omedas, P., Verschure, P., Oberg, T.D., Min, L.L., Castellano, G.: Complex network changes during a virtual reality rehabilitation protocol following stroke: a case study. In: 2019 9th International IEEE/EMBS Conference on Neural Engineering (NER), pp. 891–894. IEEE (2019)
114. Partesotti, E., Feitosa J., Manzolli, J., Castellano, G.: A pilot study evaluating brain functional changes associated to the BehCreative environments. J. Epilepsy Clin. Neurophysiol. In: Proceedings of 7th Brainn Congress, Brazilian Institute of Neuroscience and Neurotechnology, vol. 23(3), p. 12 (2020). ISSN 16176-2649
115. Borghi, A.M.: Object concepts and embodiment: why sensorimotor and cognitive processes cannot be separated. La Nuova Critica **15**(4), 447–472 (2007)
116. Rizzo, A.A., Kim, G.J.: A SWOT analysis of the field of virtual reality rehabilitation and therapy. Presence **14**(2), 119–146 (2005)
117. Magee, W.L.: Indicators and contraindications for using music technology with clinical populations: When to use and when not to use. In: Magee, W.L. (ed.) Music Technology in Therapeutic and Health Settings, pp. 83–107. Jessica Kingsley Publishers, London (2014)
118. Vartanian, O.: Neuroscience of creativity. In: Kaufman, J., Sternberg, R.J. (eds.) The Cambridge Handbook of Creativity, pp. 148–172. Cambridge University Press, Cambridge (2019)
119. MacDonald, R., Wilson, G., Baker, F.: Musical creativity and well-being. In: The Oxford Handbook of the Creative Process in Music. https://doi.org/10.1093/oxfordhb/978019063 6197.013.14

120. Cappelen, B., Andersson, A.P.: Expanding the role of the instrument. In: Proceedings of the International Conference on New Interfaces for Musical Expression (2011)
121. Sawyer, R.K., DeZutter, S.: Distributed creativity: How collective creations emerge from collaboration. Psychol. Aesthet. Creat. Arts 3(2), 81 (2009)
122. Sawyer, R.K., Mahwah, N.J.: Group Creativity: Music, Theater, Collaboration. Erlbaum, Mahwah (2003)
123. Nagler, J.: Music aesthetics, music technology, and music therapy. In: Magee, W.L. (ed.) Music Technology in Therapeutic and Health Settings, pp. 349–360. Jessica Kingsley Publishers, London (2014)
124. Cappelen, B., Andersson, A.P.: Health improving multi-sensorial and musical environments. In: Proceedings of the Audio Mostly 2016, pp. 178–185
125. Rajapakse, R., Brereton, M., Sitbon, L.: A respectful design approach to facilitate codesign with people with cognitive or sensory impairments and makers. CoDesign 17(2), 1–29 (2019). https://doi.org/10.1080/15710882.2019.1612442
126. Schaper, M.-M., Iversen, O. S., Malinverni, L., Pares, N.: FUBImethod: strategies to engage children in the co-design of full-body interactive experiences. Int. J. Hum. Comput. Stud. 132, 52–69 (2019). https://doi.org/10.1016/j.ijhcs.2019.07.008

Chapter 14
The Evolution of the Virtual Production Studio as a Game Changer in Filmmaking

Cinzia Cremona and Manolya Kavakli

Abstract This chapter analyses the key methodological and conceptual shifts in filmmaking brought about by the integrated use of VR technology, game engines, and LED walls in Virtual Production Studios (VPS). VPS combine the real time processing capabilities of game engines with LED walls and camera tracking to produce live in-camera visual effects. In this chapter, we focus not only on the elements that make an integrated VPS system, but also on what makes it qualitatively different from traditional and digital filmmaking. The methodology of the VPS transforms the relationships among display technology (screen), live action (actors), image production (cameras), and postproduction (visual effects or VFX). An understanding of how these technologies work together will form the premise for an analysis of the qualitative and performative shifts that filmmaking is undergoing. There are a number of crucial elements of an integrated VPS technology: a game engine, LED walls, a camera tracking system, a motion tracking system, Simulcam, and a Render. VP has its roots in traditional filmmaking techniques associated with combining different images in one shot, such as back projection and green screen. The back projection technique immerses performers in the environment of the action with an analogue (filmed) or a digital backlot (this technique was used, for example, in the film *Oblivion* in 2013), but brings the problem of matching the brightness of foreground and background, and, more crucially, it requires the camera on set to be still in order to avoid problems with parallax. The traditional green screen technique allows actors to perform with placeholder props and extras in an empty room lined with an even colour background. Neither of them is ideal. More recently, LED walls have replaced green and blue screen techniques, creating an active stage for actors' performance in a VPS and a safe passage from digital to virtual backlot for filmmakers. While with a green screen stage and digital backlot the final image is composed on screen, with LED walls and a virtual backlot the final image is displayed on LED walls

C. Cremona (✉)
University of Sydney and Macquarie University, 1/290 Clovelly Road, Coogee, NSW 2034, Australia
e-mail: cinziacremona@gmail.com

M. Kavakli
Aston University, Aston St, Birmingham, United Kingdom B4 7ET, United Kingdom

and updated in real time using a game engine. VFX are created in camera, making the distinction between screens and cameras superfluous, to some extent. VPS systems based on game engines have been tested since 2001, when Peter Jackson used a virtual camera in *Lord of the Rings* and Steven Spielberg used an iPad attached to a Vive controller to visualise potential shots in a CG environment updated in real time for the film *AI*. The technological components of a VPS have become progressively more sophisticated, with studios testing and adopting in-house solutions to issues of synchronisation, data management, and lighting. Everything changed when the game company *Epic Games* partnered with *Lux Machina* (LED walls), *Magnopus* (VR/AR), *Profile Studios* (mocap and VP), *Quixel* (3D graphics), and *ARRI* (cameras) to launch an out-of-the-box VPS solution at SIGGRAPH 2019 in Los Angeles with the director of photography Matt Workman. Since these developments are so recent, existing literature generally focuses on technical details and individual components of the VPS system, rather than on its impact on filmmaking. The use of LED walls in a VPS makes a radical methodological and conceptual difference to live performance and to filmmaking itself. The game changer here is the link between on-stage camera and virtual in-game-engine camera that makes changes in real time on the LED wall, as virtual camera movements are updated in sync with the movements of the on-stage camera. The analysis in this paper is supported by a selected review of publications relevant to the subject to address the domains that require further investigation. Publications that offer overviews and integrated analyses are beginning to emerge, particularly as educational institutions start to develop VFX courses that encompass VP, but an in-depth conceptual analysis of the changes in progress to capture this unique historical moment is of critical importance. We present a review of how VP and game engine technologies have evolved and converged, as well as an overview of the impact of VPS on filmmaking.

Keywords Virtual production studio · Virtual reality · LED wall · Filmmaking · Vfx · Performativity

14.1   Introduction

This chapter offers an overview of recent technological, methodological, and conceptual shifts in filmmaking driven by the introduction of tools developed in the context of game design. At the core of this shift is the integration of game design software and hardware into the Virtual Production Studio (VPS), which changes the relationship between film preproduction, performance, production, and postproduction.

As an inherently technology-based practice, filmmaking is continuously evolving. Within this evolution, we can identify key qualitative and methodological breaks—namely from analogue to digital and from digital to virtual. It is worth recalling that the term *film* refers to the material support of analogue cinema. In the late nineteenth century, the long strip of highly flammable cellulose nitrate was coated in silver salts—a blend of light sensitive minerals. When exposed to a lit environment, the silver crystals would shift in direct relation to the patterns of intensity of light reflected by the arrangement of objects in front of the camera lens. The visual correspondence between the scene and the chemical pattern on the film constitutes the analogy referred to in the expression *analogue cinema*. The emergence of electronic image capture technologies does not substantially change this relationship, as electronic analogue images are formed by patterns of charged particles on magnetic tape that imitate in two dimensions the pattern of objects in front of the camera.

The encoding of data in digital image capture technologies breaks this direct correspondence between images and objects. Visual elements in the material world are captured and encoded in the same way as sounds, words or any other content. There is no analogical relation between what is captured and what is stored, as the same pattern of digits could be decoded in different forms. Despite the more indirect relationship, in digital filmmaking the image is still created by capturing patterns of light, colour, and movement in the material world.

For decades visual art considered digital moving image practices as a separate discipline from film—video or video-film ([1, 3])—with separate funding streams, research, theories, organisations, and practitioners. In analogue filmmaking, composite images are achieved with limited optical techniques, such as double exposure and back projection or in postproduction. Conversely, digitally combining and modifying data from different captures and Computer-Generated Images (CGI) offers endless options to build realistic and believable scenes. From this perspective, it is evident that electronic and digital technologies have caused significant shifts in the methods, speed, and affordability of filmmaking. The accessibility, range, and quality of VFX have grown exponentially, making possible previously unimaginable cinematic worlds and situations [2].

It is important to recall this step to contextualise the conceptual shift introduced by virtual production in filmmaking. Virtual production is defined as 'a broad term referring to a spectrum of computer-aided production and visualization filmmaking methods', where the physical and digital worlds meet [3, 4]. This approach integrates virtual and augmented reality technologies with CGI and VFX using a game engine to enable on set production crews to capture performed scenes in real time. It is a game changer for filmmaking for the use of technologies that visualize the physics of entire three-dimensional environments and immerse actors in them. Once programmed, these technologies simulate in real time the movements of living creatures, shifts in light, water flowing, breezes playing on leaves, furs, and feathers, creating realistic and believable settings. As they change in real time, these simulations can be integrated with actors and captured in camera at a quality suitable for final release.

The technologies adopted in virtual production come into filmmaking principally from game design and from research into expanded reality (XR, encompassing virtual reality, VR, and augmented reality, AR). In the mid 1990s, the real time processing powers of game engines were first used as a form of filmmaking known as machinema, then as machinima [5]. Despite the practical links with virtual production, the history of machinima has largely been kept separate from the history of virtual production, but there are some critical crossovers worth including in this text. In VPS, tools interact in a dynamic configuration in almost real time (a lag of few tenths of a second is becoming progressively smaller as technology improves). Machinima explicitly conceptualizes these dynamics as technological performativity. It could be argued that the physics engine is *alive* and performs, that the camera performs, and that together they form an active feedback loop.

A history of XR technology is beyond the scope of this text. We will however attempt a brief history of VP and game engines so far. At time of writing, VP technology and research are in rapid development. We aim to offer a snapshot of a time when this method is becoming mainstream in filmmaking, of the unquestionable advantages it offers, and of the technological and aesthetic limitations emerging in this process. Ultimately, we want to offer a conceptual apparatus that can support a critical analysis of VP filmmaking going forward. Kipnis argues that 'film studies has been somewhat slow to come to terms with a changing apparatus, or to theorize the shifts in film language and grammar that technological change seems to have so rapidly brought about' [6]. More recently, a number of technological developments have converged into filmmaking causing a shift in practices and approaches. LED volumes in particular have shifted the performative relationship between human and non-human actors in a film studio.

An LED volume differs from an LED screen in that it transforms a film studio space into a three-sided moving image environment. It is often built in a semi-circular shape with the curve specifically calculated so that the sound doesn't echo within it. The physical floor space in front of the LED walls is painted or covered in materials that visually merge with the image on screen. Actors move within this environment, respond to it, are lit by the LED bulbs, and are followed by the camera. The perspective of the images on the LED walls that form the volume shifts in relation to the movements of the camera, avoiding parallax issues. Parallax describes how nearby objects appear to move against the background of more distant objects. In other words, camera and LED walls are in constant communication via a camera tracking system and a motion tracking system. The final composite image is computed in almost real time, and built within camera and monitor screen, making the distinction between the two more of a grey area than a clear line.

Within this system, every element has a dynamic role towards the making of the final image sequence. We argue that this requires a suitable conceptual apparatus to inform a theoretical understanding of the qualitative shifts instantiated by these technological developments and their active combination within the VPS system. This paper proposes a convergence of a number of existing theoretical tools into a conceptual vocabulary aimed at analysing the dynamics of VPS performed by its technological elements.

In this chapter, we will first offer an overview of how filmmaking methods have changed from back projection and multiple exposure via green screen to VP techniques, with examples. After reviewing how game engines evolved to allow a technological shift in film making, we explore the integrated key technological elements that make VP possible, with particular attention to the use of LED walls. A short history of the Virtual Production Studio (VPS) introduces an analysis of the conceptual changes afforded by these practices and integrated technology. We conclude by highlighting domains that require further investigation in filmmaking.

Our aim is to offer a snapshot of a field in rapid development as the book goes into print. To try and achieve an overview of the available technology, literature, research, and films, we have elected to include publications and references from more ephemeral sources such as journals for industry professionals, organizations' and companies' websites, industry digital publications, as well as presentations and conference proceedings. Moreover, more than two years of disruption due to COVID 19 has pushed a number of conferences and symposia online, and some quotes are transcribed from video presentations, talks, and interviews.

14.2 A Brief History of Composite Images

VP has its roots in traditional filmmaking techniques associated with combining different images in one shot to generate plausible or fantastic scenarios, such as double exposure, back projection, and green screen. Musser concisely states that 'major innovations include the development of the magic lantern during the 1650s, the adaptation of photography to projection ca. 1850, and the synchronization of film with recorded sound, which achieved permanent commercial standing in the late 1920s' [7]. The magic lantern is commonly acknowledged as one of the precursors of cinema as an early mechanical technology capable of generating the illusion of projected moving images. The overlapping of images projected from different glass plates offers very early experiences of composite images. In a magic lantern, different hand-coloured or printed glass plates are placed in front of sources of light and span by cranking one or more handles. In more complex systems, it is possible to fade from one plate to another by slowly covering one source of light and uncovering another pointing in the same direction. This proto-cinematic technique resembles slide projection and creates live composite images [7].

Film-based technologies derived from photography adopt comparable early techniques and additionally fix the results on a transparent support. 'Superimposition' or 'multiple exposures' describe different forms of analogue composite images created by printing portion of images from different sources to combine visual elements within the same frame [8]. A very early example of the potential of multiple exposures and superimposition can be found in the films by George Méliès. A professional stage magician, Méliès combined the fundamental illusion of movement inherent to cinema with narrative and visual illusions in his well-known films *A Trip to the Moon* (1902), *The Voyage Across the Impossible* (1904), and several others. Méliès used a number of techniques including stop motion, slow motion, dissolve, fade-out, superimposition, and double exposure. VFX artists Francis McGowan describes how director Willis Harold O'Brien used the process to produce a travelling matte for the film *King Kong* (1933). O'Brien 'used blue and yellow lighting which is then filtered onto black and white film. Then 2 strips of film would then be loaded into the camera and then you would have an in-camera composite which is known as bypacking' [9]. These methods separate actors from the visual environment in which the action is supposed to be taking place. Actors perform in costumes and with some elements of the set or on location, but only with a partial sense of the final outcome. Directors work with editors to composite and control the final visual result. Added to this, with multiple exposures the quality of the image degrades and becomes progressively grainier [10].

This radically changes when these forms of postproduction are complemented by methods that combine previously filmed backgrounds with live action. Rear screen projection, also called back projection, immerses performers in the environment of the action by back projecting a previously filmed background behind actors. Visually, this method introduces a number of difficulties including matching the brightness of foreground and background, matching image quality and, more crucially, making camera movements impossible in order to avoid problems with parallax. On the other hand, Turnock calls attention to the efficiency of back projections compared to early composite techniques as 'it could be completed immediately on the set, under the control of the producer, director, and cinematographer, instead of much later and much more slowly in postproduction' [11]. In many ways, back projection is one of the direct precursors of VPS.

The optical methods of superimposition, multiple exposures, flying matte or traveling matte evolve into the chroma key, green screen, and blue screen technologies with the advent of electronic moving image making tools. These methods use even colour backgrounds to line studio walls, floors, and ceilings. Actors perform against these colours with placeholder props and extras in otherwise empty rooms. The colours green and blue are usually chosen as they are the colour least present in the hue of human skin and this makes it easier to visually separate the performers from the backgrounds. The uniform colour is then made transparent in postproduction and replaced with previously filmed, animated or computer-generated environments. Once again, the composition of the final image and the integration of performance and visual environment happens at a later time and in a different space.

The history of video has been the focus of a number of researchers since its inception in the 1980s, and has shifted from the mythology of Nam June Pike and the Portapak [12] to an acknowledgement of parallel developments in a number of contexts [13]. The early practices of video and film remain separate in visual art, where experimental film practices are considered to be in dialogue with cinema, and video art practices with television. Purists still shudder as the mediums converge and the terms *filming* and *filmmaking* expand to encompass a variety of digital tools including mobile phones and webcams.

In cinema, integrated optical, electronic, and digital tools have combined to contribute to an expansion of storytelling possibilities. Conceptually, strong distinctions are still drawn between analogue and digital tools, as analogue electronic imaging maintains a direct correspondence with the arrangement of objects and light on either side of the lens. In practice though, for several decades, digital visual effects (VFX) and CGI have been merging seamlessly with filmed footage on location, creating composite images of plausible events, actions, and spaces, as well as more fantastic scenarios. The versatility of digital codes allows filmmakers to produce and manipulate images quickly, easily, cheaply, and extensively. Digital editing and VFX have been an essential part of major film productions since Oscar-winning editor Walter Murch used Final Cut Pro editing software on Anthony Minghella's *Cold Mountain* in 2003 [14] (the first version of the software was completed in 1998 [6]). The progressive digitisation of filmmaking reached its apex in 2012–13, when Fujifilm, a major provider of negative films for shooting and positive films for projection, decided to reduce its cinema business and Kodak was in danger of going bankrupt [15].

14.3   The Evolution of Game Engines

VP emerges at the crossroad of the shift in filmmaking from analogue to digital on the one hand, and the expansion of computer games on the other. The game engine, as opposed to the individual game, is the critical component at the heart of a VPS. A game engine is a piece of software that offers a number of modules and tools to optimize the development of a game. Layzelle defines a game engine 'as the program's core framework and the code onto which the front-end graphics sit. The functionality includes the rendering engine for the 2D and 3D graphics, the physics part of the program, including collision detection and response, sounds, animations, rendering, scripting, artificial intelligence, and memory management [16]'. Even with a superficial understanding of this definition, it is easy to see how it may converge into filmmaking. 3D graphics, physics engines, and artificial intelligence are still growing in details and sophistication, getting closer and closer to approximating filmed images. Memory and rendering management are crucial to the fast processing and archiving of moving CGI. These technological developments make VPS increasingly viable as a commercial filmmaking method.

Although in 1952 British academic A.S. Douglas developed the first piece of software that is known as a game as part of his PhD, game engines would not start evolving independently of individual games until the 1990s. *Atari*'s first attempt to engage in console gaming was not successful, but in 1972 the company introduced *Pong* with a relatively fast play and healthy profits, starting a new era in computer games. In 1983, *M.U.L.E* was released as the first multiplayer game for up to four players, demonstrating a potential for other developers to exploit the format. In 1987, *Legend of Zelda* was designed as the first fantasy role-playing game for home consoles and more adult games such as *Leisure Suit Larry*, a comedic soft porn adventure, followed. The early sport simulator *John Madden Football* was released in 1988. This and similar rather slow pixelated games helped introduce computers to a whole new section of the community.

Although the 1980s were a golden age for arcade games, there was no cross-pollination between different developers; these were all standalone titles. A game spinoff of *E.T.* (Fig. 14.1), the film directed by Stephen Spielberg, was released in 1982 by Atari, but turned out to be a commercial failure, due to the low-quality

Fig. 14.1 *E.T.* Video Game

graphics, poor collision physics, and repetitive gameplay with E.T. falling into pits [17]. This could be seen as the first crossover between games and cinema, an approach that has become very common and fertile for both contexts. This failure may also mark the first step in the development of game engines. While some game developers may have reused sections of code in their own games, none sought to license others to use their code to create games. It's this reuse of code, however, that would eventually morph into game engines, but here the official idea of a game engine is still far away.

The concept of virtual 'physics' in collision physics and physics engines is a critical aspect to enable game engines to approximate simulations of the natural world. A physics engine provides an almost real-time digital simulation of certain physical systems, such as rigid body dynamics (including collision detection), soft body dynamics, and fluid dynamics. In other words, it is possible to apply the laws of physics to digitally configured materials. This means that light, fluids, hair, skin, fabric, etc. will behave as we have come to expect them to do in the material world. The combination of graphic inputs and programmed criteria produces visual plausible results without any further human interventions. From this perspective, physics engines can be understood as performing, and this represents a crucial conceptual shift in the understanding of the role of technology within filmmaking.

In 1993, with the release of *DOOM!,* the first pioneering first-person shooter, gaming took the turn that would lead to the development of game engines, with a heavier focus on 3D graphics. The *DOOM!*'s game engine created by *id Software* emerged as a revolutionary concept in game development, isolating 'a core section of code that managed the fundamental parts of the game—such as collision physics and game assets—and leaving outer shells of the game available for modification by the developer. In this way, different characters and game intent could be built around the central coding, making the game work in the same way as other titles but have a completely different and, we would hope, 'fresh look [16].' Following this, popular first-person shooter *Quake* was released with progressive features, offering full real-time 3D rendering and early support for 3D acceleration through OpenGL, which resulted in smoother gameplay. This crucial technological element brought the aesthetics of the computer game one step closer to the smooth flow of cinematic images, and it would become an indispensable tool to make VPS not only possible, but a mainstream approach to filmmaking.

From 1993, *Quake* (Fig. 14.2) included multiplayer deathmatches, allowing online multiplayer games to become increasingly common with the *QuakeWorld* update. *DOOM!* and *Quake* inspired other game developers like *Epic Games* to develop their own game engines called *Unreal. Unreal Script* was a scripting language at the core of the *Unreal Engine*, which was easier to use and more forgiving than C++ and provided more game-oriented functionality than the visual tools. In 1996, *Tomb Raider* was released, with Lara Croft's sexual appeal and real-world adventures. This was followed by *Legend of Zelda: Ocarina of Time* introducing meaningful puzzles, and in 1998 *Half-Life* with the adventures of Gordon Freeman at a Research Facility. Eventually, *Half-Life 2* made a hit with its realistic physics engine. As a result, plenty of toolkits had been developed by the end of the 1990s.

In 1998, Tim Sweeney had created *ZZT* a game notable for its embedded creation tools, rather than its gameplay. *ZZT*'s financial success and fanbase enthusiasm led Sweeney to work on an even greater project, the *Unreal Editor* [18]. The tools Sweeney developed were revolutionary and included a wireframe editor and a texture mapper to a text editor for code functionality. Today, these are very familiar elements of game engines and allow users to automate or semi-automate the construction of virtual shapes and textures. This early version of the *Unreal Editor* gave Sweeney the ability to create virtual 3D environments, which was unheard of at the time, and thus, Unreal Editor paved the way for the modern game engine [18]. *Unreal Engine* also uniquely supported the designer with the development of game mechanics using visual programming and scripting. As Microsoft's object-oriented programming

Fig. 14.2   A still from *Quake*

language *Visual Basic* had already demonstrated in 1991, being able to visualise the logic that a program needs to run is a radical step change for programmers.

Today's main competitor of Epic Game's *Unreal Engine, Unity*, was released in 2005, aiming to make game development accessible to everyone [19]. *Unity3D* was initially released for Mac OS X and in 2006 it was a runner-up for the Best Use of Mac OS X Graphics category in Apple Design Awards [20]. The support for Microsoft Windows and Web browsers was added later. By 2008, the engine had become more sophisticated, incorporating Machine Learning Agents. This open-source software facilitated *Unity*'s connection to machine learning programs, including Google's *TensorFlow*, allowing it to be used for building lifelike virtual landscapes and self-driving cars, as well as developing robots and avatars. Because of this facility, *Unity* was used experimentally in filmmaking in two computer-generated short films, *Adam: The Mirror,* and *Adam: The Prophet*, featuring a robot escaping from prison. In 2017 *Unity* also developed a new *Cinemachine* tool and in 2019 it was used to create the backgrounds for *The Lion King*.

With *Playstation 3 / Xbox 360*, the idea of spending hundreds of thousands of hours developing your own engines became less and less fashionable [21]. *Midway*'s *Mortal Kombat* was the first to utilise *Unreal 3* for *Mortal Kombat vs. DC Universe* and in the 2011 version of the game, known simply as *Mortal Kombat*, *Midway*'s developers adapted the engine significantly. In 2014, *Capcom* started using *Unreal 4* for *Street Fighter V*, which was released two years later, and then for the *Soulcalibur* series.

At the time of writing, *Epic Games*, is the dominant game engine in filmmaking. The company has strategically partnered with VFX and film production studios large and small, it has created opportunities for independent practitioners, such as bursaries and showcases, it has organised and participated in significant engineering conferences, and generally made its presence felt. Crucially, *Epic Games* demonstrated the first proof of concept of an LED-based VPS at SIGGRAPH 2019. Although *Unity* had developed its Cinemachine tool first in 2017, by extending its software with plugins and supporting the development of individual VPS *Unreal* (Fig. 14.3) appears to have reached a dominant position in the market. The company continues to foster technological research and contribute to the development of technical refinements [22].

14.4   A Brief History of the Virtual Camera

Camera systems are used in video games to control a virtual camera or a set of virtual cameras to display a view of a 3D virtual world during the action at the best possible angle in a third-person view. The first example of an automatic camera control system was showcased in 1988 by Blinn. In 1990, a set of mappings between the user inputs and the camera behaviour was established by Ware and Osborne [23], linking the camera movements to players' behaviour. McKenna [24] demonstrated a magnetic motion sensor and a portable LCD TV at MIT, while other researchers [25, 26], were working on how to simplify camera control.

Fig. 14.3 The Virtual Production Studio at AIE Canberra. https://aie.edu.au/articles/discover-the-future-of-film-in-canberra/

Work on virtual cameras was also taking place at the University of North Carolina. Here, *The Walkthrough Project* focused on linking the camera view to physical input devices for the control of a player's movement, including dual three-axis joysticks and a billiard-ball shaped prop known as the UNC Eyeball [27]. These technical details are worth mentioning to highlight how the work developed in game design will transfer from the gaming experience to filmmaking practices. Speed of communication between physical and virtual movement, accurate controls, and multiplicity of camera angles make a virtual camera indistinguishable from a material one in their visualising performance.

As stated by Burelli [28], 'A large volume of research studies on virtual cinematography is dedicated to the analysis of robust and time-efficient techniques to place and move the camera to satisfy a set of given requirements'. Different types of shot affect player experience and the characteristics of the camera behaviour determine the impact of the visual outcomes. In video games and virtual camera systems, the movements of players require cameras to perform unpredictable movements. This is facilitated by the use of AI to select the most appropriate shots in highly interactive and changeable worlds. Already in 1996, Christianson et al. had proposed a system that automatically schedules the sequence of shots to film one or more events in a virtual environment [29]. In 2010, these systems finally reached levels of performance that allow the real-time generation of cinematographic views in these interactive and unpredictable virtual environments [30].

Resident Evil 5 was the first video game to use a combination of virtual camera and motion tracking technology [31]. In this set up, motion capture technology is used to track the position and orientation of an actor performing the movements that will be available to the game character. A virtual camera is connected with the same markers and allows the operator to intuitively move and aim a material version of the camera by simply walking about on set. The operator holds a tool that combines a tablet, motion sensors, and a support framework for recording and adjusting lens properties. This early combination was also explored by director James Cameron in the movie *Avatar* and it is a precursor of the Simulcam, one of the technological tools that has facilitated the emergence of Virtual Production Studios.

14.5   The Performance of Machinima

Marino (2004) argues that when *id Software* released *Quake*, the company's first 3D game, and included a demo recording feature, 'a critical shift occurred—*the viewpoint of the player became the viewpoint of a director*' ([32], (emphasis in original). Players used these recording features to capture multiplayer deathmatches, but they also started to construct more elaborate narrative scenarios independent from the game's progression. *Diary of a Camper* (1996) is credited with being the first short *Quake Movie*, another early name for the practice of machinima.

In this short story, one player functions as the camera as the others hit their targets in choreographed movements [32]. A community of machinima practice emerged and in 2000 Hugh Hancock founded *Machinima Inc*, a commercial entertainment enterprise later acquired by *Warner Bros* that ceased operations in 2019. However, machinima continues to thrive as a form of entertainment with dedicated platforms, and as a respected artform with specialist international festivals (for example, Rooster Teeth and the Milan Machinima Festival).

Kim Libreri—Chief Technology Officer at American video game and software developer *Epic Games*—observes that 'nowadays machinima is everywhere' and that virtual production 'is basically in-camera machinima [33]'. An early adopter of computer graphics in filmmaking, Libreri appreciates the difference between computer rendered images and the real time worlds created in game engines: 'you've basically built a simulation of the real world that can actually tell a story [33]'. Machinima and VPS, although they fulfil different purposes, build visual worlds that perform with a level of autonomy. This encompasses physics engines developed for game design that allow the computer-generated natural elements to move according to physics principles. This fascination with non-staged movement on screen can be traced back to the earliest days of cinema, when the movement of leaves in the background of a staged scene is the factor that makes cinema radically different from theatre with its painted, static sets [34].

Both machinima and VPS utilise physics engines and game technologies to build environments, characters and narratives. From this perspective, it is important to highlight this link with machinima in a text that examines the relationship between game technology and filmmaking. 'Machinima production has widened to incorporate elements of virtual performance, where human operators manipulate the game characters—like virtual puppets with in-role performance [35]'.

14.6   From Front Projection to LED Volume

The basic logistical, financial, and technical problems that film crews and filmmakers encounter have not changed substantially since 1968. The technique of front projection was famously used by Stanley Kubrick in 1968 in the opening scene of *2001: A Space* Odyssey. It consists of previously filmed images projected through a one-way mirror at 45 degrees onto a screen made of retroreflective material. The projector and camera angle have to be aligned with each other and the position of actors and props in order to avoid shadows on the screen. Front projection presents the same problems with parallax created by rear screen projections, thus making camera movements impossible. On the other hand, because of the reflective material, the background can achieve the same level of brightness as the foreground. In fact, '100 times brighter than any light image reflected from the foreground subject' [10].

Kubrick had chosen to use this technique to avoid traveling to a remote area of Southwest Africa where a location with unique rock formations had been found [36]. He also wanted to prevent the degradation of film quality derived from using blue or green screen and compositing the image with multiple exposures to combine the shots. To achieve this, he had the largest front projection screen in Europe built on a soundstage at MGM studios in Borehamwood. Still transparencies shot on location were projected on a collage of pieces of retroreflective material to increase the sense of depth. To maintain the alignment of camera, screen, and projector, Kubrik had a rotating set built and this allowed him to produce varying shots of the same scene. This set up was costly and cumbersome, and it was used again on the film *Where Eagles Dare* in the same year [36].

In 2013, for the film *Oblivion*, writer-director Joseph Kosinski also chose to use the technique of front projection as a solution. The protagonists live and work in the Sky Tower, a structure surrounded by glass high over the surface of the Earth devasted by an alien invasion. Kosinski describes the 'lighting effect and immersive quality' he wanted to achieve for actors and audiences alike [37]. He shot the skies for the background plates in Hawaii and, with director of photography Claudio Miranda, devised a system that 'used 21 projectors on the 500 feet wide screens for the Sky Tower scenes. It's the whole background and all in-camera' [38]. This allowed Miranda and Kosinski 'choices and lighting choices because you have the real thing' [38]. In this visual environment, having control of reflections was paramount. Blue and green screens create tinted reflections on the actors' skin and on any glossy surface. To achieve the futuristic feel required by the film, it was important to use chrome and glass in the structure of the Sky Tower. This would not have been possible with coloured backlots. Nevertheless, blue and green screen techniques remain in use as filmmaking, including in *Oblivion*, and often complement other techniques.

Erik Caretta—VFX supervisor at Hive Division in Treviso, Italy—was one of the film production studio VFX supervisors looking for different alternatives to blue and green screens in 2019: 'We first had a taste of a solution last year when we took care of the visual effects of *Il Talento del Calabrone* … We created the illusion of the shooting actually taking place on top of a Milan's skyscraper by first shooting the actual footage on location and then projecting the plates onto the filming stage thanks to an array of video projectors' [39]. After a year of research, the combination of *Unreal Engine* and LED walls seem to solve some of the problems that filmmakers had encountered in the making of composite images. Motion parallax was the main stumbling block to allow cameras to move on set when using rear screen projection. The link between the material camera on set and the game's virtual camera could solve this problem, while LED walls could achieve the level of background brightness that projection did not provide. Before they could apply this combined system to filmmaking, *Disney* was screening *The Mandalorian* on the *Disney Plus* channel and it was public knowledge that it had been shot in an LED volume.

From being founded by George Lucas in 1975, *ILM* has been the giant at the forefront of innovations in VFX. Turnock's recent history of the company dispels the myth of *ILM* as largely inflated by self-promotion, but the studio remains an undeniable force in the field of special effects. Cinematographer Greig Fraser first used LED screens while shooting *Rogue One A Star Wars Story* in 2015. 'Up to that point, LEDs were mostly being used for eye lights, augmentation, and as strips of lights. The new types of LED coming out around that time meant that he could dispense with gels and use the RGB palettes on the LEDs' [40]. In fact, for the 2013 film *Gravity*, 'cinematographer Emmanuel Lubezki and VFX supervisor Tim Webber created a "lightbox" structure covered with 196 panels of 4,096 LED bulbs each' [41]. "Sandy's Cage"—as it was affectionately named on set in reference to the actress Sandra Bullock who spent long hours in it—had the dual purpose of creating reflections on the actors' faces and immersing them in the environment in which their actions were supposed to take place [41].

14.7   A Brief History of the Virtual Production Studio

Variously known as StageCraft or The Volume, *ILM*'s VPS was first used in its full capacity for the *Disney Plus* television series *The Mandalorian*. Developed in 2015 for *Rogue One: A Star Wars Story*, the first version was a large cylindrical LED screen. 'All the cockpit sequences in X-Wing and U-Wing spacecraft were done in front of that LED wall as the primary source of illumination on the characters and sets' [42]. LED panels were used to light the scene and to create the right reflections, but the backgrounds were added in postproduction. Moreover, 'those LED panels had a pixel pitch of 9 mm (the distance between the centers of the RGB pixel clusters on the screen). Unfortunately, with the size of the pixel pitch, they could rarely get it far enough away from the camera to avoid moiré and make the image appear photo-real, so it was used purely for lighting purposes' [42]. Cinematographer Greig Fraser describes how they 'wanted to create an environment that was conducive not just to giving a composition line-up to the effects, but to actually capturing them in real time, photo-real and in-camera, so that the actors were in that environment in the right lighting—all at the moment of photography' [42].

Holben reports that Kim Libreri of Epic Games proposed to use Unreal engine to link the camera to the image on screen and this final step made the 'live-action Star Wars television series possible' [42]. It is this combination of game engine and LED volume that immerses actors and filmmakers in the full dynamic performance of visual environment, technology, and actors. After overcoming a number of technical difficulties, including moiré patterns (the visual effect emerging when two sets of straight or curved lines are superposed at 20 or 30 degrees with respect to one another), consistent colour calibration across the separate panels, latency in rendering, and displaying continuously updating images, the team at ILM built a prototype 35-foot-wide capture volume [43]. This eventually developed into a circle 75-foot-wide in diameter and 20-foot-high, with a ceiling and two flat walls of LED panels almost closing the volume [44].

Mike Seymour confirms that at the end of 2020 there were 'about 100–125 LED stages with another 150 due to come online beginning of next year. There is mass acceleration in building these stages because of COVID and the fact people can't travel, and it seems like a good option' [45]. Seymour observes that the consequences of the COVID-19 pandemic accelerated the development of integrated VPS systems. Historically, VP was first adopted in parallel by *ILM* (Los Angeles, US) and *Weta Digital* (Wellington, NZ) for the first time in 2001 [46]. *Weta Digital* first used a virtual camera on Lord of the Rings [4], while Steven Spielberg used an iPad attached to a Vive controller to visualise potential shots in a CG environment that updated in real-time for the film *A.I.* [47].

Film directors Robert Zemeckis and James Cameron are also credited as being early adopters of virtual production technologies and real-time rendering, working closely with VFX specialists to develop ad hoc tools [3]. In fact, in 2010 at *Weta Digital*, Cameron's film *Avatar* 'was the pivotal moment. James Cameron wanted to direct live actors on the mocap stage but view their performances in the Pandora environment' [4]. *Weta Digital* built digital assets and environments, then combined these with the digital puppets being driven by the motion capture, and fed this virtual view into the viewfinder of a camera on set. Cameron could now direct the action as if it were taking place on a live set. Without the benefit of the LED volume, actors are still performing in the vacuum created by blue and green screens, while filmmakers are still limited when it comes to the choice of materials to use on set, camera movements, speed, costs, and flexibility. Conversely, it is evident that Cameron had access to an augmented reality experience featuring an environment that was performing independently around him.

Weta, ILM and a number of smaller studios had been developing the elements that make VP possible since the early 2000s. *Weta* remains at the forefront of VPS technology and has been critical to the technological and creative developments taking place in New Zealand. In 2021, *Weta Digital* claimed to be the largest single-site VFX studio in the world, drawing artists from over 40 countries [4]. In July

2020, the company partnered with *Avalon Studios* and *Streamliner* to create a new LED stage for virtual production services also based in Wellington. The company continues to expand and build new divisions, for example by partnering with *SideFX* to create a cloud-based new software service called *WetaH* [4]. The company was acquired by *Unity* at the end of 2021, following a trend that has seen the growth in the number of hybrid companies gathering under the same ownership or management of VPS facilities with key hardware, software, display production, and distribution.

In terms of adoption of VP technology, the turning point took place in 2019 at the SIGGRAPH conference in Los Angeles, when *Epic Games* exhibited the first integrated LED VPS with *Unreal Engine* in stand 405 [48, 49]. Epic Games partnered with *Lux Machina* (LED walls), *Magnopus* (VR/AR), *Profile Studios* (mocap and VP), *Quixel* (3D graphics), *ARRI* (cameras), and director of photography Matt Workman to launch an out-of-the-box VPS solution and shoot a demo that went around the world very quickly (Fig. 14.3). In this configuration, LED walls display a portion of landscape updated in real time by the game engine as directors, directors of photography, and VFX supervisors move a camera on set, shift focus, and test shots with an actor and a motorbike [48]. This makes in camera final picture VFX and live action a reality, bypassing the need for a studio to develop in house solutions. VPS is quickly becoming the new standard in filmmaking and some similarities are starting to emerge in the digital backgrounds adopted in films [50].

Educational institutions are increasingly considering Virtual Production as a complete production and delivery system, and the majority are training their students to perform the roles emerging as VPS facilities multiply around the world. For example, in Australia, the *Academy of Interactive Entertainment (AIE)* and *AIE Institute* in Sydney is planning a dedicated VFX degree. At Edith Cowan University in Perth, Brad Nisbet is developing, as part of his doctoral research, a dedicated VP lab space and an in-house operational VPS to facilitate new and emerging technologies in media production for both students and the local industry. Bournemouth University in the UK and Chapman University in Orange, California among many private and state educational institutions include VPS technology in their plans for the near future.

14.8   Key Technological Components of a Virtual Production Studio

As we indicated above, a VPS is an integrated system of technological tools developed in different contexts. A technical understanding of the technologies involved supports our analysis of the qualitative and conceptual shift that has recently taken place in film production. Six elements of an integrated VPS system make a qualitative difference to filmmaking. We have already discussed how a game engine and LED walls combined into a volume are the core components of VPS, but the tools that connect them to each other and allow the real-time updating of rendered images on screen are equally crucial. These are a camera tracking system, a motion tracking system, a Simulcam, and finally rendering technology, as formerly discussed by Kavakli and Cremona in 2022 [51].

It is by now evident that the critical significant component that links the game engine to the processes of filmmaking is the presence of a camera. VP utilizes 'game engines that enable us to see our [CG] assets in real time', states Christina Lee Storm of the Producers Guild of America [52]. Just as it is required to do in a game, the game engine renders in real time the visual background scene that, as we have seen, can be integrated with actors' performances at different times in different ways. When displayed on LED walls, the performance can take place within the LED volume and the resulting composite image can be captured directly as final pixel (or final frame— two expressions that describe the finished visual outcome more or less unchanged). As we will see later in this chapter, it is our view that this aspect of VPS is key to the emergence of new conceptual readings of performativity in filmmaking.

Epic Games has been developing its own plug-ins and networks to extend its *Unreal Engine* and eliminate the need for customised solutions to connect game engines and LED screens. For example, nDisplay distributes the rendered images of a camera view over a number of machines. The view from the *Unreal Engine* camera is extended, rendered, and distributed onto LED panels in as close to real time as the technology progressively allows. The virtual camera films in mono (2D) and dynamically updates the background to match the perspectives that the physical camera records on set. This mechanism establishes the ground on which camera, LED panels, and the mechanisms in between can be said to be performing. The illusory depth of the set is created by the shifts images perform on the LED panels in relation to the movements of the camera, maintaining the correct perspective with the physical objects and performers on stage. To do this, and avoid problems with parallax, the LED stage needs to work in combination with the motion capture system, which is aware of the camera position and movements.

The components that track the movements of the material camera on set and link it to the virtual camera in the game engine are the hinges that make this film production method qualitatively different. The camera tracking technology in a VPS was originally developed for Virtual Reality headsets (e.g., Oculus Quest) or head mounted devices (HMD). In Virtual Reality (VR), headsets can be tracked in two ways and both ways have made their way into VPS. These can be described as looking from the outside-in or from the inside-out. When applied to VPS, in the outside-in system, sensors in the room follow the movements of the camera, while in the second kind of system the sensors are on the camera itself, watching for signals in the room. Both approaches have their own merits. Wireless systems are increasingly preferred, but sending signals wirelessly between camera and computer can add latency to the image experienced on the LED panels, and speed is a constant concern in VPS [53]. These components are part of a system known as Simulcam.

The Simulcam matches the real-life camera position to coordinates in the game engine corresponding to its internal virtual camera. With this link, the CG environment is continuously *seen* by the CG camera from the correct angle in relation to the performers on stage. The Simulcam, therefore, simulates the CG environment created for shooting from the perspective of the physical camera on set. On the LED walls, the movements of the paired cameras are rendered in almost real time, and the perspective of the scene shifts with the camera angle, avoiding parallax issues. Similarly, CG lighting and on set lighting are linked and synchronised via standardised, widely available plug ins. Frames are rendered and displayed in almost real time thanks to the game engine.

A similar motion tracking system utilises markers on performers and objects to link the movement of actors on set to CG animated characters with high-fidelity. This information is directed into the game engine and integrated into the image displayed on the LED walls with the lowest latency possible. Motion tracking systems are indispensable to integrate live on set performances with digital enhancements as well as physical and virtual objects in one visual continuum. The parallel dynamics between the motion tracking systems on camera, actors, and props create a performative loop within which all elements on set are performing and responding to each other. At present, there are two obstacles to a full performative flow that integrates and absorbs human performers—speed and power. These problems are shared by a number of data heavy technologies that require complex rendering, including VR, 3D films, and games themselves. Render farms are a partial solution to this problem. They are not specific to VPS, games or filmmaking, but have been a crucial development for the emergence of viable VPS.

The size of moving CGI files is largely determined by the number of polygons (poly-count) that forms each item—the higher the poly-count, the smoother and more photorealistic the final image and the higher the file size. Approximating filmic quality is an indispensable requirement in cinematic productions. Film history is peppered with memorable *state of the art* special effects films that, with time, reveal how far the trick was from what it was trying to simulate. Inevitably, this will happen with virtual filmmaking, and some synthetic images are already jarring for trained eyes. In this context, a high poly-count and good dynamic lighting are helpful tools, but they require complex and fast rendering. Moreover, new specific software plugins are continuously being developed to visualise realistic fur, grass, clouds, and other complex natural materials and systems that move and reflect light in unique ways. Movements, shifts, and changes in light need to be processed and displayed in real time to keep pace with camera, environment, and actors' movements in the LED volume. That's where fast processing, transferring, and storing of data become essential.

A Render farm is a cluster of individual computers connected through a network. They are used to process and render CGI, VFX shots, and still images in reasonable time. They are formed by chains of Central Processing Units (CPUs) and graphics processing units (GPUs). For a productive use of these resources, effective pipelines are indispensable. VPS pipelines need to be complemented by efficient systems of management of custom queuing, scheduling, job description, and submission systems—mechanisms that track assets in the different phases of rendering and displaying [54].

In her research on this topic, Kavakli [55] has focused on the limitations of VPS, including the problems with latency and rendering power required to update the images on the LED panels in relation to the camera movements. A delay of several frames does not sound that significant, but it is an issue in the industry and a number of companies are racing to provide solutions, with Brompton claiming to have reduced the latency to one frame [56]. The other and more significant issue is hardly ever mentioned and much more disruptive. The image on the LED panels is fully rendered and updated only in the area captured by the camera. In other words, the issue of parallax is resolved only in a limited portion of the background image. This is a material impediment to a full immersion in the environment created by the LED volume, but like many technical limitations of emerging technologies is probably temporary.

In fact, future technical research is expected to focus on these two aspects of VPS. Oakden and Kavakli [57, 58] also conducted experiments to test various hardware configurations to render in real time a variety of different scenes using the recommended graphic cards for VP. They have found that high-end and costly hardware did not offer the performance gains they had expected over the lower-end hardware. They suggest that optimising the graphics processing to split the rendering of scenes across multiple computers might offer a solution [58].

14.9   The Virtual Production Studio and Qualitative Shifts in Filmmaking

As it is to be expected in an emerging field, the terminology around VP is still fluid. The expression virtual production itself, as well as being used to reference live action filmmaking can also describe the production methods of animation, games, and VFX. In filmmaking and beyond, Previsualisation (or previs), 'techvis, postvis, motion capture, vr, ar, simul-cams, virtual cameras, and real-time rendering—and combinations of these—are all terms now synonymous with virtual production' [59]. For example, at Technicolor, VP is 'anything from green screen replacement to action in front of an LED screen' [60]. They also use the expression Real Time Production. As technologies converge, terminology converges in parallel. In-camera VFX (ICVFX) is another expression that is emerging as all the technological component of VPS systems become better integrated and make it possible to include what is captured in-camera in the final film.

As VPS increasingly becomes standard technology in filmmaking, it is worth wondering if the expression Virtual Production will be replaced by Film Studio or Production Studio. In other words, will there be retrospective terms to identify the minority of facilities that continue to use traditional filmmaking technologies? Or will the virtual, real time performative qualities of the combination of game engines and LED volumes remain a special category both linguistically and conceptually?

According to Kipnis, 'film studies has been somewhat slow to come to terms with a changing apparatus, or to theorize the shifts in film language and grammar that technological change seems to have so rapidly brought about' [6]. Srivastav and

Singh add that 'despite technological changes, the filmmaking process has remained constantly akin to the production of narrative storytelling' [61]. More recently, the development of virtual reality (VR), augmented reality (AR), and 360degree film technologies have made storytelling more challenging with their immersive and interactive nature. VP has not dramatically affected film's narrative structure, but has extended the potential for increasingly fantastic settings and characters. On the other hand, LED volumes in VPS have shifted the performative relationship between human and non-human actors in the film studio [62].

We have seen above how directors, cinematographers, DOP, and actors have sought and have started to benefit from the immersive qualities of VPS. In this final part of the chapter, we analyse and speculate on the immersive and performative qualities of this system of film production. We argue that the immersion and the radical shift in the dynamics of performativity accomplished by VPS bring a conceptual shift in filmmaking that needs to be further discussed and researched. In this framework, we propose that in the LED volume of the VPS, actors perform *with* the technology. It is indisputable that from the perspective of live performance very different mechanisms are at work here, compared to performing in a green or blue empty stage. For example, O'Shea Jackson Jr., who plays a character called Roken in the popular Disney mini-series *Obi-Wan Kenobi*, has experienced performing in the state of the art VPS at *ILM*: 'What's so crazy about The Volume is, you really forget. After a while, you forget that you're just in this studio' [63].

Here, Jackson Jr. is describing immersion, defined as 'the sensory and perceptual experience of being physically located in a non-physical, mediated, or simulated virtual environment' [64]. Although the concept and experience of immersion are usually attributed to 360degrees VR media, the effect of performing within an LED volume is certainly comparable. LED walls immerse actors in the final frame environment and VFX are created in real time for actors as well, informing their performance. The game changer here is the link between on-stage camera and virtual in-game-engine camera that makes changes in real time on the LED walls, as virtual camera movements are updated in sync with the movements of the on-stage camera. The fact that only a portion of the LED volume changes in relation to the camera movements, is relatively unimportant for actors as they seldom turn their back to the camera.

Scholar Katrina Ilmaranta [65] argues that VP is 'a real like three-dimensional environment that someone using special electronic equipment might interact in a seemingly physical way'. The ensuing sense of immersion produces 'an illusion of a spatial world' [65]. Her analysis of recent cognitive film studies and narrative studies suggests the 'co-dependency of movement and space' and proposes the concept of an enactive presence in the space 'where bodily changes lead to the forming of space-related mental images that precede the narrative' [65]. In other words, actors respond to the visual clues that surround them on LED walls with their bodies. Motivated by the muscle memory developed in the material world, they associate one with the other. In this dynamic, actors perform more directly in response to the visual environment they are immersed in, and less to the filmic narrative. From this perspective, their performance is radically altered, influenced as it is by the performance of the technology that surrounds them.

Effectively, game engines are increasingly behaving like physics engines. Once the parameters of the laws of physics have been input, all representations of materials and lighting perform to a set of instructions determined by shifting conditions. The distinction becomes important when we think in terms of performance—*Unreal* and *Unity* don't 'design' a three-dimensional world, but set coordinates and parameters for materials and conditions with known and coded behaviours to perform their roles. Once the initial settings are defined, camera movements are expressed in the behaviours of the visual qualities dictated by the conditions the same elements would *experience* in the material world. From this perspective, VP technology holds the potential to substantially transform the relationships of display technology (screen), live action (actors), image production (cameras), and postproduction (VFX) by informing their performance.

Joel Bennett examines how 'an actor's perceptual limitation to the virtual environment increases their cognitive workload and distracts their attention from a performance because they have to construct and imagine the virtual environment' [66]. This researcher focuses on Motion Capture (mocap) and on the use of XR (extended reality, including VR and AR) aids to familiarize the actor with the virtual environment before the performance. Bennett complements actors' testimonials and practice-led research with a phenomenological approach and concludes by establishing three spectra of immersion for an actor in Virtual Production: Bodily Immersion, Spatial Immersion, and Perceptual Immersion. He argues that the potential improvements that can facilitate actors' immersion include the ability to control the lighting of the virtual environment in real-time [66]. LED walls within a VPS integrated system emit light and are complemented by other sources digitally controlled within the same system. Actors are immersed in the same light as the elements represented in the visual environment that surrounds them.

Paul Debevec has conducted seminal experiments in lighting, simulating light, relighting scenes, and possible uses of arrays of LEDs for this purpose over almost 30 years. He first tested the use of LED panels to create reflections in the eyes and on the faces of actors [67]. Significantly, in 2002, Debevec published the paper *A Lighting Reproduction Approach to Live-Action Compositing*, describing the original Light Stage—a two-meter sphere of inward-pointing RGB LEDs 'focused on the actor, where each light can be set to an arbitrary color and intensity to replicate a real-world or virtual lighting environment' [68]. Most recently, his research has engaged with Neural Radiance Fields (NeRF) to explore the use of 'observed images to recover a 3D scene representation that can render the scene from novel unobserved viewpoints' [69]. It is already common practice to adopt ray tracing technology to separate the elements that form the visual environment from the dynamics of the light that will modify their appearance at different times of day and night, and from different points of view, so this technology—reminiscent of *Blade Runner* sci-fi imagination—is no longer implausible.

14.10 Conclusions

As film preproduction, production, and postproduction converge on set, virtual art department, DP, director, compositors, editors, and VFX artists tend to work together from the beginning of a film production. The majority of digital assets need to be ready in advance, available to the game engine for real time rendering and for displaying on LED walls. This advantages actors as well, as they do not need to count on vague descriptions and their imagination to visualise inexistent environments that will be added long after their performance. Light and colour grading can be controlled in almost real time on the LED walls and in camera, requiring increasingly fewer postproduction interventions. The practice of preparing iterations at different stages of production becomes obsolete as more flexible pipelines and workflows emerge.

CGI of landscapes, interiors, and a variety of environments are already available on Market Place and from a number of providers. Similarly, LUTs (Look Up Tables) that supply predetermined colour grading and VFX can be downloaded from studios and individual artists. Moreover, once a studio has digitally scanned, captured or generated any image, these can be adapted and reused an infinite number of times. These mechanisms facilitate the production of franchises and reduce costs, changing how budgeting for a production is approached in the first place. They also present a downside as an increasing number of production studios tap into the same resources— there is a danger that landscapes and visual environments start to resemble each other no matter how different the individual film's narrative.

Since traditional filmmaking requires shooting on location and in real conditions, extensive manpower and resources are involved, and thus the budget rises. With VP technology, and VPS in particular, most plates can be completed in the studio, and preproduction can be merged with production by processing characters and scenes in a game engine, blending real with virtual to achieve final pixel quality directly from assets developed in the previsualisation phase. Digital data can be distributed across networks of studios and professionals safely and rapidly for parallel development, overcoming the costs and risks of traveling. This has the potential to involve more geographically diverse practitioners and to reduce the negative impact of disabilities. It also helps the film industry greatly by reducing the cost of production, although the initial costs of setting up a VPS may be much higher than in traditional film production studios. This situation has been accelerated by the changes caused by the COVID-19 epidemic and is in continuous evolution, as off the shelf systems become available and within reach for small and independent studios.

By transposing gaming technology to filmmaking, the active point of view of the player in first person and immersive simulation games changes the position and performance of actors. Director, POD, and cinematographer are immersed in almost real time in a consistent first-person perspective that moves the narrative as camera, lights, representations, and screens perform around them materialising the rendering calculations of the game engine.

We have tried to speculate on how these performances without intentionality can be discussed, and on how they move actors to perform differently. Actors' performance is undeniably more responsive to the immersion in the active performative lighting and visual environment constructed on the LED walls and continuously updated by the game engine then to empty monochromatic rooms. Influenced by their bodily memories, performers respond to their present environment as much as to the narrative. Actors perform *with* the technology. The point of view of the player immersed in the game becomes the point of view and position of the actors—they are at the centre of the action and they are part of it. Similarly, the point of view of the filmmaker is one of the active points of view—a performing one. Director and camera traverse the set as it transforms around them, like game players. Does this also translate to the point of view of viewers?

From a conceptual perspective, this chapter offers an introduction to the idea that VPS affords a different understanding of non-human actors and machine performativity in filmmaking. Based on the technological functions of the key elements of this method, the conceptual shift derives from an understanding of the roles of each active component. A performative loop emerges in the responsiveness of each step of the process to all the others in a horizontal ontology that puts human actors at the same level of technological actors.

14.11   Domains that Require Further Investigation

In our speculations, we have made the assumption that actors are not overtly influenced by the fact that only a small portion of the LED volume in which they are immersed responds to the camera movements. As the technology evolves, this aspect will require experimental and cognitive investigation. Similarly, as the light that surrounds the actors' bodies shifts with a minute delay, more research is needed to establish if and how their performance is affected by the specific timing of the light shifts. On the other hand, it is very likely that these technical limitations will be short lived, and detailed research might not be required.

It is easy to understand the importance of integrating AI and Machine Learning in VPS, particularly in relation to the realistic representation of complex natural systems like hair, fur, water, and similar materials. This needs to become the focus of research both in the technological and theoretical fields. AI and Machine Learning are already assisting in the creation of digital humans and in the correct functioning of physics engines. Their role is still in development with regard to optimisation of graphics cards and a number of other technical improvements that can solve the remaining and emerging limitations of VPS.

Conceptually, further theoretical and experimental research is required to evaluate the ramifications of the principles introduced in this text. In particular, how performing in an LED volume VPS influences actors could be tested in the process of production. Filmmaking based on VPS methods poses a different set of questions for the researcher in film studies. We are interested in opening a conversation about definitions of performing when applied to technology.

References

1. Sandin, D.J.: Digital illusion, virtual reality and cinema. In Dodsworth, C. (ed) Digital illusion: entertaining the future with high technology. Addison-Wesley Professional: Boston, Ma. 3–26 (1998)
2. Martin, K.H.: Atomic fiction answers the call for innovation in welcome to Marwen. VFXV (2019). https://www.vfxvoice.com/atomic-fiction-answers-the-call-for-innovation-in-welcome-to-marwen
3. Kadner, N.: The Virtual Production Field Guide. Volume 1. Epic Games (2019). https://cdn2.unrealengine.com/Unreal+Engine%2Fvpfieldguide%2FVP-Field-Guide-V1.2.02-5d28ccec9909ff626e42c619bcbe8ed2bf83138d.pdf
4. Weta Digital: Virtual Production (2021). https://www.wetafx.co.nz/research-and-tech/technology/virtual-production
5. Lowood, H., Guins, R.: Debugging game history: a critical lexicon. The MIT Press Cambridge, Massachusetts London, England (2016)
6. Kipnis, L.: Film and changing technologies in the oxford guide to film studies Hill, J. and Church Gibson, P. (eds) Oxford: Oxford University Press, pp. 595–604 (1998)
7. Musser, C.: Toward a History of Screen Practice. Quart. Rev. Cinema Stud. 9(1)(Winter 1984), 59–69 (1984)
8. Rizzo, M.: The Art Direction Handbook for Film & Television, 2nd ed. Taylor & Francis Group (2014)
9. McGowan, F.: The use of rear screen projection in King Kong (2014). https://francismcgowanvfx.wordpress.com/2014/10/05/the-use-of-rear-screen-projection-in-king-kong/
10. Lightman, H.A.: Filming 2001: a space Odyssey. American Cinematographer (2018). Retrieved on 15 July 2022. https://ascmag.com/articles/filming-2001-a-space-odyssey
11. Turnock, J.: The Screen on the Set: The Problem of Classical-Studio Rear Projection. Cinema J. 51(2) (Winter 2012), 157–162. University of Texas Press on behalf of the Society for Cinema & Media Studies (2012)
12. Sherman, T.: "The Premature Birth of Video Art" University (2007). Retrieved from http://videohistoryproject.org/sites/default/files/history/pdf/ShermanThePrematureBirthofVideoArt_2561.pdf
13. Meigh-Andrews, C.: A History of video art: the development of form and function. Bloomsbury, London (2014)
14. Daly, K.M.: 2010. How cinema is digital. In Einav, G. (ed.) Transitioned media. a turning point into the digital realm. 135–147 (2010)
15. Dent, S.: Film's cinema comeback is driven by nostalgia, not logic. Engadget (2016). https://www.engadget.com/2016-02-15-films-cinema-comeback-is-driven-by-nostalgia-not-logic.html
16. Layzelle, D.: History of the Game Engine: Part 2—The 1990s (2021b). https://ultimategamingparadise.com/features/series/history-of-the-game-engine/history-of-the-game-engine-part-2/
17. Layzelle, D.: History of the Game Engine: Part 1 (2021a). https://ultimategamingparadise.com/features/series/history-of-the-game-engine/history-of-the-game-engine-part-1
18. O'Toole-Bateman, C.: The History of the Game Engine: Part 11—Unreal Engine Builds Bridges (2022). https://ultimategamingparadise.com/features/series/history-of-the-game-engine/part-11-unreal-engine/
19. McWhertor, M.: Former EA CEO John Riccitiello is now head of Unity. Polygon (2014). https://www.polygon.com/2014/10/22/7039683/electronic-arts-john-riccitiello-unity-ceo

20. Smykill, J.: Apple design award winners announced. ArsTechnica. (2006). https://arstechnica.com/gadgets/2006/08/4937
21. O'Toole-Bateman, C.: The History of the Game Engine: Part 9—Fight! (2021). https://ultimategamingparadise.com/features/series/history-of-the-game-engine/part-9-fight/
22. Niklaus, L.: Unity Vs Unreal: Which Fits You Best? Mount CG. https://mountcg.com/unity-vs-unreal/. ND
23. Ware, C., Osborne, S.: Exploration and virtual camera control in virtual three dimensional environments. ACM SIGGRAPH, **24**(2):175–183 (1990)
24. McKenna, M.: Interactive viewpoint control and three-dimensional operations. Proceedings of the 1992 Symposium on Interactive 3D Graphics (I3D '92). ACM. pp. 53–56. https://doi.org/10.1145/147156.147163
25. Phillips, C.B., Badler, N.I., Granieri J.: Automatic viewing control for 3D direct manipulation. In ACM SIGGRAPH Symposium on Interactive 3D graphics, pp. 71–74, Cambridge, Massachusetts, USA, 1992. ACM Press
26. Drucker, S.M., Zeltzer, D.: CamDroid: a system for implementing intelligent camera control. ACM Symposium on Interactive 3D Graphics and Games. (1995)
27. Brooks, F.P. et. al.: Final technical report: Walkthrough Project, Six generations of building walkthroughs. Tech. Rep. Dept. of Computer Science, Univ of North Carolina, Chapel Hill, N.C. (1992)
28. Burelli, P.: Virtual cinematography in games: investigating the impact on player experience. International Conference On The Foundations of Digital Games, Chania, Greece, pp. 1–8 (2013)
29. Christianson D.B., Anderson S.E., He L.-W., Weld D.S., Cohen M.F., Salesin D.H.: Declarative camera control for automatic cinematography. In AAAI (1996), pp. 148–155
30. Lino, C., Christie, M., Lamarche, F., Schofield, G., Olivier, P.: A real-time cinematography system for interactive 3D environments. In Eurographics ACM SIGGRAPH Symposium on Computer Animation, pp 139–148 (2010)
31. Lewinski, J.S.: Resident Evil 5 Offers Sneak Peek at Avatar's 'Virtual Camera'. Wired (2009). https://www.wired.com/2009/02/resident-evil-5-6
32. Marino, P.: 3D Game-Based filmmaking: the art of machinima. Scottsdale, AZ: Paraglyph Press (2004)
33. Harwood, T.G. Grussi, B.: Pioneers in machinima: grassroots of virtual production. Wilmington: Vernon Press (2021)
34. Shonig, J.: The Shape of Motion: Cinema and the Aesthetics of Movement. Oxford University Press (2021)
35. Carroll, J., Cameron, D.: Machinima: digital performance and emergent authorship. In changing views—worlds in play. Proceedings of DiGRA Conference. (2005) Retrieved 9 November 2021, from http://www.digra.org/dl/db/06276.32151.pdf
36. Lightman, H.A.: Front projection for 2001: a space Odyssey. In Schwam, S. (ed) The Making of 2001: A Space Odyssey. New York, Modern Library (2000)
37. Film 4: *Tom Cruise on Oblivion*. Film4 Interview Special (2013). Retrieved from https://www.youtube.com/watch?v=ielGQM63Z6o&t=252s&ab_channel=Film4
38. Definition Magazine.: Earth bound: the oblivion movie and how to avoid blue screen (2013). Retrieved on 28 June 2022. https://definitionmagazine.com/features/earth-bound-the-oblivion-movie-and-how-to-avoid-blue-screen
39. Hive Division: LED Wall Virtual Production - Bye Bye Green Screen (2020). https://www.youtube.com/watch?v=ysIOi_MP_cs&ab_channel=HiveDivision
40. Mitchell, J.: Frontier Man: Cinematographer Greig Fraser (2022). Retrieved on 15 May https://www.premiumbeat.com/blog/greig-fraser-cinematographer
41. Atkinson, S.: Interactive 'making-of' machines: The performance and materiality of the processes, spaces and labor of VFX production. In Performing Labor in the Media Industries. Fortmueller, K. (ed) Spectator **35** (2) (Fall 2015), 36–46 (2015)
42. Holben, J.: The Mandalorian: This Is the Way (2020). https://ascmag.com/articles/the-mandalorian

43. No Film School: How Cutting-Edge ILM Technology Brought 'The Mandalorian' to Life Retrieved 31 July 2022 (2020)
44. Desowitz, B.: *'Thor: Love and Thunder' Takes a Hammer to Marvel's Green Screen Problems* (2022). Retrieved on 20 June 2022. https://www.indiewire.com/2022/07/thor-love-and-thunder-stagecraft-ilm-1234738253/
45. Australian Cinematographers Society: Virtual Production - A Cinematographers' Conversation (2020). https://www.youtube.com/watch?v=3g_GhufsQ7E&ab_channel=AustralianCinematographersSociety
46. Roettgers, J.: How video-game engines help create visual effects on movie sets in real time (2019). Retrieved 7 September 2021. https://variety.com/2019/biz/features/video-game-engines-visual-effects-real-time-1203214992
47. Giardina, C.: NEP Acquires Lux Machina, Halon, Prysm Collective to Launch Virtual Production Business. Behind the Screen (2021). https://www.hollywoodreporter.com/business/digital/nep-acquires-lux-machina-halon-prysm-collective-virtual-production-business-1234991516
48. Cade, D.L.: Incredible real-time visual effects tech replaces green screens with LED walls. Peta Pixel (2019). https://petapixel.com/2019/08/22/demo-of-incredible-real-time-visual-effects-using-massive-led-walls/
49. Dalkian, S.: nDisplay Technology Whitepaper. Epic Games (2019). https://cdn2.unrealengine.com/Unreal+Engine%2Fndisplay-whitepaper-final-updates%2FnDisplay_Whitepaper_FINAL-f87f7ae569861e42d965e4bffd1ee412ab49b238.pdf
50. Baker-Whitelaw, G.: The problem with Disney's new 'Star Wars' VFX technology. 'The Volume' (2022). https://www.dailydot.com/unclick/star-wars-disney-the-volume-obi-wan/?fbclid=IwAR1IP3v8zGx8XH_hbpHS3tq3TgmMsm9RZPwKXOqpb4_uKojt2rOl5TaPTwk
51. Kavakli, M., Cremona, C.: The virtual production studio concept–an emerging game changer in filmmaking. 2022 IEEE Conference on Virtual Reality and 3D User Interfaces (VR) (2022), pp. 29–37
52. Advanced Imaging Society: Virtual Production Workflows & Insights (2020). https://www.youtube.com/watch?v=81f6FPnWMCI&ab_channel=AdvancedImagingSociety
53. Coldewey, D.: How Oculus squeezed sophisticated tracking into pipsqueak hardware. Techcrunch (2019). https://techcrunch.com/2019/08/22/how-oculus-squeezed-sophisticated-tracking-into-pipsqueak-hardware
54. Wright, M. Chambers, J. Israel.: Large Scale VFX Pipelines. ACM SIGGRAPH 2017, Talks, Article 59, pp. 1–2 (2021). https://dl.acm.org/doi/https://doi.org/10.1145/3084363.3085021
55. Kavakli, M.: How to reduce input lag in a virtual production studio. In International Conference on Human–Computer Interaction. Springer International Publishing (2022), pp. 58–65
56. Brompton: Ecreating Reality (2022). https://www.bromptontech.com/applications/virtual-production/
57. Oakden, T., Kavakli, M.: Graphics processing for real-time ray tracing using RTX architecture. 14th International Conference on Computer and Automation Engineering (ICCAE 2022) March 25–27, 2022 , pp 1–6 Brisbane, Australia (2022a). http://www.iccae.org/
58. Oakden, T., Kavakli, M.: Performance analysis of RTX architecture in virtual production and graphics processing, networked entertainment systems (NES) with ICDCS 2022 -Co-Located with the 42nd IEEE Intl. Conf. on Distributed Computing Systems (ICDCS 2022), Technically co-sponsored by the IEEE and the IEEE Computer Society, July 10 - July 13, 2022, Bologna, Italy, pp 1-11 (2022b)
59. Failes, I.: A virtual production explainer: what it is, and what it could mean for your project. Cartoon Brew (2018). https://www.cartoonbrew.com/vfx/a-virtual-production-explainer-what-it-is-and-what-it-could-mean-for-your-project-166554.html
60. Technicolor: Technicolor Genesis—Real Time Virtual Production (2019). https://www.youtube.com/watch?v=x5O2E-jvUbE&t=48s&ab_channel=technicolor
61. Srivastav, S., Singh, R.: Film Viewing Experience: Changes with Technology. Proc. 28th National Conf. Cultural Representation and Power of Media (2017). https://www.academia.edu/35899511/Film_Viewing_Experience_Changes_with_Technology

62. Latour, B.: Reassembling the social: an introduction to Actor-Network- Theory. Oxford University Press, Oxford (2007)
63. Weintraub, S.: O'Shea Jackson Jr. on How He Landed His 'Obi-Wan Kenobi' Role After Not Getting Lando and Working With The Volume Technology (2021) https://collider.com/obi-wan-kenobi-series-oshea-jackson-jr-comments-lando-the-volume-interview/
64. Zhang, C.: The why, what, and how of immersive experience. In IEEE Access **8**, 90878–90888 (2020)
65. Ilmaranta, K.: Cinematic Space in Virtual Production (2020). https://www.academia.edu/4400 2310/Cinematic_Space_in_Virtual_Production
66. Bennett, J.. Immersive performance environment: a framework for facilitating an actor in Virtual Production. PhD dissertation, Creative Industries Faculty, Queensland Univ. of Technology, Brisbane, Queensland (2020). https://eprints.qut.edu.au/203911/
67. Seymour, M.: Art of (LED Wall) Virtual Production Sets, Part Two: How you make one. VFX Guide (2020). https://www.fxguide.com/fxfeatured/art-of-led-wall-virtual-production-sets-part-two-how-you-make-one
68. Debevec, P., Wenger, A., Tchou, C., Gardner, A., Waese, J.: A lighting reproduction approach to live-action compositing. ACM Trans. Graph. (SIGGRAPH 2002) **21**(3), 547–556 (2002). https://dl.acm.org/doi/abs/https://doi.org/10.1145/566570.566614
69. Hedman, P., Srinivasan, P.P., Mildenhall, B., Barron, J.T., Debevec, P.: Baking neural radiance fields for real-time view synthesis (2021). Arxiv, abs/2103.14645, 2021 https://arxiv.org/pdf/2103.14645
70. Debevec, P., Malik, J.: Recovering high dynamic range radiance maps from photographs. Computer graphics and interactive techniques (SIGGRAPH '97). ACM Press/Addison-Wesley Publishing Co., USA, pp. 369–378 (1997) https://www.academia.edu/download/29352455/rec overing_high_dynamic_range_radiance_maps_from_photographs.pdf
71. Hu, K.: The effects of digital video technology on modern film. Master of Science Thesis, Dept. of Television Management, Drexel Univ., UK (2016)
72. Kohn, E.: 'Thor: Love and Thunder' Takes a Hammer to Marvel's Green Screen Problems (2022). Retrieved 9 July 2022. https://www.indiewire.com/2022/07/thor-love-and-thunder-sta gecraft-ilm-1234738253
73. lasermagnetic: Memories of Working On *Where Eagles Dare* (2009). Retrieved on 10 July 2022 https://www.filmboards.com/board/p/1747699
74. Montgomery, J.: Who are you: FXGuide in focus. FXGuide (2002). https://www.fxguide.com/fxfeatured/who_are_you_fxguide_in_focus/?highlight=MIKE%20SEYMOUR
75. Natale, S.: A short history of superimposition: From spirit photography to early cinema. Early Popular Visual Cult. **10**(2), 125–145 (2012)
76. Salt, B.: Film style and technology in the thirties. Film Quart. Autumn **30**(1) (Autumn, 1976), 19–32. Published by: University of California Press (1976)
77. Shields, M.: How Does Rear Projection Work? (2020) Retrieved 30 May 2022 https://filmsc hoolrejects.com/rear-projection
78. Stump, D.: Digital cinematography: fundamentals, tools, techniques, and workflows. Focal Press, Taylor & Francis Group (2014)
79. Witmer, JD., Fish, A.: ASC members offer their perspectives on what the next century has in store for the art and craft of cinematography. Future of Cinematography: The Next 100 Years (2019). https://ascmag.com/articles/the-next-100-years

Chapter 15
Interactive Film: Forking Paths to a Complete Audiovisual Experience

Bruno Mendes da Silva

Abstract The evolution of forms of immersion in the history of cinema has contributed to a paradigm shift: the narrative thread does not have to be linear and the doors to an effective interaction between the narrative and the viewer(s) are open. Nowadays, experimental film and digital media use the most advanced technologies as aesthetic strategies that seek to submerge the public, giving them the freedom to build the narrative or an aesthetic experience by interacting with it. *The Forking Paths*, a project developed by the Research Center for Arts and Communication (CIAC), focuses on different types and models of interactive films, both from a theoretical and a practical perspective. The aim of this project is to conduct original research targeting the discovery of potential new knowledge, namely through practice and through the results of this practice (practice based research). The project platform includes the films produced for the project: *Cadavre Exquis* (2019), *Valsa* (2016), *The Book of the Dead* (2015) and *Haze* (2014). Having *The Forking Paths* project as a starting point, this chapter tragets two main goals: (1) to analyse possible models and levels of film interactivity, and (2) to trace possible evolutionary paths for audiovisual language.

15.1 A Brief Preamble—From Prehistory to the Present Day

Just four years after the birth of cinema, the Cinéorama (Fig. 15.1) premières at the 1900 Universal Exhibition in Paris. This filmic artefact simulated a balloon trip exhibited by 10 projectors on a 93-m-long 360° screen. The idea of audiovisual immersiveness arrived early but with great force: the spectators watched the film at the centre of the projection, inside a simulacrum of a hot air balloon basket. In the Pavilion Gallery of Machines, the Lumière brothers also opted for a large screen, 21 m long, and with a fountain underneath it that made it transparent and, therefore,

B. M. da Silva (✉)
Research Center for Arts and Communication, University of Algarve, Faro, Portugal
e-mail: bsilva@ualg.pt

Fig. 15.1 Inside view draw of the Cinéorama, La Natur, Revue des Sciences, 1900

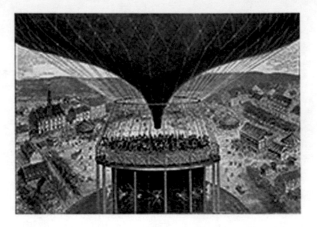

the projection was visible from both sides. A total of 150 films were projected and viewed by approximately one and a half million spectators [9, 12].

The cinematic shooting galleries also date back to the beginning of the century. Life Targets, the best known, were patented in England in 1912 (Cowan, 2018) [14]. Also considered a precursor of video games (namely first-person video games), this early target-shooting simulator consisted in a device that brought real bullets and images together. The screen was a strip of rolled paper so that the bullet holes could be easily removed. The audiovisual Pandora's box was open. Cinema could be anything. However, a certain narrative tendency quickly took hold (Thompson and Bordwell 2003). Young cinema spectators wanted to see stories, and directors like Georges Méliès understood that this medium could, in addition to telling stories, capture dreams [6].

In 1886, the Lumière had already filmed "Panorama of the departure from Ambérieu station" taken from the train. The first travelling shot in the history of cinema established an analogy, which was as simple as it was obvious, between the gaze of the train traveller and the cinema spectator: the window showing moving images is a screen. This experience was followed by different authors, and the boldness rose to a higher level [12]. Films dubbed the "Ghost Rides" genre, positioned cameras in front of the locomotive. Theis genre "inaugurated" the concept of the Subjective Camera and emphasise the passion for speed brought by transport machines.

Edwin Porter directed in 1903, "The Great Train Robbery". Porter followed the trend in train films—train robberies—and applied the morphological and syntactical innovations experienced in the complex filmic narrative "The Life of an American Fireman", released in 1902: continuity of action, mastery of the three families of the plane scale and parallel editing, a resource he perfected in "The Great Train Robbery".

Fig. 15.2 Frame from The Great Train Robbery, by Edwin Porter, 1903

The last shot of the film is a subjective close-up. Actor Justus D. Barnes, directly ribbing the audience, draws his revolver and shoots at the spectators. This interaction between the illusory moving image and the audience happens, this time, in a conscious and provocative way (Fig. 15.2).

Many films have tried to apply interactivity by defining moments of bifurcation, where the viewer chooses the path to follow among two or more possibilities, or offering different viewing options for the film narrative. One of the most successful projects is the Czechoslovakian film *Kinoautomat—one man and his house*, created in 1967 by Radúz Činčera for the World Expo in Montreal (Fig. 15.3). In this film, the audience is asked (nine times) to choose one of two given possibilities to continue the narrative. At the first screening in Montreal, the process of choice, by voting, was mediated by actors.

Several projects allow the viewer to opt for one of two endings. This is the case of the film *Mr. Sardonicus*, produced and directed by William Castle, in 1961. Before the final scene of the film, the viewers can vote using a card they are given at the beginning, with two possible drawings, just as it happened in the Roman arenas, where the gladiators fought to entertain the audience: a thumb up and a thumb down,

Fig. 15.3 Frame from the film *Kinoautomat*, by Radúz Činčera, 1967

which allows them to choose whether the character should be mercifully spared and live or be punished and die.

I'm your man, directed by Bob Bejan, in 1992, also claims the title of first interactive film in the history of cinema. Just as in previous projects, the viewers decided the unfolding of the narrative using interactive buttons installed on their chairs.

Another film announced as "the first interactive film in the history of cinema" was released in 1995, *Mr. Payback*, written and directed by Bob Gale (Fig. 15.4). Depending on the audience's interaction, this film lasted approximately half an hour. The viewers were called upon to decide at various points in the narrative, again, by using a remote which was attached to the chair. The film was not very well accepted by the critics, mainly due to the absence of a plot; nevertheless, it marked an important step in the way viewers experienced cinema, although the experience itself has been considered by many more like playing a videogame rather than watching a film.

Between 2002 and 2005, Lev Manovich devoted himself to the development of the *Soft Cinema* project, a dynamic computer-oriented installation in which the viewers can, in real time, build their own audiovisual narrative from a database containing 4 h of video and animation, 3 h of narration and 5 h of music.

Later, in 2010, the horror film *Last Call* by 13th Street, a television channel specialising in horror films, was announced as the world's first interactive horror film (Fig. 15.5). Using a software that enables voice and command recognition, one of the spectators present in the movie theatre receives a phone call from the protagonist asking for help. The protagonist wants the spectator to help her choose the best way to escape the serial killer who is chasing her. Through this technology, the film becomes unique since it depends on the instructions of the person who answers the phone.

Fig. 15.4 *Mr. Payback*, by
Bob Gale, 1995. Film poster

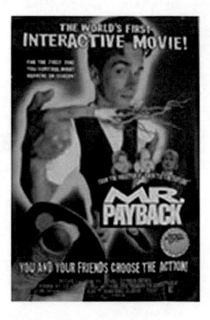

Fig. 15.5 Frame from the film *Last Call*, produced by 13th Street, 2010

Fig. 15.6 Frame from the film *The Carp and the Seagull*, by Evan Boehm, 2012

In 2012, Evan Boehm and Nexus Interactive Arts create *The Carp and the Seagull* an interactive 3D movie that takes advantage of WebGL and HTML5 technologies (Fig. 15.6). The film describes a tale of the fisherman Masato, who one day encounters the spirit Yuli-Onna that appears to him in the shape of a seagull.

However, recent examples of interactive films, such as Netflix *Black Mirror: Bandresnatch* (2018), continue to adopt the same structure used by the pioneering interactive films in the 1960s: an arborescent structure based on a simple and occasional choice made at certain moments of the narrative, where the spectator-user can choose path A or B.

15.2 The Forking Paths

The Centre for Research in Arts and Communication (CIAC) at the University of Algarve (UAlg) has been producing digital artifacts that promote the interconnection between arts and technologies, and part of the developed products are the result of projects in the field of interactive cinema. These lines of applied research, whose matrixes have served as a starting point for the emergence of several PhD projects, are based on the development and evolution of audiovisual language. On the other hand, the production of platforms whose objective is centred on the creation, dynamisation and expansion of networks of excellence in the areas of culture and digital art have been CIAC's most visible work. It is also important to remember that we live in a post-Benjaminian time [1], where the passive relationship between public and work has changed paradigm. This new relationship also embraces cinema and offers the viewer an active role of co-authorship regarding the final form of the film. It is in this context that the platform *The Forking Paths* emerges, prepared to support and host films for collective or individual viewing.

A phase of experimentation, based on 'practice-based research': original research undertaken to gain new knowledge through practice and the outcomes of that practice [2–5], resulted in the production of four interactive films:

(1) *Haze* (Fig. 15.7) shown to the public at the FILE 2015 International Festival of Electronic Art in São Paulo. "Haze" is divided into three distinct flows: one central and two lateral, one hidden on the left and the other hidden on the right. The choice of the flows will be made by the spectator-protagonist (the spectator who interferes in the narrative becoming the main character). Each flow conveys a distinct experience of the narrative. This film can also be seen on classic cinema projection screens. In this variant, the central stream is projected onto the screen and the side streams can be viewed on the mobile devices of the audience members.

(2) *The Book of the Dead* (Fig. 15.8) seeks to interact with the spectators on two levels: by controlling certain actions of the characters and by controlling the time of the narrative, allowing them to read at their own pace. When we read, we use our own reading time, we can read slower or faster. But when we listen to something being read for us, we depend on a reading time that is not ours and to which we have to adapt. The same happens when we watch a film: the viewing time is imposed by the editing rhythm, which can be faster or more contemplative. In *The Book of the Dead*, the viewers are the ones who intuitively choose the duration of each shot.

Fig. 15.7 Frame from the film *Haze*, by Bruno Mendes da Silva [10]

Fig. 15.8 Frame from the film *The Book of the Dead*, by Bruno Mendes da Silva, 2015

(3) *Waltz*, by Rui António, CIAC collaborator. Filmed with multiple cameras, *Waltz* proposes a physical interaction between man and machine, through a Kinect sensor. This film was part of the PhD project *Personagens à procura de um espet-ator*, which aims to offer the viewer control over the film editing in real time, giving them the status of co-author.

(4) *Cadavre Exquis* (Fig. 15.9) was implemented as an installation controlled by motion detection. At the time of this writing, it has been presented to the public three times: at the Literary Festival of Macau, China, in March 2019; at FICLO (Festival of cinema and literature) in Olhão, Portugal, in April 2019; at Artech 2019, in Braga, Portugal, in October of the same year. The opening scene of the film *Cadavre Exquis* is frozen (stopped in time). Three characters meet in the same room. However, the viewer (who interferes in the narrative) has the possibility to travel through the freeze-frame, getting closer to or moving away from each character. When the viewers get closer to a character, they may select him/her. That choice results in a flashback that leads up to the frozen moment. By choosing the last character, the viewer will unfreeze the opening scene, setting it in motion.

Three scriptwriters were invited to create three narratives based on a character who would meet two other characters in a room at the end of the narrative. The scripts would form a succession of sub-narratives which, just as in the *Cadavre Exquis* game played by surrealist artists, converge in the main narrative, eventually ending up in a succession of unlikely scenes. A common opening scene would be added: the scene of the three characters in the same room.

It is hereby intended a connection to the idea of automatism and to the processes that govern the unconscious: dream-condensation and dream-displacement are not perceived at the time they occur; while we are dreaming, we are not aware of the process.

The very idea of film interactivity may be regarded as an intolerable artificiality, but, at the same time, as a catalyst to the idea of collective creation, due not only

Fig. 15.9 Frame from the film *Cadavre Exquis,* by Bruno Mendes da Silva, 2015

to the possibility of co-authorship offered to the viewer, but also to the freedom the viewer is given to deconstruct the filmic structure at any time.

15.3 Models and Levels of Film Interactivity

Despite not exhausting the experiences made in the scope of interactive films, the examples we discussed describe potentially innovative strategies to make the spectator integrate the diegesis, using interfaces (remote controls, phones, tablets) or sensors that enable interaction with the filmic universe. But how can all interactive films be grouped, regardless of whether they are filmic experiences or commercial products? In an attempt to find a set of models that make it possible to encompass all films of interactive nature, the following criteria were found (Silva et al. 2019):

(a) the **arborescent model**, based on a simple one-of choice made at certain moments in the narrative, where the viewer can choose paths A or B, for which we may use the film *Last Call* (described in Sect. 15.1—A Brief preamble—from the prehistory to the present day) as an example;

(b) the **constructive model**, which involves multiple interpretation, according to a number of closed options offered to the viewer, where we can include the film *Haze* (described in Sect. 15.2—The Forking Paths);

(c) the **paired model**, which allows the incorporation of content external to the narrative, as in the film *Take This Lollipop* (Fig. 15.10), by Jason Zada (2011). The film-app requests that viewers temporarily allow access to their Facebook account to incorporate content and information from the viewer's Facebook feed that completes the narrative;

(d) the **fertile model**, whose process of interactivity between viewer and film enables the emergence of new content that was not defined a priori. This last model presents a possibility of breaking the production lines (commercial and

Fig. 15.10 Frame from the film *Take this Lollipop,* by Jason Zada, 2011

experimental) that have been taking place since the middle of the last century in the field of interactive film. Up to now, the possibilities of interactivity have always been limited to the possibilities of choice offered by each project. From the outset, they were pre-defined. However, we can only speak of an effective interactivity process when the viewer has the possibility of generating new content that was not initially foreseen. This possibility of effective man–machine interaction is complicated by nature because it depends on a process of pre-production, production and post-production. However, if we think of the hypothesis of the digital animated film conjugated with the advent of cybernetics and artificial intelligence, the emergence of narratives that enable automatic creation is very close. It is therefore important to think about the adaptability of the narrative to the user according to a new paradigm: the authorship of the film is divided between the one who thinks the film, the one who will enjoy the diegetic elements and the one who will bring the surprise factor to the narrative—the AI.

15.4   Results

The results indicate that the spectator has gained autonomy regarding control and the participation in the film narrative. However, despite technological advances in new interactive projects, such as motion-sensors, virtual reality or other interfaces, the spectator is still confined to pre-produced content. The very idea of film interactivity may be regarded as an intolerable artificiality, but, at the same time, as a catalyst to the idea of collective creation, due not only to the possibility of co-authorship offered to the viewer, but also to the freedom the viewer is given to deconstruct the filmic structure at any time. We concluded that we can only speak of true interaction when there is a reciprocal influence on the communication process. Such an interaction does not take place in any of the productions created up to the publication of this paper. The spectator is limited to certain insurmountable and predetermined options. With the imminent possibility offered by the Fertile Model, the spectator acquires creative powers beyond the author's control. The creation of unplanned content, in

Fig. 15.11 Frame from the Videogame *Telling Lies*, by Sam barlow, 2020

other words, content resulting from auto-creation, is thus enabled, whereby the idea of a meta-author becomes a reality. This will certainly be a rupture in the logical sequence of the short history of interactive film, a turning point where a film can become something it has never been before: a complete audiovisual experience. As [7] brought forward:

The typical scenario for twenty-first century cinema involves a user represented as an avatar existing literally "inside" the narrative space, (…) interacting with virtual characters and perhaps other users and affecting the course of the narrative events.

But then, how can we differentiate between an interactive film and a video game? In fact, some videogames make this line very blurry. *Telling Lies*, by Sam Barlow, 2020 (Fig. 15.11), uses a video archive of real footage to engage the user in a complex narrative plot.

We can, however, find at least one differentiating line: the interactive film, unlike most video games, does not have a concrete goal to achieve, which is, ultimately, victory within the framework of a proposed challenge. On the other hand, both aesthetics (e.g. Art Game genre) and AI are very much rooted in some video game subgenres. In the Art field, artificial intelligence has inspired numerous creative applications displayed in reference spaces such as MoMA, Ars Electronica or the Beall Center for Art + Technology.

Nonetheless, the very idea that the viewer can make a universal association of specific forms with aesthetic qualities is questionable [13]. Aesthetics is considered a human domain, which is why the intersection between AI and the discipline of aesthetics is so important. Its complexity seemed for some time incompatible with algorithmic logic. Art, aesthetics and creativity are the pinnacle of human capabilities and therefore represent an advanced stage of AI. In this sense, this complex field becomes the ultimate testing ground for the possibilities and limitations of AI [8].

15.5   A Proposal to a Final Path *of Walking on Ice*

In late November 1974, Werner Herzog received a phone call from Paris. He was told that his friend Lotte Eisner, a writer and film critic, was seriously ill and would probably die. Herzog said that this could not be. He would not allow his friend to die. He grabbed a case, a compass and a pair of new boots, and took the most direct route from Berlin to Paris, convinced that she would live if he went to her on foot (Herzog 2011).

We propose the production of an interactive film based on Herzog's journey, a film called *Of walking on ice* that portrays the harshness and solitude of that journey. While travelling from one city to the other, surprising and arbitrary encounters might take place. The user must always walk in the same direction, towards Paris. To do so, Herzog's boots will be replaced by a treadmill, set at a constant speed. The frozen landscape will be recreated three-dimensionally on the fly, i.e. while in motion, by AI and perceived through Rift glasses and VR gloves. Finally, it is important to convey how cold this late-November-journey was. Such sensory experience can be achieved through both the temperature of the exhibition venue and the binaural sound.

This proposal intends to materialize the concept of interactive film defined in the fertile model (previously mentioned in subchapter 3—Models and levels of film interactivity). It has as main reference the work *The Legible City* (1988–1991) by Jeffrey Shaw (Fig. 15.12), which allows the audience to wander, both in a virtual and interactive way, through a city that reveals itself through words. Those words are used to replace spaces and forms of architecture and are meant to be read by the spectator: "Using the ground plans of actual cities—Manhattan, Amsterdam and Karlsruhe—The Legible City completely replaces the existing architecture of these cities with text formations written and compiled by Dirk Groeneveld" (Shaw, n.d). In this work, we find the ideas of immersiveness, space reconfiguration and dissolution of the boundaries between the physical and the digital, taking the viewers into the work, as if it was their (new) place. Technically, the work resorts to position sensors activated by the interface of the pedals and the steering wheel of the bicycle. On the other hand, we have a projector and an LCD monitor that together allow the visualization of images in real time. The simulation happens when the spectators, while pedalling a bicycle previously equipped with position and speed sensors, choose their route from a small LCD monitor to be able to view the projected and always updated images.

In the proposal *Of walking on ice* (Fig. 15.13), the user's ascent to the treadmill is perceived by a sensor that triggers a narrative introduction: a voice over that contextualises the situation, a sort of audible open caption. Then the path begins. The users' steps are marked by the binaural sound: sometimes users step on water, others on ice, they might also step on snow, or even on mud. Nature works as a whole and reacts to the behaviour of each user. Random animals can run away, attack, allow themselves to be petted or follow the user. The outcome depends on the way the users interact with the diegetic elements they casually encounter along the way: snowstorms, fog, hailstorms, rain, ice, deer, trees, crow flocks, northern goshawks, pheasants, noises, goats with rattles, cackling roosters, a tractor that leaves furrows,

Fig. 15.12 *The Legible City,*
by Jeffrey Shaw, 1988–1991

Fig. 15.13 A draft for the
film *Of Walking on snow*

bells, hills of fog, dogs, an old woman with crooked legs and a bicycle, two men of
African descent, a train at standstill and a bus stop.

We conclude by saying that producing this pioneering film as an example of
The fertile model is the final goal of *The Forking Paths* project. *Of Walking on Ice*
proposes to be the first interactive film to use AI, seeking the way to the complete
audiovisual aesthetic experience.

References

1. Benjmin, W.: A Obra de Arte na Era da sua Reprodução técnica. In Geada, E. (Ed.) Estéticas
   do Cinema, Publicações Dom Quixote (1985)
2. Candy, L., Edmonds, E.: Practice-based research in the creative arts: foundations and futures
   from the front line. *Leonardo* **51**(1), 63–69. https://doi.org/10.1162/LEON_a_01471
3. Candy, L.: *Inter-disciplinary art and technology practice-based research and the creative arts.*
   NiTRO (2019). https://bit.ly/3EQvZdU. Accessed 10 December 2021

4. Candy, L.: Practice-based research: a guide. In: Creativity and Cognition Studos. CCS Report: 2006 V1.0 November (2006). https://www.creativityandcognition.com/resources/PBR%20G uide-1.1-2006.pdf
5. Dewey, J.: *Arte* como experiência. Martins Fontes, São Paulo (2010)
6. Malthête-Méliès, M.: Georges Méliès l'Enchanteur. La Tour Verte, Condé-sur-Noireau (2011)
7. Manovich, L.: The Language of New Media. The MIT Press, Massachusetts (2011)
8. Manovich, L.: Arielli, E.: Artificial Aesthetics: A Critical Guide to AI, Media and Design (2011)
9. Marechal, G.: Le Cinématographe. La Nature, No **1427**, 295–283 (1900)
10. Mendes da Silva, B., Tavares, M., Araujo, A.: Cadavre exquis—a motion-controlled interactive film. In: Proceedings of Artech 2019, 9th International Conference on Digital and Interactive Arts, Braga, 23–25 Oct 2019
11. Shaw. J.: Legible City. *Jeffrey Shaw Compendium*. Retrieved https://www.jeffreyshawcomp endium.com/portfolio/legible-city/ (N/D). Accessed 2 Dec 2021
12. Smith, R.: The Last Machine, BBC. Television documentary (1995)
13. Specker, E. et al.: Warm, lively, rough? Assessing agreement on aesthetic effects of artworks. PLOS One (2020). https://doi.org/10.1371/journal.pone.0232083
14. Voorhees, G.: Chapter 31: shooting. In: Perron, B. (ed.) The Routledge Companion to Video Game Studies, pp. 251–258. Taylor & Francis, London (2014)

Chapter 16
Design for Science: Proposing an Interactive Circular 2-Level Algorithm

Bruno Azevedo⬚, Pedro Branco⬚, and Francisco Cunha⬚

Abstract Finding important scientific results in the actual glut of information have never been more challenging. In the context of digital scientific libraries, there is a gap in the design of exploratory interfaces to support the discovery and visualization of scientific knowledge, as well as new ways to foster cooperation between the main actors in science, namely researchers. Science is defined by a social dimension where communities of researchers cooperate with each other. Researchers play a key role as curators of important scientific articles by sharing with their peers what they read. This kind of behavior already happens on social platforms, however without a logic and medium that supports and maps these social dynamics. Circular layouts are well-known structures to visualize relations between entities, and as opposed to force-directed layouts, are a more effectively and user-friendly solution for visualize relations. This article presents a novel 2-level circular structure to map and support cooperation in science communities. The 2-level circular layout was designed and implemented using D3.js and it is able to portray the relations between two different entities, the scientific community that falls within a specific knowledge domain (outer circle), and the scientific articles shared by this community (inner circle). The 2-level circular structure provide an overview plus details of the most read and shared scientific articles within a community. The preliminary results provide interesting results and pave the way for future research on the topic. The link to access the visualization and code can be found at the end of the article.

B. Azevedo (✉)
ALGORITMI Research Center – engageLab, University of Minho, Guimarães, Portugal
e-mail: d7447@algoritmi.uminho.pt

P. Branco
Department of Information Systems (DSI), ALGORITMI Research Center – engageLab, University of Minho, Guimarães, Portugal
e-mail: pbranco@dsi.uminho.pt

F. Cunha
University of Minho, Braga, Portugal
e-mail: a84059@alunos.uminho.pt

© The Author(s), under exclusive license to Springer Nature Switzerland AG 2023
A. L. Brooks (ed.), *Creating Digitally*, Intelligent Systems Reference Library 241,
https://doi.org/10.1007/978-3-031-31360-8_16

445

Keywords Chord diagram · Circular layouts · Information design · Information visualization · Interaction design · Node-link diagram · Social cooperation

16.1 Introduction

Researchers are always trying to get the most important scientific knowledge, and finding accurate results in the current science deluge have never been more critical than now. The pandemic crisis emphasized even more the information overload problematics in science. Finding important scientific results in the actual glut of scientific publications have never been more challenging. In May 2020 more than 4000 scientific articles related with Covid-19 were publish every week [1]. Researchers described a tsunami of scientific publications that was impossible to handle and filter [2, 3]. It is important to emphasize that this is not a problem concerning our contemporary society because all previous Eras were information Eras [4]. However, if in previous times there was an effort in collecting and storing information, the digital information deluge and the urgency in developing efficient strategies to cope with the exponential growth of information defines the main challenge of the current digital information society [4, 5]. The exponential growth of scientific articles [6] makes searches time-consuming and the support that is provided in the search process is very limited, despite the significant advantages of science digital libraries mechanisms, as for instance the use of ranking metrics to provide better results in the search process. Researchers normally select the most important according to a non-transparent metric (e.g., relevance and ranking metrics), and therefore the probability of selecting the remaining scientific publications is very reduced [7–10]. This perspective imposes a shift from the actual approaches that are based on digital libraries interfaces defined by advanced search strategies (e.g., using Boolean operators), to a find paradigm [11], based on a process of social filtering [12] supported by a novel visualization *medium.*

The organization, presentation, and communication of scientific knowledge embody an urgent and continuous challenge [13]. In science digital libraries context, there is a lack of interfaces that better support researchers and the research process. Therefore, it's important to responsibly design effective and efficiently solutions that simplify the digital systems we use, namely the design of novel ways to visualize and interact with scientific knowledge. In fact, overwhelming systems/interfaces are everywhere. The cultural evolution of information systems [4] validates the advantages of the digitization processes however, some downsides portray this evolution. Some examples are the data collections inaccessibility, mainly due to the implemented technologies; a lack of truly open data policies in some fields (e.g., science field); and interfaces or websites unfriendly designed that difficult the access to information and knowledge (e.g., open data repositories, government websites, science digital libraries) [14]. For this reason, is inevitable and urgent to promote the design of exploratory interfaces that enhance the presentation and communication of scientific knowledge. An approach that is distinctive from other digital libraries (e.g.,

Google Scholar, Scopus, and PubMed) that returns as visual outputs massive list of results with hundreds of pages that remain unknown [15].

A growing number of interfaces have emerged, but with no practical results for the main actors in science: that are researchers [13, 16]. This scenario imposes a challenge that is described by the fact that it is not possible to have a complete overview and go deeper in a complex topic in a *glimpse of an eye*. In fact, science digital libraries interfaces are not an optimal solution when dealing with complex topics (e.g., Covid-19 scientific articles flood). In fact, researchers are drowned in scientific outputs and missing important scientific results have a big impact in the research process (e.g., literature review process) [3].

A recent article proposes a new type of interface or visualization tool as a data scope, and compares our current capacity to understand and visualize large sets of data to the 17th-century astronomy evolution accomplished by Galileo [14]. In the past, researchers didn´t have powerful tools to capture the different layers of our complex, dynamic and multidimensional world. Drawing was the primary medium to communicate observations, ideas and conceptual theories, as for instance John Snow and the cholera map of London (1854), Charles-Joseph Minard's map of Napoleon's flawed Russian campaign of 1812, Florence Nightingale and the Nightingale Rose among other information visualization drawings/milestones [17–21]. A more detailed discussion about this topic is beyond the scope of the current article. Currently, and through the use of our contemporary tools and technologies it is possible to explore the actual information landscapes[1] with more resolution. These new tools give rise to a new medium and a new way of understanding complex phenomena, while providing a new form of communication [14, 22]. Hidalgo [14] *datascopes* concept also defines a richer relationship between the visualization and the audience who plays an explorer role instead of a passive observer role, arguing that this implies a change on the form but also on the function. This concept also defines an approach that can be used in the search for new understanding, rather than the study of a single "snapshot" of a specific topic/problem. The *DataViva*[2] project is an example of a *datascope* that aims to democratize access to public records with a more powerful resolution. Thus, based on the concept of datascope and considering the context of scientific knowledge deluge, we propose a two-level circular layout that will be integrated in a future interface as an exploratory *knowledgescope* for science, where researchers play a collaborative and curatorial role [13].

The described problematic defines two main challenges:

The first challenge is that science is described by a social dimension[3] [23], where researchers as curators share scientific publications based on their personal experience. This hypothetical scenario portrays a filtering process where researchers are curators and share what they read, and provide a research synthesis based on their personal experience [13]. Therefore, the goal is to design an exploratory interface as

---

[1] Information Landscapes: Muriel Cooper at the TED5 Conference.

[2] http://dataviva.info/en/.

[3] Science Dimensions: Social dimension: people, institutions; Institutional dimension: rules, funding, metrics, indicators; and Cognitive dimension: texts, journals.

a *knowledgescope* that maps and supports this scenario. However, it is fundamental to establish a theoretical background to understand the mechanics underlying this social cooperation behavior [23–29]. Is important to emphasize that a detailed discussion about this topic is beyond the scope of the current article. Thus, the described context discloses fundamental clues about the shape/topology of science social cooperation [30–32].

Science is a community-based social activity, that equally happens in the digital space. In fact, there are communities of researchers who are willing to share and assist each other's. An example of science social cooperation is when researchers request scientific publications on a specific topic. Figure 16.1 portrays an example of social cooperation on Twitter, and Fig. 16.2 provides a scenario, according to Hidalgo post, where researchers share scientific articles but without a logical and structured communication process. The cooperation behavior presented in Fig. 16.1 suggests a relation triggered by an individual request, however, it is driven by a social cooperation process [25], namely when the scientific community satisfies the researcher's request, and shares scientific publications that match the researcher's interest. The curatorial and sharing process takes place in the social dimension of science, yet the communication process is not structured as stated in Fig. 16.2. This scenario emphasizes that the curatorial process must be supported and mapped out through a structured communication process.

This premise makes clear the second challenge, and emphasizes the main goal which is to design novel exploratory visualization interfaces to support and map science social collaborative and curatorial processes [13]. Information visualization

Fig. 16.1 Elizabeth Page-Gould Tweet, 2019

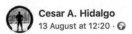

Cesar A. Hidalgo   •••
13 August at 12:20 · 🌐

PSA: To my academic colleagues. When you post about your new paper or
book, please, tell us what the paper is about. Everyday, I see a posts along
the lines: "New paper published. Wohoo! [LINK]," or "Paper out in XYZ
Journal!! Yay! " without any attempt to describe the question, result, or
contribution, of the paper. I think that many of us use social media as a
channel to learn about what's new in our communities, and just sharing
"paper accepted," is bad communication. Telling people that your paper is
published is not informative (everyday thousands of papers are published).
Telling people what your paper is about is much more useful. If I know what
the paper is about I am more likely to remember it, or to click on it to learn
more.

I understand that academic publishing has created a culture in which
getting published seems more important than what is published. But as
scholars, it is important for us to remember that at the end of the day, we
are working to create new knowledge, not to get our name on print in this or
that journal. Trust me that, if you lead with your story, it will more easily find
its audience.

Thanks!

End PSA.

Fig. 16.2   Cesar A. Hidalgo Post, 2019

framework and workflow design [33, 34] transforms data into information and information into knowledge through the use of a visual abstraction process (e.g., points
and lines) and spatial principles (e.g., position) [35].

The main goal is to mitigate the cognitive load and time-consuming research
process and to map/support the social cooperation process through interactive structure's that visual translate the relations between researchers/communities, knowledge
domains, sub-domains, topics and scientific articles. The current proposal is a work
in progress [13], and is an extended version of the article Designing an Interactive 2-Level Circular Algorithm to Visualize and Support Collaboration in Science
published and presented at the EAI ArtsIT 2022 conference.[4] It presents one of the
structures for modeling and visualizing science social cooperation according to the
described context. The goal is to turn these interactions into "natural interfaces".
To address this challenge, we propose one of the structures to be incorporated in a
future exploratory *knowledgescope*, [36], namely a novel two-level circular layout
based on a dummy dataset. Circular layouts are a distinct category of techniques to
visualize networks. A contextualization on this topic is presented in the next section.
The remaining article is divided in the following sections: Sect. 16.3 describes the
design and implementation process of the novel interactive two-level circular layout/
algorithm in D3.js, and Sect. 16.4 provides some preliminary results. The last section
presents a brief discussion and future directions. The Observable link to access the
visualization and code can be found at the end of the article.

---

[4] EAI ArtsIT 2022—11th EAI International Conference: ArtsIT, Interactivity and Game Creation.

16.2   Circular Layouts: Definition and Contextualization

There are two distinct categories of techniques to visualize networks, namely node-link diagrams, and matrices [37]. A further discussion about matrices layouts is beyond the scope of the current article. Node-link diagrams techniques fall into two main sub-categories, namely force-directed layouts [38–41], and fixed layouts that have three sub-categories, namely grid, linear and circular/radial [34, 42–44]. The main focus of this section is a sub-category of node-link diagrams, namely circular layouts, where the node's fixation placement determines the layout strategy [43, 45]. Fixed layouts are compact layouts because the nodes have a fixed position according to a set of defined attributes (e.g., ranking, alphabetical order). This strategy differs from force-directed layouts where nodes are randomly placed in space. Force-directed layouts are a non-deterministic algorithm that generate more complex layouts, and where the nodes placement depends on forces [46, 47]. In a fixed-layout strategy, it's harder to identify the topological structure of networks in contrast to force-directed layouts. However, the advantage of fixed-layout strategies is that it is simpler to identify large or small nodes through interactive approaches [44], and the division of the layout into different sections allows the grouping of nodes into different categories or classes.

The use of graphical components, namely nodes and links, defines the mapping encoding strategy. Another strategy used in fixed layouts is the assignment of quantitative values and the consequent proportional scaling of these graphical components. The use of hues to represent different categories allows easy comparisons and differentiation between clusters.

Circular layouts are an optimized approach for portraying complex information and are a common approach among social network and ontologies visualization approaches. The process of efficiently visualizing complex relational structures between entities, and also between different categories, results from a simple design approach: the nodes placement occurs at specific locations, namely nodes are placed equidistantly along the circle, so it is possible to have a cleaner and uncluttered layout, allowing nodes and links to have more visibility [48]. The link placement usually happens in the center of the circle and represents the relations between nodes. A disadvantage of circular layouts is the number of links displayed in the center of the circle, specifically when there are a considerable number of links. One mitigation strategy to overcome this problem, among other approaches, and achieve readability is the use of interaction techniques, such as the edge bundling technique that will be presented and discussed later in this section.

An early example of a circular layout is the circular relation map proposed by [49]. The goal of previous approaches was to find better and more efficient circular drawing techniques to minimize the number of edge crossings [50–52]. The work of [53] point out important references of early circular layouts or layouts with a connected ring pattern, and emphasizes important features about circular layouts: "Nodes positioned around circumference of ring; line segments connecting nodes; additional nodes optionally appear in ring's interior".

Therefore, it is important to reference a set of contributions in the context of circular layouts to identify and understand important approaches.

The above work presented in Fig. 16.3 [45] aims to reduce the edge length based on the node placement strategy by routing the outer edges through the outer part of the circle. The goal of the algorithm is to optimally select a set of externally routed edges by coupling groups of edges as bundled splines that share part of their path. The goal is to reduce visual clutter, density, and edge crossings.

The hierarchical edge bundling technique aim is to reduce the circle inner clutter as presented in Fig. 16.4b. It uses curved edges instead of straight edges, and the goal is to reduce the visual clutter based on a lower number of crossings between edges. The technique is used to visualize implicit adjacent edges between parent nodes that are the result of explicit adjacent edges between their respective child nodes, this means that it groups the lines that follow the same direction [54].

Instead of drawing straight lines between two points, the line follows a path based on an interpolation method between the points in the hierarchical structure, that bundles together the lines that follow similar paths. The path follows the control points in order to draw the curved lines. Therefore, the computed structure is based on control points that create the spline [55]. However, a problem with this technique is that edges that follow similar paths and pass through the same path are occluded.

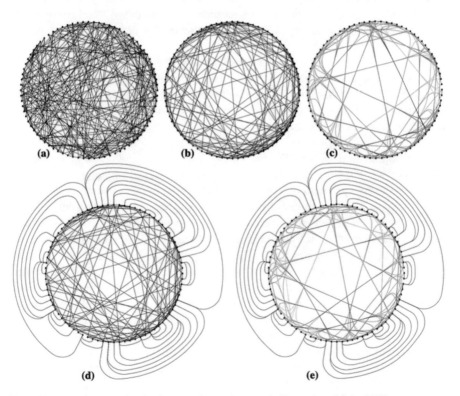

Fig. 16.3 Variations on circular layouts of a random graph (Reproduced from [45])

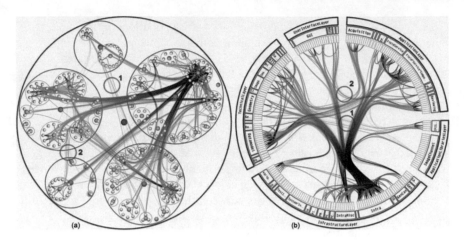

Fig. 16.4 Hierarchical edge bundling technique (Reproduced from [54])

It's important to note that the hierarchical circular structure in Fig. 16.4 uses a parent–child relation approach between data items, and therefore uses a hierarchical dataset structure. This means that the edge-bundling technique can be used in conjunction with tree visualization structures, such as radial-tree layouts [55]. Succinctly, the goal is to group adjacent edges together in order to mitigate visual clutter.

The Circos package uses a circular representation of matrix data (relational data), namely a chord diagram [56] as show in Fig. 16.5. The software package turns tabular data (matrix data structures) into a circular layout, namely a chord diagram to depict links between objects or between positions. The algorithm represents a network structure with a circular layout, and at the same time it is possible to visualize weighted relationships between various entities.

The advantage of the Chord Diagram is that it reduces visual clutter in the center of the circle, when compared to force-directed layouts. Scalability and nodes distribution is a common problem on force-directed layouts. In fact, when force-directed layouts compute a large number of nodes and links, the algorithm degenerates into a hairball of visual clutter, which consequently decreases the effectiveness of the layout. This makes more difficult to understand the overall relational structure, or understand the individual links/paths between nodes [47]. In the chord diagram layout, the links distribution and arrangement are more efficient, because entities are equidistant placed and arranged in a circular configuration. The thickness of the links can also represent attributes, namely the extent or magnitude of the relational structure between elements (e.g., degree of similarity, traffic between elements, among other variables). A different number of hues can be used to distinguish categories or entities, and identify patterns, which makes the perception/recognition process easier. Another point is that when relations have a direction, links can be colored by source or target element. Another advantage is the easier decoding of the links

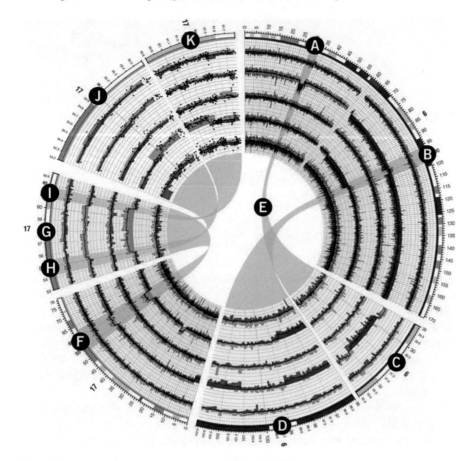

Fig. 16.5   Circos software package (Reproduced from [56])

because curved lines facilitate eye movement, as opposed to sharp angles (straight lines) that make it more difficult to trace a specific path between nodes [56].

The TreeNetViz [57] combines in a single circular layout a tree structure and a network structure as presented in Fig. 16.6. It is possible to visualize and interact with different levels and sectors of the tree structure and simultaneously visualize the relations between the various entities. It uses a radial space-filling (RSF) technique (Sunburst) to represent a tree structure. The circular arcs are generated by an RSF layout, and by a circular layout that represents the network structure. It also uses an edge bundling technique adapted from [54] to reduce the inner visual clutter. Another point is the use of an algorithm to improve node placement considering a set of constraints. The slider lets control which scale a network should be aggregated and displayed, and also it is possible to get a cross-scale view of a network over a tree structure by expanding or collapsing a node. It is also possible to increase or decrease the width of a node by using the mouse wheel. Two disadvantages define this

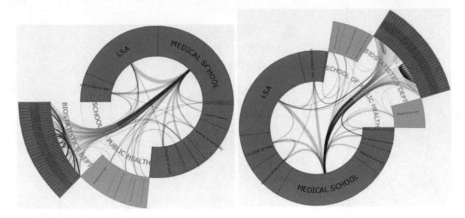

Fig. 16.6 TreeNetViz (Reproduced from [57])

approach: scalability, and the structure does not support changes in the hierarchical structure because it is not dynamic.

The multiple view visualization tool [48] is a hierarchical and relational approach that allows an exploratory interaction as presented in Fig. 16.7. A hierarchically connected circle, an indented tree, and a node-link diagram are the main visualization structures. This approach allows for exploratory interaction from different viewpoints. A non-hierarchical linking approach is used, which draws the connections within de circle using the edge bundle technique for visual clarity.

The above approach is an interactive two-level circular layout combined with a force-directed layout in D3.js and JQuery as presented in Fig. 16.8 [58]. Each node is a song, and the nodes are scaled based on popularity, and colored by artist. Links indicates that two songs are similar to one another. Is important to emphasize that the author "hijacked" the D3 force-directed layout using an entity called *radial placement*, where a function called *radial location* converts the polar coordinates to a radial location.

In this work it is referenced a set of important works, and it discussed the edge crossing issue [59]. It is proposed a technique aimed to edge crossing optimization, a circular visualization of biconnected graphs on a single embedding circle, and also the merging of the force-directed layout with a circular layout in order to reduce the number of edges crossings as presented in Fig. 16.9.

The ClusterVis is an approach that is applied to the field of science, namely to a co-authorship network and is implemented in D3.js as presented in Fig. 16.10. The user study on satisfaction metrics shows good results [60].

This work aims to help users better understand their consumption profiles based on two approaches. It is used a chord diagram as presented in Fig. 16.11 and a bar chart [61]. The authors conducted a user-centered study to evaluate the effectiveness of the chord diagram for improving human decision making. The goal was to reveal the regions of the recommendation space that are unknown to the users. The study

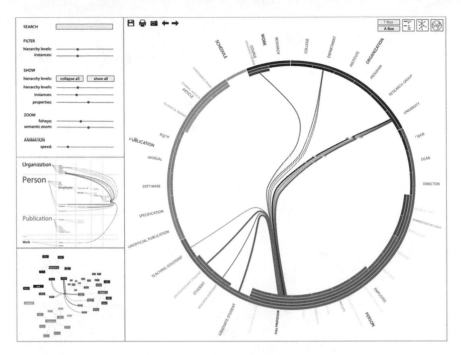

Fig. 16.7 Multiple view visualization tool (Reproduced from [48])

Fig. 16.8 Circular layout combined with a force-directed layout (Reproduced from [58])

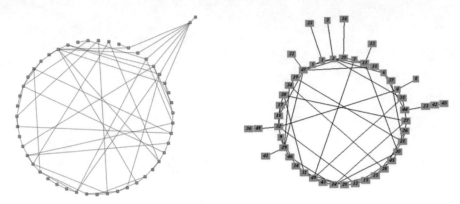

Fig. 16.9 Circular Algorithm with forces (Reproduced from [59])

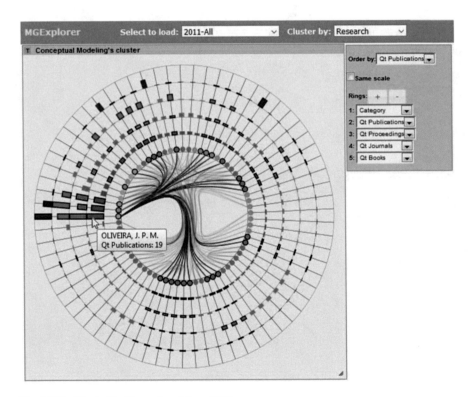

Fig. 16.10 ClusterVis (Reproduced from [60])

suggest that users can understand the proposed chord diagram and that the structure is effective in helping users to identify the blind-spots in the recommendation space.

The above work is an ongoing study that proposes basic capabilities of forming and combine visual shapes in polar coordinates in order to design node-link-group

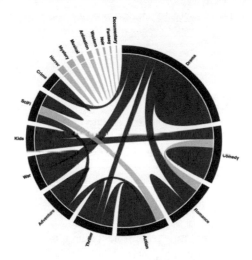

Fig. 16.11 Chord diagram used in the usability study (Reproduced from [61])

diagrams [62] as presented in Fig. 16.12. The authors also highlight an important problem related to the term's usage in the data visualization and graph drawing fields, namely the usage of "circular" and "radial" terms. Both terms refer to the polar coordinates system, where the starting point is the center of the circle. However, and according to the authors, a structure based on polar coordinates comprises two dimensions: a circular scale and angle that determines the position of a visual mark along the circular line, and a linear scale (radius) that determines the distance between the elements and the center. Succinctly, the authors propose a set of visual notions between data and their relationships, namely an outline of their application to design node-link-group diagrams.

This work provides an analysis of a set of interaction techniques for chord diagrams and presents a novel technique based on the hierarchical edge bundling technique [44]. The authors present a summary table of techniques and approaches of important works such as edge optimization, edge clustering, data-and segment-centric approaches, animation features, filtering techniques, and identifies the use of hierarchical dataset structures as we can see in Fig. 16.13.

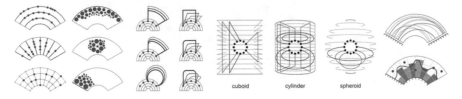

Fig. 16.12 Spatial allocation (Reproduced from [62])

Works	Edge Optimization	Edge Bundling	Data Centered	Segment Centered	Animation	Filtering	Hierarchical
Holten [1]		✓				✓	✓
Gansner and Koren [3]	✓	✓					
Meyer et al. [25]		✓		✓		✓	
Krzywinski et al. [26]						✓	
Bae and Lee [27]						✓	
Bostock et al. [4]		✓	✓		✓	✓	✓
Zhao et al. [28]		✓	✓	✓	✓	✓	✓
Gou and Zhang [29]		✓		✓		✓	✓
Kuhar and Podgorelec [30]		✓				✓	
Alsallakh et al. [31]						✓	
Borkin et al. [32]			✓	✓		✓	
Zeng et al. [22]		✓					✓
Peixoto [33]		✓					
Crnovrsanin et al. [21]		✓				✓	✓
Nicholas et al. [23]		✓				✓	
Etemad et al. [24]		✓				✓	
Wang et al. [34]			✓	✓		✓	✓
Cava et al. [16]		✓	✓		✓	✓	✓
Papp and Kunkli [15]	✓						
Ren et al. [35]							

Fig. 16.13 Summarization of related work (Reproduced from [44])

It also presented a new interaction technique for circular layouts based on the hierarchical edge bundling technique [54], namely a deformation technique for interactive exploration of connected data in a chord diagram, see Fig. 16.14. This approach uses a hierarchical dataset structure and proposes a set of interaction techniques based on dynamic selection, rotation and expandable layout for segment focus. The user can select individual segments of the circular layout and expand it to maximize the structure to a more detailed view. Another feature is a sketch-based and multivariate filtering technique based on multiple connections, visualization of direction and order, namely a method for tracking calls across multiple connections.

The Information Eye [63] uses a tree and a network structure in a single layout, as presented in Fig. 16.15. It embeds a radial tree inside the circular layout and it uses the concept of force, as it is based on the idea of the force-directed algorithm. The layout represents the relations without edges, and the center of the layout displays attributes-related information. The layout is divided into two sections, the circular layout and the center tree layout. An attraction matrix is used to optimize the distance from the node to the center of the circle by displaying labels. Instead of using edge attributes (e.g., thickness), it relies on the distance between points to reduce visual clutter.

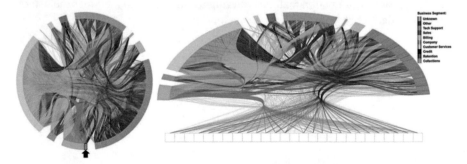

Fig. 16.14 Deformation technique for a detailed view of individual nodes and multi-variate brushing for filtering (Reproduced from [44])

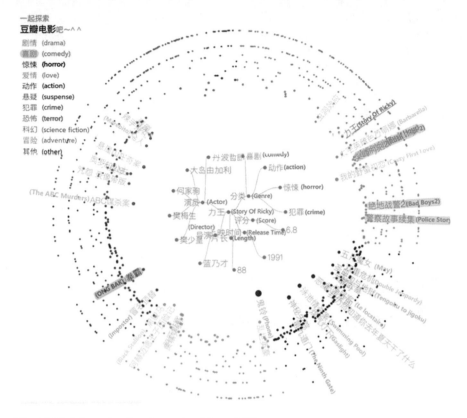

Fig. 16.15 Information Eye (Reproduced from [63])

Unlike node-link diagrams, the schema represents the relationships without showing the edges. In this approach the relationships are based on the distance between the nodes. Therefore, when the node of interest is placed in the center of the circle the relation between the nodes is presented based on the distance between the central node and the outer nodes. This is a very interesting approach because position is a very effective channel (i.e., position at common scale or position and unaligned scales) given the effectiveness principle and the ranking of the visual channel's accuracy [64], however, we have no link to access and experiment this approach.

It is important to highlight that the main goal of this section was to cover node-link layouts, specifically circular layouts/chord diagrams that are a subcategory of fixed layouts. Another point is that the present analysis excluded radial tree layouts [65] or sunburst layouts [66] because these structures are subcategories of node-linked tree layouts and are based on hierarchical dataset structures (e.g., root, parent, child and leaves). Relational datasets structures are the basis of the proposed two-level circular structure.

The analysis summary presented in Table 16.1 provided some important cues and interaction techniques, such as the edge bundling technique to reduce the inner clutter, density and the number of crossings. The examples based on hierarchical dataset structures denotes a wide adoption of the hierarchical edge bundling technique, however there are some approaches that uses hierarchical data and relational data in a single layout. Another important point is that relational dataset structures with weighted links are used in chord-diagrams. One exception is the work of [44] that uses a chord-diagram with a hierarchical dataset.

Table 16.1 Circular layouts analysis summary

Authors	Layout	Technique	Relational data	Hierarchical data	D3.js	AntV	Usability study
Holten [54]	Circular layout	Hierarchical edge budling		●			
Gansner and Koren [45]	Circular layout	External edge routing		●			
Krzywinski et al. [56]	Circular layout (chord diagram)		●				
Gou and Zhang [57]	Circular layout	Radial space-filling (RSF) technique, and hierarchical edge budling	●	●			
Kuhar and Podgorelec [48]	Circular layout	Hierarchical edge budling	●	●			
Vallandingham [58]	Force-directed layout, and circular layout		●		●		
Tollis [59]	Force-directed layout, and circular layout		●				
Cava et al. [60]	Circular layout		●	●	●		●
Tintarev et al. [61]	Circular layout (chord diagram)		●				●
Rees et al. [44]	Circular layout (chord diagram)	Hierarchical edge budling and edge optimization		●			
Yang et al. [63]	Circular layout, radial tree, and force-directed layout		●	●		●	

16.3 Designing and Implementing a 2-Level Circular Layout

We propose a two-level circular interactive layout/algorithm to visualize and explore science social cooperation [13, 36] as presented in Fig. 16.16. The 2-level structure, see Fig. 16.17, was implemented in D3.js which is a JavaScript visualization library.

The data-driven concept implies that the dataset should have a specific structure, namely a relational dataset structure (network) in order to compute and draw the links between the articles and the researchers. Two main circles define the circular structure, researchers are in the outer circle, and the articles grouped by scientific domain and by different shades are in the inner circle. It was used a dummy dataset in JSON according to the described context and scenario, namely a standard user profile (personas) and a set of shared scientific articles. Two items define the dataset, the scientific articles in the object category (nominal variable), and the researchers in a specific knowledge domain in the person item category (nominal variable). A set of attributes defines both items. For more details, see the dummy dataset in the Observable page. It is important to note that the items researchers and items scientific articles are at the same level in the data structure.

In the mapping strategy, visual marks are used to represent data items, points and lines (nodes and links). In the visual encoding strategy, namely the visual channels, color intensity, hue, and position are used. In the outer circle structure, it is used text instead of the visual mark circle (node). The objective is to provide a fast distinction between two different items, namely researchers and scientific articles.

The scientific articles are sorted according to a hypothetical score that defines the number of articles placed in the inner and outer circle (e.g., weighting factor) [13]. The same approach takes place in the researcher's circle (outer circle). A further discussion about the score/ weighting factor is beyond the scope of the current article [67].

The circles are based on the `circle_gen` function to generate the position of items, persons and objects, based on cartesian coordinates (x, y) rather than polar coordinates. The function performs the nodes computation and the equidistant distribution along the circumferences based on the x and y coordinates. Succinctly, the function uses a set of attributes, namely, `data`, `gen_radius`, `transform_text`, `scale_color`, `inc_angle`, and `type`.

The `data` attribute is related to the data needed to draw the circle, namely the item persons (researchers) which is sorted alphabetically and the item objects (articles) which are grouped by knowledge domain clusters and a hypothetical score; the `gen_radius` defines the radius of the circles (e.g., outer circle slider button: max 1000 px, min 600 px; inner circle slider button: max 900 px, min 200 px) see Fig. 16.19; `transform_text` is the text size and padding parameterization; the `scale_color` defines the nodes color scale, however has not been used, because every item in the dataset contains itself a color; `inc_angle` defines the division of the circumference by the number of equidistantly distributed items, and also the node angle on the circles using the following function:

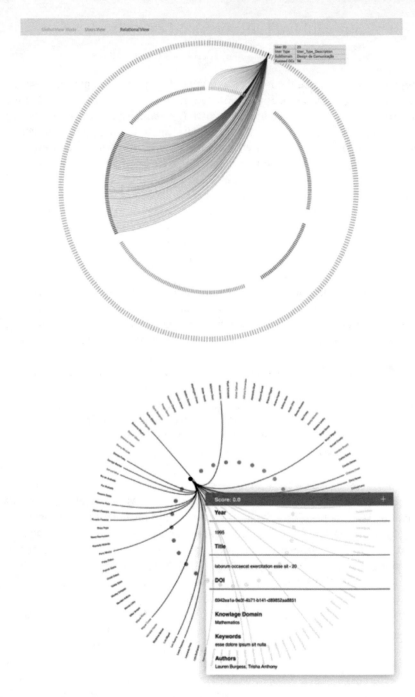

Fig. 16.16 User interface: community web page and the two-level circular structure

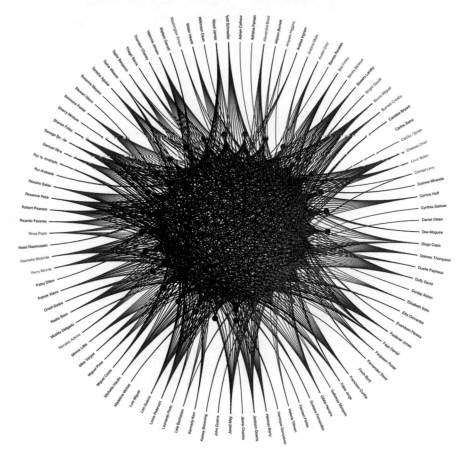

Fig. 16.17 D3.js Bi-level layout: researchers are represented in the outer circle and the scientific articles are represented in the inner circle

2*PI/n (e.g., `array(person; array(object).length;`

The attribute `type` performs the distinction between items, namely objects (articles) and persons (researchers).

The distribution of the items based on cartesian coordinates is performed using `object_id` that is the identifier of each item and makes the distribution of the nodes equidistant. This value is multiplied by `inc_angle` to determine the relative angle of the node on the circle. This converts the node angle and radius values to cartesian coordinates based the following functions:

`new_object.x = gen_radius * Math.cos(object_id * inc_angle)`

`new_object.y = gen_radius * Math.sin(object_id * inc_angle`

The Cartesian coordinates are thus applied to the following function that calculate the positions of the nodes in order to generate the circular structures:

```
node.append('circle').attr('cx', d = > d.x).attr('cy', d = >
d.y).
```

However, to obtain the attributes required for the links computation, the map type called `coor` (coordinates) associated with the items objects and persons is used. A set of attributes define the value map, namely the cartesian coordinates and the `uid` performs the link identification.

The connections between the items are based on two attributes, namely `source` and `target`. An array with links was implemented to store attributes related to the cartesian coordinates of the `source` and `target` attributes (persons and objects). The nodes placement it is based on arrays, namely array `object` and array `person`. This solution provides scalability to the structure. Adding new items with their respective array makes it possible increment the number of levels/circles.

To reduce internal visual clutter, the technique edge bundling without hierarchy was used. This approach differs from the hierarchical edge bundling technique [54], because it uses a relational dataset structure rather than a hierarchical dataset. The non-hierarchical edge bundling implementation was based on the following code[5]:

```
.attr("d", function(d){
                    var lineData = [
                      {
                        x: Math.round(d.target_x),
                      y: Math.round(d.target_y)
                        x: Math.round(d.target_x) -
Math.round(d.target_x)/2,
                        y: Math.round(d.target_y) -
Math.round(d.target_y)/2
                      },{
                        x: Math.round(d.source_x) -
Math.round(d.source_x)/2,
                        y: Math.round(d.source_y) -
Math.round(d.source_y)/2
                      },{
                        x: Math.round(d.source_x),
                        y: Math.round(d.source_y)
                      }];
re-
turn`M${lineData[0].x},${lineData[0].y}C${lineData[1].x
},${lineData[1].y},$
{lineDa-
ta[2].x},${lineData[2].y},${lineData[3].x},${lineData[3
].y}`;})
```

---

[5] https://stackoverflow.com/questions/34263110/d3-js-edge-bundling-without-hierachy.

Color was used to simplify category identification, and is based on Lch/HCL color space interpolation that is both intuitive and perceptually uniform [68]. The transformation from RGB to Lch/HCL interpolation space, was performed in the Leonardo color webtool,[6] and the color scheme was generated in Coolors[7] webtool. A further discussion about the color topic is beyond the scope of the current article.

16.3.1   Interaction Techniques

Interactivity plays an important role in exploratory information visualization approaches. The users can use the mouse wheel to change the scale of detail trough the zoom technique; it is used two slide buttons to change the outer and inner circle radius; the *connections on* button activates all the links and provides the context overview as shown in Fig. 16.18, and the *connections off* button clears the structure. Is important to highlight that the buttons shape isn't the final design and not the final option. In future work new buttons will be designed.

A mouseover event triggers a tooltip/infotip to display detailed information/ metadata related to researchers and scientific articles (e.g., Title, Authors, Year among other metadata) as presented in Fig. 16.19. Another visual feature is the transparency of the tooltips to prevent users from losing sight of the context. A click on a node provides only the links between scientific articles and researchers, and also removes the tooltip and the remaining nodes. The outer nodes (researcher's nodes) are arranged alphabetical and clockwise from the central axis of the circular structure. The reference line with the characters A and Z is used to indicate the start and end point of the circular structure. The aim is to reinforce a clockwise orientation/reading process of the structure as shown in Fig. 16.19.

An important technique is to visually "deactivate" the context to reduce visual clutter. Opacity with a value of 0.07 was used to make the context transparent, as shown in Fig. 16.18. This technique makes it possible to focus on a node and its respective links without letting users lose sight of the context, and thus provide a continuous and predictable experience as the interface context change.

We believe that a clear design as well as the implemented interaction techniques will better support the user's exploratory activity.

16.4   Preliminary Results

This section presents the preliminary results of the two-level circular structure [13, 36]. Section two provided important guidance, namely novel interaction techniques, and clutter mitigation strategies, such as the hierarchical edge bundling technique to

---

[6] https://leonardocolor.io.

[7] https://coolors.co.

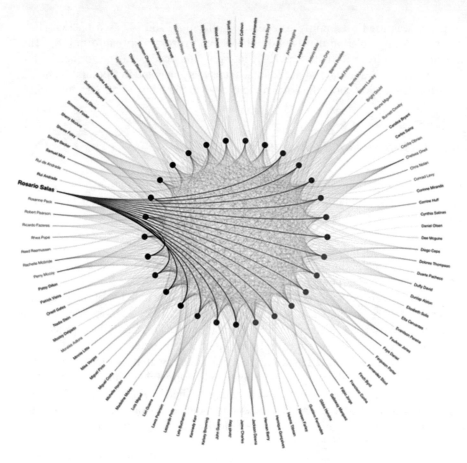

Fig. 16.18 Graying out the context

reduce visual clutter [54]. Our approach used a dummy relational dataset structure with 100 nodes in the outer circle, 29 nodes in the inner circle, making 963 links. It was used a dataset with a relational structure, and the items persons and objects are at the same level. D3.js library was used to design and implement the structure, and the layout structure and links were based on Marchese's[8] D3.js approach published in July 2020, which is also based on a relational dataset structure. This approach solved some problems at the functional level, specifically with the node distribution and arrangement. This work was the basis for the circular and equidistant distribution of the nodes in the two circles. However, our approach differs from Marchese's in terms of aesthetics, functionality, and interaction.

The non-hierarchical edge bundling technique reduced the visual clutter between the outer circle and the inner circle. However, with a high number of connections, the inner circle remains cluttered. The first objective was to test whether the technique

---

[8] https://observablehq.com/@juanmarchese/chord.

Connections Of   Connections On

This slider input with the name "Outter Circle Size" allows the user change the radius of the external circle

Outter Circle Size

649

This slider input with the name "Inner Circle Size" allows the user change the radius of the internal circle

Inner Circle Size

394

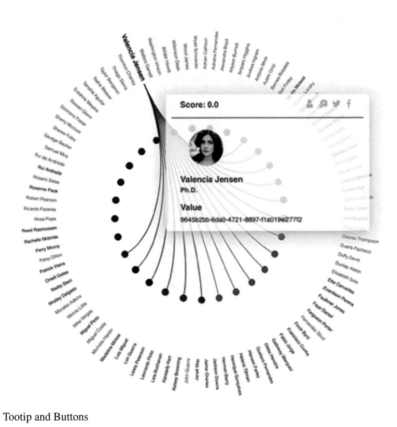

Fig. 16.19 Tootip and Buttons

could handle a high number of connections and crossings in a two-level circular structure, as shown in Fig. 16.17. The second goal, and in future work, is to improve the technique and achieve better readability by avoiding the overlapping between nodes and links. However, more research is needed on this point. In future work the number of researchers and articles will be limited according to a score, namely a weighting factor.

Succinctly, the two-level structure can interactively and aesthetically accommodate social cooperation based on a curatorial process of scientific articles. Interaction techniques allow users to interact and navigate, but also mitigate visual clutter.

The results are still at a preliminary stage, and the goal is to conduct a future usability study to test whether the layout is an efficient exploratory framework.

16.5   Conclusion and Future Work

This article presents a design approach to information, and the goal is to address the flood of scientific knowledge. The proposed two-level interactive circular structure is an exploratory visualization that aims to support a collaborative and curatorial process based on the wisdom of science communities. This kind of behavior already happens on social platforms (e.g., Twitter, Facebook, among others), however without a logic and *medium* that supports and maps this social dynamics.

The structure uses two circles, the inner circle represents the most shared scientific articles, and the outer circle represents the researchers who read and share them. The computed links provide the connections between these two levels. Is important to emphasize the use of a non-hierarchical edge bundling technique to reduce visual clutter, however further research directions are needed regarding this point. The social cooperation flows of science are presented metaphorically, as an animation is used, however the flows do not represent any variables, so further research directions are also needed regarding this topic.

In future work, we will design a metadata filter panel and implement a set of attributes to reduce visual clutter (filtering mechanisms and ponderation factors). It is important to conduct a user study/evaluation based on performance metrics and satisfaction metrics to understand how researchers will interact and navigate in the two-level circular layout. Important artistic works should also be considered in a future literature review, the work of [69] provides important cues on how to transform digital information into an aesthetic of participative processes.

The visualization and code can be found in the following link: https://observabl ehq.com/@brunomiguelam/2-level-circular-layout.

Acknowledgements This work has been supported by FCT—Fundação para a Ciência e Tecnologia within the R&D Units Project Scope: UIDB/00319/2020.

References

1. Brainard, J.: Scientists are drowning in COVID-19 papers. Can new tools keep them afloat? Science **1979** (2020). https://doi.org/10.1126/science.abc7839
2. Kupferschmidt, K.: 'A completely new culture of doing research.' Coronavirus outbreak changes how scientists communicate. https://www.sciencemag.org/news/2020/02/completely-new-culture-doing-research-coronavirus-outbreak-changes-how-scientists?fbclid=IwAR0q Tbw5A_EPg5lNJxhsiz4nXB37ajUybrvivauVxlHAmOAy6N8c1d9MtyA
3. Goulard, C.: Mapping the Covid19 research landscape. https://www.youtube.com/watch?v= 5dmzvlBSDJE&list=PLaoyTABsSAEEsFdgqZnOENu6YCtG-jYxT&index=6&t=56s
4. Wright, A.: Glut: Mastering Information Through The Ages. Cornell University Press, Ithaca, United States (2008)
5. Thackara, J.: In the Bubble. Designing in a Complex World. MIT Press (2006)
6. Johnson, R., Watkinson, A., Mabe, M.: The STM Report: An overview of scientific and scholarly Journal Publishing. https://www.stm-assoc.org/2018_10_04_STM_Report_2018. pdf
7. Seglen, P.O.: Why the impact factor of journals should not be used for evaluating research. BMJ **314**, 497–497 (1997). https://doi.org/10.1136/bmj.314.7079.497
8. Hicks, D., Wouters, P., Waltman, L., de Rijcke, S., Rafols, I.: Bibliometrics: the Leiden Manifesto for research metrics. Nature **520**, 429–431 (2015). https://doi.org/10.1038/520429a
9. Castillo, C.: Fairness and transparency in ranking. ACM SIGIR Forum. **52**, 64–71 (2019). https://doi.org/10.1145/3308774.3308783
10. Muller, J.: The Tyranny of Metrics (2018)
11. Bonsiepe, G.: Design as tool for cognitive metabolism : from knowledge production to knowledge presentation. In: International Symposium on the Dimensions of Industrial Design Research Ricerca+Design, pp. 1–14 (2004)
12. Reingold, H.: Smart Mobs: The Next Social Revolution. Perseus Basic Books (2002)
13. Azevedo, B.: New Paradigm Applied To A Digital Repository of Scientific Publications: Design of a Collaborative Structure For Visualization And Filtering Information. https://collections. pomona.edu/labs/record/?pdb=3737
14. Hidalgo, C., Ali, A.: The Data-Visualization Revolution—Scientific American. www.scientifi camerican.com/article/the-data-visualization-revolution/
15. Jeschke, J.M., Lokatis, S., Bartram, I., Tockner, K.: Knowledge in the dark: scientific challenges and ways forward. FACETS. **4**, 423–441 (2019). https://doi.org/10.1139/facets-2019-0007
16. Azevedo, B.M., e Sa, J.O., Baptista, A.A., Branco, P.: Information visualization: conceptualizing new paths for filtering and navigate in scientific knowledge objects. In: 2017 24° Encontro Português de Computação Gráfica e Interação (EPCGI), pp. 1–8. IEEE, Guimarães (2017)
17. Tufte, E.R.: Envisioning Information. Graphics Press (1990)
18. Tufte, E.R.: Visual Explanations: Images and Quantities. Graphics Press, Evidence and Narrative (1997)
19. Tufte, E.R.: The Visual Display of Quantitative Information. Graphics Press, 2nd edn (2001)
20. Tufte, E.R.: Beautiful Evidence. Graphics Press (2006)
21. Friendly, M., Denis, D.: Milestones in the history of thematic cartography, statistical graphics, and data visualization. Seeing Sci. Today Am. 15–56 (2008)
22. Viégas, F.: Data Visualization as Exploratory Medium: From Scientific Insight to Artistic Expression. https://www.youtube.com/watch?v=N2ddykLU7EE
23. Scharnhorst, A., Börner, K., van den Besselaar, P.: Models of Science Dynamics. Springer, Berlin Heidelberg, Berlin, Heidelberg (2012)
24. Leydesdorff, L.: The Challenge of Scientometrics: The Development, Measurement, and Self-Organization of Scientific Communications. Universal Publishers/uPUBLISH.com, USA (2001)
25. Castellano, C., Fortunato, S., Loreto, V.: Statistical physics of social dynamics. Rev. Mod. Phys. **81**, 591–646 (2009). https://doi.org/10.1103/RevModPhys.81.591

26. Chang, Y.-F.: Social synergetics, social physics and research of fundamental laws in social complex systems. Physics (College Park Md) **468**, 11 (2009)
27. Szolnoki, A., Wang, Z., Perc, M.: Wisdom of groups promotes cooperation in evolutionary social dilemmas. Sci. Rep. **2**, 576 (2012). https://doi.org/10.1038/srep00576
28. Adami, C., Hintze, A.: Thermodynamics of evolutionary games. Phys. Rev. E. **97**, 062136 (2018). https://doi.org/10.1103/PhysRevE.97.062136
29. Fortunato, S., Bergstrom, C.T., Börner, K., Evans, J.A., Helbing, D., Milojević, S., Petersen, A.M., Radicchi, F., Sinatra, R., Uzzi, B., Vespignani, A., Waltman, L., Wang, D., Barabási, A.-L.: Science of science. Science **1979**, 359 (2018). https://doi.org/10.1126/science.aao0185
30. Sun, X., Kaur, J., Milojević, S., Flammini, A., Menczer, F.: Social dynamics of science. Sci. Rep. **3**, 2–7 (2013). https://doi.org/10.1038/srep01069
31. Weng, L., Ratkiewicz, J., Perra, N., Gonçalves, B., Castillo, C., Bonchi, F., Schifanella, R., Menczer, F., Flammini, A.: The role of information diffusion in the evolution of social networks. In: Proceedings of the 19th ACM SIGKDD International Conference on Knowledge Discovery and Data Mining—KDD '13, p. 356. ACM Press, New York, New York, USA (2013)
32. Sarrafzadeh, B., Vtyurina, A., Lank, E., Vechtomova, O.: Knowledge graphs versus hierarchies. In: Proceedings of the 2016 ACM on Conference on Human Information Interaction and Retrieval - CHIIR '16, pp. 91–100. ACM Press, New York, New York, USA (2016)
33. Fry, B.: Visualizing Data: Exploring and Explaining Data with the Processing Environment. (2008)
34. Börner, K., Polley, D.: Visual Insights: A Practical Guide to Making Sense of Data. MIT Press (2014)
35. Manovich, L.: What is visualisation? Vis. Stud. **26**, 36–49 (2011). https://doi.org/10.1080/147 2586X.2011.548488
36. Azevedo, B., Baptista, A.A., Oliveira e Sá, J., Branco, P., Tortosa, R.: Interfaces for science: conceptualizing an interactive graphical interface. In: Brooks, A., Brooks, E., Sylla, C. (eds) 7th EAI International Conference, ArtsIT 2018, and 3rd EAI International Conference, DLI 2018, ICTCC 2018, Braga, Portugal, October 24–26, 2018, Proceedings, pp. 17–27. Springer, Cham (2019)
37. Henry, N., Fekete, J.: Matrix explorer: a dual-representation system to explore social networks. IEEE Trans. Vis. Comput. Graph. **12**, 677–684 (2006). https://doi.org/10.1109/TVCG.200 6.160
38. Gibson, H., Faith, J., Vickers, P.: A survey of two-dimensional graph layout techniques for information visualisation. Inf. Vis. **12**, 324–357 (2013). https://doi.org/10.1177/147387161 2455749
39. Kobourov, S.G.: Force-directed drawing algorithms. In: Chapman and Hall/CRC (ed.) Handbook of Graph Drawing and Visualization, p. 862. Roberto Tamassia (2013)
40. Hua, J., Huang, M.L., Wang, G.: Graph layout performance comparisons of force-directed algorithms. Int. J. Perform. Eng. **14**, 67–76 (2018). https://doi.org/10.23940/ijpe.18.01.p8.6776
41. Cheong, S.-H., Si, Y.-W.: Force-directed algorithms for schematic drawings and placement: a survey. Inf. Vis. **19**, 65–91 (2020). https://doi.org/10.1177/1473871618821740
42. Meirelles, I.: Design for Information: An Introduction to the Histories, Theories, and Best Practices Behind Effective Information Visualizations. Rockport Publishers (2013)
43. Bertini, E.: Visualizing Networks and Trees. https://docs.google.com/presentation/d/11ApQe 7ODFH4J3BLRvPYLeGUQon07VWVZhhEsZqTTiHM/edit#slide=id.gaee402ba2_0_107
44. Rees, D., Laramee, R.S., Brookes, P., D'Cruze, T.: Interaction techniques for chord diagrams. In: Proceedings of the International Conference on Information Visualisation, pp. 28–37 (2020). https://doi.org/10.1109/IV51561.2020.00015
45. Gansner, E.R., Koren, Y.: Improved circular layouts. In: Graph Drawing, pp. 386–398. Springer Berlin Heidelberg, Berlin, Heidelberg (2007)
46. Ghoniem, M., Fekete, J.-D., Castagliola, P.: A comparison of the readability of graphs using node-link and matrix-based representations. In: IEEE Symposium on Information Visualization, pp. 17–24. IEEE (2004)
47. Munzner, T.: Visualization Analysis and Design. A K Peters/CRC Pres (2014)

48. Kuhar, S., Podgorelec, V.: Ontology visualization for domain experts: a new solution. In: 2012 16th International Conference on Information Visualisation, pp. 363–369. IEEE (2012)
49. Salton, G., Allan, J., Buckley, C., Singhal, A.: Automatic analysis, theme generation, and summarization of machine-readable texts. In: Information Retrieval and Hypertext, pp. 51–73. Springer US, Boston, MA (1996)
50. Six, J.M., Tollis, I.G.: Circular Drawings of Biconnected Graphs. In: Lecture Notes in Computer Science (including subseries Lecture Notes in Artificial Intelligence and Lecture Notes in Bioinformatics), pp. 57–73 (1999)
51. Six, J.M., Tollis, I.G.: A framework for circular drawings of networks. Lecture Notes in Computer Science (including subseries Lecture Notes in Artificial Intelligence and Lecture Notes in Bioinformatics) **1731**, 107 116 (1999). https://doi.org/10.1007/3-540-46648-7_11
52. Doğrusöz, U., Madden, B., Madden, P.: Circular layout in the graph layout toolkit. In: Lecture Notes in Computer Science (including subseries Lecture Notes in Artificial Intelligence and Lecture Notes in Bioinformatics), pp. 92–100 (1997)
53. Draper, G.M., Livnat, Y., Riesenfeld, R.F.: A survey of radial methods for information visualization. IEEE Trans. Vis. Comput. Graph. **15**, 759–776 (2009). https://doi.org/10.1109/TVCG.2009.23
54. Holten, D.: Hierarchical edge bundles: visualization of adjacency relations in hierarchical data. IEEE Trans. Vis. Comput. Graph. **12**, 741–748 (2006). https://doi.org/10.1109/TVCG.2006.147
55. Holten, D.: Visualization of graphs and trees for software analysis (2009)
56. Krzywinski, M., Schein, J., Birol, I., Connors, J., Gascoyne, R., Horsman, D., Jones, S.J., Marra, M.A.: Circos: an information aesthetic for comparative genomics. Genome Res. **19**, 1639–1645 (2009). https://doi.org/10.1101/gr.092759.109
57. Gou, L., Zhang, X.: TreeNetViz: revealing patterns of networks over tree structures. IEEE Trans. Vis. Comput. Graph. **17**, 2449–2458 (2011). https://doi.org/10.1109/TVCG.2011.247
58. Vallandingham, J.: How to Make an Interactive Network Visualization. http://projects.flowingdata.com/tut/interactive_network_demo/
59. Six, J.M., G. Tollis, I.: Circular drawing algorithms. In: Handbook of Graph Drawing and Visualization: Discrete Mathematics and Its Applications, p. 862. Chapman and Hall/CRC (2013)
60. Cava, R., Freitas, C.M.D.S., Winckler, M.: ClusterVis. In: Proceedings of the Symposium on Applied Computing, pp. 174–179. ACM, New York, NY, USA (2017)
61. Tintarev, N., Rostami, S., Smyth, B.: Knowing the unknown: visualising consumption blind-spots in recommender systems. In: Proceedings of the ACM Symposium on Applied Computing, pp. 1396–1399 (2018). https://doi.org/10.1145/3167132.3167419
62. Guchev, V., Buono, P., Gena, C.: Towards intelligible graph data visualization using circular layout. Proceedings of the Workshop on Advanced Visual Interfaces AVI. **1**, 15–17 (2018). https://doi.org/10.1145/3206505.3206592
63. Yang, L., Zhou, F., Li, X.Y.: Information eye: a hybrid visualization approach of exploring relational information space. IEEE Trans. Syst. Man Cybern. Syst. 3753–3759 (2020). https://doi.org/10.1109/SMC42975.2020.9282899
64. Mackinlay, J.: Automating the design of graphical presentations of relational information. ACM Trans. Graph. **5**, 110–141 (1986). https://doi.org/10.1145/22949.22950
65. Ka-Ping Yee, Fisher, D., Dhamija, R., Hearst, M.: Animated exploration of dynamic graphs with radial layout. In: IEEE Symposium on Information Visualization, INFOVIS 2001. pp. 43–50. IEEE (2001)
66. Stasko, J., Catrambone, R., Guzdial, M., Mcdonald, K.: Evaluation of space-filling information visualizations for depicting hierarchical structures. Int. J. Hum. Comput. Stud. **53**, 663–694 (2000). https://doi.org/10.1006/ijhc.2000.0420
67. Azevedo, B., Silva, H., Tortosa, R.: Harnessing User Knowledge in the Construction of Rating Flows: the Design of a Collaborative System Applied to Academic Repositories. In: Libro de Actas (ed) Systems & Design: Beyond Processes and Thinking (IFDP - SD2016), pp. 780–792. Universitat Politècnica València, Valencia (2016)

68. Ware, C.: Information Visualization: Perception for Design. Morgan Kaufmann Publishers (2012)
69. Strauss, W., Fleischmann, M.: The art of the thinking space—a space filled with data. Digit. Creat. **31**, 156–170 (2020). https://doi.org/10.1080/14626268.2020.1782945

Chapter 17
Children's Generative Play with a Combination of Analogue and Digital Resources

Eva Brooks⊙ and Anders Kalsgaard Møller⊙

Abstract This chapter focuses on children's play and making activities when involved in creative processes, in particular focusing on how different resources and materials foster their agency. The chapter will include empirical data where children (3–9 years of age) are engaged with both digital and analogue resources, (LEGO, Minecraft, and Virtual Reality). The analysis was based on data in the form of ethnographic observations, and informal conversations. The analysis focused on how children engaged with the different resources and how they unified or separated them in the design activity. We narrowed this focus into four main categories: problem scoping, idea generation, design monitoring, and design experience. The chapter contributes to further specifying dimensions of digital creation opening new avenues of play and creation with a unifying combination of analogue and digital resources. This has implications on educational and societal instances by renewing our understanding of children's contemporary play.

Keywords Generative play · Analogue and digital resources · Multimodal social semiotics · Meaning potentials · Affordances · Communication · Representation

17.1 Introduction

In this chapter, we outline a perspective on digital creation as a play-oriented and generative activity, beyond an understanding of creation as (only) a digital endeavour. We can notice that children's play and creation by means of a combination of analogue and digital resources has become more recognised in the education literature. It is described in terms of 'hands-on play through design' [5], 'digital play' [31], 'converged play' [9], 'digital convergence' [2], 'trans-media' [15] and playscapes

E. Brooks (✉) · A. K. Møller
Department of Culture and Learning, Xlab, Aalborg University, Aalborg, Denmark
e-mail: eb@ikl.aau.dk

A. K. Møller
e-mail: ankm@ikl.aau.dk

[1]. New kinds of physical-digital ('phygital') play technologies [23] are emerging in the form of smart toys and digitally augmented play spaces [3], and making its way into children's digital play and creation activities. This chapter presents findings from design-oriented research in which children's generative play with a combination of analogue and digital resources was a central premise.

Play as a concept has several connotations [25, 27, 30, 34]. Children themselves consider almost anything as an opportunity for play and hence have a relatively unrestrained view of play [13]. This is aligned with Sutton-Smith's [30] view on play as characterised by its quirkiness, redundancy, and flexibility. However, the traditional understanding of play has changed in recent years because of the development of play-oriented digital technology, which has resulted in new play environments for play. Play processes including a combination of analogue and digital resources pose new pedagogical challenges in educational settings. Traditionally, different kinds of play are known to support learning, for instance through exploration, problem-solving, and social interaction [26]. To this end, play including a combination of analogue and digital resources has emerged only within the last decade, since the entrance of the iPad™ in 2010, which enabled children to engage with digital technology [10]. In this regard, designers have created digital experiences mirroring traditional play contexts to increase children's analogue and social interactivity of play [23]. Currently, the internet of things, robots, wearable technologies, digital agents, and voice-activation are more commonly used, which offer children further experiences of combined analogue–digital experiences [16, 31]. With an exception to robots, children's play with analogue and digital resources is still largely centred on screen-based interactions [31]. Due to that this kind of combined play in educational settings is new, the implications for children's play, learning, and teaching are still scarce [9, 31].

Like play, design activities promote curiosity and exploration. While design typically leads to concrete solutions for specific situations, play can be seen as an open-ended natural asset where people imaginatively interact with the world. Similarly, designers are curious to ask the question 'how' in order to get to the intended solution, whereas playing humans imaginarily ask the question 'what if'. Yet, both questions reflect a wonder about the creation of a possible future [29]. Stappers [29] refers to the act of designing as the locus where new ideas get created by confronting them with the world [35 p. 493]. Zimmerman et al. [35] state that designers in this way have skills to create "a product that transforms the world from its current state to a preferred state". In that context and from a social semiotics perspective, design can be considered as a multimodal assertation of an individual's interest in participating in the world and an insistence to shape this interest through the design of representations with resources available in a specific situation [20]. In this way, design is prospective by projecting an individual's interest into their world to make meaning of and influence the future. To meet gaps identified in research, we consider a design-oriented approach based on such overarching configurations offering potentials to open new avenues of play and creation with a combination of analogue and digital resources. This is also accompanied by educational and societal implications such as renewing our understanding of children's contemporary play.

Theoretically, this chapter approaches generative play from a multimodal social-semiotic perspective specifically focusing on the concepts of meaning making, communication, and representation [20, 32]. Within this overall approach, design workshops, called *The World of Scale: Design—Scan—Experience*, were designed drawing inspiration from generative ontologies to the study of generative play with a combination of analogue (LEGO blocks) and digital resources (Minecraft and Virtual Reality–VR). We set out to explore and illustrate the intersections of generative play across analogue and digital domains in a semi-formal learning environment. In particular, we focused on how analogue–digital resources of generative play enabled school children (6–9 years of age) to unify materially and digitally mediated resources to develop ideas, express knowledge and communicate with others.

The chapter is structured as follows: we first outline the conceptual background, describing the main aspects and concepts of the applied multimodal social-semiotic theory. We briefly summarize previous research that depicts the relatively recent shift in children's play and learning landscape. This is followed by a description of the included case study. Moreover, we provide examples to illustrate some of the features of play and creation in cross-modal activities. Finally, the chapter presents a discussion including takeaway points for further development that we anticipate as helpful when designing activities for digital play and learning that cross-over modes of actions, which is followed by conclusive comments.

17.2 Generative Play—Playing with Meaning Potentials

The concept of play can have many different meanings. Generally, play is seen as an open-ended natural asset where people imaginatively play with the world. Play is often associated with social engagement, discovery and inquiry occupying a child's play experience, rather than a specific outcome [5, 24, 28].

In play, children naturally explore, creatively imagine, and invent new ways of doing things. According to Sutton-Smith [30] this involves different dimensions of unpredictability promoting play as children's very own activity promoting flexible and adaptable meaning-making processes. Hence, it can be stated that play actions involve exploration by means of imagination and creativity.

> *It may very well be the ability to take leave of the world-as-it-is that provides the space through which new possibilities for thought emerge. Relinquishing the constraints of convention in order to explore in the mind's eye the unconventional might provide one of the most important arenas in which creativity itself could be generated* [11, p. 46].

Considering this quote by Eisner, children's digital creation can be seen as generative play actions, where children are modelling with ideas and meaning making, e.g. by reasoning, testing, and questioning. Expressed differently, through acts of meaning making, children can make their own choices and decisions and, thereby, find their ways of dealing with what they already know and their creative imagination of what could be. For example, asking children to provide explanations is to prompt them to generate inferences, i.e. to make meaning, that moves beyond given information [6], which as such requires reasoning abilities deducted from prior knowledge [4]. According to a multimodal social-semiotic perspective, meaning making in children's generative play arises in social environments and in social interaction [20]. This is how we in this chapter emphasise play as a generator of meaning, which as such always arises out of the interest and need of the player(s).

According to Kress [22], a meaning can move across modes and the affordances that each mode offers. In the context of this chapter, this means that meaning originating from children's play with analogue resources (LEGO blocks) moves when children transfer their LEGO block constructions across to the digital mode. The digital mode includes Minecraft and a Minecraft plugin to VR, which transfer the LEGO blocks into a VR representation. The question is what actions and processes exist in these two modes, and what relations between the modes that can be implied? For example, the three-dimensional and tactile features of LEGO blocks and their constructions afford meanings about balance and tactility, something that is less available, or completely different in Minecraft or VR. Generative play processes thus relate to a play with meaning potentials, where meaning is considered as a resource. Considering this, we draw on Kress' [20] take on the concept of affordances, which refers to their distinct meaning potentials. Here, the concept of 'aptness' is relevant to bring in as it focuses on 'fitness for its purpose'; i.e. the best fit for a specific purpose [20, p. 156]. Thus, aptness refers to the resources that are available in a meaning making process (e.g. LEGO blocks and VR) [20]. Further, Kress states that what individuals identify as being most apt depends on their interests in context. This means that we apply the analytical concept of *aptness* to identify the affordances that children use in their generative play. As such, we investigate questions about affordances in terms of: '-apt for whom? -apt to what? and -when apt?'. This concerns how affordances function in social environments. On the other hand, this is also a question of how affordances rest on the materiality of a certain resource. This will be further elaborated in the following section.

17.3   Multimodal Communication and Representation

As a theory, social semiotics draws on Halliday's [14] supposition that language emerges from social processes and that the resources of a language are shaped by the functions developed by people in specific contexts to satisfy their needs [16, 20, 32]. Hodge and Kress [16], van Leeuwen [33], and Kress [20] as further developed Halliday's socially framed perspective of language to include all semiotic resources, i.e. a theory of multimodal sign-making. As a concept, multimodality has attracted researchers from different disciplines concerned with meaning and communication [18]. Kress [22] initiated an investigation of multimodal meaning making, studying young children's use of affordances of materials they had at hand, or techniques they mastered on the basis of their interest. The findings exposed that children used what were of pivotal importance for them at a given moment [33]. Further Kress wrote [22 p. 13]: *As children are drawn into culture, 'what is to hand' becomes more and more that which the culture values and therefore makes readily available.* This can be related to the study of affordances and learning potentials of different semiotic modes [21] and why different modes of communication can be considered as resources for representation and interaction [33].

Kress' [20] perspective of representation is grounded in an individual's interest and engagement in the world and an urge to give material realisation of this interest. He conceptualises this wish to realise e.g. ideas of interest as communication, which he argues is distinct from the process of representation. Expressed differently, communication focuses on the need to make a representation available to others while interacting with them. In the context of this study, when children have a wish of a certain representation, they ask themselves what it is they want to represent and how this can be materially expressed?

By contrast, when children communicate, or realise, their creation of ideas, they put their understanding of the world in a material shape as a sign into an interaction with others, like into an audience [20]. Kress further elaborates on this by stating that both representation and communication are social processes. While representation focuses on the idea of an individual, shaped by his or her social histories and a certain focus to give material form, communication focuses on social interaction including the individual together with others in a specific social environment. This is to say that representation *takes place* in a social environment and communication *constructs* a social environment. In this study then, the participating children becomes sign makers, or designers, articulating and expressing their design ideas and solutions. They do so by choosing the material that best fits their ideas and solutions and make use of them in relation to their interests and previous experiences. In other words, they re-create what they already know (what is) into new representations and communications (what could be). Kress [20] explains that manipulation with digital technology and other expressive resources (e.g. LEGO blocks) afford communication.

17.4 Creating Cross-Over Modalities—A Case Study

The case study included in this chapter is based on data generated from three design workshops including a total of 66 children (35 girls and 31 boys) between 6–9 years of age. The study was part of a project, 'Architecture is Play', where children's interaction with a combination of analogue and digital technology targeting architectural learning (math, science, and design skills) through play were investigated. Data were gathered from observations and informal conversations with the participating children, and their teachers. The workshops took place at a Design Centre in Northern Jutland, Denmark and were led by two facilitators from the centre and to some extent supported by the participating schoolteachers.

The workshops in question were titled *The World of Scale: Design—Scan—Experience*, where the children explored the world of scale in LEGO, Minecraft, and VR.

Through this exploration, the children become familiar with some of an architect's many tools and methods for designing, researching, and communicating. In these workshops, the children should create high buildings; a skyscraper (in two of the workshops) and an observation tower (in one of the workshops).

The workshop started with a discussion about what a skyscraper/observation tower is, what functions it has, and experiences of standing at the top of the building. Then, one of the facilitators introduced the children to the concepts of scale, module, and model, which led to discussions about the meaning of scale and the relationship between size of a model and size in reality (Figs. 17.1, 17.2). In general, the introduction, which lasted 20 min, had a conversational character and generated lots of questions to and from the children.

Fig. 17.1 What is a scale? © Eva Brooks

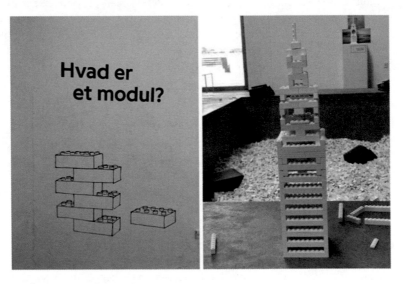

Fig. 17.2 What is a module? © Eva Brooks

After this workshop introduction, the children started by creating a high building with white LEGO (Fig. 17.2) and concretely examined the concept of scale by creating sketches, modules, and models. They worked in groups (3–4 participants in each group), which were formed beforehand by their teachers. This part of the workshop lasted 60 min.

Next phase of the workshop was to use a CAPTOR 3-D Scanner to transfer the groups' LEGO creations into Minecraft,[1] where they could experience how their creations were transformed into Minecraft-blocks as well as continuing to create digitally on their buildings.

Finally, the groups of children could experience their skyscrapers/observation towers in a 3-D VR mode using a Minecraft VR plugin for Oculus Quest 2 Head Mounted Display (HMD). In addition, the children could experience climbing to the top of some of the world's highest building and experience these kinds of designs through the HMD. This phase lasted 45 min.

The workshop ended with a joint review of the workshop, which included a re-cap of the concepts in relation to the concrete design creations.

The study was subject to research-ethical principles of transparency in the research process and quality of documentation as well as the protection of sources and individuals [8, 12]. Teachers and parents were informed about the study in writing.

All parents confirmed that their child could participate in the study by signing informed consent forms including approval to use videos and photos for scientific purposes.

The United Nations convention on the rights of the child [7] was fully respected and participating children were carefully informed before verbal consent was negotiated with them ahead of and during every workshop.

The data provided rich insights into children's generative play in the context of analogue and digital resources to solve a design problem.

In this chapter, we focus on how children engaged with the different resources (LEGO, Minecraft, and VR) and how they potentially unified or separated them in the design activity. We narrowed this focus into four main categories: problem scoping, idea generation, design monitoring, and design experience. This will be further elaborated in the next section.

17.5 Analysis

Following the research question, we focused on trying to understand how and when analogue–digital resources enabled children's generative play to develop ideas, express knowledge and communicate with others. Also, how children's ways of

---

[1] Minecraft is a sandbox digital game developed by Mojang Studios. Minecraft could be considered an online, updated version of the classic LEGO building blocks. Lego blocks are connected and assembled to create a practically limitless variety of structures. The same is true for Minecraft, except that instead of handling building blocks, users operate in a virtual world, using pixelated cubes [19].

Table 17.1 Categories for children's generative play using analogue and digital resources

Problem scoping	Idea generation	Design monitoring	Design experience
Exploring boundaries of the design problem	*Situations of imagining, brainstorming, and planning*	*Assessing goal completion of the design*	*Experiencing the outcome of the design*
• LEGO promoting collaboration • LEGO made creation possible for joint attention • LEGO offered choices of directions towards design goals	• Minecraft made creation possible for individual actions • LEGO promoted joint decision making • LEGO forced problem-solving	• Minecraft providing information about progress towards goal • LEGO making play interdependent	• Fast paced experiences in VR turned children into conversation with each other • VR made design experiences possible for both individual and joint attention

engaging in generative play, may differ considering the kind of resources they used. This resulted in four main categories: (1) Problem scoping; (2) Idea generation ; (3) Design monitoring; and (4) Design experience (Table 17.1).

Problem scoping activities included children's identification of opportunities and constraints in relation to the design problem as well as familiarising with the resources/materials and assigning roles (who should do what). This happened mainly when the children were creating with LEGO.

After being introduced to the design problem, the children should start creating their preferred high building (a skyscraper or a tower). Each of the groups (approximately four participants in each of the group) started by dividing the problem into different tasks. This most often resulted in that two of the group members had the task to sketch how the creation should look like and the other two started to gather LEGO blocks. The children were completely focused on collectively solving the design problem.

To create a stable high building different modules should be put on top of each other to form a recurring pattern (Fig. 17.3). This is illustrated in the below Excerpt 1, which is from the first author's observation notes.

17.5.1 Excerpt 1 (June 19th, 2019)

The facilitator shows the children how to put together different modules in systems and patterns; not too high or low and not too broad to keep the stability of the creation. She shows examples of models that in different ways are put together in modules, systems, and patterns. After this introduction, the children start to create their high buildings. The children are engaged and explore how they can begin building their own creations. However, some of the children do not fully understand what a model or module is and how they can be put together to form a system and pattern. They need help from their teachers and from the facilitators. The children also help each other by showing how they have solved the

Fig. 17.3 Different LEGO
modules forming a recurring
pattern © Eva Brooks

module, system, pattern problem. The teachers encourage the children to learn from each other, "Look at Martin and how he has put together the LEGO blocks." One of the children has tried several times to create modules and to put them together, which resulted in the following conversation:

David: "And how should we put together all this? It is very hard to put together modules when we shall Create a skyscraper, It can just fall apart."

Axel: "Let's do like this [showing how they can put modules on top of each other], I think this can work".

David: "Oh, yes, this way, yes, It is very smart."

While exploring the scope of the problem as well as the boundaries of the LEGO, the collaboration was an attempt to develop a shared understanding of how to create a stable high building with the LEGO blocks. This happened when David arrived at a shared understanding in his conversation with Axel (see the end of Excerpt 1). The material (LEGO) and the way the design problem was designed, invited the children to manage joint attention and to be able to collaborate in meaningful ways, i.e. to fulfil the design goal of creating a stable high building.

When asking the teachers what they thought the children learnt from the LEGO creation, one of them answered: *"That they can create with LEGO without a manual when they have learnt what scales and modules are. To explore, discover, and reflect fits so well with how children learn."*

Even though some children initially were confronted with difficulties in understanding the concepts of scale, module, system, and pattern, they were focused on the design goal and had strong preferences about how the high building should look.

This was prompted by processes of exploring and discovering, which promoted the children to choose which pathways to take to reach their preferred design goal.

Idea generation encompassed processes of brainstorming, and planning where decision making, and problem-solving were central to the children's activities. This happened both when children created with LEGO and with Minecraft. However, the processes emerged in different ways. While most actions with LEGO were collaborative, actions with Minecraft were more individual. Minecraft promoted children's awareness of their own performance towards meeting a design goal and did so by giving contingent feedback about a child's performance. In this way, Minecraft provided contingent information about the child's progress status. Also, Minecraft offered opportunities to test certain changes without creating too many problems if the change did not work out. The repair process in Minecraft was rather effortless compared to LEGO, where a mistake could cause a complete re-build of the creation. Most often one or two group members were occupied by planning for certain improvements of their creation in Mindstorm while the other two or three children walked around to see the other groups' creations.

The analysis showed that children's decision making while creating with LEGO was complex and influenced by their ability to reason about certain decisions. When the children were brainstorming and imagining about their design creations, their reasoning were characterised by information gathering. When they have studied other high LEGO creations made by other children, they reasoned about how a certain decision would influence their own creation; would it break or would it hold. This reasoning required the children to make certain choices towards their specific design goal. It also forced the children to focus on solving the problem at hand. We identified that the children through reasoning moved from an initial problem state to a goal state. This was visible when the groups jointly solved problems. In this way, LEGO required the children to engage in collaborative problem-solving. In the below Excerpt 2 from the first author's observation notes, a group of three girls have decided to create an innovative high building and they have divided the tasks between themselves.

17.5.2   Excerpt 2 (December 12th, 2018)

A group of three girls have started creating their high building (a tower). Two of them have started building special parts of the tower. The third girl picks up LEGO blocks. The girls sort out types of blocks and organise them into different piles. They have built a base in the shape of a square, which they then build up in height. The facilitator comes by: "it looks exciting what you are building - how you are building the corners of the tower. There are corners there, aren't there?" The girls inform the facilitator how they imagine their tower to be, and it appears that they have aesthetics in mind rather than the robustness of the creation. Their purpose is to build around the square base with "holes" that should be possible to see through (Fig. 17.4). One of the girls points to the corners and says that the facilitator is right: "we don't have any ties to connect the corners." The girls test and touch the different freestanding walls that are not built together. It turns out to be a little unstable when they wiggle them. They talk about stability. The girls now start connecting the corners. The girls talk about how nice their tower is going to be. The girls' attempt to make their tower more stable does not go quite as they want to, it continues to be unstable. Two of the girls try to fix it. The third girl builds an ornament as a decoration for the tower - like windmill blades. She shows this to the other two, but they are in a hurry and does not notice. The group's creation grows in height. They build carefully so that each layer builds on the other. One of the girls is very concentrated. The other two girls are building separately. The facilitator points out that a layer sticks out a little further than other layers in one of the corners. One girl immediately tries to repair and remove blocks to create balance. Now two of the girls are building exterior decorations for the tower - one with windmill blades and the other girl is building something that looks like a fan. Finally, they put the decorations on their tower, but they have problems with this and start reasoning about how to solve the problem. They get help from a group of boys and eventually succeed.

Excerpt 2 illustrates how the group of three girls had imagined their construction to be as well as how they planned to do it; it should not only become an efficient construction, but an aesthetically nice creation. They kept to their imagined creation even though they ran into construction problems. Finally, they were assisted by another group and jointly they solved the main stability problem.

Design monitoring was observed when the children assessed the entirety of their designs. Design monitoring occurred most frequently towards the completion of an

Fig. 17.4 The start of a design creation (tower) where a square base with "holes" and additional decorations should shape a high building where it was possible to see through. © Eva Brooks

activity, which happened within both the LEGO and the Minecraft activities. We identified that children's creation with Minecraft provided them with insights into their entire creation and based on this they could decide whether or which changes they wished to do. The interaction with Minecraft was mostly individual, which resulted in that the child who was monitoring the creation tried out different changes, but by the end kept to the original version. Minecraft tended to afford more individual creation by children, and it facilitated less collaboration compared to LEGO. Excerpt 3 illustrates how the joint attention and collaboration turned individual in the Minecraft activity.

17.5.3   Excerpt 3 (December 12th, 2018)

Anna: [Asking Peter] "What are you doing?"

> *Peter: "We have chosen this model." [pointing to his group's model in Minecraft]*
> *Anna: "Oh, should we do like that? Shouldn't we just play?"*
> *Peter: "No, we shouldn't."*

However, some of the children could identify that their scanned LEGO models were too compact and not transparent, which one of the children expressed: "*it is not possible to see what is behind the walls in our skyscraper. I have to change it; it must be possible to see behind the wall. I remove [Minecraft] blocks.*"

The design monitoring by means of LEGO was of another kind compared to with Minecraft. LEGO fostered children's collaborative monitoring, which generated an interdependency among the children. We noticed how the children needed each other to achieve the goals they set out with their LEGO creations. This is illustrated in Excerpt 2, where two of the girls tried to fix the stability problem and the third girl was building an exterior decoration. By the end the outcome of the one girl's creation directly linked to the outcome of the two other girls' creation; the different creations merged into a single shared creation. This resulted in an increased engagement and pride among the girls, which the following observation note illustrates: "*The girls are very proud and say: 'No other towers have windmill blades, ours is special and it's so nice'*". This kind of joint experience of the entirely creation was an important factor supporting both creative and social actions.

Even though the design problem of creating a high building was set by the facilitators, the question of how this skyscraper or tower should look like was up to the children to decide. The analysis showed that this open-endedness or ambiguity of the design problem prompted the children to customise their creations to their own desires and interests.

Design experience was prevalent primarily in the VR activity. The children's interaction with the VR environment was HMD-based and therefore only one of the group members at a time could experience their creation in the VR. However, the analysis revealed that the VR made the children engage vigorously in communication. The children communicated both verbally and non-verbally with lively gestures. While the ones outside the VR asked what the person with the HMD could see or if (s)he could see a certain feature from their creation, the person with the HMD energetically referred to what (s)he could experience. Irrespective of the children wore an HMD or not, they were jumping up and down when communicating. Consequently, the children experienced an individual as well as a joint attention. VR promoted a fast-paced experience, which meant that also the conversation was carried out at a fast pace. LEGO, on the other hand, facilitated a slower paced interaction, which turned the children collaborative in a way where they could apply their reasoning and negotiation skills. Children's design experiences with Minecraft tended to put their attentive space between themselves as an individual child and Minecraft. This shows that attentive spaces was experienced both as individual, collective, and collaborative in the holistic design experience including both LEGO, Minecraft, and VR.

17.6 Discussion

Following the theoretical framework, and based on the categorisation of the findings, we further developed design guidelines aligned with how analogue–digital resources mediated generative play to develop ideas, express knowledge and communicate with others. We identified three affording design guidelines: (1) Action potentials; (2) Characteristics of analogue and digital play; and (3) Aligning design goal, resource, and play actions. These design guidelines will frame the discussion and are unfolded in the following.

17.6.1 Action Potentials

The analysis revealed that the different kinds of affordances generated by analogue and digital technology could vary in the ways they invited or guided children's actions. That is, how the affordances offered distinct meaning potentials. We suggest that the material of LEGO invited children to play with action potentials, which in turn opened for joint attention, as well as making strategic choices to advance their design creations. Likewise, Minecraft invited children to play with action potentials directed towards monitoring of a child's individual actions while trying out different changes of the design creation. While LEGO promoted action potential for children's joint attention, Minecraft invited to apt action potentials toward individual attention by providing information about the child's own performance. This is aligned with Sutton-Smith's [30] theory suggesting that different dimensions of unpredictability promote generative play by inviting children to flexible and adaptable meaning-making processes. In this way, action potentials suggest that meaning is considered as a resource with inherent affordances, which are apt for specific purposes [20].

It was noticeable, that the design of the activity guided the children to make specific choices of how they should solve problems and reach specific design goal. In this way, the design of an activity can reduce children's action potentials and direct them towards certain goal-bound actions. In line with [20] we suggest that the affordances of LEGO were apt to the children's joint and collaborative interests. On the other hand, the aptness of Minecraft was limited and related to an individual child's interest while waiting for experiencing the design creation in VR. Thus, the function of affordances related to action potentials were distinctly related to the social environment. Based on this, we suggest that resources' action potentials can facilitate children to engage in a creative design activity, where children's actions can become both enabling and constraining depending on the design of the activity. In this way, the chapter contributes to further specify dimensions of the design of analogue–digital activities to either enable or constrain action potentials.

17.6.2   Characteristics of Analogue and Digital Resources

While action potentials afforded children's apt engagement in an analogue–digital design activity, the characteristics of analogue and digital resources contributed to the children's generative play. It was clear that generative play was important in relation to individual monitoring, decision making, and problem-solving. Additionally, generative play and characteristics of the resources were essential for collaboration. By means of direct feedback, Minecraft provided information about progress towards set goals, which enhanced children's awareness of, for instance, which choices to make. The analysis showed that analogue characteristics of the resources were critical for decision making as they provided children with different choices of direction to move towards specific goals. The decision making was prevalent within children's reasoning about what would happen if they carried on in one way or another. Characteristics promoting problem-solving were related to the framing of the design activity, which promoted children to engage in collaborative problem-solving to progress within the activity. For example, the characteristics of LEGO forced problem-solving, in particular within idea generation processes. It was evident that children through reasoning moved from an initial problem state to a goal state, which was notable when the groups jointly solved problems. Moreover, the characteristics of VR provided the children with a holistic view of their analogue and digital creations, which promoted children's joint play with meaning potentials.

Consistently with [20], our analysis demonstrates that both analogue and digital characteristics can promote generative play. In doing so, we emphasise the aspects of communication and representation. Communication was established when children through reasoning constructed a social environment including a wish or need to realise ideas, i.e. to put their understanding of the problem at hand in a material shape as a sign into an interaction with others.

To a large extent, the children became sign-makers using LEGO by articulating and expressing their ideas and solutions. Minecraft offered some of the individual children to re-create their designs, which fostered reflective processes of what they had done into potentially new representations. VR tied together the LEGO and Minecraft creations offered prospective design experiences. This adds to the literature suggesting that an activity based on a design approach, where analogue and digital resources are combined, can project children's engagement and interest into new avenues and collective meaning making. However, the analysis showed that the design of the activity must seriously consider the alignment of this combination.

17.6.3   Aligning Design Goal, Resource, and Play Actions

The aligning of an activity's design goals and resources can be considered as a structure of possible play actions, which can afford children's generative play. The analysis showed that aligning design goals and resources focused on the knowledge the children jointly and individually generated, which offered a variety of analogue and digital play actions.

We found that children engage in aligned play actions in ways that are similar to conceptualisation of design as well as play, for example by promoting exploration, imagination, and discovery. While design goals prompted the children to come up with concrete solutions for specific problems, the resources opened for exploration and imaginary pathways. In this way, the alignment nurtured play actions, which afforded questions like 'how', 'what if', and 'what could be'. The analysis illustrated well the importance of aligning design goal and resources for developing open-ended play actions.

Our findings are consistent with previous research referring to acts of play and design as the locus where new ideas get created [5, 29, 35]. Our study shows that the affordances of resources as well as design goals are relevant to understand the links between children's generative play when it is afforded by interest rather than learning [5, 20].

17.7   Conclusion

This chapter aimed to gain a better understanding of how analogue–digital resources of generative play enabled school children (6–9 years of age) to unify materially- and digitally mediated resources to develop ideas, express knowledge and communicate with others. In particular, the study examined how a structure of analogue and digital resources promoted children's generative play, that is their digital creation. The analysis focused on how the children engaged with the different resources (LEGO, Minecraft, and VR) and how they potentially unified or separated them in the design activity We will now turn to conclusive comments related to these matters.

We identified three affording design guidelines that promoted engagement in generative play actions by means of a combination of analogue and digital resources:

(1) The combination offered action potentials such as individual and collective monitoring of actions, collaboration, decision making, meaning making, and problem-solving, (2) Characteristics of analogue and digital play contributed to children's generative play, and (3) The importance of aligning design goal, resource, and play actions. As an overall conclusion, we suggest that unifying analogue and digital resources can engage children in digitally enhanced creation beneficial for the development of meaningful generative play.

We found that the alignment of an activity's design goals and resources can be considered as a structure of possible play actions inviting and requiring children to individual as well as collaborative interactions. However, this should be framed by a design perspective on generative play as it forms an interest-based connection between the analogue and digital resources.

This chapter contributes to further specify dimensions of digital creation, in particular through its multimodal social semiotic approach as it offers potential to open new avenues of play and creation with a unifying combination of analogue and digital resources. This also has implications on educational and societal instances by renewing our understanding of children's contemporary play.

Acknowledgements The authors direct their sincere thanks to all the children and teachers who participated in and contributed to the study. We also thank the Utzon Center, Aalborg, Denmark, for all support. The study was carried out within the project 'Architecture is Play', supported by Nordea-fonden, Denmark.

References

1. Abrams, S.S., Roswell, J., Merchant, G.: Virtual convergence: Exploring culture and meaning in playscapes. Teach. Coll. Rec. **119**, 120301 (2017). https://doi.org/10.1177/016146811711901208
2. Bajovic, M.: Playing and learning across concrete and digital realms: A new context for the new learners. International Journal of Play **7**(2), 199–209 (2018). https://doi.org/10.1080/21594937.2018.1496002
3. Barr, R., Linebarger, D.N.: Media exposure during infancy and early childhood. Springer (2016)
4. Brod, G.: Generative learning: Which strategies for what age? Educ. Psychol. Rev. **33**, 1295–1318 (2021). https://doi.org/10.1007/s10648-020-09571-9
5. Brooks, E.: Designing as play. In E. Brooks, S. Dau, & S. Selander (Eds.), Digital learning and collaborative practices. Lessons from inclusive and empowering participation with emerging technologies. Routledge (2021)
6. Chi, M. T. H.: Self-explaining expository texts: The dual processes of generating inference and repairing mental models. In Glaser, R. (Ed.), Advances in instructional psychology (pp. 161–238). Lawrence Erlbaum Associates (2000). https://doi.org/10.1177/1046878102238607
7. Convention on the rights of the child.: Treaty no. 27531. United Nations Treaty Series, 1577, pp. 3–178 (1989). Available at: https://treaties.un.org/doc/Treaties/1990/09/19900902%2003-14%20AM/Ch_IV_11p.pdf (Accessed 19 September 2022)
8. Danish code of conduct for research integrity.: Ministry of higher education and research, copenhagen, Denmark. Retrieved March 7th, 2021 from: https://ufm.dk/en/publications/2014/files-2014-1/the-danish-code-of-conduct-for-research-integrity.pdf (2014)

9. Edwards, S., Mantilla, A., Grieshaber, S., Nuttall, J.: Converged play characteristics for early childhood education: multi-modal, global-local, and traditional-digital. Oxf. Rev. Educ. **46**(5), 637–660 (2020). https://doi.org/10.1080/03054985.2020.1750358

10. Edwards, S., Nuttall, J., Grieshaber, S., Wood, E.: New play: A pedagogical movement for early childhood education. In: Whitebread, D., Pino-Pasternak, D. (eds.) The Sage handbook of developmental psychology and early childhood education, pp. 227–287. SAGE (2019)

11. Eisner, E. W.: The role of art and play in children's cognitive development. In E. Klugman, & S. Smilansky (Eds.), Children's play and learning: Perspectives and policy implications. Teachers' College Press (1990)

12. General data protection regulations (GDPR) (2016/679). Official Journal of European Union. Retrieved March 7th, 2021 from: https://our lex.europa.eu/legal-content/EN/TXT/PDF/?uri-CELEX:32016R0679

13. Glenn, N.M., Knight, C.J., Holt, N.L., Spence, J.C.: Meanings of play among children. Childhood **20**(2), 185–199 (2013). https://doi.org/10.1177/0907568212454751

14. Halliday, M.A.K.: Language as social semiotic: The social interpretation of language and meaning. Arnold (1978).

15. Herr-Stephenson, B., Alper, M., & Reilly, E., & Jenkins, H.: T is for transmedia: Learning through transmedia play. USC Annenberg Innovation Lab. The Joan Ganz Cooney Center at Sesame Workshop (2013)

16. Hodge, R. & Kress, G.: Social semiotics. Polity (1988)

17. Holloway, D., Green, L.: The internet of toys. Communication Research and Practice **2**(4), 506–519 (2016). https://doi.org/10.1080/22041451.2016.1266124

18. Jewitt, C.: An Introduction to Multimodality. In Jewitt, C. (Ed.), The routledge handbook of multimodal analysis, 2nd ed. Routledge (2014)

19. Karsenthi, P. T., Bugmann, J., & Gros, P. P.: Transforming education with Minecraft? Results of an exploratory study conducted with 118 elementary-school students. Canada Research Chair on Technologies and Education. CRIFPE (2017)

20. Kress, G.: Multimodality. Routledge, A social semiotic approach to contemporary communication (2010)

21. Kress, G.: Literacy in the new media age. Routledge (2003)

22. Kress, G.: Before writing: Re-thinking the paths to literacy. Routledge (1997)

23. Lupetti, M. L., Piumatti, G. & Rossetto, F.: Phygital play HRI in a new gaming scenario. In *Proceedings of the 7th International Conference on Intelligent Technologies for Interactive Entertainment (INTETAIN'15)* (pp. 17–21) (2015)

24. Ortlieb, E.T.: The pursuit of play within the curriculum. J. Instr. Psychol. **37**(3), 241–246 (2010)

25. Pellegrini, A.D.: Research and policy on children's play. Play Policy **3**(2), 131–136 (2009)

26. Pyle, A., DeLuca, C., Danniels, E.: A scoping review of research on play-based pedagogies in kindergarten education. Review of Education **5**(3), 311–351 (2017). https://doi.org/10.1002/rev3.3097

27. Schwartzman, H.B.: The anthropological study of children's play. Annu. Rev. Anthropol. **5**, 289–328 (1976)

28. Sherwood, S.A., Reifel, S.: The multiple meanings of play: Exploring preservice teachers' beliefs about a central element of early childhood education. Journal of Early Childhood Teacher Education **31**(4), 322–343 (2010). https://doi.org/10.1080/10901027.2010.524065

29. Stappers, P. J.: Doing design as a part of doing research. In Michel, R. (Ed.) *Design research now. Board of International Research in Design*, 81–91. Birkhäuser Basel. (2007) https://doi.org/10.1007/978-3-7643-8472-2_6

30. Sutton-Smith, B.: The ambiguity of play. Harvard University Press (1997)

31. Torres, P.E., Ulrich, P.I.N., Cuciat, V., Cukurova, M., Fercovic De la Presa, M.C., Luckin, R., Carr, A., Dylan, T., Durrant, A., Vines, J., Lawson, S.: A systematic review of physical-digital play technology and developmentally relevant child behaviour. International Journal of Child-Computer Interaction **30**, 100323 (2021). https://doi.org/10.1016/j.ijcci.2021.100323

32. van Leeuwen, T.: Introducing social semiotics. Routledge (2005)

33. van Leeuwen, T.: Multimodal literacy. *Viden om Literacy*, 21. Nationalt Videncenter for Læsning (National Knowledge Center for Reading) (2017)
34. Wood, E.: Conceptualizing a pedagogy of play: International perspectives from theory, policy and practice. In: Kuschener, D.S. (ed.) From children to red hatters: Diverse images and issues of play, pp. 166–190. University Press of America (2009)
35. Zimmerman, J., Forlizzi, J. & Evenson, S.: Research through design as a method for interaction design research in HCI. In *Proceeding of the SIGCHI conferences on Human factors in computing systems*, San Jose, California, USA, 493–501. ACM. (2007). https://doi.org/10. 1145/1240624.1240704

Chapter 18
Empowering Creativity and Feedback: Lessons Learned from the Development of an App to Assist Game-Based Learning Activities

Maria Helena Reis, Ana Margarida Almeida, and Catarina Lelis

Abstract The chapter aims to underline the importance of considering 'Empowerment of creativity and feedback', one of the dimensions in Chou's Octalysis framework, when developing digital learning game-based activities. A proposal of an app which aims at cataloguing and evaluating digital educational games is described, emphasising the design process used during its creation. The chapter is organised in four parts: the first underlies the theme and addresses the benefits of game-based learning in teaching and the obstacles encountered by teachers when selecting games for using during classes; the second seeks to give an overview of the Octalysis framework and highlights its dimension 'Empowerment of creativity and feedback'; in the third part, the steps of the Communicative Design Paradigm are explored, which allowed, through a participatory design approach, the collection of the functional requirements for the proposed model; finally, it is clarified how 'Empowerment of creativity and feedback' was used to scaffold the proposal of an app for cataloguing and evaluating digital educational games, highlighting the role that such an app can have in surpassing teachers' barriers when selecting educational games.

M. H. Reis · A. M. Almeida (✉) · C. Lelis
Department of Communication and Art/DigiMedia, University of Aveiro, Aveiro, Portugal
e-mail: marga@ua.pt

M. H. Reis
e-mail: hsreis@ua.pt

C. Lelis
e-mail: lelis@ua.pt

491

18.1   Introduction

One of the advantages, for the various actors in the educational process (teachers and students), of using digital educational games in the classroom is related to a good receptiveness from younger people, as they often already bring game playing practices from home. Playing games is quite attractive for most teenagers and youngsters [7]. Incorporating games in the educational context seems to have a positive impact on the motivation, learning, behaviour, and attention of students [7, 40]. When articulated with the most traditional methodologies, educational games seem to favour a more effective transmission and produce excellent results in essential learning [28]. From another perspective, Serious Games allow players to connect with the situation, which may lead them to assume a position of self-criticism. On the other hand, games may also bring a relational approach, favouring the relationships between peers (students) and between student and teacher [25].

However, once the advantages of using Game-Based Learning (GBL) as a teaching strategy are perceived, it is important to improve the conditions given to teachers in the game selection process. There are some obstacles which can lead to demotivation when the teacher is considering the use of educational games in the classroom: peers not recognising the adopted strategy, the full compliance of school programs prescribed by the government or education-related authorities, which affects the necessary time to prepare classes in innovative ways, the process of choosing the game that best fits both the educational programme and the intended users (teacher and students), the technical and technological conditions, among others. For teachers, it is complex to quickly identify how a particular game may be relevant to a curriculum theme, as well as the quality and accuracy of existing content within the game [44].

Thus, the app proposed, and presented here, is directed to teachers who intend to use a didactic game in the context of their classroom. It aims at promoting a more conscious use of GBL so that it can reproduce in students a greater motivation towards the taught contents and learning. When planning the lesson, the teacher, employing the output resulting from the app we shall propose, can make a better selection of the game, thus ensuring the fulfilment of the outlined objectives, and reducing some constraints that come from the use of an inappropriate or misappropriated resource for the class.

Therefore, the creation of a digital application (henceforth "app") to support the cataloguing and evaluation of educational games—based on practical and real cases and supported by evaluation models recognised by the scientific community—may transform the process of choosing the game into a more robust one. When designing the app, it is important to keep the end-user involved, since only then can the application respond appropriately to their needs and interests throughout time [9]. Hence, developing a feasible and innovative application that can solve a problem in the educational community became a reason of interest and motivation to the research presented in this chapter.

A multidisciplinary team of experts was created aiming at enriching the domain under study and bringing their empirical knowledge to the design of the model. The experts, organised in groups, allowed the identification of requirements for the development of the wireframes and for motivating future users-teachers to use the digital application, considering its genesis—the gamification principles of the Octalysis framework [9]. Thus, it is undeniable that there is a distinction between GBL and gamification. While the former (GBL) refers to a game context, the latter (Gamification) allows the implementation of strategies and activities that use tools associated with the game, but in a 'non-game' environment. The main objective of gamification is to support and motivate users to perform a set of tasks [15].

18.2 Octalysis Framework

The theoretical-methodological Octalysis framework proposed by Chou [9] reinforces that gamification goes beyond points, medals and rankings. This framework helps to unravel the motivational impulses obtained through gamified activities. It can be used to design gamified applications and is supported in eight core drives of human motivation. It reports to the concept of human-centred gamification where the main goal is to provide the user with the best possible experiences. The human dynamics that prompt the user and lead them to continue using an application are not always the same. Some are motivation-based but others are grounded on manipulation and obsession. This understanding of human dynamics may be converted into changes and bring transformations in the current practices, leading to enriched experiences [30]. Chou's [9] tool is developed into eight dimensions (Fig. 18.1):

1. Epic meaning and calling
2. Development and accomplishment
3. Empowerment of creativity and feedback
4. Social influence and relatedness
5. Unpredictability and curiosity
6. Loss and avoidance
7. Scarcity and impatience
8. Ownership and possession.

The importance given to each segment will expand or contract its side of the octagon [9, 17].

Epic meaning and calling is about assigning tasks or functions to the user so that they feel part of a mission where they feel valued. This dynamic involves the user, making them think they are participating in something great or that they have been chosen to accomplish something.

Development and accomplishment implies winning, awarding rewards through challenges that keep users engaged. It is an internal dynamic to advance and develop skills, and succeed in the face of challenges. It is the easiest to conceive and often resorts to badges, points and leaderboards.

Empowerment of creativity and feedback refers to the assignment of skills that may increase the user's personal fulfilment. Users are involved in the creative process

Fig. 18.1 Octalysis
framework Chou (Adapted
from Chou [9])

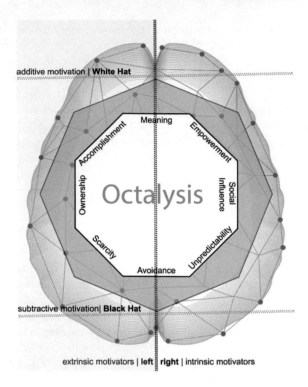

through which they must solve problems repetitively and by different combinations. In addition to expressing their creativity, they also need to receive feedback from the application regarding said creativity.

Social influence and relatedness considers the interaction with other people (or simply knowing what they like, think, or feel), since it may engage the user. This dynamic includes all the social aspects that motivate users, such as mentoring, cooperation, as well as competition and envy. When a friend reaches a certain level and particular perks, the user tends to be motivated to achieve the same. There is also an interest in belonging to a group. On the other hand, contacting with something that appeals to past memories, increases the desire to acquire/achieve it, for example a song, voucher for an event.

Unpredictability and curiosity implies that not everything should be controlled and/or regulated, since the discovery piques the user's interest. This dynamic is associated with wanting to know what will happen in the next moment. The unknown makes the user think about it. It is one of the factors that leads to addiction and is associated with sweepstakes.

Loss and avoidance presupposes events that should be avoided because they cause unpleasant situations and may carry negative consequences. Missed opportunities are part of this dynamic. In this way, in order not to miss the opportunities, users act immediately.

Scarcity and impatience entails the appreciation of a particular activity by the user because it is associated with a good of limited access. Thus, this dynamic refers to something that is desired and that cannot be obtained. The user may have access at any given time, which will lead them to return to the application.

Ownership and possession involves activities that enable the user to acquire something. Users are motivated because they feel they have something. When you feel that, the tendency is to want to improve what you have and get even more. It is the dynamic that provokes the desire to accumulate wealth. In digital applications, it refers to virtual goods. When a user spends time customising their avatar or profile, they get the feeling of owning a certain property. This dynamic is also fun because it leads to the collection of badges or to the completion of jigsaw-like activities [9].

The nature of the different dimensions differs due to their approach to the axis:

- *upper horizontal*—White Hat (additive motivation), when it comes to rewards or pleasant feelings; if something motivates the user and they are allowed to use their creativity, if the user feels capable due to their skills or due to feeling important, powerful.
- *lower horizontal*—Black Hat (subtractive motivation), when we report for feelings of fear, uncertainties or anxiety [9, 17]; if the user performs without knowing what will happen, they are afraid of failing or afraid of never being able to achieve something. This may result in a disgruntled and frustrated user that will abandon the task as soon as they have the opportunity; these dynamics are not meant to be bad since its inherent motivational factors can be used to obtain positive and productive results.
- *right vertical*—the right side of the brain, usually associated with creativity and socialisation, calls for intrinsic motivation (inner strength—emotional side); the activity itself rewards the individual, which means that users find the motivation in themselves.
- *left vertical*—the left side of the brain, usually related to logic, calculation, possession and analytical thinking, concerns extrinsic motivation and the need to achieve a goal, an object (rational side) [9].

18.2.1 Empowerment of Creativity and Feedback

Empowerment of Creativity and Feedback is expressed when users are engaged in a creative process where they repeatedly figure new things out and try different combinations. Thus, its main characteristic lies in the ability to continuously involve the user. Through this dimension, it is possible that the user expresses their creativity, experiences some kind of fun and satisfaction, potentially turning imagination and ideations into practice. The user also receives the results of such activities and reviews their own efforts.

In the Octalysis framework, this dimension is found in the upper right corner. It means that it is located in the White Hat, where emotions are positive in the long-term, as well as on the right side of the brain, where intrinsic motivation is emphasised.

It is the only dimension that refers to positive emotion and comes from invisible motivation. In other words, it promotes feelings of satisfaction, control, certainty and firmness [9].

"Motivation refers to processes that instigate and sustain goal-directed activities" ([37], p. 1), whilst intrinsic motivation, in a professional setting such as the one we focus on, means that individuals see work as a reward in itself, being confident in making decisions for themselves in regard to how they want to act and what to choose so they can achieve their goals [8]. The subject enjoys the task itself, rather than for the expectation of achieving some external goal or reward.

Several authors have been demonstrating the existence of a strong positive relationship between intrinsic motivation and creativity [2, 45]. Intrinsic motivation includes both cognitive and affective components, in which autonomy and competency are factors that determine cognitive intrinsic motivation [12], whilst several affective components, such as interest, excitement and the optimal experience associated to deep task involvement [11], and happiness, surprise, and fun [33], are also determinant to this kind of impetus.

As a consequence, when individuals are sufficiently engaged, interested, comfortable and confident with the performing of their tasks, the more active they are in the sense of developing different and effective approaches and of solving the constraints the task may impose, hence resulting in increased creative performance [18]. In fact, according to Guo-liang [19], intrinsic motivation is of greater relevance in innovation performance, with positive impact on both idea generation and development. Ensuring that there is more than one way for a task to be completed enables users to continue to be creative, as many possibilities for action are generated and motivation is nurtured. Additionally, the feedbacks obtained through experience are expected to retain users for a longer time. Of course, it is predicted that creative, engaged, empowered, and excited teachers to be the greatest conductors of equally creative, engaged, empowered, and excited students.

Hence, when Empowerment of Creativity and Feedback is designed, it is important to create a configuration in which the user is prompted with a goal, as well as with a variety of tools and actions that can lead to the achievement of said goal. To make Empowerment of Creativity and Feedback more evident, it is possible to include some game techniques to involve users [9]:

- *boosters*—at a given time and for a provisional state, the user receives privileges; boosters allow that a certain level is achieved more quickly and leads users to develop new strategies;
- *milestone unlocks*—a level is reached to gain something, and new skills are tested; feedback is received indicating points that can be obtained by performing the task; if the points are not reached, there will always be new possibilities and the user can improve their performance and try again;
- *choice perception*—indicates that having options to choose from might be better than not having them; even if it is not the most suitable technique, as it can lead to user frustration when faced with meaningless options, it is most commonly used by designers (as opposed to *meaningful choices*);

- and finally, *meaningful choices*—options that are available to the user so that he/
  she can achieve the same goal, as much as other users, whilst creating their own
  personalised path; the user is provided with meaningful and uncompromising
  choices that can be revealed as personal preferences, allowing the adjustment of
  their strategies; it is about giving the user a sense of freedom and control.

This chapter explores in detail how *Empowerment of Creativity and Feedback* can
be found in the creative process of designing an app for cataloguing and evaluating
educational games. This approach was particularly inspiring to envision the development of the app as Empowerment of Creativity and Feedback awakens the desire
to create and empowers the users in directing their own way of "playing", giving
them the possibility to configure the game environment, to some extent according to
their imagination.

18.2.1.1 Empowering Creativity

Creativity is regarded as an essential target of the educational system [10] and is
one of the competencies for individuals in the twenty-first century. The Partnership
for 21st Century Skills[1] is a consortium that gathers some of the biggest and most
relevant corporations and brands in the world (i.e., Apple, Blackboard, Dell, Ford,
Lego, Microsoft, Pearson, Sesame Street, just to name a few) with the support of non-
governmental entities such as the OCDE. They propose a framework for the development of competencies that are seen as relevant in the current times, being Creativity
one of them. Despite the increased awareness of the relevance of creativity and of its
well-known benefits for individuals, the development of creativity as a competence
in education has been far from being a priority, both in regard to enhancing students'
creativity, and in what relates to those facilitating the conditions for such creativity
to be enhanced: the teachers.

This frames the empowerment of creativity in a very specific context: the one of
work. Several authors [3, 16] have been extensively demonstrating that creativity
in the context of a professional activity (which may naturally include Education) is
immensely reliant on the worker's competencies, but also, on the work characteristics
and conditions, and contextual/environmental factors that stimulate creative thinking,
in which technology is included.

According to Shabalina et al. [38] the discussion around creativity in digital
environments makes people think first in digital games, since it is fairly accepted
that games stimulate creativity, and that the success of a player is grounded on
their creative attitude—because they are expected to make decisions that open up
the opportunity to change the course of game-related narratives, and because they
can personalise the game environment [31]. It is worth noting that the output to

---

[1] http://www.21stcenturyskills.org/.

be proposed in this chapter, aiming at supporting teachers in implementing game-based learning activities in their classes, is developed following the principles of gamification.

Gamified approaches reorganise work practices in the sense that they induce engagement, sense of playfulness and desired outcomes [26], being creativity one of them. According to Amabile and Pratt [3] creativity can be seen as an outcome when considered as a work method or process, but also when looking into the potential of the ideas and solutions produced by an individual (or a team, for that matter). The authors also defend that employee motivation is the greatest driver for creative performance, whilst [21] suggest that work gamification improves motivation by making profession-related tasks more meaningful, pleasurable and, simultaneously, challenging [6], which seems to fully align to Csikszentmihalyi's [11] optimal experience, grounded on intrinsic motivation. Moreover, Ikhide et al. [23] state that, in the context of games, a creative attitude is associated with independence, curiosity, interest in the task, further suggesting that, as per Mumford [29], accompanying these intrinsic calls with rewards and feedback is an important motivating strategy.

18.2.1.2   Provide Feedback

Many were the progresses of the Feedback dimension in application design; currently iterative interactions with the user are far from the only simple text contact line that indicated indicates "contact request sent" [14]. Feedback is one of the most valued elements in gamification. It is referred to as one of the most important elements by the fact of directing, giving return and enabling a systematic view of the situation in which the user is positioned [46]. This dimension helps decision-making and allows the user to define which strategy to adopt. On the other hand, feedback contributes to desired behaviour and can be the key to effective motivation [5]. The user receives bonuses in real time, is driven, and instantly gets a reaction of their actions [46].

Kapp [24] also announces that, in the course of involving users, the challenges, defined by rules, must have interactivity and feedback and that they must result in quantifiable responses, in which they cause emotional reactions to emerge. In this case, the feedback is informative because the return is on the actions, either correct or incorrect, without indicating the path to correction.

All the elements underlying gamified systems—points, leaderboards, badges—have a feedback function [36], as they inform the participant of the actions won and lost.

In addition to the Octalysis framework, other models approach the feedback mechanism as an important dimension that does not concern the user's intrinsic motivation within the gamified design process. We highlight the Motivational Design Lenses by Deterding [13], that addresses a feedback dimension so-called Feedback Lenses. This emerges as a dimension that depends on the context, in which affordances are used, aimed at informing users of their progress and the next available actions or challenges. The dimension presents itself in:

- *Clear and Immediate Feedback*—the systems inform users immediately of any changes in an easy and graspable way.
- *Actionable Feedback*—the user is informed about how next actions and improvements are to be available.
- *Graspable Progress*—the system informs the user of where they are and which way to go to progress [42].

In a gamified system, instant feedback increases user motivation, but with targeted and appropriate feedback that allows for an overview of performance and condition, the participant recognises which way to go for their progression and can make decisions whilst defining strategies [5].

18.3   A Research Roadmap Based on a Communicative Design Paradigm

Considering this emphasis on *Empowerment of creativity and feedback*, a qualitative methodological approach, using participatory design, targeting the interpretation and analysis of the contributions of a group of eight experts, was used to gather and refine the functional requirements of an app to be proposed. A focus group was composed of GBL and Gamification experts (all secondary school teachers from a school located in central Portugal) from different subjects (such as Geography, Sociology and STEAM—Science, Technology, Engineering, Arts and Mathematics).

Combining a team with different backgrounds increased the chances of obtaining a more distinctive digital app, since, as proposed by Design Thinking practices, each element examined the problem in a different perspective (IDEO 2009), neither imposing nor feeling constrained by any specific specialism or perceived expertise [4]. It is important to establish a dialogue between those who design the app and those who will use the digital application [27]. This participatory design approach increases the chance of acceptance of the app at the end of the process [39].

Considering the research objectives—which are (1) to propose a digital application that supports game evaluation and selection, (2) to determine the components and dimensions to be considered for the specification of the app, (3) to prototype and validate the app, (4) to assess the suitability of the Octalysis gamification framework to represent the dimensions and components of the app—the methodological design of this study is strongly influenced by the approaches of Educational Design Research [1].

The functional requirements of the app were gathered according to the Communicative Design Paradigm (CDP) [20], aiming to identify the indicators of the proposed app, using the Octalysis gamification framework. In the context of CDP, consensus prevails among the professionals involved (experts) throughout the process of developing an educational solution [43].

In this study, throughout the six stages of CDP (Fig. 18.2), the dynamics were established among the experts of the focus group (consisting of the eight secondary school teachers) and a Portuguese publishing house, well-known for developing educational contents. The latter agent makes it possible to have a business vision, linked to education, and brings into the discussion its vast experience in the production of digital educational resources.

Moreover, in Educational Design Research [1] it is common to use multiple sources of data, particularly considering the need to enquire end-users and experts. Grounded on the opinions of the experts, the focus group was used to gather the requirements for the digital app to be developed, considering its main objective: to be used as a gamified resource. This exchange of ideas among all enabled the consensual improvement of the proposal under discussion, as detailed below, following the procedures of the CDP framework.

On the 1st stage—Ideas Platform—the focus group's elements participated in an online explanatory session (Zoom™ meeting), where the problem was defined, its restrictions and fundamental assumptions were identified as the existing digital solutions were demonstrated.

Fig. 18.2 CDP framework Gustafson et al. (Adapted from Gustafson et al. [20])

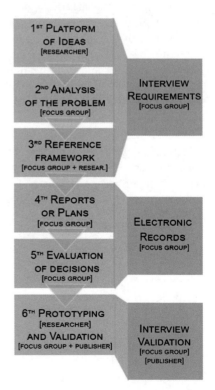

Fig. 18.3 Excerpt of one of
the produced mind maps

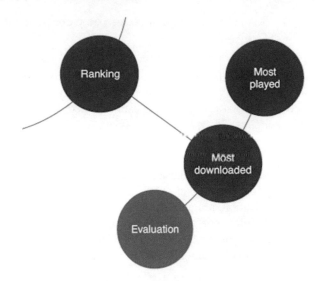

On the 2nd and 3rd stage—Analysis of the Problem and Reference Framework—
the experts were encouraged to share their views in a forum of the institution's
Learning Management System (LMS) platform, and, working in pairs, were invited
to create mind maps (Fig. 18.3) representing the app requirements, where the development of different solution scenarios was encouraged. They considered that these
requirements should be included in the application (4th stage—Development of
Reports or Plans).

After sharing the electronic records of the mind maps in the LMS platform, moments of discussion and validation took place among the experts (5th
stage—Evaluation of Decisions).

Finally, the 6th stage—Prototyping and Validation—involved not only the
expected prototyping of the proposed app, but also the creation and discussion of
new ideas, until a coherent proposal was obtained. The validation of the app was
performed by the teachers in the focus group and by the Portuguese publishing
house represented by its Director of Digital Education. The focus group and interview
survey script constructed for this purpose was segmented into three blocks–"Thinking
Aloud" [22] which enabled free and individual manipulation and collection of critical
comments (aloud); "Cognitive Walkthrough" [32] which consisted of the execution
of manipulation tasks in a guided and shared way; and "Retrospective Think Aloud"
[41] which fostered discussion towards the improvement of the app design.

18.4   An App for Supporting the Selection of Educational Games

The analysis of the requirements for the design of the digital application was based on the mind maps and interactions created during the focus group with the experts on the LMS platform. From the data obtained, the indicators were grouped into six main domains [34, 35]:

1. Domain Game

    - subject area—science, mathematics, technologies, languages, visual arts, social sciences and humanities, sport, citizenship and development, sexual education, religion.
    - specifications—synopsis, release date, degree of interactivity, free of charge, usage tips, mobile or desktop format, tutorial, rules, operating system type, technical requirements such as memory/disk space, accessibility and related games.
    - level of education of the target audience—pre-school, primary school and secondary school.
    - category—quiz, strategy, memory/reasoning, simulator, board, family, jigsaw puzzles, mime, crossword, motricity, colouring and letter soup.
    - number of players—single or multiplayer.
    - target audience age group—3–4 years, 5–6 years—two-year grouped levels up to 18 years.
    - sorting—most played, newest, most popular, alphabetically A-Z and Z-A.
    - search—by game name, subject area, teaching cycle and release date.
    - evaluation survey—according to the dimensions, motivation, user experience and learning. At any given time, the user can choose whether to change the questions in the game evaluation questionnaire, in the learning component.

2. Domain News panel

    - News/Updates with the possibility for the user to insert comments—launch of new games, updates of existing games, sharing of experiences by users, key users, promotions, events related to the games, websites of interest, scientific publications in the GBL area.

3. Domain Interaction

    - public/private chat and discussion forum.
    - pairing—suggestion of users with a similar profile for Private Chat.
    - notification.

4. Domain User

    - identification—name, email, username, avatar construction / profile picture, age and gender.

- interests—areas of interest, subject group, education levels, favourite games, followers and following.
- Domain Status *Star Club*—beginner/junior/senior/expert user rating, resulting from their interaction on the platform, badges awarded as, for example, the user of the month. The user who publishes the most, the user who shares the most, prizes such as discounts on the purchase of games, or tickets for events on games, unlocking content, custom settings of the application.

5. Domain Ranking (Games)

- scale—global, motivation, user experience, learning. It is worth mentioning that the four scores mentioned result from the evaluation of the game made by the students after they use it in the classroom. In the proposed model, the student has an active role in the evaluation of the game. Thus, after the game session, the teacher invites their students to fill out a questionnaire to evaluate the game according to the three indicators of motivation/UX/acquired learning. The player's vision is thus captured and valued. This opinion, once scored, leads to the creation of several rankings for the existing games in the model.
- overall scale—not recommended/reasonable/optimal/good/excellent based on the overall assessment of the three dimensions under study.
- interaction scale—most rated game, most commented game.
- scale of emotion regarding the effectiveness of the game applied in the classroom—went well/badly, liked/did not like.

6. Domain Language

- Portuguese, English, Spanish, French and German.

After analysing the requirements collected, each domain was graphically represented in 47 wireframes.

Activities that awaken the user are good triggers for intrinsic motivation, either by appealing to novelty, challenge, or aesthetic value. By assigning differentiated statuses (Silver, Gold and Diamond) to the user, through the degree of interaction, the teacher receives different privileges and, among others, can configure the application in a personalised way, building a digital environment with which he/she best identifies, creating a feeling of belonging and control. The possibility of choosing and perceiving that there is room for self-determination contributes to intrinsic motivation because it promotes in the user the feeling of autonomy. This aspect is mirrored in the personalised definition of the questions of the learning indicator in the evaluation of the game, which will serve as a stimulus to the imagination and creativity of the teacher. There is a choice between adopting the predefined questionnaire of the application or the personalised mode with the possibility of being creative in the definition of questions and in the interactivity generated with the respondents.

Immediate feedback, assumed as a reward and as a privileged communication channel, leads to feelings that facilitate intrinsic motivation. By receiving immediate feedback from the app, engagement and productivity can be unlocked [9]. In the proposed model, feedback is boosted in several domains, whether through notifications, discussions in the forum or real-time communication through a chat (Fig. 18.4). This connection with others plays a fundamental role in maintaining and developing intrinsic motivation. Additionally, the user is invited to comment on the publications made by the other teachers and gets a sense of the preferences (through the Like feature) allocated to both games and news (Fig. 18.5).

In their personal profile, users receive feedback on their position as members, as well as the benefits they may achieve, namely the possibility of customisation and configuration of the app (Fig. 18.6).

Through these interaction methods aiming the users' empowerment, they understand how they can reach the next level of privileges and move on from Silver

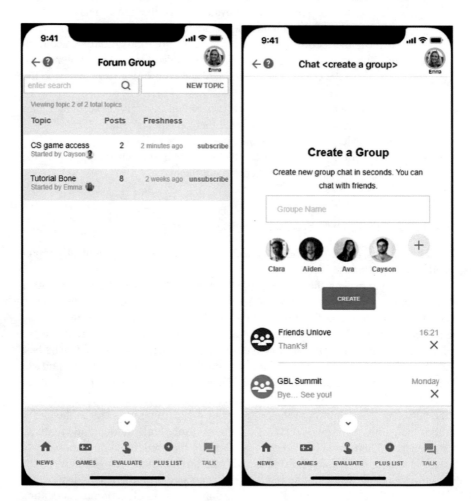

Fig. 18.4 Domain interaction—discussion forum and public chat

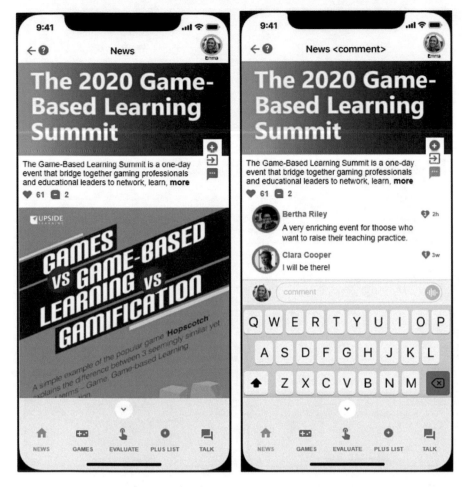

Fig. 18.5 Domain news panel—comments and likes

membership (the initial level) to Gold and, eventually, to Diamond (*milestone unlocks*).

The navigation paradigms used on the proposed app favour different strategies to obtain points or perform an action, meaning that there is no unique, right way for a task to be successfully performed. This diversity promotes empowerment since the user can act independently according to his/her preferences. To obtain member points (those that give access to benefits), the user has a range of possibilities, clearly explicit in their profile. The options are divided into sharing news on social networks, inviting friends, adding events and games to their favourites, activating notifications, and asking students to evaluate a game (Table 18.1). The user thus exercises his/her power of free will, in a personalised way (*meaningful choices*) and without realising of being driven.

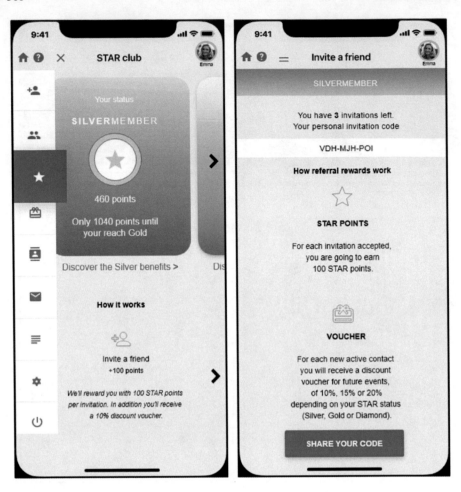

Fig. 18.6 Domain User (Star Club)—status (badge, points and benefits)

Through the benefits achieved and the feedback obtained, users take control (real-time control), which gives them access to the application settings, where colours and the environment image can be changed, or even where they can configure notification alerts, choose events for the app's vouchers, receive notifications from users with similar profiles, and choose the friends they want to invite to join as members.

Table 18.1 Domain User (Star Club)—points and privileges for status member Silver, Gold and Diamond

	Silver	Gold	Diamond
Unlocked status	300–1499 points	1500–2499 points	>2500 points
Invite a friend	+100 points		
Invitations	3 friends	5 friends	10 friends
Benefits	Colour Layout—set up the Games4Class APP with the colours you like		Colour Layout—set up the Games4Class APP with the colours or background you like
Voucher	10% discount	15% discount	20% discount
Priority news alert	Email alerts of new games and events		
Sign up for newsletter	+100 points for your first subscription		
Recommended user	–		With the same interests
Share in social media	+50 points		
Adds in plus list (favourite games and events)	+10 points		
Evaluate game	+50 points		

18.5 Final Considerations

According to Chou [9], once Empowerment of Creativity and Feedback has been achieved, it is easier to reach the dimensions *Development and Accomplishment*, *Social influence and relatedness* and *Unpredictability and Curiosity* of the Octalysis framework, which is why this is such a worthy and significant dimension.

In the Octalysis framework, *Development and Accomplishment* is indispensable for making progress, developing skills, achieving mastery, and ultimately overcoming challenges. Getting a badge without overcoming a challenge is not meaningful at all. A challenging and meaningful task is a justification for a badge or award, and it is within this dimension that most of the points, badges, leader boards can be found. The users are driven by a sense of growth and a need for accomplishment of targeted goals. *Development and Accomplishment* is an enthusiasm generator and leads to a commitment towards learning new skills. At a given moment, the user will be able to enjoy differentiating privileges and perform actions that set him apart from the remaining users. We realize that, in the prototype and in the proposed app, elements such as points, leader boards and badges act as informational feedback. Given the typology of feedback, these features focus on the result of a given task and on their own correction, but are also centred on the individual, for self-regulation.

On the other hand, *Social influence and relatedness* incorporate all the social elements that motivate people, including: mentorship, social acceptance, social feedback, companionship, and even competition. When we realise that someone in our circle of relationships is competent in some area, he or she becomes a model and induces us to behave equally. The same happens with events or products when these provoke feelings of nostalgia and increase the possibility of acquisition or replication. Also, communication between app users connects teachers and creates a community that enhances the concepts of this dimension. Assigning status to members refers to template actions that can be replicated by other users.

Finally, the dimension *Unpredictability and Curiosity* refers to a constant involvement because it is not known what will happen next. It represents the main force behind the user's passion with experiences that are uncertain and involve chance. In this situation, the brain is alert and there is no attention left to the unexpected. In the app, this dimension can be increased with the launch of prizes with limited reach through sweepstakes.

In the proposed model, the game techniques indicated for *Empowerment of Creativity and Feedback* are represented, however empowerment can be further stimulated through the use of boosters, and the feedback associated with creative processes can be enhanced.

Although there are external rewards in the app, what really matters to the teacher, and retains them as users in the app, is the process of formulating and optimising strategies in the evaluation of games. For example, upon receiving the answers to the questionnaire applied to students, the teacher can review the question design, outline a new strategy and innovate next time they need to apply to questionnaire again.

Assessing the cognitive and affective components of intrinsic motivation in teachers resorting to these kinds of resources might be an interesting future avenue of research. Equally, monitoring the development of self-perceived creative skills would be a relevant add-on, since teachers are necessarily expected to develop their creativity in order to better support the development of creative skills in their students.

References

1. van den Akker, J., Bannan, B., Kelly, A.E., Nieveen, N., Plomp, T.: Educational Design Research—Part A: An introduction. SLO—Netherlands Institute for cuurriculum development, Enshede (2013)
2. Amabile, T.M., Hill, K.G., Hennessey, B.A., Tighe, E.: The work preference inventory: assessing intrinsic and extrinsic motivational orientations. J. Pers. Soc. Psychol. **66**(5), 950–967 (1994)
3. Amabile, T.M., Pratt, M.G.: The dynamic componential model of creativity and innovation in organizations: Making progress, making meaning. Res. Organ. Behav. **36**, 157–183 (2016). https://doi.org/10.1016/j.riob.2016.10.001
4. Brown, T.: Change by Design: How Design Thinking Transforms Organizations and Inspires Innovation. HarperBusiness, New York, USA (2009)
5. Burgos, D., van Nimwegen, C., van Oostendorp, H., Koper, R.: Game-based learning and the role of feedback: a case study. Adv. Technol. Learn. **4**, 188–193 (2007)
6. Cardador, M.T., Northcraft, G.B., Whicker, J.: A theory of work gamification: something old, something new, something borrowed, something cool? Hum. Resour. Manag. Rev. **27**, 353–365 (2017). https://doi.org/10.1016/j.hrmr.2016.09.014

7. Carvalho, A., Zagalo, N., Araujo, I.: From games played by secondary students to a gamification framework. Soc. Inf. Technol. Teach. Educ. Int. Conf. **2015**, 737–744 (2015)
8. Chen, C.-A., Chen, D.-Y., Xu, C.: Applying self-determination theory to understand public employee's motivation for a public service career: an East Asian case (Taiwan). Public Perform. Manag. Rev. **41**, 365–389 (2018). https://doi.org/10.1080/15309576.2018.1431135
9. Chou, Y.: Actionable Gamification—Beyond Points, Badges and Leaderboards. Octalysis Media, CA (2016)
10. Craft, A.: Creativity in schools: tensions and dilemmas. Creat. Sch. Tens. Dilemmas. (2005). https://doi.org/10.4324/9780203335/965
11. Csikszentmihalyi, M.. Flow: The Psychology of Optimal Experience (1990)
12. Deci, E.L., Ryan, R.M.: The general causality orientations scale: self-determination in personality. J. Res. Pers. **19**, 109–134 (1985). https://doi.org/10.1016/0092-6566(85)90023-6
13. Deterding, S.: Situated motivational affordances of game elements : a conceptual model (2011)
14. Deterding, S.: The lens of intrinsic skill atoms: a method for gameful design. Hum. Comput. Interact. **30**, 294–335 (2015). https://doi.org/10.1080/07370024.2014.993471
15. Deterding, S., Dixon, D., Khaled, R., Nacke, L.: From game design elements to gamefulness. In: Proceedings of the 15th Internayional Academic MindTrek Conference Envisioning Futur. Media Environ.—MindTrek '11(9) (2011). https://doi.org/10.1145/2181037.2181040
16. Engen, M., Magnusson, P.: Exploring the role of front-line employees as innovators. Serv. Ind. J. **35** (2015). https://doi.org/10.1080/02642069.2015.1003370
17. Ferreira, M., Miranda, G., Morgado, L.: Análise das funcionalidades de gamificação nos ambientes de aprendizagem Classcraft e Moodle à luz da framework Octalysis, http://hdl.handle.net/10400.2/7294 (2018)
18. Gong, Y., Huang, J.-C., Farh, J.-L.: Employee learning orientation, transformational leadership, and employee creativity: the mediating role of employee creative self-efficacy. Acad. Manag. J. **52**, 765–778 (2009). https://doi.org/10.5465/AMJ.2009.43670890
19. Guo-liang, Z.: The relationship between work motivation and individual innovation behavior. Soft Sci. (2007)
20. Gustafson, K., Visscher-Voerman, I., Plomp, T.: Educational design and development: an overview of paradigms. In: Springer—Science+Business Media, B.V. (ed.) Design Approaches and Tools in Education and Training, pp. 15–29. Kluwer Academic Publishers (1999)
21. Hammedi, W., Leclercq, T., Poncin, I., Alkire (Née Nasr), L.: Uncovering the dark side of gamification at work: impacts on engagement and well-being. J. Bus. Res. **122**, 256–269 (2021). https://doi.org/10.1016/j.jbusres.2020.08.032
22. Hartson, R., Pardha S.P.: The UX Book (2nd edn.). Morgan Kaufmann (2019)
23. Ikhide, J.E., Timur, A.T., Ogunmokun, O.A.: The potential and constraint of work gamification for employees' creative performance. Serv. Ind. J. **42**, 360–382 (2022). https://doi.org/10.1080/02642069.2022.2045278
24. Kapp, K.M.: The Gamification of Learning and Instruction: Game-based Methods and Strategies for Training and Education. Wiley (2012)
25. Kirriemuir, J., Mcfarlane, A.: Literature review in games and learning. Futurelab. **3**, 39 (2004). https://doi.org/10.1111/j.1541-0072.1974.tb01308.x
26. Koivisto, J., Hamari, J.: The rise of the motivational information systems: a review of gamification research. Int. J. Inf. Manage. **45**, 210 (2019). https://doi.org/10.1016/j.ijinfomgt.2018.10.013
27. Krueger, R.A., Casey, M.A.: Focus Groups: A Practical Guide for Applied Research. Sage Publication - International Educational and Professional Publisher, London (2015)
28. Lisenbee, P.S., Ford, C.M.: Engaging students in traditional and digital storytelling to make connections between pedagogy and children's experiences. Early Child. Educ. J. **46**, 129–139 (2018). https://doi.org/10.1007/s10643-017-0846-x
29. Mumford, M.D.: Managing creative people: strategies and tactics for innovation. Hum. Resour. Manag. Rev. **10**, 313–351 (2000). https://doi.org/10.1016/S1053-4822(99)00043-1
30. Oliveira, S., Cruz, M.: The gamification octalysis framework within the primary english teaching process: the quest for a transformative classroom. Rev. Lusofona Educ. **41**, 63–82 (2018). https://doi.org/10.24140/issn.1645-7250.rle41.04

31. Ott, M., De Gloria, A., Arnab, S., Bellotti, F., Kiili, K., Freitas, S., Berta, R.: Designing serious games for education: from pedagogical principles to game mechanisms. In: Proceedings of the European Conference on Games-based Learning (2011)
32. Polson, P.G., Lewis, C., Rieman, J., Wharton, C.: Cognitive walkthroughs: a method for theory-based evaluation of user interfaces. Int. J. Man. Mach. Stud. **36**, 741–773 (1992). https://doi.org/10.1016/0020-7373(92)90039-N
33. Reeve, J., Cole, S.G., Olson, B.C.: Adding excitement to intrinsic motivation research. J. Soc. Behav. Personal. **1**, 349–363 (1986)
34. Reis, M., Almeida, A.: Designing an application to support game-based learning: gathering functional requirements from a qualitative approach. In: 2021, C. (ed.) 16ª Conferência Ibérica de Sistemas e Tecnologias de Informação (CISTI), 23 e 26 de junho de 2021. CISTI 2021, Chaves (2021)
35. Reis, M.H.: Prototyping an app to assist game-based activities: co-design using a qualitative approach. CHItaly 2021 Jt Proc. Interact. Exp. Dr. Consort. - CEUR Work. Proc. **2892**, 41–45 (2021)
36. Sailer, M., Hense, J.U., Mayr, S.K., Mandl, H.: How gamification motivates: An experimental study of the effects of specific game design elements on psychological need satisfaction. Comput. Hum. Behav. 371–380 (2017)
37. Schunk, D.H., DiBenedetto, M.K.: Motivation and social cognitive theory. Contemp. Educ. Psychol. **60**, 101832 (2020). https://doi.org/10.1016/j.cedpsych.2019.101832
38. Shabalina, O., Mozelius, P., Vorobkalov, P., Malliarakis, C., Tomos, F.: Creativity in digital pedagogy and game-based learning techniques; theoretical aspects, techniques and case studies (2015)
39. Shneiderman, B.: Designing the User Interface: Strategies for Effective Human-Computer Interaction. Addison-Wesley (1998)
40. Syal, S., Nietfeld, J.L.: The impact of trace data and motivational self-reports in a game-based learning environment. Comput. Educ. **157**, 103978 (2020). https://doi.org/10.1016/j.compedu.2020.103978
41. Technology, T.: Guidelines for Using the Retrospective Think Aloud P rotocol with Eye Tracking (2009)
42. Tondello, G.F., Kappen, D.L., Mekler, E.D., Ganaba, M., Nacke, L.E.: Heuristic evaluation for gameful design. In: CHI Play 2016—Proceedings of Annual Symposium on Computer Interaction in Play Companion, pp. 315–323 (2016). https://doi.org/10.1145/2968120.2987729
43. Visscher-Voerman, I., Gustafson, K.L.: Paradigms in the Theory and Practice of Education and Training Design. https://www.learntechlib.org/p/165942/ (2004)
44. Yousefi, B.H., Mirkhezri, H.: Toward a game-based learning platform : a comparative conceptual framework for serious games. In: Proceeding of 2019 Internmational Serious Games Symposium ISGS, pp. 74–80 (2019). https://doi.org/10.1109/ISGS49501.2019.9046979
45. Zhou, J., Hoever, I.: Research on workplace creativity: a review and redirection. Annu. Rev. Organ. Psychol. Organ. Behav. **1**, 333–359 (2014). https://doi.org/10.1146/annurev-orgpsych-031413-091226
46. Zichermann, G., Cunningham, C.: Gamification by Design: Implementing Game Mecha-Nics in Web and Mobile Apps. [S.l.]: "O'Reilly Media, Inc." (2011)

Chapter 19
Virtual Reality Prosumers on YouTube and Their Motivation on Digital Design Students

Alejandra Lucía De La Torre Rodríguez, Ramón Iván Barraza Castillo, David Cortés Sáenz, Tayde Edith Mancillas Trejo, and Anahí Solís Chávez

Abstract There is no denying that in our digital culture, prosumers have an impact on brand positioning and purchase decisions with their audience. However, the actions and discourse of the prosumers can also exert other types of motivation in the academic field. Thus, we wanted to assess if the YouTube prosumer movement has had any impact in the desire of digital design students towards video game development. In this chapter we first introduce the concept of motivation, the different types, and how it relates to the design process. Though motivation theory is not the focus of the research, it is relevant to better understand what the driving force behind the students' video game projects proposals is. Through the literature review we explore the work of authors on the topic of impact on motivation prosumers have had on students. Finally, we present the results of two surveys conducted amongst students of the Digital Design for Interactive Media (DDIM) undergraduate program. Subjects had to have some video game and Virtual Reality (VR) design and development experience to participate in this research. The first survey laid foundation about students' motivation in relation to YouTube prosumers. Second survey focused on finding if the opinion of prosumers regarding VR, motivated the students in pursuing school projects that use this technology. Both questionnaires were based on a five-point Likert scale and dichotomous questions. The information gathered, allowed a

A. L. De La Torre Rodríguez · R. I. Barraza Castillo (✉) · D. Cortés Sáenz ·
T. E. Mancillas Trejo · A. Solís Chávez
Architecture, Design and Art Institute, Ciudad Juárez Autonomous University, Ciudad Juárez, Chihuahua, México
e-mail: ramon.barraza@uacj.mx

A. L. De La Torre Rodríguez
e-mail: lucia.delatorre@uacj.mx

D. Cortés Sáenz
e-mail: david.cortes@uacj.mx

T. E. Mancillas Trejo
e-mail: tayde.mancillas@uacj.mx

A. Solís Chávez
e-mail: asolis@uacj.mx

© The Author(s), under exclusive license to Springer Nature Switzerland AG 2023   511
A. L. Brooks (ed.), *Creating Digitally*, Intelligent Systems Reference Library 241,
https://doi.org/10.1007/978-3-031-31360-8_20

better understanding of the context of video games and VR from the perspective of YouTube content creators. In addition, the influence they have on DDIM students.

Keywords Prosumer · Digital design · Digital influence · Digital society

19.1   Introduction

Communication mediated by technology gives rise to changes in the consumption habits of the digital society. Within this group of digitally connected people, it is common practice to search for reviews and opinions through videos or blogs made by prosumers to share experiences about a product or service they want to acquire. Thanks to the reach of communication on digital platforms and social networks, the reviews made by prosumers have an impact on the digital consumer, they evoke reactions and entice purchases. The goal of this chapter is to assess if Digital Design students are motivated or get ideas to create innovative video game projects from seeing the content shared by important YouTube prosumers.

The term prosumer, coined by Toffler [16] in his book, refers to individuals who would merge the roles of producer and consumer to create new personalized products. We live more than ever in an era of digital prosumers, the advent of social media platforms such as Facebook,[1] Instagram,[2] YouTube,[3] Twitch,[4] TikTok,[5] among many others, allows users to produce high-quality content for others to consume. This has also brought about the idea of social media influencers, referring to people who have amassed millions of subscribers to their channels and, according to Freberg et al. [4], "represents a new type of independent outside sponsor shaping the attitudes of the community audience through blogs, tweets, and the use of other social networks" (p. 90).

Gaming industry has attracted interest due to the amount of economic income generated, according to [14]. "It is estimated that the global gaming market will amount to 268.8 billion U.S. dollars annually in 2025, up from 178 billion U.S. dollars in 2021" (para. 1). Therefore, video game platforms continue to invest through consoles, smartphones, and computers.

According to Skwarczek [15], during the pandemic "both Microsoft and Sony recently posted record growth figures for their gaming revenue streams, with the console sector alone earning more than \$45 billion in 2020" (p. 2).

Van Dreunen [18] argues that video games are becoming more important in the industry due to the accessibility that they now have through different mobile devices.

---

[1] https://www.facebook.com/.

[2] https://www.instagram.com/.

[3] https://www.youtube.com/.

[4] https://www.twitch.tv/.

[5] https://www.tiktok.com/.

Over the past fifteen years, video games have transitioned from the fringes of the entertainment business to become one of the biggest and fastest-growing segments during a period characterized by two developments in particular: the widespread adoption of consumer broadband and the popularization of smartphones (p. 4).

Determining whether a video game will be successful is no easy task, there are various internal and external factors to consider, however, the Internet opens the door for developers to have a direct communication channel with their target audience, making it easier for them to both produce content for users to watch and to receive feedback on it.

YouTube is a web platform that offers registered users the possibility to upload, freely share and even monetize their video content. This has enabled an increasing community of online personalities to become leaders in positioning brands, as they can reach a sizable number of subscribers and influence their perception with their audiovisual content.

Technology such as VR has been used by YouTube content creators to demonstrate a wide spectrum of products, from real estate, home decor, retail, tourism, video games, among others. They share with their audience, the overall experience with the product, as well as the tech specs of their VR equipment.

Designing video games from a Digital Design teaching perspective is a challenge that currently arises in higher education, there are different barriers that must be overcome, starting with technological equipment, trained personnel, the design process that must be followed, as well as the creative processes for generating innovative proposals by students.

19.2 Motivation

This chapter focuses on investigating whether the YouTube prosumer can motivate the design student to propose video games so that they become popular on current digital platforms. For this reason, motivation is a determining factor that must be present throughout the progress of the work plan until it is completed. Below are theoretical concepts of motivation and how it influences people to achieve their stated goals.

Motivation in higher education is a phenomenon that occurs to a greater or lesser extent; however, it is essential to stand out through completed goals that reflect the quality and effort of the student to achieve their goals. Huitt [7] mentions that "most motivation theorists assume that motivation is involved in the performance of all learned responses; that is, a learned behavior will not occur unless it is energized" (p. 4).

However Ryan and Deci [13] assure that "Motivation is often treated as a singular construct, even superficial reflection suggests that people are moved to act by very different types of factors, with highly varied experiences and consequences" (p. 69).

Designing and developing a video game is an activity that involves extensive knowledge on different topics that the video game goes through, for that reason the motivation is present from the moment the project is planned, at the moment when the ideas flow and are reflected on paper.

Conscious purpose, determination to achieve goals, as well as intimate thoughts are factors that the human being acquires throughout his growth, which lead to search for sources of motivation independently of the learning acquired in the classroom. Carrillo et al. [1] highlight that motivation arises when "The interest in an activity is "aroused" by a need, which is a mechanism that incites the person to action, and which can be of a character physiological or psychological origin" (p. 21).

According to Liu [9] "Motivation theory suggests two types: (1) intrinsic motivation, and (2) extrinsic motivation. Most past studies relating to motivation still use extrinsic and intrinsic motivation as a starting point to examine the effects motivation has on learning performance" (p. 2). This research does not focus on working with the theory of motivation directly with students, however it is considered necessary to investigate what type of motivation is produced from observing the content positioned by the YouTube prosumer, which leads to proposing ideas. for the design and development of video games.

In this sense, motivation is a factor that can be promoted in different scenarios, however, for design students who work directly with creativity, innovation that results in learning through extensive projects such as video games. Through the questionnaires that were applied to the students, it can be determined if the YouTube prosumer has the ability to generate motivation to propose video games as well as identify the type of motivation, whether extrinsic or intrinsic.

Digital Design for Interactive Media (DDMI) is an undergraduate academic program offered at Ciudad Juárez Autonomous University, whose mission is to "Prepare professionals who use technology to create innovative digital design solutions to meet the needs of digital consumption and provide students with a body of generic skills and competencies that will help them stay in the labor market" [17]. The program includes subjects in PC, console, and mobile game design and development, with an emphasis on the inclusion of emerging technologies like Extended Reality.

Being a young program, few generations have graduated as of 2022, yet there are already some digital products that have been designed and developed by DDMI students as one of the requisites to obtain their bachelor's degree. They are video game projects that they carried out as evidence of applied knowledge. Throughout their studies, theoretical knowledge is accessed, however, students must apply that knowledge and present innovative project ideas that aim to be released on known digital distribution platforms.

Figure 19.1a shows the main screen of the short film that was produced as an interactive immersive VR experience, the project took over a year to complete. Figure 19.1b depicts part of the production process, running in the Unity[6] video game engine that served as the core platform for its creation.

---

[6] Unity Technologies 2022. https://unity.com/es.

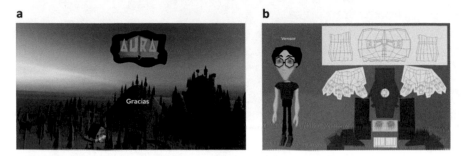

Fig. 19.1 **a** Aura, a short film proposal in interactive immersive virtual reality, **b** Design of the main character of the project. Recovered from: https://heronstd.com/Portafolio/Aura/corto.html

The video game shown in Fig. 19.2 is named after Huitzilopochtli, a digital project that aims to recall the history of México's mythology through scenes inspired by ancient Tenochtitlan. The students decided on a 3D action platform genre and a 2D cell-shaded art style for the two main characters Huitzilopochtli and Coatlicue.

The projects presented are evidence that students learn to design and develop video games, as well as other products using VR technology. However, the question of knowing where the inspiration for designing creative digital projects comes from arises. Are these ideas something that the students have had in their heads for many years or are they triggered by external factors such as watching content creators talk about or play current video games?

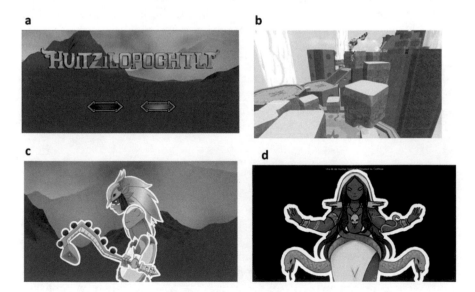

Fig. 19.2 **a** Videogame name, **b** Videogame scene, **c** Huitzilopochtli, **d** Coatlicue

19.3 Problem Statement

Ferchaud et al. [3] argues that "The most popular video-sharing platform today is YouTube" (p. 2), thus, this leads to the community that constantly produces videos and makes their channel subscribers share and comment on the content, reach popularity and be seen by important brands that sponsor them, positioning them as prosumers. Position themselves as prosumers on a digital platform as popular as YouTube, where they are seen by millions of people among whom they have the ability to exert influence, therefore, it is considered necessary to determine through this research if the influence leads to the motivation to propose digital projects such as video games.

There are currently recognized universities with design programs where include courses in development and interaction design for videogames. However, motivation is an essential ingredient that is needed to achieve objectives that are involved in a project such as the design and development of video games. Motivation is what leads the student to set goals at the end of the video game, a goal can be to launch the video game independently and to achieve popularity among the gamer society. This research is conducted to find out if the YouTube prosumer through its video game-themed content provides motivation and ideas to the student when designing a video game. Students learn theoretical concepts, procedures and tools for design and development in the classroom, but what happens with the innovative ideas, inspiration and motivation that is necessary for the proposal of technological projects.

Our research objective is to identify if the YouTube prosumer plays an important role in the motivation or production of ideas in Digital Design for Interactive Media students to propose new video game projects. In order to investigate whether the digital dialogue between students and YouTube prosumers influences the decision not only to design video games, but also to include the use of VR technology. Within the study program there is no subject called virtual reality, the student will learn to design and develop projects with different technological outputs through different subjects such as programming languages, video game engines and interface design. For this reason, it is important to analyze how students take learning and present technological projects.

19.4 Literature Review

The digital community of prosumers is of research interest due to the impact they can have on their viewers, the number of subscribers they add every day, the communication space they can have with their followers, and the number of views their content can have. These factors allow research proposals to be made from the academic field. In this section, a literature review of the articles that have recently been published on YouTube prosumers and their impact on society is presented.

Delgado et al. [2], carried out a study to find out how digital journalism develops in a certain region, and how it promotes the emergence of prosumers among students of the Faculty of Communication Sciences, Tourism and Archeology of the National University San Luis Gonzaga of Ica located in the country of Peru. The research was descriptive in nature and followed an experimental design methodology. Some of the conclusions of the investigation were regarding digital journalism which promotes the emergence of content producers as well as prosumers by the students of the Faculty of Communication Sciences, Tourism and Archeology.

Zimmerman et al. [20], presents a cross sectional quantitative study where they assess the analytical and critical evaluations young people have about YouTube videos on political and social issues and the effect they have based on age and gender.

Murillo-Zamorano et al. [10], focuses on understanding the collective collaboration of gamification in higher education from the development of a theoretical framework and its application to a real case in a digital configuration, this means that through design they motivate the participants with the goal of searching economic gains towards an activity that is intrinsically motivating.

Velásquez García [19], carries out a study to trace the social and cultural profile of young prosumer communication students in the Medellín communication schools in the zone of Aburrá Valley. In this study they focus their interest on observing the analysis of the impacts of communication on mobility, on society in general, and on the way in which young communicators are trained. In the results they were able to observe that they are in a scenario where the classic media face a crisis that has transformed them and where hybridizations are the order of the day, the appearance and configuration of new media and new narrative forms transform cultural consumption.

In the work by Ribeiro et al. [12], it is an evaluation on the impact of YouTuber with the brand that promote, as well as the credibility and congruence on consumers' attitudes. Results indicate that consumers assume that digital influencers have ideas and interests similar to theirs, which leads to a positive reception towards the brand managed by the Youtuber, therefore, credibility is produced towards the influencer and a close relationship with his followers.

Gonzalez and Velasquez [5], examined the influence of the most recognized YouTubers in Ecuador and Colombia. To determine the ranking of the Youtubers, the authors of the article performed an analysis of quantifiable variables regarding their activity on the platform and based on the Social Blade[7] statistical tool they identified the five YouTubers with the greatest influence in these two countries.

Ladhari et al. [8], explored the influencing effects a vlogger had concerning purchase decisions within the context of the beauty industry. The authors examined how homophily, emotional attachment, and credibility influenced the vlogger's popularity.

Hortigüela-Alcalá et al. [6] carried out a study where they explore the effect that social networks have on student motivation and involvement in a university

---

[7] https://socialblade.com/ Social Blade tracks user statistics for YouTube, Twitch, Instagram, and Twitter.

subject called Physical Education and its Didactics through comparisons between an experimental group identified as Twitter and Instagram users and a control group where they did not use social networks. The subject is part of the University Master's degree in primary education for future teachers in physical education. The results they obtained were that there is evidence that supports the idea that the use of social networks can serve as an educational component that produces greater motivation in students.

19.5   Materials and Methods

Our study group focused on digital design students at a public university. They are characterized by having skills and an inclination for emerging technology, they have experience in video game design and development as part of their class and research projects.

Two surveys were conducted, the first one helped gain insight on whether digital design students followed prosumers with a high number of subscribers on YouTube and whether they considered them a source of motivation to design and develop video games. The second survey was carried out to assess whether digital design students consider making innovative and creative proposals that use VR technology since it has now become the focus of attention thanks in part to the Metaverse announcement by Facebook.[8]

Both surveys are mostly 5-point Likert scale questions, but also include dichotomous and multiple-choice items, answers were collected using Google Forms.[9]

19.6   Procedure

First questionnaire provided the authors an overall understanding of the YouTube prosumer scene, and thus was broad in nature. Game content creators on the platform publish different video types when reviewing a particular video game, some rely purely on a spoken review, others include gameplay footage, while others do their live playthrough with commentaries.

Twenty-three questions were asked regarding prosumers' recommendations for creative video game design. Questions included age, gender, educational level, and items relating to whether the students identify the prosumer as a determining factor in the success of a video game due to their popularity on the platform. One hundred

---

[8] The metaverse is the next step in the evolution of social connections. https://about.facebook.com/ltam/meta/.

[9] Collect and organize all kinds of information with Google Forms. And free. https://www.google.com/intl/es_mx/forms/about/.

and seventeen advanced level students with experience in video game design and development responded.

19.6.1 Video Game Design Survey

1. What is your age?
2. What is your gender?
3. What semester are you current enrolled?
4. How much do you agree with a YouTube prosumer reviewing and rating a video game?
5. How much do you agree that a YouTube prosumer can motivate you to design a video game?
6. Would you follow recommendations from a prosumer for the design of a video game?
7. What characteristics should a prosumer have to motivate you to design a video game?
8. How much do you agree that a famous YouTube prosumer's content is for entertainment only?
9. How much do you agree that a YouTube prosumer motivated you to develop a digital design project?
10. If you were to design a video game, would you agree to send it to a YouTube prosumer to use within their content?
11. Would you produce a video of your video game just to have it posted on a prosumer's YouTube channel?
12. Would it be a personal goal to get a YouTube prosumer to publish a video game designed by you on their content?
13. Do you agree with the idea that famous prosumers on YouTube set trends in the video game industry?
14. How much do you agree with a YouTube prosumer testing video games so that they become popular among the fan community?
15. How much do you agree that the YouTube prosumer has the final say in determining whether a video game will be successful?
16. How much do you agree that there are many hurdles for a YouTube prosumer to be able to use a video game designed by you?
17. How much do you agree that for a video game to remain in trend, it must constantly be used by a prosumer on YouTube?
18. Do you agree that the video games produced today are famous thanks to the comments of a prosumer?
19. Have you played a video game recommended by a YouTube prosumer?
20. Do you agree that a YouTube prosumer is the most important source of information to determine the impact of a video game?
21. Do you agree that prosumers must have specific characteristics so that they rate and put out a review for video game?

22. Do you agree that the YouTube prosumer is a reliable source regarding their opinions towards a video game?
23. Have you gotten design ideas for your video game projects from watching content and following a YouTube prosumer that posts video game playthroughs?

The second questionnaire was oriented towards knowing how familiarized with VR technology digital design students were and if they have considered incorporating it in their video game design and development. Therefore, this survey focused more on asking about VR itself and YouTube prosumers that use this technology in their content.

The thirteen-question survey was responded to by forty-one advanced level students with previous exposure to VR technology.

19.6.2   VR Technology Survey

1. Do you know the difference between VR and Augmented Reality?
2. Which companies that produce VR equipment do you know?
3. Have you played any video game with VR technology?
4. Which VR equipment do you consider the best to play?
5. Have you seen videos on YouTube where video games are played with VR technology?
6. Have you seen any YouTube influencer, use VR technology?
7. Do you consider the opinion of a YouTube content creator important when choosing a VR equipment?
8. Do you think it is important to develop VR video games?
9. Have you considered developing a video game using VR technology?
10. Have you been influenced by any YouTube content creator to design a video game?
11. Do you think that a video game is better enjoyed if it uses VR technology?
12. If you were to develop a video game, do you consider the VR platform and equipment to be a determining factor?
13. If you were to develop a video game, what VR platform and equipment would you chose?

19.7   Results and Discussion

This section presents the results of the surveys. They were used to understand if DDMI students identify YouTube prosumers as an external source for their knowledge. It is knowledge outside the traditional context of the university classroom, the dialogue of the prosumer is translated or impregnated in their academic projects.

The first survey had a sample size $N = 117$ (ages 18–41, mean $= 21.96$, mode $=$ 23, median $= 22$, standard deviation $= 3.17$), 60.68% ($n = 71$) were male, 35.89% female ($n = 42$), and 3.41 ($n = 4$) preferred not to answer.

Question four weighs on the level of agreement students have on a YouTube prosumer scoring and reviewing a video game through their gameplay experience. This helps to establish if the students perceive the prosumer's activity as meaningful. The results were that 83 students agreed that a prosumer can in fact review and score a videogame, eight responded that they do not agree prosumers should do it, while 26 remain neutral (see Fig. 19.3).

Question five deals with the motivation a prosumer might have on students regarding video game development. Results in Fig. 19.4 where 87 students feel that a prosumer could motivate them, only eight said they do think they could, and 22 felt impartial. This might reflect on the students' desire to create something that an important content creator and their followers could enjoy.

Question ten further delves into the previous idea, asking students if they would agree to send their video game to an important YouTube prosumer for them to play, comment, and review. Again, most of the responses, 88, said that they would agree; 9 stated they would not, whilst the remaining 20 felt neutral towards the idea (see Fig. 19.5).

In question twelve, students were asked if they would consider a goal for them to have their video game featured on a prosumer's YouTube channel, 70.94% ($n = 83$) people agreed, while 29.06% ($n = 34$) did not, (see Fig. 19.6) This gave the authors insight into determining if the motivation that students have when developing a video game is extrinsic to their formal education and that achieving this goal comes from the students' self-determination by being immersed in an environment

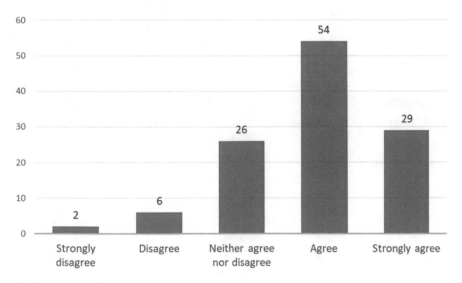

Fig. 19.3 Results for question four

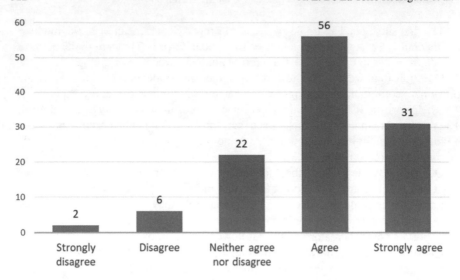

Fig. 19.4 Results for question five

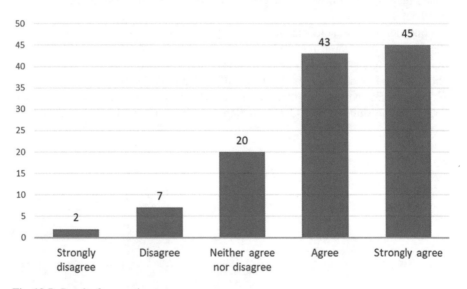

Fig. 19.5 Results for question ten

completely dominated by digital platforms and technology as stated by Pourbaix [11], "Generation C are people born after 1990, who reach maturity after 2000, commence their studies, and then enter the labor market. These are people who function predominantly in virtual reality, and the world of digital media is their standard" (p. 92).

The results of the first questionnaire indicate that the activity of the YouTube prosumer within the gaming community exerts a motivation towards the design

Fig. 19.6 Results for
question twelve

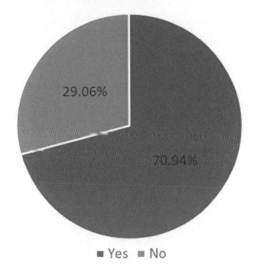

■ Yes ■ No

and development of video games. This is mentioned from the findings obtained in the responses of the students. Where it is observed that the prosumer is a media figure who identifies himself as an expert in the management and use of video games. The creativity in the narrative with which the prosumer communicates attracts the attention of the digital society, well-known brands, consumers and future designers of leisure and entertainment video games. With this knowledge, the second questionnaire was geared towards discovering if the same held true for VR projects.

Second survey, demographic information was not collected, as the first survey saw no significant differences between the groups. The sample size was reduced to N = 41 as it is representative of the population of advanced level students with previous exposure to VR technology in the classroom. It was applied to investigate whether the digital design student follows the content of prosumers who use video games with VR technology. With the objective of knowing if the student has the desire to design and develop a video game in VR, as it is a technology that is increasingly positioned in the entertainment and communications market. It is a questionnaire that focuses on VR; therefore, questions were asked regarding the selection of technology brands to develop a video game, and if the prosumer influenced the use of VR in video games.

In this section, the most important findings that were obtained from 13 questions that were asked to the students are discussed. The survey focuses on two lines: the technological brands of VR and the motivation exerted by the prosumer to develop a video game with this technology.

Question seven was concerning the importance of the prosumers' opinion on a particular VR equipment, to which 73.17% ($n = 30$) students answered that it was important and 26.83% ($n = 11$) that it was not. This is interesting, as many content creators do sponsor advertisement videos for brands in exchange for their products and thus their opinions might be biased. Bigger YouTube channels tend to keep this to a minimum or only accept to do products reviews if they are allowed to publish a completely honest analysis.

For the second survey, demographic information was not collected, as the first survey saw no significant differences between the groups. The sample size was reduced to $N = 41$ as it is representative of the population of advanced level students with previous exposure to VR technology in the classroom. From the previous survey, the authors concluded that YouTube prosumers did in fact influence and motivated students to pursue the development of video game projects. With this knowledge, the second questionnaire was geared towards discovering if the same held true for VR projects.

Question seven was concerning the importance of the prosumers' opinion on a particular VR equipment, to which 73.17% (n = 30) students answered that it was important and 26.83% (n = 11) that it was not. This is interesting, as many content creators do sponsor advertisement videos for brands in exchange for their products and thus their opinions might be biased. Bigger YouTube channels tend to keep this to a minimum or only agree to do products reviews if they are allowed to publish a completely honest analysis.

In question eight, the authors wanted to know if it is important for students to develop a video game that uses VR technology, 85.37% ($n = 35$) of the responses were affirmative and only 14.63% ($n = 6$) negative. This would suggest they understand the current and future trends of the industry and are keen to learn and adopt them in their academic projects.

Question ten, allows us to contrast that although students enjoy the videos that prosumers on YouTube upload to the platform. Some extent considers important the opinion they may have about games and VR equipment; they have not affected their interest in designing and developing a video game with VR technology since only 36.59% (n = 15) reported to have been influenced against 63.41% (n = 26) that have not.

19.8   Conclusions

There are currently few video game prosumers on YouTube who would consider showcasing games made by a student on their channel. Findings in this research prompt the authors with the idea of creating an online repository and a YouTube channel where students can upload their VR projects and share the experience. Thus, effectively becoming prosumers themselves, having the opportunity to gain the recognition of their peers, and perhaps influence others into creating innovative projects.

From the results, it can be understood that the students consider the YouTube prosumer as an important public figure and that the motivation it exerts on them towards the creation of video game projects is extrinsic, since they want to achieve a reward, this reward can be either economic or to gain popularity on the digital platform.

Likewise, it can be observed that digital design students are up to date with respect to emerging technologies such as VR and are eager to use them in their projects. It is important to highlight that the prosumers are a driving force behind new technologies and that their media presence help paves the way for new generations of video game designers to incorporate said technologies into academic projects.

References

1. Carrillo, M., Padilla, J., Rosero, T., Villagómez, M.S.: La motivación y el aprendizaje. Alteridad **4**, 20–33 (2009)
2. Delgado, E., Oré, M., Pecho, Y., Soto, G., Cordero, E.: Digital journalism and the rise of prosumers in the public university. In: ICFET 2021, 2021 The 7th International Conference on Frontiers of Educational Technologies, Bangkok, 4–7 June 2021
3. Ferchaud, A., Grzeslo, J., Orme, S., LaGroue, J.: Parasocial attributes and YouTube personalities: exploring content trends across the most subscribed YouTube channels. Comput. Hum. Behav. **80**, 88–96 (2018). https://doi.org/10.1016/j.chb.2017.10.041
4. Freberg, K., Graham, K., McGaughey, K., Freberg, L.A.: Who are the social media influencers? A study of public perceptions of personality. Public Relat. Rev. **37**, 90–92 (2011). https://doi.org/10.1016/j.pubrev.2010.11.001
5. Gonzalez-Criollo, M.J., Velasquez-Benavides, A.V.: YouTubers and its digital influence. Case study: ecuador and Colombia. In: 2019 14th Iberian Conference on Information Systems and Technologies (CISTI), Coimbra, 19–22 June 2019
6. Hortigüela-Alcalá, D., Sánchez-Santamaría, J., Pérez-Pueyo, Á., Abella-García, V.: Social networks to promote motivation and learning in higher education from the students' perspective. Innov. Educ. Teach. Int. **56**, 412–422 (2019). https://doi.org/10.1080/14703297.2019.1579665
7. Huitt, W.: Motivation to learn: An overview. Educational Psychology Interactive. http://www.edpsycinteractive.org/topics/motivation/motivate.html (2011). Accessed 27 Feb 2022
8. Ladhari, R., Massa, E., Skandrani, H.: YouTube vloggers' popularity and influence: The roles of homophily, emotional attachment, and expertise. J. Retail. Consum. Serv. **54**, 102027 (2020). https://doi.org/10.1016/j.jretconser.2019.102027
9. Liu, I.: The impact of extrinsic motivation, intrinsic motivation, and social self-efficacy on English competition participation intentions of pre-college learners: differences between high school and vocational students in Taiwan. Learn. Motiv. **72**, 101675 (2020). https://doi.org/10.1016/j.lmot.2020.101675
10. Murillo-Zamorano, L.R., López-Sánchez, J.Á., Bueno-Muñoz, C.: Gamified crowdsourcing in higher education: a theoretical framework and a case study. Think. Skills Creat. **36**, 100645 (2020). https://doi.org/10.1016/j.tsc.2020.100645
11. Pourbaix, P.: Prosumer of the XXI century-new challenges to commerce and marketing. Acta Sci. Pol. Oeconomia. **15**, 89–97 (2016)
12. Ribeiro, M.I.B., Fernandes, A.J.G., Lopes, I.M.: The impact of YouTubers' credibility and congruence in consumers' attitude towards the brand. In: Rocha, Á., Reis, J.L., Peter, M.K., Cayolla, R., Loureiro, S., Bogdanović, Z. (eds.) Marketing and Smart Technologies. Smart Innovation, Systems and Technologies, vol. 205, pp. 645–656, Springer, Singapore (2021). https://doi.org/10.1007/978-981-33-4183-8_51
13. Ryan, R.M., Deci, E.L.: Self-determination theory and the facilitation of intrinsic motivation, social development, and well-being. Am. Psychol. **55**, 68–78 (2000). https://doi.org/10.1037/0003-066X.55.1.68

14. STATISTA: Global video game market value from 2020 to 2025. https://www.statista.com/statistics/292056/video-game-market-value-worldwide/#:~:text=Video%20gaming%20market%20size%20worldwide%202020%2D2025&text=It%20is%20estimated%20that%20the,growth%20in%20the%20Asian%20region (2021). Accessed 20 April 2022

15. Skwarczek, B.: How The Gaming Industry Has Leveled Up During The Pandemic. Forbes. https://www.forbes.com/sites/forbestechcouncil/2021/06/17/how-the-gaming-industry-has-leveled-up-during-the-pandemic/?sh=3ec6e84297c5 (2021). Accessed 9 Mar 2022

16. Toffler, A.: The Third Wave: The Classic Study of Tomorrow. Bantam, New York (1984)

17. UACJ: Licenciatura en Diseño Digital de Medios Interactivos. https://www.uacj.mx/oferta/IADA_LDDMI.html (2022). Accessed 28 Feb 2022

18. Van Dreunen, J.: One Up: Creativity, Competition, and the Global Business of Video Games. Columbia University Press, New York (2020)

19. Velásquez-García, G.: Perfil social y cultural de los jóvenes prosumidores en las facultades de comunicación de Medellín. Rev. Aportes la Comun. y la Cult. 55–64 (2020)

20. Zimmermann, D., Noll, C., Gräßer, L., Hugger, K.U., Braun, L.M., Nowak, T., Kaspar, K.: Influencers on YouTube: a quantitative study on young people's use and perception of videos about political and societal topics. Curr. Psychol. (2020). https://doi.org/10.1007/s12144-020-01164-7

Chapter 20
Design of a Bespoke Web-Based 3D Virtual Venue and Video Streaming Event Platform

Ramón Iván Barraza Castillo, Alejandra Lucía De La Torre Rodríguez, Iris Iddaly Méndez-Gurrola, and Anahí Solís Chávez

Abstract As technology advances, enterprises, retail stores, museums, universities, and other organizations have learned to adapt to the ever-changing challenge of drawing public attention to their offerings. This has become even more important in the current era we are living in, where not having a digital presence severely hinders their possibilities to reach a broader market. This issue has drawn even more attention in the past two years as the COVID-19 pandemic spiked the need to hold remote seminars, conferences, classes, and other events through videoconferencing platforms. As with other organizations, universities have been using their websites to promote academic programs, campus installations, events, and amenities and since have expanded to social media platforms and mobile applications to offer more features to both the students and visitors. Colleges and universities across the world have been integrating multimedia resources such as panoramic photos, 360-degree videos, interactive maps, and guided tours of their campuses either from their website or mobile app. To further expand on this idea, and in an effort to reproduce the experience of attending an event in a physical location, albeit, to a certain extent; an interdisciplinary team of digital design, interior design, graphic design, and computers science students, along with professors at Ciudad Juárez Autonomous University, developed a virtual venue that will be accessible from the university web site. Though the concept is not entirely new, the approach and developing process aims to produce a modern, configurable, and flexible platform that allows more freedom for the end users. Attendees will be able to interact with the environment,

R. I. Barraza Castillo · A. L. De La Torre Rodríguez · I. I. Méndez-Gurrola (✉) · A. S. Chávez
Architecture, Design and Art Institute, Ciudad Juárez Autonomous University, Ciudad Juárez, Chihuahua, México
e-mail: iris.mendez@uacj.mx

R. I. Barraza Castillo
e-mail: ramon.barraza@uacj.mx

A. L. De La Torre Rodríguez
e-mail: lucia.delatorre@uacj.mx

A. S. Chávez
e-mail: asolis@uacj.mx

© The Author(s), under exclusive license to Springer Nature Switzerland AG 2023
A. L. Brooks (ed.), *Creating Digitally*, Intelligent Systems Reference Library 241,
https://doi.org/10.1007/978-3-031-31360-8_21

move freely from one live conference to another, or to any showroom available at that event. By leveraging the experience of the experts throughout the entire process, from interior space design, 3D modeling, texturing, illumination, signage, brand identity, and programming, the final product will provide a more engaging experience than a traditional virtual setting, without requiring any specialized viewing equipment.

Keywords Digital design · Virtual campus · Design process · Metaverse · 3D modeling · Digital twin

20.1 Introduction

As the COVID-19 pandemic enters its third year, the world has dramatically changed in many ways. People around the world have learned to adapt and live in accordance with new rules and regulations imposed by the various levels of government. It is palpable that this global event has disrupted all facets of social life.

As the result of the stay-at-home mandates, the workplace, school, retail, entertainment and almost every industry had to resort to the use of video conferencing software such as Zoom,[1] Google Meet,[2] Microsoft Teams,[3] and other platforms to conduct their meetings and everyday communication, [1] explains this led to a spike in the use of this type of Software. According to [1], this trend is likely to continue in the near future, and even overtake the whole in-person meeting experience in a significant percentage of businesses.

Universities, like other institutions, had to act swiftly to adapt to the circumstances, switching to a remote learning environment during the onset of the pandemic, and slowly shifting back to a hybrid and flexible setting as conditions allow. But academic life is more than just attending seminars and lectures at school.

Conferences play a significant role in the life of students, teachers, and researchers. According to Rowe, "they can be found in all of the major disciplines of science and academia, many trades and profession" [2]. Donlon recognizes them as "a central component of academic life, providing important opportunities for dissemination of research, professional networking, and extending scholarship" [3]. Similarly, [4] agrees that attendance at this type of event is a key factor for early-career academics, as it provides experience and helps develop skills and confidence needed for a future in research.

There is no denying the social benefits that the traditional face-to-face conference format inherently has, yet there is also the other side of the coin that must be considered. Most events require participants to pay high registration fees [5], along with the expenses related to international travel and accommodation, which will in turn discourage those of less privileged backgrounds from attending. [6] also highlights

---

[1] https://zoom.us/.

[2] https://apps.google.com/meet/.

[3] https://www.microsoft.com/en-us/microsoft-teams/group-chat-software.

the drawbacks, for people with special travel needs due to physical disabilities or other health related problems, as their travel expenses increase.

Event organizers have been relying on different digital technologies to accommodate for the drawbacks of face-to-face conferencing, by providing access to other means of communication such as websites forums, social media channels, and video streams, to those who cannot physically attend the events.

However, the restrictions brought upon by the pandemic have forced organizers either to cancel, postponed or reimagine the events as online experiences to remain afloat. Donlon asserts how "Remarkable creativity and variation have been witnessed in converting traditional face-to-face conferences to online events" [3]. Of course, all of this comes with its own challenges, the technical and logistical side of it, the cost of the infrastructure and software needed for live streaming, and the interaction and visual appeal of the event. As a result, academic institutions have resorted to makeshift solutions to cope with the need; instead of fully embracing the opportunities to create a cohesive and enjoyable online experience for virtual attendees.

The aim of this chapter is to provide an overview of the creation and deployment of a virtual venue platform, which serves as an extension to Ciudad Juárez Autonomous University online conferencing offering, not only to differentiate and position itself in the market, but to make it more appealing to a wider audience. It follows a multidisciplinary approach, which includes interior, graphical and digital designers.

20.2 Literature Review

Ever since the Internet became mainstream back in the mid to late '90s, there have been multiple attempts to create virtual environments where people all around the world could come together and interact. Each generation of technological ecosystems was made possible and limited by the advances in technology, whereas the firsts were restricted to text only interchange such as message boards the most recent ones involve Virtual Reality (VR) headsets to fully immerse the user in the experience.

20.2.1 Moving Towards the Metaverse

During the Connect2021 conference, Mark Zuckerberg the CEO of former company Facebook, Inc., introduce Meta, according to him it "brings together our apps and technologies under one new company brand" [7]. Along with this announcement, Zuckerberg presented their vision of a metaverse, a term that was first coined by Neal Stephenson in his sci-fi novel Snow Crash [8], where he envisioned the future of the Internet as a vast virtual world where users interacted using avatars.

According to Meta Platforms, Inc.,[4] their metaverse is sort of a hybrid of their current social media experiences "sometimes expanded into three dimensions or projected into the physical world. It will let you share immersive experiences with other people even when you can't be together — and do things together you couldn't do in the physical world" [7].

Though this might seem surreal, the truth is, that it has been attempted a few times before with various degrees of success, the clearest example being Second Life,[5] created by Linden Lab. Back in its peak in 2007, it reached upwards of 1.1 million active users per month [9]. The company servers hosted and simulated environment and provided client software with a set of tools for users to create content and communicate with others. This quickly became a flourishing community, where participants held diverse types of events in their virtual venues. Most recently, game platforms such as Minecraft,[6] Sandbox,[7] Roblox,[8] and Animal Crossing[9] have become the go to place for enthusiast that want to get into the metaverse action, as they provide the tools needed for creating interactive environments that can be then shared and enjoyed with others.

In article by Park and Kim, the authors dive into the taxonomy, components, and open challenges of what they define as the new type of metaverse, and how it differs from earlier conceptions. They identify three key aspects: the advances in deep learning and computer vision technology, which "enables a more immersive environment and natural movement" [10], the ubiquitous access to the metaverse, and the integration of coding tools within the virtual world itself. They also differentiate Augmented Reality (AR) and VR from the metaverse, as they view them as desirable but not required technologies.

Contrastingly, the idea of the metaverse in the work of Lee et al., is "characterized by perpetual, shared, concurrent, and 3D virtual spaces concatenating into a perceived virtual universe" [11], and their technological framework includes the use of Extended Reality (XR) technologies.

From the standpoint of Zhao et al., the metaverse development is still in its early stages, therefore lacks the frameworks needed for convenient visual construction, in their paper, they "propose a framework that summarizes how graphics, interaction, and visualization techniques support the visual construction of the metaverse and user-centric exploration" [12].

---

[4] https://about.facebook.com/.

[5] https://secondlife.com/.

[6] https://www.minecraft.net/en-us.

[7] https://www.sandbox.game/en/.

[8] https://www.roblox.com/.

[9] https://www.animal-crossing.com/.

20.2.2 From Virtual Tours to Virtual Campuses and Beyond

The academic world has tried to keep up with technological advances, implementing online courses, video on demand conferences, 360-degree video tours of their facilities, all the way to fully fledge VR campuses.

As with other organizations, universities have been using their websites to promote academic programs, campus installations, events, amenities, etc., and since have expanded to social media platforms and mobile applications to offer more features to both the students and visitors. Another now common sight on college websites are their virtual campus tours as alternatives to on-campus tours and information sessions. The Princeton Review lists 884 colleges and universities across the US that offer virtual campus tours, many of which were created using YouVisit platform, allowing for panoramic photos, 360-degree videos, and interactive elements to be added to the experience [13].

Going back to the Second Life era, on a survey by [14], they searched for educational institutions that met certain criteria to consider they had presence in that world, and found that at least 170 accredited colleges, universities, and schools did. According to Second Life's destination website, as of 2022, there are currently only 13 universities listed, searching within the client program itself, brings 39 sites that match the query "university".

In the case of Minecraft, aside from the Education Edition which is a version specifically designed for use in the classroom as it supports collaboration, assessment, and pre-built lessons across K-12 curriculum, many universities have seen their campuses recreated using the Java or Bedrock editions, by their own students as a personal project, associations within the school, or even as a sponsored project by the institution itself.

With the help of the COVID-19 Creative Community Response Grant in 2020, a couple of students at Stanford University started a project with the aim "to engage new students as they encounter Stanford through the digital world. It also brings comfort to current students who are no longer on campus" [15].

Jeffery Yu, a student at MIT started a home project and connected with other members of the community through a Discord server and began the ambitious task of building a replica of the campus on Minecraft, as more students joined and the project grew, a MIT volunteer computing group known as the Student Information Processing Board now hosts the server [16].

Boston University has now a growing team of students that support the once small CraftBU[10] project that was started back in 2019 by the BU Gaming Club, according to their website, they now have practical uses for the virtual campus, including virtual tours, events, plan housing, among others.

Similar projects have appeared across other institutions, such as Carnegie Mellon University, University of California Irvine, University of Pennsylvania, University of Minnesota, University of Washington, and the University of California Berkley has

---

[10] https://craftbu.com/.

his aptly named project Blockeley,[11] where they even hosted the first ever Minecraft graduation ceremony for the class of 2020.

In 2020, as part of the induction process for new students at Teesside University in the UK, a research team devised an "interactive induction activity that would allow students to get to know peers as well as staff" [17], they developed a challenge that had the teams solve a task in their virtual campus.

This trend was not limited to the US universities only, students from around the world had to abide to their country's lockdown mandates. Feeling they no longer had a "place" to attend their classes and hang out with friends, they took matters into their own hands. As reported by [18], students belonging to Tecnológico de Monterrey collages across four states in México, joined in an effort to build and connect their campuses, so any student could come and visit the installations.

On a local level, Ciudad Juárez Autonomous University had its Engineering and Technology Institute (IIT) and Architecture, Design and Art Institute (IADA) reproduced in Minecraft during the beginning of the pandemic. What started out as an individual effort from a computer science undergraduate sophomore student, quickly grew to a team of over 50 students from other academic programs [19].

During the autumn of 2021, Stanford's Virtual Human Interaction Lab launched its first VR class as part of an ongoing study, 263 students used their own VR headsets to attend the lectures, engage in group activities and build their own virtual worlds using the ENGAGE[12] platform.

On December 2021, Meta released his VR online videogame called Horizon Worlds, which according to its website is "a social experience where you can create in extraordinary ways. In Worlds, you are not just a visitor. You're a part of what makes it great" [20]. Furthermore, a partnership between Meta Platforms, Inc., and VictoryXR[13] has been announced, to create 10 university campuses across the US by 2022, according to XRToday "Victory XR will use digital twin technologies to create highly-detailed and accurate versions of preexisting campuses, allowing remote students to immerse in interactive campus environments" [21].

Along with VictoryXR, which offers a variety of VR solutions aimed mostly towards schools, there is also the ENGAGE platform. It combines media, file sharing, interactive tools such as whiteboards and sticky notes, with content creation tools to create MetaWorlds that are accessible using both VR headsets and traditional desktops and mobile platforms through a proprietary client application.

Literature review showed how the academic world has not only accepted virtual learning environments in the classroom but embrace virtual and online technologies as alternatives during the pandemic confinement. From online videoconferencing classes, virtual tours, recreated school campuses on games, to reimagined academic events and conferences. Although there are integrated software solutions that help with the virtual creation of school installations, and conference venues, they do not accurately depict the real buildings, furniture, lighting, and overall feel. There is also

---

[11] https://blockeley.com/.

[12] https://engagevr.io/.

[13] https://www.victoryxr.com/.

the cost to consider, as these solutions tend to quickly add up for large groups. The rest of this chapter focuses on creating a modern, configurable, and reusable backend platform alongside an engaging 3D environment frontend of an iconic university building that will be accessible through a web browser.

20.3   Proposed Methodology

The project was born as an idea from the Information and Technology department at Ciudad Juárez Autonomous University, they wanted to leverage their videoconferencing infrastructure in the wake of the pandemic. Seeing how most, if not all face-to-face events and conferences at the campus were canceled, postponed, or in the best-case scenario converted to an online format, they wanted to bring back part of the experience of attending a live event.

Having seen and analyzed the platforms that other universities have used to bring their campus to the virtual world, such as YouVisit,[14] Ricoh360,[15] Floorfy,[16] as well as projects in Minecraft and SecondLife by students and alumni, and other online services that offer the creation of virtual galleries, exhibitions, and venues like Artsteps,[17] Kunstmatrix,[18] Ikonospace,[19] Hexafair,[20] Curat10n,[21] and VR All Art,[22] among others; it was clear that none of these solutions fitted the overall requirements.

While some of the online services offer a free tier plan, they are not without limitations, mainly in regards of venue size, quality of the models, event type, user quantity, to name a few. Their paid plans expand their capabilities but incur a monthly subscription fee. Moreover, if the user requires a bespoke environment different from the available templates, the price increases significantly as they must build it from reference material, a process that can take weeks if not months to complete.

Thus, the IT department recruited the help of a multidisciplinary group of designers and developers from the Architecture, Design and Art Institute who shares the same campus and suffered from the same problems during lockdowns. The team was composed of teachers and senior students from three different undergraduate programs. Figure 20.1 shows a high abstraction view of the development process and the main tasks for the teams. Each component and their methodological approach will be addressed in more detail in the following subsections.

---

[14] https://www.youvisit.com/.

[15] https://www.ricoh360.com/tours/.

[16] https://floorfy.com/.

[17] https://www.artsteps.com/.

[18] https://www.kunstmatrix.com/en.

[19] https://www.ikonospace.com/.

[20] https://www.hexafair.com/.

[21] https://curat10n.com/.

[22] https://vrallart.com/.

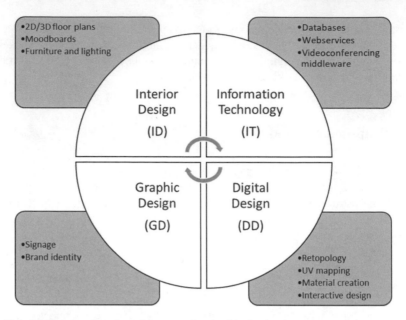

Fig. 20.1 High abstraction view of the teams involved in the development process

20.3.1   Interior Design

According to Dodsworth and Anderson, "Good interior design adds a new dimension to a space. It can increase our efficiency in the way we go about our daily lives, and it adds depth, understanding, and meaning to the built environment" [22]. Therefore, the first step towards recreating one of the iconic off campus buildings that would serve as the virtual venue for conferences, exhibitions, and other online events, was to model the structure itself. Since the team already had the 2D floor plans, the decision to not recruit the help of an architect was taken. Instead, they relied on the Interior Design (ID) group to undertake the task.

Though there are specific design processes available, [22] emphasize that regardless of the one chosen, it should be flexible to adapt to the unique nature of the project. Figure 20.2 illustrates the simplified process diagram followed by the team.

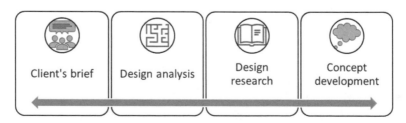

Fig. 20.2 Interior Design simplified process diagram

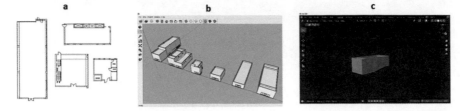

Fig. 20.3 **a** 2D floorplan, **b** 3D floorplan, **c** 3D model

20.3.2 2D/3D Floor Plan

After the initial brief with the IT department, the team set out to digitize the existing 2D floor plans using AutoCAD.[23] Once the master files were created, they were imported into SketchUp[24] to begin creating the 3D version of the floor plan. This step only included walls, floors, ceilings, and other structural building elements. The last step of this stage was to export the 3D model as an obj file that will be used by the Digital Design team. Figure 20.3 shows an example of three different iterations of the building.

20.3.3 Mood Boards, Furniture, and Lighting

With the 3D space constructed, it was time to analyze and survey the space to begin conceptualizing the project. This is where the imagination of the ID team was given free rein to envision and sketch their ideas.

During this process they had to think about the function each space was to fulfill, how it interacts with the rest of the spaces, and most importantly, how people are going to connect with it. Things like dimension, scale, proxemics, and ergonomics were taken into consideration when designing each area.

After successfully defining the 3D space, the important, but sometimes misunderstood task of deciding on a decorative scheme was tackled. Dodsworth and Anderson state that it is "the decorative scheme that adds those elements that complete the sensory experience. Interior design gives purpose to decoration. It adds texture, light, and color" [22].

Through the initial design analysis where the function of the space was identified, furniture such as counters, coffee tables, sofas, and chairs, along with other elements for exhibitions like room dividers, modular panels, pedestals, canvas, and stands were proposed. The mood boards created by the team for each of the building areas included a distribution layout, color scheme, a detailed view of each element as well as the materials chosen. Lastly, it is also important to note that lighting was taken

---

[23] https://www.autodesk.com/products/autocad/overview.

[24] https://www.sketchup.com.

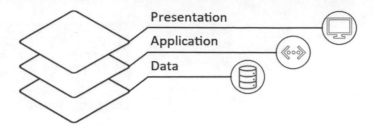

Fig. 20.4 Three-tier architecture

into consideration when choosing the style of luminaries as well as the colors and material finishes.

20.3.4 Information Technology

The IT team responsibility was twofold, the backend deployment and management, as well as the middleware for the video streaming services. Figure 20.4 shows the well-established three-tier layer architecture. The main benefit of this architecture is "its logical and physical separation of functionality. Each tier can run on a separate operating system and server platform" [23].

20.3.4.1   Database Backend

Levering on the IT department access to database servers running Microsoft SQL Server[25] as well as MongoDB,[26] the team created and populated the databases needed to host the data that will feed the client application and store the state of the virtual environment. Figure 20.5 shows a sequence diagram for the initial request made by the client application, every time a new user logs and launches the webpage it triggers the action.

The client application uses this initial request to get all the information needed to dynamically create and populate assets for the different exhibition areas when the user navigates between buildings. Similarly, when the user goes through a door, the application will notify the server back so it can keep track of the number of avatars in each room.

---

[25] https://www.microsoft.com/en-us/sql-server.

[26] https://www.mongodb.com.

Fig. 20.5 Initial request
sequence diagram

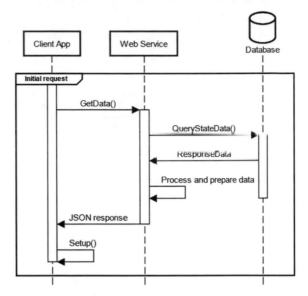

20.3.4.2 Videoconferencing Middleware

Along with the database server and the webservice that feeds the client application, a media server was deployed. Using the open-source runtime environment
Node.JS[27] and the Node-Media-Server[28] project, the IT team can publish live streams
using FFmpeg[29] suit of libraries and programs or any streaming software such as
OBS.[30] This middleware solution was implemented to help with the integration of the
multiple videoconferencing platforms, live feeds, and local streams. Once published,
the streams can be accessed through RTMP or HTTP protocols using HTTP-FLV,
WS-FLV, HLS, or DASH adaptive bitrate streaming technologies.

As with the initial request setup, whenever a user navigates to the auditorium,
a new request is triggered to get the current information for any live streams that
might be available. Although there is only one conference room, there could be
many video feeds available for the user to select. Figure 20.6 illustrates the complete
infrastructure diagram.

---

[27] https://nodejs.org.

[28] https://github.com/illuspas/Node-Media-Server.

[29] https://ffmpeg.org/.

[30] https://obsproject.com.

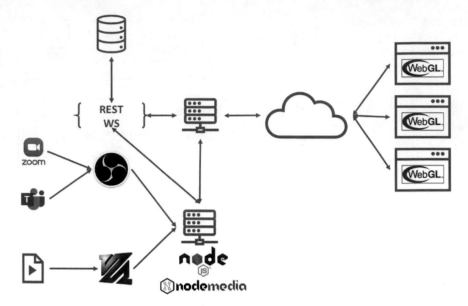

Fig. 20.6 Application backend and middleware server infrastructure

20.3.5 Digital Design

The Digital Design (DD) group had two main responsibilities in the project. The first one was to take the ideas and models produced by the ID team and using the open-source 3D computer graphics software Blender,[31] create the materials and textures for the buildings and objects. Their second task was to develop all the interaction the user can have with the virtual environment, as well as the code needed to communicate with the back end. The team followed the incremental build model, which takes the traditional waterfall model and combines it with the iterative nature of the prototype-based software development one [24].

20.3.5.1 Retopology

During the process of creating the texture maps, the team quickly realized that one of their initial fears came true. The software used by the ID group (Sketchup), produced models with unnecessarily complex 3D meshes, like the sofa seen in Fig. 20.7 that has 12,679 triangles. The team agreed that the models, though perfectly usable in the Unity[32] game engine, were not suitable for the practical needs of the application, as they will have a negative impact on the performance and size of the final product.

---

[31] https://www.blender.org/.
[32] https://unity.com.

Fig. 20.7 Complex 3D
mesh example from
Sketchup

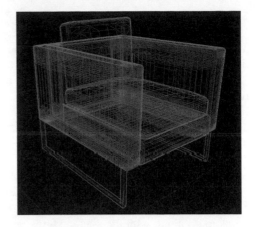

Therefore, they had to rework some of the meshes through a process known as retopology to simplify the topology of a mesh to make it lighter and easier to work. This can be achieved by manually manipulating the geometry of the object or using automated methods available in Blender's Poly build tool. The DD team used a combination of manual manipulation and Quad remeshing, which according to Blender's manual "uses the Quadriflow algorithm to create a Quad based mesh with few poles and edge loops following the curvature of the surface" [25]. Figure 20.8a shows the original object (702 faces), while Fig. 20.8b depicts the same model after going through the retopology process (228 faces).

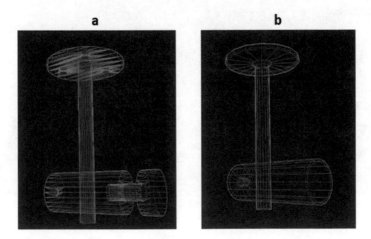

Fig. 20.8 **a** Original model, **b** Reworked model

20.3.5.2 UV Mapping and Material Creation

The last step for the 3D assets was to create the materials defined earlier in the process by the ID team. To do so, a technique called UV mapping was employed, which takes a 2D image texture and wraps it onto a 3D mesh. The importance of the retopology process becomes clear as a clean and well proportionate mesh will yield better results when marking the mesh seams.

After unwrapping the meshes, the team went back to the mood boards to get the color and texture references for each of the materials that they had to be created. Using real material samples like leather, steel, wood, and glass as well as photographic references, they created the varied materials seen in Fig. 20.9 using the built-in Blender tools as well as Substance 3D Painter.

As part of the modern development art pipeline, the team created the normal maps to go along with the main textures, as they "allow you to add surface detail such as bumps, grooves, and scratches to a model which catch the light as if they are represented by real geometry" [26]. This technique was especially useful for this case, as it allowed us to add back details that the retopology process took away from the model without adding in extra geometry. Figure 20.10a shows the unwrapped UV texture and Fig. 20.10b the normal map for the sofa seen in Fig. 20.7.

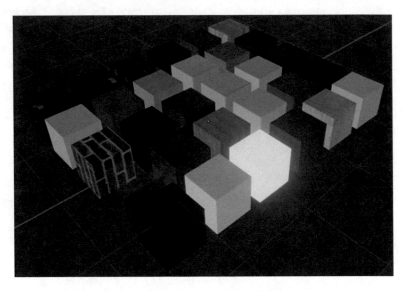

Fig. 20.9 Cubes with materials applied from one of the exhibit rooms

a

b

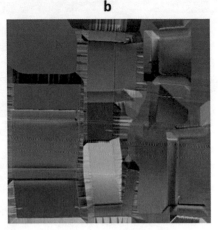

Fig. 20.10 **a** UV texture map, **b** Normal map

20.3.5.3 Interactive Design

With all the objects now ready and following the 2D and 3D floorplans, the team recreated and populated every room and auditorium as a real-life scale virtual environment in Unity game engine. The first iteration of the project had the following functional requirements (FR) and non-functional requirements (NFR).

- FR1: The user can launch the application in a browser.
- FR2: The user can quit the application.
- FR3: The user can move around the environment.
- FR4: The user can look around the environment.
- FR5: The user can click to select a certain object.
- FR6: The user can sit in a chair in the auditorium.
- FR7: The user can select the live video feed that he wants to watch.
- FR8: The user cannot go through walls, furniture, and other users.
- FR9: The user cannot manipulate objects.
- FR10: The user cannot manipulate illumination.
- NFR1: The virtual environment should be a 1:1 representation of the real building.
- NFR2: The virtual environment should be realistic.
- NFR3: Application should work on Firefox and Chromium-based browsers.
- NFR4: The application should be light weighted.

The original idea was not to have each user navigate the virtual environment on its own, but rather have the different areas filled with other users that are currently connected, as if they were attending a face-to-face event. Rather than implementing a full-blown real time multiplayer application, the team used Unity's navigation system.

Fig. 20.11 NavMesh component and door trigger points

By defining a NavMesh, the team described the walkable surfaces and points of interest on the environment. When a new user launches the application, it spawns a NavMeshAgent in the lobby, and as it moves around the exhibition rooms, it triggers callbacks to the server to inform of their activity. Every user then periodically queries this data to approximate the number of avatars it should encounter and the areas where they are congregating. Figure 20.11 shows the NavMesh and door colliders used as trigger points for server callbacks.

As defined by the functional requirements, the user can interact with certain elements in the environment, for this purpose, visual cues will appear on screen to let the user know when and how to do it.

To integrate the live video feeds into the virtual environment, the team had two main ideas. The first consisted of having the user's avatar take a seat on the auditorium and switch to a full screen HTML/JavaScript player, though effective, it might break the user's immersion. The second idea was to use a plug-in that would allow the video feed to be used as a texture for an object inside the virtual environment, to this end, the team chose AVPro Video by RenderHeads.[33] Once the user takes a seat in the auditorium, the client application sends a request to the server, the client then parses the response JSON, and configures the player with the information needed to connect to the media server.

20.3.6 Graphic Design

To aid the user navigate through the virtual environments, the team recruited the help of a Graphic Designer (GD), whose responsibilities included the integration of signage and wayfinding as well as the graphical identity for the project.

---

[33] https://renderheads.com/products/avpro-video/.

20.3.6.1 Signage

Just like in the physical world where we need to find our way through it, it is important that users in a virtual environment know where and how they can navigate and reach certain areas. In modern Environmental Graphic Design (EGD) the "activity involves the development of a systematic, informationally-cohesive, and visually unified graphic communication system for a given site within the built environment" [27].

Since the project is a replica of an existing part of the university's campus, a signage and wayfinding system already exists. Instead of creating a new scheme that would cause dissonance to those familiar with the buildings, the Graphic Designer reproduced the same signs, symbols, diagrams, arrows, and other graphical elements and adapted them for the virtual environment.

20.3.6.2 Brand Identity

Just like with signage, it was important that the brand identity for the project kept in line with what already exists. According to Wheeler "Brand identity fuels recognition, amplifies differentiation, and makes big ideas and meaning accessible. Brand identity takes disparate elements and unifies them into whole systems" [28]. Furthermore, the brand is more than just a logo and a catch phrase [29], and special attention should be given when working on it.

Ciudad Juárez Autonomous University has a guide that defines basic design rules pertaining to the elements of its graphical identity and their correct usage. Based on these rules, the designer took particular care with every institutional reference within the virtual environment.

20.4 Results

The last step of the process was for the DD team to integrate all the components into a single Unity project and build the solution so they could assess if it meets all functional requirements as well as evaluate the nonfunctional ones. One of the goals of the project was to avoid the need for the end user to download and install anything in their devices, as not all users know how to do so, and those who do, might not want to do it for security, privacy, or other reasons. The team also wanted to avoid the need for special hardware such as VR/AR headsets to appeal to a broader audience.

Even though most of the development and testing was done from the Unity editor and a Windows standalone build, the team took particular care to only use components and libraries that would be compatible with a modern Web build. Unity now targets WebGL 1.0/2.0 for its Web builds and relies on WebAssembly and the.NET Standard 2.0 to run high-performance applications on a browser. Combined with the optimized

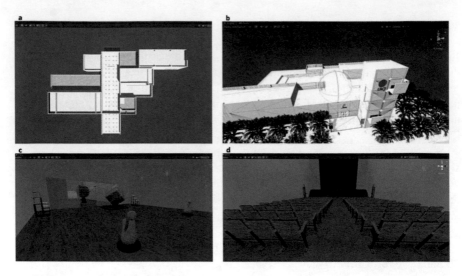

Fig. 20.12 **a** Top view of the complex, **b** outside view of the complex, **c** exhibition room, **d** auditorium

3D models, detailed texture maps, normal maps, and baked global illumination, the images shown in Fig. 20.12 are examples of the first full project build.

Even with the care taken when optimizing the 3D models, textures, and illumination, the build exceeded the file size the team considered to be appropriate for a Web experience. After reviewing the build log from Unity, it was clear that the meshes and textures were still the ones contributing the most to the size.

Therefore, the DD team went back to the editor and tweaked the platform specific setting for the texture compression to RGB Compressed DXT1, changed the Max Size to 1024, and enabled Crush Compression with a 75-quality value, this yielded a project 52% the original sized without noticeable impact in the quality of the textures, as seen in the original brick pattern on Fig. 20.13a and the optimized version on Fig. 20.13b.

The designers took a similar approach in regards of the 3D meshes, Unity provides two ways to achieved this, Vertex and Mesh compression, the latter was chosen as it can be set on individual meshes and it has the most impact on size. After tunning the settings and rebuilding the project, the file size was only 31.7% of the original, making it acceptable for a modern Web experience.

The final assessment came in the form of a limited private beta test, as the team behind the project has not officially launched it to the academic community. This is partly because the official launch is expected to be accompanied by a socialization campaign, in addition to the fact that the team seeks to integrate some new features that emerged during development. Though this chapter will not cover the beta test result details, it is worth to mention that the overall impression was positive, with testers only citing the lack of avatar creation tools and voice chat, while the former is a feature the team will work to integrate into the final build, the latter is not.

a   b

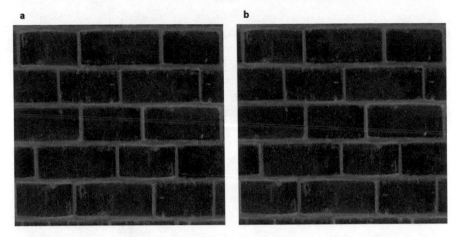

Fig. 20.13 Texture quality/size comparison **a** original texture (2.7MiB) **b** optimized texture (197KiB)

20.5 Conclusion

With the pandemic, the world is now a different place, we grown accustomed to video-conferencing, virtual campus tours, and online academic conferences, and thus, it has become evident that once the restrictions ease, is not only important but necessary to continue investing in developing technological alternatives to be prepared for similar future events and to take advantage of the economic and social benefits they bring.

This chapter gave insight on how different members of the academic guild have used the available tools to create virtual environments to adapt to the changing circumstances brought upon by the traveling restrictions around the world. It also highlighted the growth and interest of companies that offer virtual meeting and event services, ranging from simple 360-dregree video/photo walkthroughs to massive immersive environments that require specialized VR/AR equipment.

After researching the work other universities have done and analyzing the pros and cons of the available platforms for creating virtual environment that would be an extension and to a certain degree, a digital twin of an iconic campus building, it was clear that it had to be tailored made.

An interdisciplinary approach between teams of specialist in different areas of Design and Technology, lead to the creation of a modern, configurable, and flexible Web-based platform for both the management team and the end users. The virtual environment is accessible from a Web browser and does not require any installation, it allows attendees to move freely from one live conference to another, or to any showroom available in the buildings, as if attending a live event. By levering a discrete trigger system and intelligent agents instead of a real time multiplayer approach, the system can represent and manage crowds of users making their way through the buildings.

As the project continues to evolve, the team already has ideas they want to implement for future version. They want to address the avatar creation tool comment brought up by the beta test group, the creation a Web based management tool that would allow them to have a grid-like view of each of the building floorplans, so they can arrange the rooms with furniture and exhibition equipment and have it dynamically populated in the virtual world. There is also interest in expanding the online experience by adding new buildings, 3D elements, and audiovisual content using Unity's AssetBundles and Addressables, as to keep the project file size manageable and load additional content on demand.

References

1. Karl, K.A., Peluchette, J.V., Aghakhani, N.: Virtual work meetings during the COVID-19 pandemic: the good, bad, and ugly. Small Gr. Res. (2021). https://doi.org/10.1177/104649642 11015286
2. Rowe, N.: 'When you get what you want, but not what you need': the motivations, affordances and shortcomings of attending academic/scientific conferences. Int. J. Res. Educ. Sci. **4**, 714–729 (2018). https://doi.org/10.21890/ijres.438394
3. Donlon, E.: Lost and found: the academic conference in pandemic and post-pandemic times. Irish Educ. Stud. **40**, 367–373 (2021). https://doi.org/10.1080/03323315.2021.1932554
4. Timperley, C., Sutherland, K.A., Wilson, M., Hall, M.: He moana pukepuke: navigating gender and ethnic inequality in early career academics' conference attendance. Gend. Educ. **32**, 11–26 (2020). https://doi.org/10.1080/09540253.2019.1633464
5. Niner, H.J., Johri, S., Meyer, J., Wassermann, S.N.: The pandemic push: can COVID-19 reinvent conferences to models rooted in sustainability, equitability and inclusion? Socio-Ecological Pract. Res. **2**, 253–256 (2020). https://doi.org/10.1007/s42532-020-00059-y
6. De Picker, M.: Rethinking inclusion and disability activism at academic conferences: strategies proposed by a PhD student with a physical disability. Disabil. Soc. **35**, 163–167 (2020). https://doi.org/10.1080/09687599.2019.1619234
7. Meta: The Facebook Company Is Now Meta. https://about.fb.com/news/2021/10/facebook-company-is-now-meta/ (2021). Accessed 03 Dec 2021
8. Stephenson, N.: Snow crash: A novel. Bantam Books, New York, NY (1992)
9. New World Notes: Second Life Usage Now Higher Than Its 2007 Hype Level Period?. https://nwn.blogs.com/nwn/2020/06/second-life-sl-linden-lab-mau-concurrency.html (2020). Accessed 03 Dec 2021
10. Park, S.-M., Kim, Y.-G.: A metaverse: taxonomy, components, applications, and open challenges. IEEE Access. **10**, 4209–4251 (2022). https://doi.org/10.1109/ACCESS.2021.314 0175
11. Lee, L.-H., Braud, T., Zhou, P., Wang, L., Xu, D., Lin, Z., Kumar, A., Bermejo, C., Hui, P.: All One Needs to Know about Metaverse: A Complete Survey on Technological Singularity, Virtual Ecosystem, and Research Agenda. arxiv.org (2021)
12. Zhao, Y., Jiang, J., Chen, Y., Liu, R., Yang, Y., Xue, X., Chen, S.: Metaverse: Perspectives from graphics, interactions and visualization. Vis. Informatics. (2022). https://doi.org/10.1016/j.vis inf.2022.03.002
13. The Princeton Review: Virtual Campus Tours. https://www.princetonreview.com/college-adv ice/virtual-tours (2022). Accessed 10 Jan 2022
14. Jennings, N., Collins, C.: Virtual or virtually U: Educational institutions in Second Life. Int. J. Soc. Sci. **2**, 180–186 (2007)

15. Aquilanti, A.: Stanford, but in Minecraft – Stanford Arts: A profile of one of the COVID-19 Creative Community Response Grant projects. https://arts.stanford.edu/stanford-but-in-minecraft/ (2020) Accessed 10 Jan 2022
16. Tran, S.: Building and reconnecting MIT in Minecraft. https://news.mit.edu/2020/building-and-reconnecting-mit-minecraft-0407 (2020) Accessed 10 Jan 2022
17. Tidy, H., Carney, H., Wood, A., Anderson, C.: Using Minecraft Education Edition to Create Meaningful Student Interactions. Paper presented at the 3 Rivers Conference, Teesside University, Middlesbrough, 25 June (2021)
18. Sierra, E.: Estudiantes construyen al TEC en Minecraft. https://tec.mx/es/noticias/san-luis-potosi/arte-y-cultura/estudiantes-construyen-al-tec-en-minecraft-fotogaleria (2020) Accessed 10 Jan 2022
19. Luna, K.: Llega UACJ a Minecraft. https://netnoticias.mx/juarez/llega-uacj-a-minecraft/ (2020) Accessed 10 Jan 2022
20. Lewis, J.E., Trojovsky, M., Jameson, M.M.: New Social Horizons: Anxiety, Isolation, and Animal Crossing During the COVID-19 Pandemic. Front. Virtual Real. 2, (2021). https://doi.org/10.3389/frvir.2021.627350
21. Greener, R.: Meta, VictoryXR to Launch 10 Metaverse Campuses - XR Today. https://www.xrtoday.com/virtual-reality/meta-victoryxr-to-launch-10-metaverse-campuses/ (2021) Accessed 24 Jan 2022
22. Dodsworth, S., Anderson, S.: The Fundamentals of Interior Design. Bloomsbury (2019)
23. IBM Cloud Education: What is Three-Tier Architecture | IBM. https://www.ibm.com/cloud/learn/three-tier-architecture (2022) Accessed 05 Feb 2022
24. Douglass, B.P.: What Are Agile Methods and Why Should I Care? In: Agile Systems Engineering. pp. 41–84. Morgan Kaufmann (2016)
25. Blender Foundation: Retopology — Blender Manual. https://docs.blender.org/manual/en/latest/modeling/meshes/retopology.html (2022) Accessed 12 Feb 2022
26. Unity: Unity - Manual: Normal map (Bump mapping). https://docs.unity3d.com/2022.1/Documentation/Manual/StandardShaderMaterialParameterNormalMap.html (2022) Accessed 12 Feb 2022
27. Calori, C., Vanden-Eynden, D.: Signage and wayfinding design. John Wiley & Sons Inc., Hoboken, New Jersey (2015)
28. Wheeler, A.: Designing brand identity: an essential guide for the whole branding team. John Wiley & Sons Inc., Hoboken, New Jersey (2013)
29. Mogaji, E.: Brand Identity. In: Brand Management. pp. 85–122. Palgrave Macmillan, Cham (2021)